Methods in Molecular Biology™

Series Editor
John M. Walker
School of Life Sciences
University of Hertfordshire
Hatfield, Hertfordshire, AL10 9AB, UK

For further volumes:
http://www.springer.com/series/7651

mTOR

Methods and Protocols

Edited by

Thomas Weichhart

Division of Nephrology and Dialysis, Department of Internal Medicine III,
Medical University Vienna, Vienna, Austria

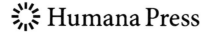

Humana Press

Editor
Thomas Weichhart
Division of Nephrology and Dialysis
Department of Internal Medicine III
Medical University of Vienna
Vienna, Austria
thomas.weichhart@meduniwien.ac.at

ISSN 1064-3745 e-ISSN 1940-6029
ISBN 978-1-61779-429-2 e-ISBN 978-1-61779-430-8
DOI 10.1007/978-1-61779-430-8
Springer New York Dordrecht Heidelberg London

Library of Congress Control Number: 2011941581

Printed on acid-free paper

Humana Press is part of Springer Science+Business Media (www.springer.com)

Preface

Since its discovery, the mammalian target of rapamycin (mTOR) has been shown to regulate many critical molecular processes in eukaryotes, such as metabolism, growth, survival, aging, synaptic plasticity, memory, and immunity. Exciting findings in these areas explain the enormous scientific interest in the mTOR pathway and extend the potential therapeutic applications of mTOR inhibitors, which are currently employed in the clinic as immunosuppressive and anticancer agents. The book is primarily intended for basic and translational scientists in the fields of molecular cell biology, energy metabolism, immunology, transplantation, and cancer research. This volume unites current biochemical assays and *in vitro* as well as *in vivo* models to cover the divergent fields of mTOR research ranging from the analysis of signal transduction events within a cell to the assessment of complex human diseases, such as metabolic disorders or cancer.

Vienna, Austria *Thomas Weichhart*

Contents

Contributors

KOICHI ARAKI • *Department of Microbiology and Immunology, Emory Vaccine Center, Emory University School of Medicine, Atlanta, GA, USA*

JOSEPH AVRUCH • *Diabetes Unit and Medical Services and Department of Molecular Biology, Massachusetts General Hospital, Boston, MA, USA; Department of Medicine, Harvard Medical School, Boston, MA, USA*

PEARL BAKHRU • *Department of Pathology and Laboratory Medicine, University of Texas Medical School, Houston, TX, USA*

MANUELA BATTAGLIA • *San Raffaele Telethon Institute for Gene Therapy, Milan, Italy; San Raffaele Diabetes Research Institute, Milan, Italy*

PHILIP BERGMAN • *Developmental and Molecular Pathways, Novartis Institutes for BioMedical Research, Cambridge, MA, USA*

RAKHMET I. BERSIMBAEV • *Department of Natural Sciences, The L.N. Gumilyov Eurasian National University, Astana, Kazakhstan*

OLGA BULGAKOVA • *Department of Molecular and Cellular Oncology, University of Texas M. D. Anderson Cancer Center, Houston, TX, USA; Department of Natural Sciences, The L.N. Gumilyov Eurasian National University, Astana, Kazakhstan*

JAE WON CHANG • *Department of Cancer Biology, Dana Farber Cancer Institute, Boston, MA, USA; Department of Biological Chemistry and Molecular Pharmacology, Harvard Medical School, Boston, MA, USA*

LOUIS-MARIE CHARBONNIER • *Institute for Medical immunology, Université Libre de Bruxelles, Gosselies, Belgium*

CLAUDIO J. CONTI • *Department of Molecular Carcinogenesis, The University of Texas M. D. Anderson Cancer Center, Smithville, TX, USA*

JENNIFER L. COTTON • *Program in Molecular Medicine, University of Massachusetts Medical School, Worcester, MA, USA*

MEGAN CULLY • *Signal Transduction Laboratory, Cancer Research UK London Research Institute, London, UK*

NADINE CYBULSKI • *Biozentrum, University of Basel, Basel, Switzerland*

GREG M. DELGOFFE • *Sidney-Kimmel Comprehensive Cancer Research Center, Johns Hopkins University School of Medicine, Baltimore, MD, USA*

JULIAN DOWNWARD • *Signal Transduction Laboratory, Cancer Research UK London Research Institute, London, UK*

LEIF W. ELLISEN • *Massachusetts General Hospital Cancer Center, Harvard Medical School, Boston, MA, USA*

QI-WEN FAN • *Departments of Neurology, Pediatrics, Neurological Surgery and Brain Tumor Research Center, and Helen Diller Family Comprehensive Cancer Center, University of California, San Francisco, CA, USA*

Leonard J. Foster • *Department of Biochemistry and Molecular Biology, Centre for High-Throughput Biology and University of British Columbia, Vancouver, BC, Canada*
David A. Fruman • *Molecular Biology & Biochemistry, University of California, Irvine, CA 92617, USA*
Lorenzo Galluzzi • *INSERM, U848, Villejuif, France; Institut Gustave Roussy, Villejuif, France; Université Paris-Sud, Le Kremlin-Bicêtre, France*
Irma B. Gimenz-Conti • *Department of Molecular Carcinogenesis, The University of Texas M. D. Anderson Cancer Center, Smithville, TX, USA*
Ivan Gout • *Department of Structural and Molecular Biology, Institute of Structural and Molecular Biology, University College London, London, UK*
Nathanael S. Gray • *Department of Cancer Biology, Dana Farber Cancer Institute, Boston, MA, USA; Department of Biological Chemistry and Molecular Pharmacology, Harvard Medical School, Boston, MA, USA*
Kun-Liang Guan • *Department of Pharmacology and Moores Cancer Center, University of California San Diego, La Jolla, San Diego, CA, USA*
David A. Guertin • *Program in Molecular Medicine, University of Massachusetts Medical School, Worcester, MA, USA*
Dongjun Guo • *Department of Neurology, The Hope Center for Neurological Disorders, Washington University School of Medicine, St. Louis, MO, USA*
Michael N. Hall • *Biozentrum, University of Basel, Basel, Switzerland*
Markus Hengstschläger • *Medical Genetics, Medical University of Vienna, Vienna, Austria*
Sungki Hong • *Department of Molecular and Integrative Physiology, Life Sciences Institute, Ann Arbor, MI, USA; Division of Nephrology in the Department of Internal Medicine, University of Michigan Medical School, Ann Arbor, MI, USA*
Jingxiang Huang • *Department of Pathology, National University Hospital of Singapore, Singapore*
Wooyoung Hur • *Department of Cancer Biology, Dana Farber Cancer Institute, Boston, MA, USA; Department of Biological Chemistry and Molecular Pharmacology, Harvard Medical School, Boston, MA, USA*
Valentina Iadevaia • *School of Biological Sciences, University of Southampton, Southampton, UK*
Ken Inoki • *Life Sciences Institute, Department of Molecular and Integrative Physiology, University of Michigan Medical School, Ann Arbor, MI, USA; Division of Nephrology in the Department of Internal Medicine, University of Michigan Medical School, Ann Arbor, MI, USA*
Chinnaswamy Jagannath • *Department of Pathology and Laboratory Medicine, University of Texas Medical School, Houston, TX, USA*
Matthew R. Janes • *Drug Discovery, Intellikine Inc., La Jolla, CA 92037, USA*
Seong A. Kang • *Whitehead Institute for Biomedical Research, Cambridge, MA, USA*
Oliver Kepp • *INSERM Institut Gustave Roussy, U848, Villejuif, France; Institut Gustave Roussy, Villejuif, France; Université Paris-Sud, Le Kremlin-Bicêtre, France*

MAHMOUD KHALIL • *Department of Structural and Molecular Biology, Institute of Structural and Molecular Biology, University College London, London, UK*
JINHEE KIM • *Department of Molecular Carcinogenesis, The University of Texas M. D. Anderson Cancer Center, Smithville, TX, USA*
BOGUMILA T. KONIECZNY • *Department of Microbiology and Immunology, Emory Vaccine Center, Emory University School of Medicine, Atlanta, GA, USA*
GUIDO KROEMER • *INSERM Institut Gustave Roussy, U848, Villejuif, France; Metabolomics Platform, Institut Gustave Roussy, Villejuif, France; Centre de Recherche des Cordeliers, Paris, France; Pôle de Biologie, Hôpital Européen Georges Pompidou, AP-HP, Paris, France; Université Paris Descartes, Sorbonne Paris Cité, Paris, France*
ALAIN LE MOINE • *Institute for Medical immunology, Université Libre de Bruxelles, Gosselies, Belgium*
DONG LI • *Department of Cell and Developmental Biology, Institute for Genomic Biology, University of Illinois at Urbana-Champaign, Urbana, IL, USA*
HUAWEI LI • *Program in Molecular Medicine, University of Massachusetts Medical School, Worcester, MA, USA*
JAMES K. LIAO • *Vascular Medicine Research Unit, Brigham and Women's Hospital and Harvard Medical School, Cambridge, MA, USA*
QINGSONG LIU • *Department of Cancer Biology, Dana Farber Cancer Institute, Boston, MA, USA; Department of Biological Chemistry and Molecular Pharmacology, Harvard Medical School, Boston, MA, USA*
MARIA CHIARA MAIURI • *INSERM Institut Gustave Roussy, U848, Villejuif, France; Institut Gustave Roussy, Villejuif, France; Université Paris-Sud, Le Kremlin-Bicêtre, France; Dipartimento di Farmacologia Sperimentale, Università degli Studi di Napoli Federico II, Napoli, Italy*
ARISTOTLE M. MANNAN • *Department of Molecular and Integrative Physiology, Life Sciences Institute, Ann Arbor, MI, USA; Division of Nephrology in the Department of Internal Medicine, University of Michigan Medical School, Ann Arbor, MI, USA*
ANDREW MARKHARD • *Whitehead Institute for Biomedical Research, Cambridge, MA, USA*
HIROYUKI MORI • *Department of Molecular and Integrative Physiology, University of Michigan, Ann Arbor, MI, USA*
EUGENIA MORSELLI • *INSERM Institut Gustave Roussy, U848, Villejuif, France; Institut Gustave Roussy, Villejuif, France; Université Paris-Sud, Le Kremlin-Bicêtre, France*
LEON O. MURPHY • *Developmental and Molecular Pathways, Novartis Institutes for BioMedical Research, Cambridge, MA, USA*
DUYU NIE • *Department of Neurology, The F.M. Kirby Neurobiology Center, Children's Hospital Boston, Harvard Medical School, Boston, MA, USA*
BEAT NYFELER • *Developmental and Molecular Pathways, Novartis Institutes for BioMedical Research, Cambridge, MA, USA*

NANCY OTTO • *Department of Molecular Carcinogenesis, The University of Texas M. D. Anderson Cancer Center, Smithville, TX, USA*
ANGELA PAPAGEORGIOU • *Diabetes Unit and Medical Services and Department of Molecular Biology, Massachusetts General Hospital, Boston, MA, USA; Department of Medicine, Harvard Medical School, Boston, MA, USA*
JONATHAN D. POWELL • *Sidney-Kimmel Comprehensive Cancer Research Center, Johns Hopkins University School of Medicine, Baltimore, MD, USA*
CHRISTOPHER G. PROUD • *School of Biological Sciences, University of Southampton, Southampton, UK*
NICHOLAS R. RENSING • *Department of Neurology, The Hope Center for Neurological Disorders, Washington University School of Medicine, St. Louis, MO, USA*
MARGIT ROSNER • *Medical Genetics, Medical University of Vienna, Vienna, Austria*
DAVID M. SABATINI • *Whitehead Institute for Biomedical Research, Cambridge, MA, USA; Department of Biology, Howard Hughes Medical Institute, Massachusetts Institute of Technology, Cambridge, MA, USA; Koch Center for Integrative Cancer Research at MIT, Cambridge, MA, USA*
MUSTAFA SAHIN • *Department of Neurology, The F.M. Kirby Neurobiology Center, Children's Hospital Boston, Harvard Medical School, Boston, MA, USA*
DOS D. SARBASSOV • *Department of Molecular and Cellular Oncology, University of Texas M. D. Anderson Cancer Center, Houston, TX, USA; The University of Texas Graduate School of Biomedical Sciences at Houston, Houston, TX, USA*
TATTYM SHAIKEN • *Department of Molecular and Cellular Oncology, University of Texas M. D. Anderson Cancer Center, Houston, TX, USA*
TAEBO SIM • *Department of Cancer Biology, Dana Farber Cancer Institute, Boston, MA, USA; Department of Biological Chemistry and Molecular Pharmacology, Harvard Medical School, Boston, MA, USA*
ANGELA STABILINI • *San Raffaele Diabetes Research Institute, Milan, Italy*
CARSON C. THOREEN • *Department of Cancer Biology, Dana Farber Cancer Institute, Boston, MA, USA; Department of Biological Chemistry and Molecular Pharmacology, Harvard Medical School, Boston, MA, USA*
LOURDES TORAL-BARZA • *Pfizer Oncology Research Unit, Pearl River, NY, USA*
ELEONORA TRESOLDI • *San Raffaele Telethon Institute for Gene Therapy, Milan, Italy*
DOUANGSONE D. VADYSIRISACK • *Massachusetts General Hospital Cancer Center, Harvard Medical School, Boston, MA, USA*
ILIO VITALE • *INSERM Institut Gustave Roussy, U848, Villejuif, France; Institut Gustave Roussy, Villejuif, France; Université Paris-Sud, Le Kremlin-Bicêtre, France*
CHERYL L. WALKER • *Department of Molecular Carcinogenesis, The University of Texas M. D. Anderson Cancer Center, Smithville, TX, USA*
CHAO-YUNG WANG • *Second Section of Cardiology, Department of Medicine, Chang Gung Memorial Hospital and Chang Gung University College of Medicine, Taoyuan, Taiwan*
FEI WANG • *Department of Cell and Developmental Biology, Institute for Genomic Biology, University of Illinois at Urbana-Champaign, Urbana, IL, USA*

JINHUA WANG • *Department of Cancer Biology, Dana Farber Cancer Institute, Boston, MA, USA; Department of Biological Chemistry and Molecular Pharmacology, Harvard Medical School, Boston, MA, USA*
XUEMIN WANG • *School of Biological Sciences, University of Southampton, Southampton, UK*
THOMAS WEICHHART • *Division of Nephrology and Dialysis, Department of Internal Medicine III, Medical University Vienna, Vienna, Austria*
WILLIAM A. WEISS • *Departments of Neurology, Pediatrics, Neurological Surgery and Brain Tumor Research Center, and Helen Diller Family Comprehensive Cancer Center, University of California, San Francisco, CA, USA*
CHRISTOPHER J. WILSON • *Developmental and Molecular Pathways, Novartis Institutes for BioMedical Research, Cambridge, MA, USA*
MICHAEL WONG • *Department of Neurology, The Hope Center for Neurological Disorders, Washington University School of Medicine, St. Louis, MO, USA*
ZHONG YAO • *Department of Biological Regulation, Weizmann Institute of Science, Rehovot, Israel; Department of Biochemistry and Molecular Biology, Centre for High-Throughput Biology, University of British Columbia, Vancouver, BC, Canada*
AMÉNA BEN YOUNES • *INSERM Institut Gustave Roussy, U848, Villejuif, France; Institut Gustave Roussy, Villejuif, France; Université Paris-Sud, Le Kremlin-Bicêtre, France*
KER YU • *Cancer Pharmacology, Fudan University, Shanghai, PR, China*
JIANMING ZHANG • *Department of Cancer Biology, Dana Farber Cancer Institute, Boston, MA, USA; Department of Biological Chemistry and Molecular Pharmacology, Harvard Medical School, Boston, MA, USA*
JIAXI ZHOU • *Department of Cell and Developmental Biology, Institute for Genomic Biology, University of Illinois at Urbana-Champaign, Urbana, IL, 61801, USA*
VITTORIA ZINZALLA • *Biozentrum, University of Basel, Basel, Switzerland*

Chapter 1

Mammalian Target of Rapamycin: A Signaling Kinase for Every Aspect of Cellular Life

Thomas Weichhart

Abstract

The mammalian (or mechanistic) target of rapamycin (mTOR) is an evolutionarily conserved serine-threonine kinase that is known to sense the environmental and cellular nutrition and energy status. Diverse mitogens, growth factors, and nutrients stimulate the activation of the two mTOR complexes mTORC1 and mTORC2 to regulate diverse functions, such as cell growth, proliferation, development, memory, longevity, angiogenesis, autophagy, and innate as well as adaptive immune responses. Dysregulation of the mTOR pathway is frequently observed in various cancers and in genetic disorders, such as tuberous sclerosis complex or cystic kidney disease. In this review, I will give an overview of the current understanding of mTOR signaling and its role in diverse tissues and cells. Genetic deletion of specific mTOR pathway proteins in distinct tissues and cells broadened our understanding of the cell-specific roles of mTORC1 and mTORC2. Inhibition of mTOR is an established therapeutic principle in transplantation medicine and in cancers, such as renal cell carcinoma. Pharmacological targeting of both mTOR complexes by novel drugs potentially expand the clinical applicability and efficacy of mTOR inhibition in various disease settings.

Key words: mTOR, Rapamycin, Immunity, Cancer, Kinase

1. Identification of mTOR

When the antifungal drug rapamycin was isolated from the soil bacterium *Streptomyces hygroscopicus* on Easter Island (Rapa Nui) in the 1970s, nobody could have imagined back in those days that this drug would be fundamental for the identification of a signaling network that regulates so many different aspects of cellular life (1). Initially, rapamycin was developed as antifungal agent, but soon afterward it was found that rapamycin possesses immunosuppressive and antiproliferative properties (2). Yeast genetic screens discovered that rapamycin inhibits two genes called target of rapamycin 1 and 2 (TOR1 and TOR2) and later the mammalian homolog mammalian

Thomas Weichhart (ed.), *mTOR: Methods and Protocols*, Methods in Molecular Biology, vol. 821,
DOI 10.1007/978-1-61779-430-8_1, © Springer Science+Business Media, LLC 2012

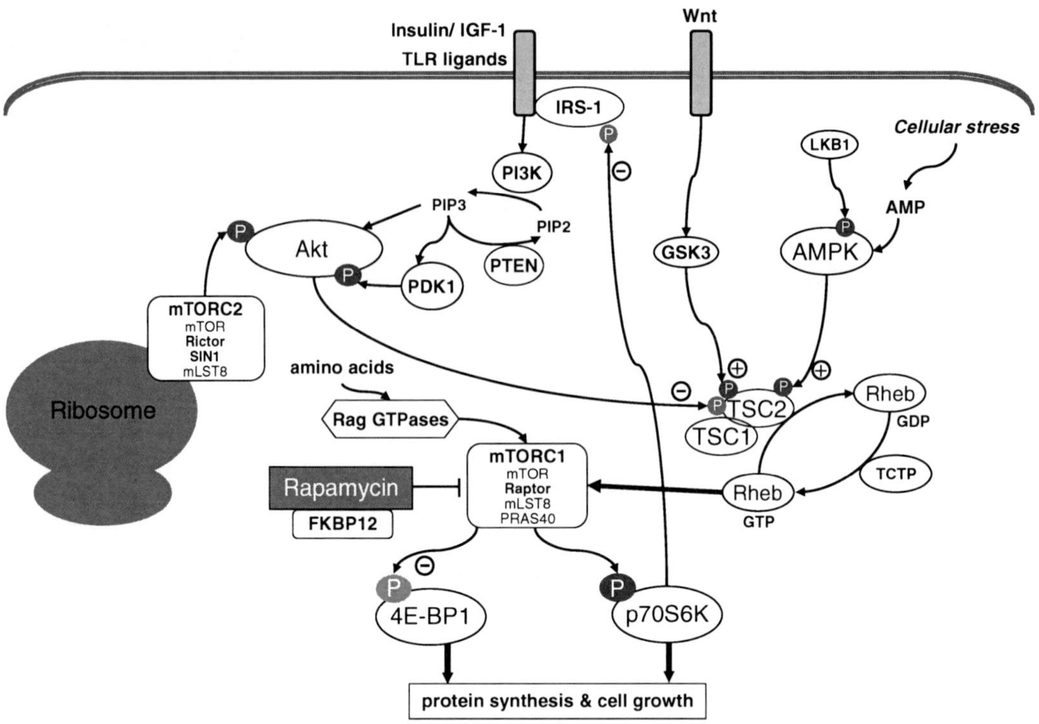

Fig. 1. The current understanding of mTOR signaling. All pathway members, which are discussed in the text, are shown. Please note that additional proteins in the mTOR pathway have been described that are comprehensively reviewed by Zoncu et al. (3) or Yang and Guan (1).

(or mechanistic) TOR (mTOR; also known as FRAP1 or RAFT1) was identified (1). Rapamycin does not directly inhibit mTOR, but instead binds the FK506-binding protein 12 (FKBP12), and it is this complex, which inhibits mTOR (Fig. 1) (1). mTOR is a signaling kinase that affects broad aspects of cellular functions, including metabolism, growth, survival, aging, synaptic plasticity, immunity, and memory (3). It is an atypical serine-threonine protein kinase belonging to the phosphatidylinositol kinase-related kinase (PIKK) family with a predicted molecular weight of 290 kDa. mTOR is the catalytic subunit of two distinct complexes called mTOR complex 1 (mTORC1) and mTORC2.

2. The mTOR Signaling Pathway

The mTOR signaling pathway is highly conserved from yeast to humans and is activated by a variety of divergent stimuli. mTOR senses cellular energy levels by monitoring cellular ATP:AMP levels via the AMP-activated protein kinase (AMPK), growth factors such as insulin and insulin-like growth factor 1 (IGF-1) via the insulin receptor and the IGF-1 receptor respectively, amino acids via Rag GTPases, and signals from the Wnt family via glycogen synthase kinase 3 (GSK3) (1, 4). In the immune system, stimulation of antigen receptors (T and B cell receptors), cytokine receptors (e.g., Interleukin [IL]-2 receptor), or toll-like receptors (TLRs) (1, 5–8) all lead to the activation of mTOR (Fig. 1). As the archetypical and best-documented example, triggering of the insulin receptor activates tyrosine kinase adaptor molecules at the cell membrane leading to the recruitment of the class I family of phosphatidylinositol-3 kinases (PI3K) to the receptor complex (9). Following receptor engagement, PI3K phosphorylates phosphatidylinositol 4,5-bisphosphate (PIP2) to generate phosphatidylinositol-3,4,5-trisphosphate (PIP3) as a second messenger to recruit and activate downstream targets, including Akt (Fig. 1). The serine-threonine kinase Akt, also termed protein kinase B (PKB), connects PI3K and mTOR (10). There are three highly homologous isoforms of Akt encoded in the genome (Akt1, Akt2, and Akt3) representing some of the most important survival kinases involved in regulating a similarly wide array of cellular processes as mTOR, including metabolism, growth, proliferation, and apoptosis (11). The tumor suppressor phosphatase and tensin homolog deleted on chromosome 10 (PTEN) is a lipid phosphatase and dephosphorylates PIP3 to negatively regulate PI3K signaling (12). A main effector of Akt is the tuberous sclerosis complex (TSC) protein 2 (TSC2) (1). TSC2 is a tumor suppressor that forms a heterodimeric complex with TSC1. Mutations in TSC1 or TSC2 give rise to the hamartoma syndrome TSC and the proliferative lung disorder lymphangioleiomyomatosis (13). TSC2 is phosphorylated and inactivated by Akt leading to a loss of suppression of mTOR by the TSC1–TSC2 complex (1). Moreover, downstream of TSC1–TSC2, the GTPase Ras homolog enriched in brain (Rheb) is essential for mTOR activation (14). Amino acid-induced activation of the Rag GTPases promotes the translocation of mTORC1 to lyosomal compartments that contain Rheb (4, 15–17). mTOR controls protein synthesis through the direct phosphorylation and inactivation of a repressor of mRNA translation, eukaryotic initiation factor 4E-binding protein 1 (4E-BP1), and through phosphorylation and activation of S6 kinase (S6K1 or p70S6K), which in turn phosphorylates the ribosomal protein S6 (18). Cytokines, growth factors, amino acids, insulin, or TLR ligands activate mTOR and increase the phosphorylation status of 4E-BP1 and S6K1 in a rapamycin-sensitive manner (18). Importantly, a negative feedback loop has been described in the insulin receptor pathway involving the insulin receptor substrate-1 (IRS-1). Activation of mTORC1 promotes an inhibitory phosphorylation of IRS-1 via S6K1 that leads to inactivation of PI3K and Akt (Fig. 1). Conversely, inhibition of mTORC1 should lead to hyperactivation of Akt, which indeed has been documented in some rapamycin-treated cancer patients (3). Loss of mTOR function leads to an arrest in the G1 phase of the cell-cycle along with a severe reduction in protein synthesis in many cells.

mTOR exists in two distinct complexes in the cell. mTORC1 consists of the regulatory-associated protein of mTOR (Raptor), a conserved 150 kDa protein, which recruits S6K1 and 4-E-BP1, and the adaptor protein mLST8 (19). Rapamycin inhibits mTORC1 activity by blocking its interaction with Raptor (20). mTORC2 is composed of mTOR, mLST8 and the adaptor proteins Rictor (rapamycin-insensitive companion of mTOR) and Sin1 (21). mTORC2 is usually rapamycin-insensitive and is thought to regulate actin cytoskeleton dynamics. mTORC2 directly phosphorylates Akt at serine 473 via cotranslational phosphorylation and through a direct interaction with the ribosome (22–24). The functional role of mTORC2, which is upstream of Akt and mTORC1, is not well understood within this signaling circuit (25). Interestingly, long-term treatment with rapamycin (>18 h) alters the mTORC1: mTORC2 equilibrium resulting in reduced mTORC2 levels and, hence, impaired Akt signaling in some cells (26). Collectively, receptor engagement leads to a coordinated activation of PI3K, Akt, TSC1–TSC2 and Rheb, which are integrated at the level of mTORC1 (1, 27). Interestingly, mTORC1 seems to be predominantly cytoplasmic, whereas mTORC2 is abundant in the cytoplasm and the nucleus in human primary fibroblasts (28). The detailed functions of the mTOR complexes in these compartments are currently unknown. Further aspects of mTOR signaling are comprehensively reviewed, e.g., by Zoncu et al. (3), Yang and Guan (1).

3. Cell Growth and Proliferation

mTORC1 controls growth and proliferation by modulating mRNA translation through phosphorylation of the 4E-BP1, 2, and 3 and the S6K1 and 2 (29). More specifically, the three 4E-BPs do not regulate cell size, but they block cell proliferation by inhibiting the translation of messenger RNAs that encode proteins involved in proliferation and cell cycle progression (29). In T lymphocytes, mTOR controls cell cycle progression from the G1 into S phase in IL-2-stimulated cells (30). The cyclin-dependent kinase (Cdk) enzymes, when associated with the G1 cyclins D and E, are rate-limiting for the entry into the S phase of the cell cycle. IL-2 activates Cdk by causing the elimination of the Cdk inhibitor protein p27Kip1, a process that is prevented by rapamycin in T cells (31). Moreover, mTORC2 activates the serum- and glucocorticoid-induced protein kinase-1 (SGK1), which in turn phosphorylates p27Kip1. Once phosphorylated, p27Kip1 is retained in the cytoplasm rendering it incapable of blocking Cdk1 or Cdk2 activation and therefore allowing entry into the cell cycle (32, 33).

4. Autophagy

Autophagy is a starvation-induced degradation of cytosolic components ranging from individual proteins (microautophagy) to entire organelles (macroautophagy). Autophagy is critical in providing substrates for energy production under conditions of limited nutrient supply (3). mTORC1 actively suppresses autophagy and, conversely, inhibition of mTORC1 strongly induces autophagy (34). In *S. cerevisiae*, TOR-dependent phosphorylation of autophagy-related 13 (Atg13) disrupts the Atg1–Atg13–Atg17 complex that triggers the formation of the autophagosome (3). The mammalian homologs of yeast Atg13 and Atg1, ATG13 and ULK1, bind to the 200 kDa FAK family kinase-interacting protein (FIP200; a putative ortholog of Atg17) and the mammalian-specific component ATG101 (35). mTOR phosphorylates ATG13 and ULK1 to block autophagosome initiation (35).

5. Developmental Aspects of mTOR Signaling

As the mTOR pathway is critical for many basic aspects of cell biology, it is not surprising that mTOR also plays a prominent role in development. For example, embryonic homozygous deletion of mTOR leads to a developmental arrest at E5.5 (36). Moreover, mTOR−/− embryos show a defect in inner cell mass proliferation consistent with an inability to establish embryonic stem cells from mTOR-deficient embryos (36). The catalytic function of mTOR is critical for the embryonic development as knock-in mice carrying a mutation in the catalytic domain of mTOR die before embryonic day 6.5 (37). Rheb, the essential upstream regulator of mTORC1, is likewise important for embryonic development (38, 39). Interestingly, in contrast to mTOR or Raptor mutants, the inner cell mass of Rheb−/− embryos differentiate normally (38). Moreover, embryonic deletion of Rheb in neural progenitor cells abolishes mTORC1 signaling in the developing brain and increases mTORC2 signaling. While embryonic and early postnatal brain development appears grossly normal in these mice, there are defects in myelination (39). These results suggest that mTORC1 signaling plays a role in selective cellular adaptations but is not decisive for general cellular viability (39).

6. Nerve Function and Epilepsy

TSC is an autosomal dominant disease caused by mutations in either TSC1 or TSC2. TSC is characterized by the presence of benign tumors called hamartomas, which within the brain are known as cortical tubers. Neurological manifestations in TSC patients include epilepsy, mental retardation, and autistic features (40). In mice, Tsc2 haploin sufficiency causes aberrant retinogeniculate projections and the TSC2–Rheb–mTOR pathway controls axon guidance in the visual system (41). In addition, dorsal root ganglial neurons (DRGs) in the peripheral nervous system activate mTOR following damage to enhance axonal growth capacity (42). Hence, the mTOR pathway has a central role in axon guidance, regeneration, and growth (43).

7. mTOR and Pluripotency in Stem Cells

Human embryonic stem cells (hESCs), derived from blastocyst-stage embryos, can undergo long-term self-renewal and have the remarkable ability to differentiate into multiple cell types in the human body (44). A role for mTOR in these processes has recently been appreciated (45). mTOR integrates signals from extrinsic pluripotency-supporting factors and represses the transcriptional activities of a subset of developmental and growth inhibitory genes in hESCs. Repression of the developmental genes by mTOR is necessary for the maintenance of hESC pluripotency (46). A similar mechanism is operative in human amniotic fluid stem cells (47). On the other hand, it has been proposed that mTOR-mediated activation of S6K1 induces differentiation of pluripotent hESCs (48). In that line, mTORC1 activation is detrimental to stem cell maintenance in spermatogonial progenitor cells (SPCs) (49, 50).

8. Regulation of β-Cell Function and Obesity by mTOR Signaling

mTOR regulates the metabolism, growth and survival of β-cells, the cardinal cells in the pancreas that produce insulin, a hormone that controls the level of glucose in the blood to regulate food intake (51). Studies in S6K1 knockout mice demonstrate a central positive role of mTOR/S6K1 signaling in β-cell growth and function (51). Indeed, these mice develop glucose intolerance despite increased insulin sensitivity. This is associated with depletion of the pancreatic insulin content, hypoinsulinemia and reduced β-cell mass suggesting that lack of S6K1 activity impairs β-cell growth and function (51). In addition, it has been shown that obesity develops in older hypothalamic Tsc1 knockout animals; however, young animals display a prominent gain-of-function β-cell phenotype prior to the onset of obesity (52). Young hypothalamic Tsc1 knockout animals display improved glycemic control due to mTOR-mediated enhancement of β-cell size and insulin production. Thus, mTOR disseminates a dominant signal to promote β cell/islet size and insulin production, and this pathway is crucial for β-cell function and glycemic control (52).

9. Myeloid Phagocytes Activate mTOR to Limit Proinflammatory Responses

The immune system is a complex network of cells that protect against disease by identifying and killing pathogens and tumor cells, but it is also implicated in homeostatic mechanisms like tissue remodeling and wound healing (53).

A growing body of evidence indicates that in myeloid phagocytes (monocytes, macrophages, and myeloid DC; mDC) mTOR is crucially implicated in TLR signaling and might serve as a decision maker to control the cellular response to pathogens by modulating cytokines, chemokines, and type I interferon responses (54, 55). Inhibition of mTOR by rapamycin in these cells promotes IL-12 and IL-23 production via the transcription factor NF-κB but blocks the release of IL-10 via Stat3 (7, 8, 56, 57). These results have been confirmed in kidney transplant patients *in vivo*. The most prominent transcriptional alterations in peripheral blood from rapamycin-treated kidney transplant recipients affect the innate immune cell compartment and hyperactivation of NF-κB-mediated proinflammatory pathways (58). Moreover, kidney transplant patients on rapamycin display an increased inflammatory and immunostimulatory potential of myeloid monocytes and dendritic cells *in vivo* compared with patients on calcineurin inhibitors (59, 60). Moreover, rapamycin can augment inflammation and pulmonary injury by enhancing NF-κB activity in the lung of tobacco-exposed mice (61). In dendritic cells, autophagy facilitates the presentation of endogenous proteins on MHC class I and class II molecules. This leads to the activation of CD4+ T cells and connects autophagy in innate immune cells with enhanced adaptive immune responses. For example, in Mycobacterium tuberculosis-infected DCs, rapamycin-induced autophagy enhances the presentation of mycobacterial antigens (62).

10. mTOR Mediates Type I Interferon Production in Plasmacytoid Dendritic Cells

Plasmacytoid DCs (pDCs) constitute a specialized cell population that produce large amounts of type I interferons (IFN) in response to viral infection via the activation of cytoplasmic receptors or TLRs (63). The mTOR pathway is important for the regulation of type I IFN production in murine pDCs (6, 64). Inhibition of mTOR or its downstream mediators S6K1 and S6K2 during pDC activation block the phosphorylation and nuclear translocation of the transcription factor IRF-7, which results in impaired IFN-α and β production (6). In addition, translation of IRF-7 in pDCs is negatively regulated by the 4E-BP pathway downstream of mTOR (65). Hence, mTOR via its two downstream effectors, 4E-BP and S6K, controls translation and activation of IRF-7.

11. CD4+ T Helper Differentiation and the Role of mTOR

Peripheral CD4+ T helper (Th) lymphocytes are critical in regulating immune responses as well as autoimmune and inflammatory diseases. Upon activation, naïve CD4+ Th cells differentiate into distinct effector subsets depending on the cytokine milieu (66). Recent data show that mTOR-deficient naïve CD4+ T cells are unable to differentiate into Th1, Th2, and Th17 cells, but preferentially develop into induced regulatory T (Treg) cells, which can potently suppress adaptive immune responses (67). In line, rapamycin is able to enrich Treg cells *in vitro* (68). mTORC2 is important in these processes as Rictor-deficient CD4+ T cells are unable to differentiate into Th2 cells demonstrating that mTORC2 is critical for Th2 differentiation (69, 70). On the other hand, Rheb-deficient T cells fail to generate Th17 responses *in vitro* and *in vivo* (67). The role of mTORC1 and mTORC2 for Th1 differentiation is currently under debate.

12. mTOR Controls CD8+ T Lymphocyte Migration and Memory

Another insight how mTOR regulates adaptive immunity *in vivo* can be deduced from recent experiments showing that mTOR influences the migratory properties of murine CD8+ T lymphocytes and the differentiation of CD8+ memory T cells (71, 72). The migratory properties of naïve CD8+ T lymphocytes into the lymph nodes crucially depends on the constitutive expression of the chemokine receptor 7 (CCR7) and L-selectin (CD62L), which is controlled by the transcription factor Krüppel-like factor 2 (KLF2).

mTOR negatively regulates KLF2, and therefore controls migration of activated CD8+ T lymphocytes *in vivo* (71). Memory CD8+ T cells are a critical component of protective immunity, and inducing effective memory T-cell responses is a major goal of vaccines against chronic infections and tumors (72). Recently, it was demonstrated that rapamycin promotes the generation of memory CD8+ T cells after viral or bacterial infection *in vivo* (72, 73). Importantly, mTOR acts cell intrinsically to regulate memory T-cell differentiation (72).

13. mTOR and Longevity

The molecular and cellular mechanisms that regulate aging are currently under scrutiny because aging is linked to many human diseases. The TOR pathway is emerging as a key regulator of aging and inhibition of mTOR by rapamycin or genetic deletion has been shown to expand life-span of invertebrates, including yeast, nematodes, and fruit flies (74). Even more strikingly, rapamycin extends median and maximal life span of male and female mice when feeding began at 600 days of age (75). Rapamycin is the only pharmacological substance so far, which has been shown to expand life span in a mammal species. Calorie restriction extends life in rhesus monkeys (76) and inhibits mTOR signaling, however the underlying mechanism, how rapamycin exerts its life-extending effects is currently unknown but potentially involves S6K (74, 77). It should be noted that the concentrations used in these mice (~80 ng/ml) were about ten times higher than the concentrations, which are currently used in human transplantation medicine.

14. Current Applications of mTOR Inhibitors

In 1999, rapamycin was approved by the US Food and Drug Administration for the prevention of kidney allograft transplant rejection (78). Currently, rapamycin (sirolimus) and its derivative RAD0001 (everolimus) are mostly evaluated and used as alternative treatments in all organ and bone-marrow transplantations in lieu of calcineurin inhibitors (e.g., cyclosporine) which cause chronic renal allograft damage. Rapamycin-eluting stents are used for the prevention of in-stent restenosis (a thrombosis in a blood vessel) after percutaneous coronary revascularization in patients with coronary artery lesions (79) Three rapamycin analogs, CCI-779, RAD001, and AP23573 are currently in advanced clinical trials for the treatment of cancers (80). These include renal cell carcinomas, bone sarcomas, glioblastomas, mantle cell lymphomas, and endometrial carcinomas. Patients with TSC or sporadic lymphangioleiomyomatosis often develop benign renal neoplasms called angiomyolipomas, which impair renal function. In initial clinical trials, inhibition of mTOR showed some promise in leading to a partial angiomyolipoma regression (13, 81).

15. Novel Inhibitors Targeting mTORC1 and mTORC2

Rapamycin was initially seen as a holy grail for cancer therapy; however, the potency of rapamycin as an anticancer drug in clinical trials was limited to the tumor types described above (3). The constricted success of rapamycin and the appreciation that mTORC1 has both rapamycin-sensitive and rapamycin-insensitive substrates (82, 83) as well as the finding that mTORC2 activation is largely unaffected or even increased by acute cellular exposure to rapamycin promoted the development of novel active-site mTOR inhibitors that fully block both mTOR complexes. These novel ATP-competitive inhibitors have improved anticancer activity compared to rapamycin in a variety of solid tumor models *in vitro* and *in vivo* and at least three candidate compounds have entered clinical trials (84–88). They show a consistently potent effect against tumors that are driven by PI3K–Akt; however, the clinical effectiveness of these novel mTOR inhibitors remain to be determined.

16. Conclusion

In the last years, there has been a tremendous amount of novel data establishing how the mTOR pathway is regulated by various environmental and intracellular molecules and how mTOR controls many different processes implicated in health and disease. Inhibition of mTOR is currently established in allogeneic transplantation and in certain forms of cancer. Novel applications for mTOR inhibitors, such as the generation of high number of Treg cells ex vivo for immunotherapy or the improvement of vaccines by the promotion of memory CD8+ T-cell responses, are currently evaluated. The identification of novel signaling pathways important for the control of mTOR in different tissues of the body may open further clinical applications of inhibiting mTOR in human disease.

Acknowledgments

TW is supported by the Else-Kröner Fresenius Stiftung. I apologize to those authors whose primary work I did not reference directly in the text.

References

1. Yang, Q., and Guan, K. L. (2007) Expanding mTOR signaling. *Cell Res* **17,** 666–81.
2. Abraham, R. T., and Wiederrecht, G. J. (1996) Immunopharmacology of rapamycin. *Annu Rev Immunol* **14**, 483–510.
3. Zoncu, R., Efeyan, A., and Sabatini, D. M. (2011) mTOR: from growth signal integration to cancer, diabetes and ageing. *Nat Rev Mol Cell Biol* **12**, 21–35.
4. Avruch, J., Long, X., Ortiz-Vega, S., Rapley, J., Papageorgiou, A., and Dai, N. (2009) Amino acid regulation of TOR complex 1. *Am J Physiol Endocrinol Metab* **296**, E592–602.
5. Donahue, A. C., and Fruman, D. A. (2007) Distinct signaling mechanisms activate the target of rapamycin in response to different B-cell stimuli. *Eur J Immunol* **37**, 2923–36.
6. Cao, W., Manicassamy, S., Tang, H., Kasturi, S. P., Pirani, A., Murthy, N., et al. (2008) Toll-like receptor-mediated induction of type I interferon in plasmacytoid dendritic cells requires the rapamycin-sensitive PI(3) K-mTOR-p70S6K pathway. *Nat Immunol* **9**, 1157–64.
7. Schmitz, F., Heit, A., Dreher, S., Eisenacher, K., Mages, J., Haas, T., et al. (2008) Mammalian target of rapamycin (mTOR) orchestrates the defense program of innate immune cells. *Eur J Immunol* **38**, 2981–92.
8. Weichhart, T., Costantino, G., Poglitsch, M., Rosner, M., Zeyda, M., Stuhlmeier, K. M., et al. (2008) The TSC-mTOR signaling pathway regulates the innate inflammatory response. *Immunity* **29**, 565–77.
9. Fruman, D. A. (2004) Towards an understanding of isoform specificity in phosphoinositide 3-kinase signalling in lymphocytes. *Biochem Soc Trans* **32**, 315–9.
10. Guertin, D. A., and Sabatini, D. M. (2005) An expanding role for mTOR in cancer. *Trends Mol Med* **11**, 353–61.
11. Brazil, D. P., Yang, Z. Z., and Hemmings, B. A. (2004) Advances in protein kinase B signalling: AKTion on multiple fronts. *Trends Biochem Sci* **29**, 233–42.
12. Deane, J. A., and Fruman, D. A. (2004) Phosphoinositide 3-kinase: diverse roles in immune cell activation. *Annu Rev Immunol* **22**, 563–98.
13. Paul, E., and Thiele, E. (2008) Efficacy of sirolimus in treating tuberous sclerosis and lymphangioleiomyomatosis. *N Engl J Med* **358**, 190–2.
14. Stocker, H., Radimerski, T., Schindelholz, B., Wittwer, F., Belawat, P., Daram, P., et al. (2003) Rheb is an essential regulator of S6K in controlling cell growth in Drosophila. *Nat Cell Biol* **5**, 559–65.
15. Kim, E., Goraksha-Hicks, P., Li, L., Neufeld, T. P., and Guan, K. L. (2008) Regulation of TORC1 by Rag GTPases in nutrient response. *Nat Cell Biol* **10**, 935–45.
16. Sancak, Y., Bar-Peled, L., Zoncu, R., Markhard, A. L., Nada, S., and Sabatini, D. M. (2010) Ragulator-Rag complex targets mTORC1 to the lysosomal surface and is necessary for its activation by amino acids. *Cell* **141**, 290–303.
17. Sancak, Y., Peterson, T. R., Shaul, Y. D., Lindquist, R. A., Thoreen, C. C., Bar-Peled, L., et al. (2008) The Rag GTPases bind raptor and mediate amino acid signaling to mTORC1. *Science* **320**, 1496–501.
18. Hay, N., and Sonenberg, N. (2004) Upstream and downstream of mTOR. *Genes Dev* **18**, 1926–45.
19. Kim, D. H., and Sabatini, D. M. (2004) Raptor and mTOR: subunits of a nutrient-sensitive complex. *Curr Top Microbiol Immunol* **279**, 259–70.
20. Oshiro, N., Yoshino, K., Hidayat, S., Tokunaga, C., Hara, K., Eguchi, S., et al. (2004) Dissociation of raptor from mTOR is a mechanism of rapamycin-induced inhibition of mTOR function. *Genes Cells* **9**, 359–66.
21. Jacinto, E., Loewith, R., Schmidt, A., Lin, S., Ruegg, M. A., Hall, A., et al. (2004) Mammalian TOR complex 2 controls the actin cytoskeleton and is rapamycin insensitive. *Nat Cell Biol* **6**, 1122–8.
22. Zinzalla, V., Stracka, D., Oppliger, W., and Hall, M. N. (2011) Activation of mTORC2 by Association with the Ribosome. *Cell* **144**, 757–68.
23. Oh, W. J., Wu, C. C., Kim, S. J., Facchinetti, V., Julien, L. A., Finlan, M., et al. (2010) mTORC2 can associate with ribosomes to promote cotranslational phosphorylation and stability of nascent Akt polypeptide. *EMBO J* **29**, 3939–51.
24. Sarbassov, D. D., Guertin, D. A., Ali, S. M., and Sabatini, D. M. (2005) Phosphorylation and regulation of Akt/PKB by the rictor-mTOR complex. *Science* **307**, 1098–101.
25. Huang, J., Dibble, C. C., Matsuzaki, M., and Manning, B. D. (2008) The TSC1-TSC2 complex is required for proper activation of mTOR complex 2. *Mol Cell Biol* **28**, 4104–15.
26. Sarbassov dos, D., Ali, S. M., Sengupta, S., Sheen, J. H., Hsu, P. P., Bagley, A. F., et al. (2006) Prolonged rapamycin treatment inhibits mTORC2 assembly and Akt/PKB. *Mol Cell* **22**, 159–68.

27. Rosner, M., Hanneder, M., Siegel, N., Valli, A., and Hengstschlager, M. (2008) The tuberous sclerosis gene products hamartin and tuberin are multifunctional proteins with a wide spectrum of interacting partners. *Mutat Res* **658**, 234–46.
28. Rosner, M., and Hengstschlager, M. (2008) Cytoplasmic and nuclear distribution of the protein complexes mTORC1 and mTORC2: rapamycin triggers dephosphorylation and delocalisation of the mTORC2 components rictor and sin1. *Hum Mol Genet* **17**, 2934–48.
29. Dowling, R. J., Topisirovic, I., Alain, T., Bidinosti, M., Fonseca, B. D., Petroulakis, E., et al. (2010) mTORC1-mediated cell proliferation, but not cell growth, controlled by the 4E-BPs. *Science* **328**, 1172–6.
30. Morice, W. G., Brunn, G. J., Wiederrecht, G., Siekierka, J. J., and Abraham, R. T. (1993) Rapamycin-induced inhibition of p34cdc2 kinase activation is associated with G1/S-phase growth arrest in T lymphocytes. *J Biol Chem* **268**, 3734–8.
31. Nourse, J., Firpo, E., Flanagan, W. M., Coats, S., Polyak, K., Lee, M. H., et al. (1994) Interleukin-2-mediated elimination of the p27Kip1 cyclin-dependent kinase inhibitor prevented by rapamycin. *Nature* **372**, 570–3.
32. Hong, F., Larrea, M. D., Doughty, C., Kwiatkowski, D. J., Squillace, R., and Slingerland, J. M. (2008) mTOR-raptor binds and activates SGK1 to regulate p27 phosphorylation. *Mol Cell* **30**, 701–11.
33. Rosner, M., Freilinger, A., Hanneder, M., Fujita, N., Lubec, G., Tsuruo, T., et al. (2007) p27Kip1 localization depends on the tumor suppressor protein tuberin. *Hum Mol Genet* **16**, 1541–56.
34. Noda, T., and Ohsumi, Y. (1998) Tor, a phosphatidylinositol kinase homologue, controls autophagy in yeast. *J Biol Chem* **273**, 3963–6.
35. Jung, C. H., Jun, C. B., Ro, S. H., Kim, Y. M., Otto, N. M., Cao, J., et al. (2009) ULK-Atg13-FIP200 complexes mediate mTOR signaling to the autophagy machinery. *Mol Biol Cell* **20**, 1992–2003.
36. Gangloff, Y. G., Mueller, M., Dann, S. G., Svoboda, P., Sticker, M., Spetz, J. F., et al. (2004) Disruption of the mouse mTOR gene leads to early postimplantation lethality and prohibits embryonic stem cell development. *Mol Cell Biol* **24**, 9508–16.
37. Shor, B., Cavender, D., and Harris, C. (2009) A kinase-dead knock-in mutation in mTOR leads to early embryonic lethality and is dispensable for the immune system in heterozygous mice. *BMC Immunol* **10**, 28.
38. Goorden, S. M., Hoogeveen-Westerveld, M., Cheng, C., van Woerden, G. M., Mozaffari, M., Post, L., et al. (2011) Rheb is essential for murine development. *Mol Cell Biol* **31**, 1672–8.
39. Zou, J., Zhou, L., Du, X. X., Ji, Y., Xu, J., Tian, J., et al. (2011) Rheb1 is required for mTORC1 and myelination in postnatal brain development. *Dev Cell* **20**, 97–108.
40. Haidinger, M., Hecking, M., Weichhart, T., Poglitsch, M., Enkner, W., Vonbank, K., et al. (2010) Sirolimus in renal transplant recipients with tuberous sclerosis complex: clinical effectiveness and implications for innate immunity. *Transpl Int* **23**, 777–85.
41. Nie, D., Di Nardo, A., Han, J. M., Baharanyi, H., Kramvis, I., Huynh, T., et al. (2010) Tsc2-Rheb signaling regulates EphA-mediated axon guidance. *Nat Neurosci* **13**, 163–72.
42. Abe, N., Borson, S. H., Gambello, M. J., Wang, F., and Cavalli, V. (2010) Mammalian target of rapamycin (mTOR) activation increases axonal growth capacity of injured peripheral nerves. *J Biol Chem* **285**, 28034–43.
43. Park, K. K., Liu, K., Hu, Y., Smith, P. D., Wang, C., Cai, B., et al. (2008) Promoting axon regeneration in the adult CNS by modulation of the PTEN/mTOR pathway. *Science* **322**, 963–6.
44. Li, D., Zhou, J., Wang, L., Shin, M. E., Su, P., Lei, X., et al. (2010) Integrated biochemical and mechanical signals regulate multifaceted human embryonic stem cell functions. *J Cell Biol* **191**, 631–44.
45. Zhou, J., Su, P., Wang, L., Chen, J., Zimmermann, M., Genbacev, O., et al. (2009) mTOR supports long-term self-renewal and suppresses mesoderm and endoderm activities of human embryonic stem cells. *Proc Natl Acad Sci USA* **106**, 7840–5.
46. Lee, K. W., Yook, J. Y., Son, M. Y., Kim, M. J., Koo, D. B., Han, Y. M., et al. (2010) Rapamycin promotes the osteoblastic differentiation of human embryonic stem cells by blocking the mTOR pathway and stimulating the BMP/Smad pathway. *Stem Cells Dev* **19**, 557–68.
47. Siegel, N., Rosner, M., Unbekandt, M., Fuchs, C., Slabina, N., Dolznig, H., et al. (2010) Contribution of human amniotic fluid stem cells to renal tissue formation depends on mTOR. *Hum Mol Genet* **19**, 3320–31.
48. Easley. C.A.t., Ben-Yehudah. A., Redinger. C.J., Oliver. S.L., Varum. S.T., Eisinger. V.M., et al., (2010). mTOR-mediated activation of p70 S6K induces differentiation of pluripotent human embryonic stem cells. *Cell Reprogram.* 12, 263–73.
49. Yilmaz, O. H., Valdez, R., Theisen, B. K., Guo, W., Ferguson, D. O., Wu, H., et al. (2006)

Pten dependence distinguishes haematopoietic stem cells from leukaemia-initiating cells. *Nature* **441**, 475–82.

50. Hobbs, R. M., Seandel, M., Falciatori, I., Rafii, S., and Pandolfi, P. P. (2010) Plzf regulates germline progenitor self-renewal by opposing mTORC1. *Cell* **142**, 468–79.
51. Leibowitz, G., Cerasi, E., and Ketzinel-Gilad, M. (2008) The role of mTOR in the adaptation and failure of beta-cells in type 2 diabetes. *Diabetes Obes Metab* **10 Suppl 4**, 157–69.
52. Mori, H., Inoki, K., Opland, D., Muenzberg, H., Villanueva, E. C., Faouzi, M., et al. (2009) Critical roles for the TSC-mTOR pathway in {beta}-cell function. *Am J Physiol Endocrinol Metab* **297**, 1013–22.
53. Janeway, C. A., Jr. (2001) How the immune system works to protect the host from infection: a personal view. *Proc Natl Acad Sci USA* **98**, 7461–8.
54. Saemann, M. D., Haidinger, M., Hecking, M., Horl, W. H., and Weichhart, T. (2009) The multifunctional role of mTOR in innate immunity: implications for transplant immunity. *Am J Transplant* **9**, 2655–61.
55. Weichhart, T., and Saemann, M. D. (2009) The multiple facets of mTOR in immunity. *Trends Immunol* **30**, 218–26.
56. Yang, C. S., Song, C. H., Lee, J. S., Jung, S. B., Oh, J. H., Park, J., et al. (2006) Intracellular network of phosphatidylinositol 3-kinase, mammalian target of the rapamycin/70 kDa ribosomal S6 kinase 1, and mitogen-activated protein kinases pathways for regulating mycobacteria-induced IL-23 expression in human macrophages. *Cell Microbiol* **8**, 1158–71.
57. Ohtani, M., Nagai, S., Kondo, S., Mizuno, S., Nakamura, K., Tanabe, M., et al. (2008) Mammalian target of rapamycin and glycogen synthase kinase 3 differentially regulate lipopolysaccharide-induced interleukin-12 production in dendritic cells. *Blood* **112**, 635–43.
58. Brouard, S., Puig-Pey, I., Lozano, J. J., Pallier, A., Braud, C., Giral, M., et al. (2010) Comparative Transcriptional and Phenotypic Peripheral Blood Analysis of Kidney Recipients under Cyclosporin A or Sirolimus Monotherapy. *Am J Transplant* **10**, 2604–14.
59. Haidinger, M., Poglitsch, M., Geyeregger, R., Kasturi, S., Zeyda, M., Zlabinger, G. J., et al. (2010) A versatile role of mammalian target of rapamycin in human dendritic cell function and differentiation. *J Immunol* **185**, 3919–31.
60. Weichhart, T., Haidinger, M., Katholnig, M., Kopecky, C., Poglitsch, M., Lassnig, C., et al. (2011) Inhibition of mTOR blocks the anti-inflammatory effects of glucocorticoids in myeloid immune cells. *Blood* **117**, 4273–83.
61. Yoshida, T., Mett, I., Bhunia, A. K., Bowman, J., Perez, M., Zhang, L., et al. (2010) Rtp801, a suppressor of mTOR signaling, is an essential mediator of cigarette smoke-induced pulmonary injury and emphysema. *Nat Med* **16**, 767–73.
62. Jagannath, C., Lindsey, D. R., Dhandayuthapani, S., Xu, Y., Hunter, R. L., Jr., and Eissa, N. T. (2009) Autophagy enhances the efficacy of BCG vaccine by increasing peptide presentation in mouse dendritic cells. *Nat Med* **15**, 267–76.
63. Colonna, M., Trinchieri, G., and Liu, Y. J. (2004) Plasmacytoid dendritic cells in immunity. *Nat Immunol* **5**, 1219–26.
64. Kaur, S., Lal, L., Sassano, A., Majchrzak-Kita, B., Srikanth, M., Baker, D. P., et al. (2007) Regulatory effects of mammalian target of rapamycin-activated pathways in type I and II interferon signaling. *J Biol Chem* **282**, 1757–68.
65. Colina, R., Costa-Mattioli, M., Dowling, R. J., Jaramillo, M., Tai, L. H., Breitbach, C. J., et al. (2008) Translational control of the innate immune response through IRF-7. *Nature* **452**, 323–8.
66. Zhu, J., Yamane, H., and Paul, W. E. (2010) Differentiation of effector CD4 T cell populations (*). *Annu Rev Immunol* **28**, 445–89.
67. Delgoffe, G. M., Kole, T. P., Zheng, Y., Zarek, P. E., Matthews, K. L., Xiao, B., et al. (2009) The mTOR kinase differentially regulates effector and regulatory T cell lineage commitment. *Immunity* **30**, 832–44.
68. Battaglia, M., Stabilini, A., and Roncarolo, M. G. (2005) Rapamycin selectively expands CD4+CD25+FoxP3+ regulatory T cells. *Blood* **105**, 4743–8.
69. Lee, K., Gudapati, P., Dragovic, S., Spencer, C., Joyce, S., Killeen, N., et al. (2010) Mammalian target of rapamycin protein complex 2 regulates differentiation of Th1 and Th2 cell subsets via distinct signaling pathways. *Immunity* **32**, 743–53.
70. Weichhart, T., and Saemann, M. D. (2010) T helper cell differentiation: understanding the needs of hierarchy. *Immunity* **32**, 727–9.
71. Sinclair, L. V., Finlay, D., Feijoo, C., Cornish, G. H., Gray, A., Ager, A., et al. (2008) Phosphatidylinositol-3-OH kinase and nutrient-sensing mTOR pathways control T lymphocyte trafficking. *Nat Immunol* **9**, 513–21.
72. Araki, K., Turner, A. P., Shaffer, V. O., Gangappa, S., Keller, S. A., Bachmann, M. F., et al. (2009) mTOR regulates memory CD8 T-cell differentiation. *Nature* **460**, 108–12.
73. Pearce, E. L., Walsh, M. C., Cejas, P. J., Harms, G. M., Shen, H., Wang, L. S., et al. (2009) Enhancing CD8 T-cell memory by modulating fatty acid metabolism. *Nature* **460**, 103–7.

74. Hands, S. L., Proud, C. G., and Wyttenbach, A. (2009) mTOR's role in ageing: protein synthesis or autophagy? *Aging (Albany NY)* **1**, 586–97.
75. Harrison, D. E., Strong, R., Sharp, Z. D., Nelson, J. F., Astle, C. M., Flurkey, K., et al. (2009) Rapamycin fed late in life extends lifespan in genetically heterogeneous mice. *Nature* **460**, 392–5.
76. Colman, R. J., Anderson, R. M., Johnson, S. C., Kastman, E. K., Kosmatka, K. J., Beasley, T. M., et al. (2009) Caloric restriction delays disease onset and mortality in rhesus monkeys. *Science* **325**, 201–4.
77. Selman, C., Tullet, J. M., Wieser, D., Irvine, E., Lingard, S. J., Choudhury, A. I., et al. (2009) Ribosomal protein S6 kinase 1 signaling regulates mammalian life span. *Science* **326**, 140–4.
78. Miller, J. L. (1999) Sirolimus approved with renal transplant indication. *Am J Health Syst Pharm* **56**, 2177–8.
79. Moses, J. W., Leon, M. B., Popma, J. J., Fitzgerald, P. J., Holmes, D. R., O'Shaughnessy, C., et al. (2003) Sirolimus-eluting stents versus standard stents in patients with stenosis in a native coronary artery. *N Engl J Med* **349**, 1315–23.
80. Faivre, S., Kroemer, G., and Raymond, E. (2006) Current development of mTOR inhibitors as anticancer agents. *Nat Rev Drug Discov* **5**, 671–88.
81. Bissler, J. J., McCormack, F. X., Young, L. R., Elwing, J. M., Chuck, G., Leonard, J. M., et al. (2008) Sirolimus for angiomyolipoma in tuberous sclerosis complex or lymphangioleiomyomatosis. *N Engl J Med* **358**, 140–51.
82. Thoreen, C. C., Kang, S. A., Chang, J. W., Liu, Q., Zhang, J., Gao, Y., et al. (2009) An ATP-competitive mammalian target of rapamycin inhibitor reveals rapamycin-resistant functions of mTORC1. *J Biol Chem* **284**, 8023–32.
83. Thoreen, C. C., and Sabatini, D. M. (2009) Rapamycin inhibits mTORC1, but not completely. *Autophagy* **5**, 725–6.
84. Yu, K., Shi, C., Toral-Barza, L., Lucas, J., Shor, B., Kim, J. E., et al. (2010) Beyond rapalog therapy: preclinical pharmacology and antitumor activity of WYE-125132, an ATP-competitive and specific inhibitor of mTORC1 and mTORC2. *Cancer Res* **70**, 621–31.
85. Yu, K., Toral-Barza, L., Shi, C., Zhang, W. G., Lucas, J., Shor, B., et al. (2009) Biochemical, cellular, and in vivo activity of novel ATP-competitive and selective inhibitors of the mammalian target of rapamycin. *Cancer Res* **69**, 6232–40.
86. Chresta, C. M., Davies, B. R., Hickson, I., Harding, T., Cosulich, S., Critchlow, S. E., et al. (2010) AZD8055 is a potent, selective, and orally bioavailable ATP-competitive mammalian target of rapamycin kinase inhibitor with in vitro and in vivo antitumor activity. *Cancer Res* **70**, 288–98.
87. Janes, M. R., and Fruman, D. A. (2010) Targeting TOR dependence in cancer. *Oncotarget* **1**, 69–76.
88. Janes, M. R., Limon, J. J., So, L., Chen, J., Lim, R. J., Chavez, M. A., et al. (2010) Effective and selective targeting of leukemia cells using a TORC1/2 kinase inhibitor. *Nat Med* **16**, 205–13.

Chapter 2

Biochemical and Pharmacological Inhibition of mTOR by Rapamycin and an ATP-Competitive mTOR Inhibitor

Ker Yu and Lourdes Toral-Barza

Abstract

The mammalian target of rapamycin (mTOR) is the catalytic subunit of two multiprotein complexes, mTOR complex-1 (mTORC1) and mTOR complex-2 (mTORC2). Clinically used rapamycin and rapalogs are FKBP12-dependent allosteric inhibitors of mTORC1. The recently discovered WYE-125132 and related drugs represent a new generation of ATP competitive and highly specific inhibitors targeting mTOR globally. As mTORC1 and mTORC2 mediate diverse sets of both redundant and distinctive cellular pathways of growth, nutrient and energy homeostasis, rapamycin and WYE-125132 elicit both overlapping and distinctive pharmacological properties with important implications in treating cancer, metabolic, and age-related degenerative diseases. Detailed methods are described for the determination of mTOR inhibition by rapamycin and WYE-125132 in assays of recombinant mTOR enzyme, immunprecipitated native mTOR complexes, growth factor- and amino acid-induced cellular phosphorylation cascades as well as the PI3K/AKT/mTOR hyperactive breast tumor model *in vitro* and *in vivo*. The methods have been particularly useful in discovery and biochemical characterization of mTOR inhibitors, cellular and *in vivo* mTOR substrate phosphorylation analysis, and in deciphering novel biomarkers of mTORC1 and mTORC2 signaling pathways.

Key words: mTOR, Rapamycin, WYE-125132, mTOR inhibitor, DELFIA

1. Introduction

Mammalian target of rapamycin (mTOR) is the founding member of a family of unconventional phosphoinositide-3-kinase (PI3K)-related kinases (PIKKs) that are uniquely large molecular weight (>250 kDa) serine/threonine kinases with catalytic sites homologous to those of PI3K but different from those of the conventional protein kinases (1, 2). mTOR resides in at least two functional multiprotein complexes, mTOR complex-1 (mTORC1) and mTOR complex-1 (mTORC2), mediating diverse signals from

Thomas Weichhart (ed.), *mTOR: Methods and Protocols*, Methods in Molecular Biology, vol. 821,
DOI 10.1007/978-1-61779-430-8_2, © Springer Science+Business Media, LLC 2012

growth factors, nutrients, and energy supply (3, 4). mTORC1 is well known for its critical roles in protein synthesis and growth, which involves phosphorylation of the ribosomal protein S6 kinase 1 (S6K1) and the eukaryotic translation initiation factor eIF4E-binding protein 1 (4EBP1). mTORC2 phosphorylates and activates AKT, a key regulator of cell growth, metabolism, and survival. For more than a decade we have relied largely on the use of rapamycin to study mTOR function but it now appears that rapamycin is a partial inhibitor of mTOR through allosteric inhibition of mTORC1 but not mTORC2. Recent development of highly potent and specific active-site mTOR inhibitors, such as WYE-125132 (5), not only offers new therapeutic potentials but also provides invaluable tools for deciphering novel insights into the signaling network of mTOR complexes (6, 7).

Traditionally, biochemical assessment of mTOR kinase activity has been challenging due to its exceedingly large molecular weight and intricate multiprotein complexes that are often sensitive to buffer conditions. Employing a truncated and catalytically active form of mTOR expressed in and affinity-purified from HEK293 cells, we illustrate a high-capacity Dissociation-Enhanced Lanthanide Fluorescence Immunoassay (DELFIA) for the determination of inhibition kinetic mechanism of rapamycin and WYE-125132. Native mTORC1 and mTORC2 are immunoprecipitated from cytoplasmic extracts and it is shown that mTORC1-specific substrate phosphorylation of S6K1 and mTORC2-specific phosphorylation of AKT are differentially inhibited by rapamycin and WYE-125132. Likewise in cellular models, signaling biomarkers of mTORCs induced by insulin-like growth factor 1 (IGF-1), amino acids, or by PI3K hyperactivation in breast cancer cells are only partially inhibited by rapamycin but are globally targeted by WYE-125132.

2. Materials

2.1. Assay of Recombinant mTOR via Dissociation-Enhanced Lanthanide Fluorescence Immunoassay

1. The FLAG-tagged recombinant mTOR enzyme and the kinase assay substrate His6-S6K1 are obtained as described (8) and stored in –80°C freezer.
2. Assay buffer: 10 mM HEPES (pH 7.4), 50 mM NaCl, 50 mM β-glycerophosphate, 10 mM $MnCl_2$, 0.5 mM dithiothreitol (DTT), 0.25 μM Microcystin LR, and 50 μg/mL bovine serum albumin (BSA) (see Note 1).
3. Adenosine-5′-triphosphate (ATP): 20 mM stock in distilled water and store in –20°C in small single use aliquots.
4. Inhibitor chemicals: Rapamycin and WYE-125132 (Wyeth Research) dissolved in dimethyl sulfoxide (DMSO) as 20 mM

stock solutions. His6-FKBP12 is obtained as described (8) or equivalent.

5. Reaction stop buffer: 20 mM HEPES (pH 7.4), 20 mM ethylenediaminetetraacetic acid (EDTA), and 20 mM ethylene glycol tetraacetic acid (EGTA).
6. Inhibitor dilution plate: Polypropylene 96-well round bottom plate (Costar #3359).
7. Assay plate (for carrying kinase reaction): Nontreated 96-well round bottom polystyrene plate (Costar #3795), plate adhesive film (VWR), and plate shaker (Lab-Line Instruments).
8. DELFIA detection plate: MaxiSorp plate 96-well (Nunc #44-2404-21).
9. Dulbecco's phosphate-buffered saline without calcium and magnesium (DPBS, Mediatech).
10. DELFIA assay buffer (PerkinElmer).
11. Eu-P(T389)-S6K1 antibody: P(T389)-S6K1 polyclonal antibody labeled with europium-N1-ITC (Eu) (Cell Signaling, PerkinElmer). Dilute Eu-P(T389)-S6K1 antibody to 40 ng/mL with DELFIA assay buffer.
12. DELFIA wash buffer: DPBS with 0.05% Tween-20 (Bio-Rad).
13. DELFIA enhancement solution (PerkinElmer).
14. VICTOR Model Plate reader (PerkinElmer).

2.2. Immune-Complex Kinase Assay of mTOR complexes

1. HEK293 cells from American Type Culture Collection (ATCC).
2. Dulbecco's Modified Eagle's Medium (DMEM) supplemented with 10% fetal bovine serum (FBS), 100 μg/mL penicillin, and 50 μg/mL streptomycin (Gibco/Invitrogen).
3. Cytosolic fraction extraction buffer: 5 mM HEPES (pH 7.4), 5 mM β-glycerophosphate, 100 μg/mL digitonin, 1 mM DTT, 1 mM $MgCl_2$, 50 nM Microcystin LR, 1× protease inhibitor cocktail (Roche).
4. Immunoprecipitation (IP) antibody: Anti-FRAP (N19) goat antibody and control goat IgG (Santa Cruz Biotechnology).
5. Agarose beads: Protein A/G PLUS-Agarose (Santa Cruz Biotechnology).
6. Wash buffer: extraction buffer supplemented with 50 mM NaCl and 0.5% NP40 (Roche).
7. Assay buffer (see Subheading 2.1, item 2).
8. Inhibitor chemicals: WYE-125132 and CCI-779 (Wyeth Research) dissolved in DMSO as 20 mM stock solutions, His6-FKBP12 (8) or equivalent.

9. Substrate mix: dilute the His6-S6K1 to 0.6 μM and the inactive His6-AKT1 to 0.34 μM in assay buffer.
10. ATP mix: dilute ATP to 200 μM in assay buffer.
11. NuPAGE LDS sample buffer (4×) (Invitrogen) to be used as kinase reaction stop buffer.
12. SDS-PAGE and Western blotting system: All materials for SDS-PAGE and Western blotting are purchased from Invitrogen unless otherwise specified (see Note 2).
13. NuPAGE Sample Reducing Agent.
14. Precast gels in a 26-well midi format, 4–12% NuPAGE Bis–Tris and 3–8% NuPAGE Tris–acetate.
15. Running buffer: NuPAGE MOPS SDS (20×) and NuPAGE Tris–acetate SDS (20×). Dilute to 1× with distilled water.
16. Molecular weight markers: dilute 100 μL Sharp Pre-Stained Protein Standard 90 μL 1× sample buffer and add 10 μL MagicMark XP Western Standard.
17. Transfer buffer: 2× NuPAGE transfer buffer, 0.2% NuPAGE Antioxidant, and 20% methanol.
18. Transfer system: iBlot Dry Blotting System and iBlot Gel Transfer Stacks in nitrocellulose or comparable manual transferring unit.
19. Membrane wash buffer: TBS-T (Tris-buffered saline). Dilute 100 mL of 10× TBS (Bio-Rad) with 900 mL distilled water plus 1 mL Tween-20.
20. Blocking buffer: 5% nonfat dry milk (Bio-Rad) in TBS-T.
21. Primary antibodies: mTOR, Raptor, P-S6K1(T389), AKT, P-AKT (S473) (Cell Signaling), Rictor (Novus Biologicals), Anti-polyhistidine (clone-1) (Sigma).
22. Dilution buffer for primary antibodies: 3% BSA in TBS-T supplemented with 0.05% sodium azide.
23. Secondary antibodies: goat antirabbit IgG-HRP and goat antimouse IgG-HRP (Santa Cruz Biotechnology).
24. Detection: Enhanced chemiluminescent (ECL) reagents and Amersham Hyperfilm for ECL (GE Healthcare).

2.3. Analysis of mTOR Activity in Cultured Cells

1. Rat1 and HEK293 cells (ATCC) grown in DMEM growth medium (see Subheading 2.2, item 2).
2. MDA361 cells (ATCC), grown in Minimum Essential Medium (MEM) supplemented with 10% fetal bovine serum, 100 μg/mL penicillin and 50 μg/mL streptomycin, 1 mM Na pyruvate (Gibco/Invitrogen).
3. Low serum DMEM, replace FBS concentration to 0.1%.

4. Inhibitor chemicals WYE-125132, rapamycin, 17-hydroxy-wortmannin (Wyeth Research) prepared as 1,000× concentrate in DMSO.
5. Recombinant human IGF-1 (R&D Systems), reconstituted according to the manufacturers' recommendation. Store in aliquots at −80°C.
6. Amino acid (AA)-free medium: Earle's balanced salt solution (EBSS) 10×, MEM vitamins (100×), MEM nonessential amino acids (NEAA) (100×), 7.5% sodium bicarbonate, 45% D(+)-glucose solution, dialyzed FBS. To make 100 mL of AA-free medium: 10 mL EBSS, 1 mL vitamins, 1 mL NEAA, 2.7 mL sodium bicarbonate, 1 mL glucose, 10 mL FBS and bring to 100 mL with distilled water. Filter sterilized.
7. Amino acid solution (50×) (Sigma #M5550).
8. Cell lysis buffer: NuPAGE LDS sample buffer (4×) diluted to 1.1× with distilled water and stored at room temperature.
9. Protein concentration determination: Bio-Rad RC/DC protein Assay.
10. SDS-PAGE (see Subheading 2.2, items 13–16).
11. Total protein staining Ponceau S Solution (Sigma).
12. Western blotting (see Subheading 2.2, items 17–24).
13. Additional primary antibody: P-AKT (T308), AKT, FKHRL1, P-FKHRL1 (T32), GSK3, P-GSK3 (S21/9) (Cell Signaling).

2.4. mTOR inhibition and antitumor activity by WYE-125132 against Tumor Xenografts in Nude Mice

1. MDA361 cells (ATCC).
2. MDA361 xenograft tumors in female nude mice (5).
3. Formulation vehicle: 5% ethanol, 2% Tween-80, 5% polyethylene glycol (PEG)-400.
4. WYE-125132 (Wyeth Research).
5. Tumor lysis buffer: 25 mM HEPES, pH 7.5, 100 mM NaCl, 20 mM β-glycerophosphate, 1.5 mM $MgCl_2$, 0.5 mM EGTA, 0.25 mM EDTA 1% NP-40, 10 mM Na_3VO_4, 10 μg/mL aprotinin, 10 μg/mL leupeptin, 1 mM PMSF, 1 μM microcystin LR, and 0.1% 2-mercaptoethanol.
6. Probe homogenizer (Ultra-Turrax T8, IKA Labortechnic Staufen).
7. SDS-PAGE (see Subheading 2.2, items 13–16).
8. Western blotting (see Subheading 2.2, items 17–24).
9. Additional primary antibody: Cleave PARP (Cell Signaling) and β-actin (Chemicon International).

3. Methods

3.1. In Vitro Biochemical Assays of mTOR

3.1.1. Assay of Recombinant Flag-mTOR via DELFIA

For full details on the production of recombinant mTOR enzyme and substrate His6-S6K1, readers are referred to Toral-Barza et al. (8). The protocol assumes assays in 96-well plate format with reagent volumes intended for each assay well. The kinase assay and DELFIA detection protocols are carried out at room temperature.

1. Prepare inhibitor serial dilution in DMSO in the inhibitor dilution plate according to the desired dose range to be tested. In general, prepare inhibitors as 50× concentrate so that they will be diluted 50× in the final assay (see Note 3). Rapamycin serial dilution was prepared in DMSO, while FKBP12 was prepared in the assay buffer (this is to keep the concentration of DMSO in the assay at 2%).
2. Prepare enzyme mix: dilute FLAG-mTOR enzyme to 12 nM in assay buffer.
3. Prepare substrate-ATP mix: dilute the His6-S6K1 to 2.5 μM and ATP to 200 μM in assay buffer.
4. Dispense 12 μL enzyme mix to assay plate (see Note 4). The assay buffer alone is used as background control.
5. Add 0.5 μL of inhibitor or DMSO to assay plate (see Note 5). Seal the plate with adhesive film and incubate for 10 min on a plate shaker (500 rpm) to thoroughly mix enzyme and inhibitor.
6. Initiate kinase reaction by adding 12.5 μL of substrate–ATP mix. The reaction is continued for 2 h with shaking.
7. Stop kinase reaction by adding 25 μL of reaction stop buffer and return the plate to plate shaker for 10 min.
8. Dispense 60 μL DPBS to DELFIA detection plate. Add 40 μL of the stopped reaction mixture and let the peptide substrate to bind to the plate for 2 h on plate shaker.
9. Aspirate the binding mixture from the DELFIA plate and wash the plate (see Note 6) once with DPBS.
10. Dilute Eu-P(T389)-S6K1 antibody to 40 ng/mL with DELFIA assay buffer. Add 100 μL of Eu-P(T389)-S6K1 antibody. Incubate for 1 h on plate shaker.
11. Aspirate and wash the plate 4× with DELFIA wash buffer. Add 100 μL DELFIA enhancement solution and incubate for 30 min without shaking.
12. Measure DELFIA TF fluorescence in VICTOR Model Plate reader.
13. Calculate the inhibitor activity as percent of DMSO control with background (without enzyme) value subtracted.

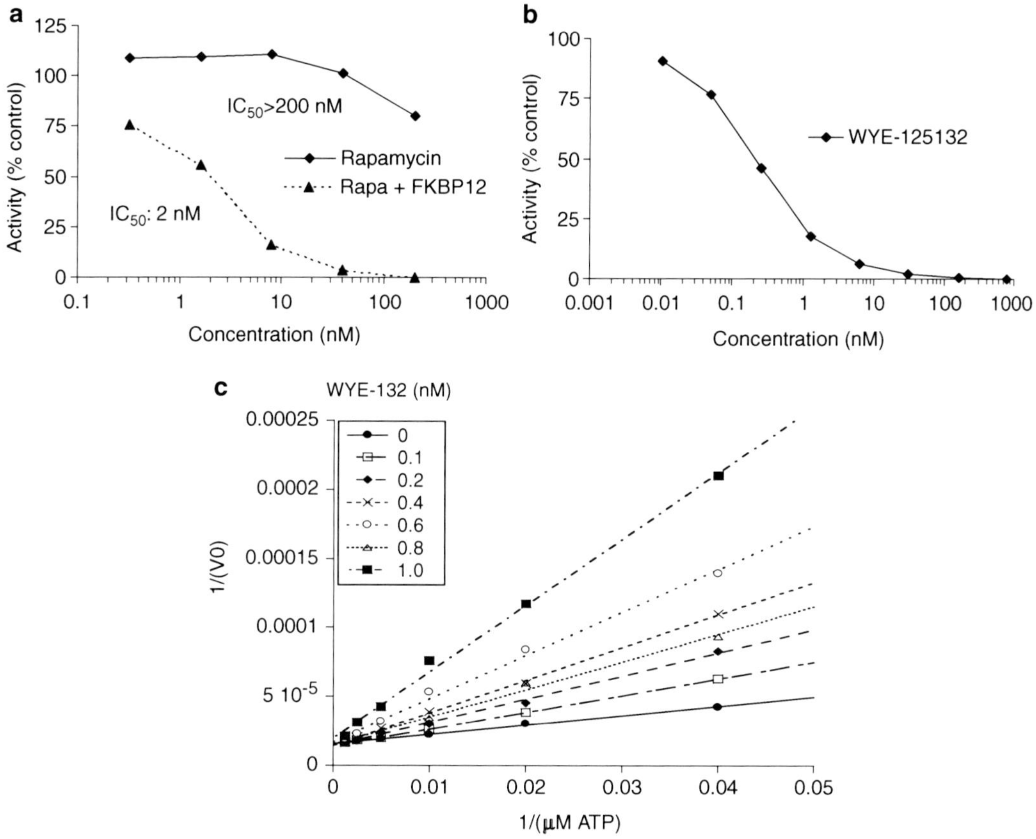

Fig. 1. *In vitro* kinase assay of recombinant mTOR by DELFIA. Dose–response curves of rapamycin alone, rapamycin plus FKBP12 (**a**), or an ATP-competitive inhibitor WYE-125132 (**b**). Note that rapamycin–FKBP12 complex but not rapamycin alone potently inhibits mTOR, whereas WYE-125132 alone inhibits mTOR. (**c**) WYE-125132 versus ATP matrix competition assay. The initial enzyme rates (V_0) against 0, 0.1, 0.2, 0.4, 0.6, 0.8, and 1 nM WYE-125132 at various concentrations of ATP were measured to generate the double-reciprocal plot (**c** reproduced from ref. 5 with permission from the American Cancer Society).

14. Determine the half maximal inhibitory concentration (IC_{50}) of the inhibitor by constructing inhibitor dose–response curves. Examples of the results produced are shown in Fig. 1a, b.

3.1.2. Inhibitor Versus ATP Matrix Assay

Follow methods in Subheading 3.1.1, steps 1–12 with modifications as follows.

1. Prepare inhibitors as 50× concentration of 0.5, 1, 2, 3, 4, and 5 times the inhibitor's IC_{50} value. For example, for an inhibitor with IC_{50} value of 2 nM, prepare 50× concentration of 1, 2, 4, 6, 8, and 10 nM.
2. Prepare enzyme mix: dilute FLAG-mTOR enzyme to 20 nM in assay buffer.

3. Prepare substrate–ATP mix: dilute His6-S6K1 to 2.5 μM with varying concentrations of ATP (50, 100, 200, 400, 800, and 1,600 μM) in assay buffer.
4. Reduce kinase reaction time to 30 min.
5. Transfer 15 μL of the stopped reaction mix to DELFIA detection plate with predispensed 85 μL DPBS.
6. Generate a Lineweaver–Burk analysis by plotting the reciprocal of the initial enzyme rate (V_0) versus the reciprocal of ATP. An example of results produced is shown in Fig. 1c.

3.2. Immune-Complex Kinase Assay of mTORC1 and mTORC2

1. HEK293 cells were grown in 150 mm culture dish until 80% confluence. Cells are freshly fed 1 day before experiment.
2. Prepare cell extraction buffer and keep on ice.
3. Aspirate growth medium from plate and wash 2× with cold DPBS and place plate on ice.
4. Add 3 mL cell extraction buffer to the plate and incubate for 15 min.
5. Collect the supernatant (cytosolic extract) and centrifuge it for 5 min at 1,000 rpm to remove cell debris.
6. Carefully transfer the extracted protein into a new tube and determine the protein concentration by the Bradford method (Bio-Rad).
7. To immunoprecipitate (IP) mTOR complexes, prepare a 1 mg/mL protein extract in a microcentrifuge tube and supplement with NaCl to final 50 mM. Add 2 μg anti-FRAP/mTOR (N19) or control IgG and incubate for 1.5 h while rocking at 4°C. Add 20 μL of a 50% slurry of agarose beads and incubate for 2 additional hours at 4°C.
8. Spin down the beads at 3,000 rpm for 5 min. Wash the beads 3× with 0.5 mL wash buffer and once with 0.5 mL assay buffer. Remove as much of the assay buffer without disturbing the beads.
9. Resuspend the beads in 24 μL substrate mix and add 1 μL of 50× inhibitor. Preincubate for 10 min at 25°C with shaking in a microcentrifuge mixer (see Note 7).
10. Initiate kinase reaction by adding 25 μL ATP mix. Incubate the reaction for 1 h at 25°C in a microcentrifuge mixer.
11. Stop the reaction by adding 17 μL 4× LDS sample buffer and proceed with SDS-PAGE (see Note 8).
12. Add 7 μL 10× reducing agent and denature samples at 70°C for 10 min.
13. Spin the sample for 5 s at 14,000 rpm and transfer the supernatant to a new microcentrifuge tube.

14. Run 5 μL of molecular weight markers and samples in a 26-well 3–8% Tris–acetate gels using Tris–acetate SDS running buffer to resolve mTOR, Rictor, and Raptor. His6-AKT and His6-S6K1 are separated in a 26-well 4–12% gels using MOPS SDS running buffer.
15. Remove gel from the cassette and soak in the transfer buffer for 5 min on an orbital shaker.
16. Place the gel on top of the iBlot nitrocellulose anode stack. Soak an iBlot filter paper in the transfer buffer and place it on top of the gel. Remove any trapped air bubbles by using a roller. Then place the iBlot cathode stack on top and remove air bubbles.
17. Transfer using the iBlot dry blotting system (see Note 9).
18. Incubate the membrane in 50 mL blocking buffer for 1 h in a rocking platform at room temperature.
19. Discard the blocking buffer and rinse membrane 3× for 2 min each with TBS-T. Incubate the membrane with various primary antibodies of mTOR, Rictor, Raptor, P-S6K1 (T389), P-AKT (S473), and His6 (1:1,000 dilution) overnight on a rocking platform at 4°C.
20. Remove the primary antibody and wash the membrane 3× for 5 min each with TBS-T. Add the secondary antibody diluted to 1:5,000 in blocking buffer and incubate for 1 h at room temperature on a rocking platform.
21. Discard the secondary antibody and wash 3× for 5 min each with TBS-T and 2× with TBS only.
22. Prepare the ECL reagent in just enough volume to completely cover the membrane. Discard the final wash and add the ECL reagent and incubate for 1 min.
23. Place the membrane between sheets of Saran wrap and remove excess ECL reagent with a roller and Kim-Wipes. Tape the corners of the wrapped membrane in a film cassette.
24. Expose to ECL film from 1 s to several minutes to obtain desired exposure for each antigen. An example of the results produced is shown in Fig. 2.

3.3. Pharmacological Inhibition of mTOR in Cultured Cells

3.3.1. Inhibition of IGF-1-Induced mTOR Activation

1. Plate Rat1 cells in six-well culture dish in growth medium for 1–2 days.
2. Cells (60% confluence) are washed 1× with serum-free DMEM and incubated in 3 mL low serum DMEM for 24 h.
3. Serum-starved Rat1 cells are pretreated with 3 μL inhibitor (1,000× dilution from the prepared stocks) for 1 h followed by IGF-1 stimulation (100 ng/mL) for 30 min.
4. Cells are washed 1× with DPBS and lysed in 200 μL 1.1× NuPAGE LDS sample buffer followed by 1 min of water-bath

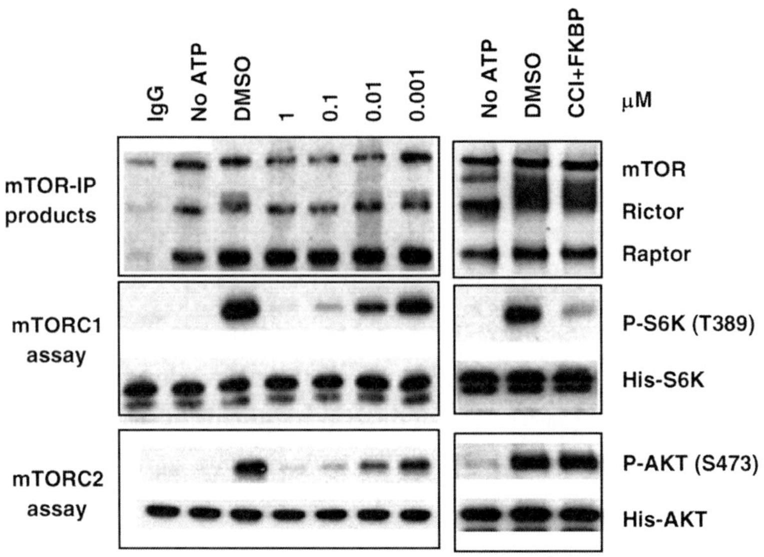

Fig. 2. Immune-complex kinase assay of mTORC1 and mTORC2. Native mTOR complexes immunoprecipitated from HEK293 cytosolic lysates were assayed for substrate phosphorylation of S6K (for mTORC1 activity) or AKT (for mTORC2 activity) without or with inhibitors. The assay products were immunoblotted. Note that CCI-779 (a rapamycin analog) in complex with FKBP12 only inhibits mTORC1, whereas WYE-125132 inhibits both mTORC1 and mTORC2 (the *left* and *right panels* reproduced from refs. 5 and 10, respectively, with permission from the American Cancer Society).

sonication in a cold room. Keep samples on ice or store at –80°C for later use.

5. Perform Western blotting (Subheading 3.2, steps 14–24) for P-S6K1(T389), P-AKT (S473), and P-AKT (T308). An example of the results produced is shown in Fig. 3a, left panel.

3.3.2. Inhibition of Amino Acid-Induced mTOR Activation

1. Plate HEK293 cells in six-well culture dish in growth medium and let cells to reach 70–80% confluence.
2. Very carefully wash the cells once with amino acid-free medium and incubate in 3 mL amino acid-free medium for 2 h.
3. Amino acid-starved cells are pretreated with 3 μL inhibitors (1,000× dilution from the prepared stocks) for 2 h. Cells are then stimulated with amino acids (add 60 μL of the 50× solution) for 1 h.
4. Prepare total cell lysate as described for the IGF-1 experiment.
5. Perform Western blotting (Subheading 3.2, steps 14–24) for P-S6K1(T389), P-AKT (S473), and P-AKT (T308). An example of the results produced is shown in Fig. 3a, right panel.

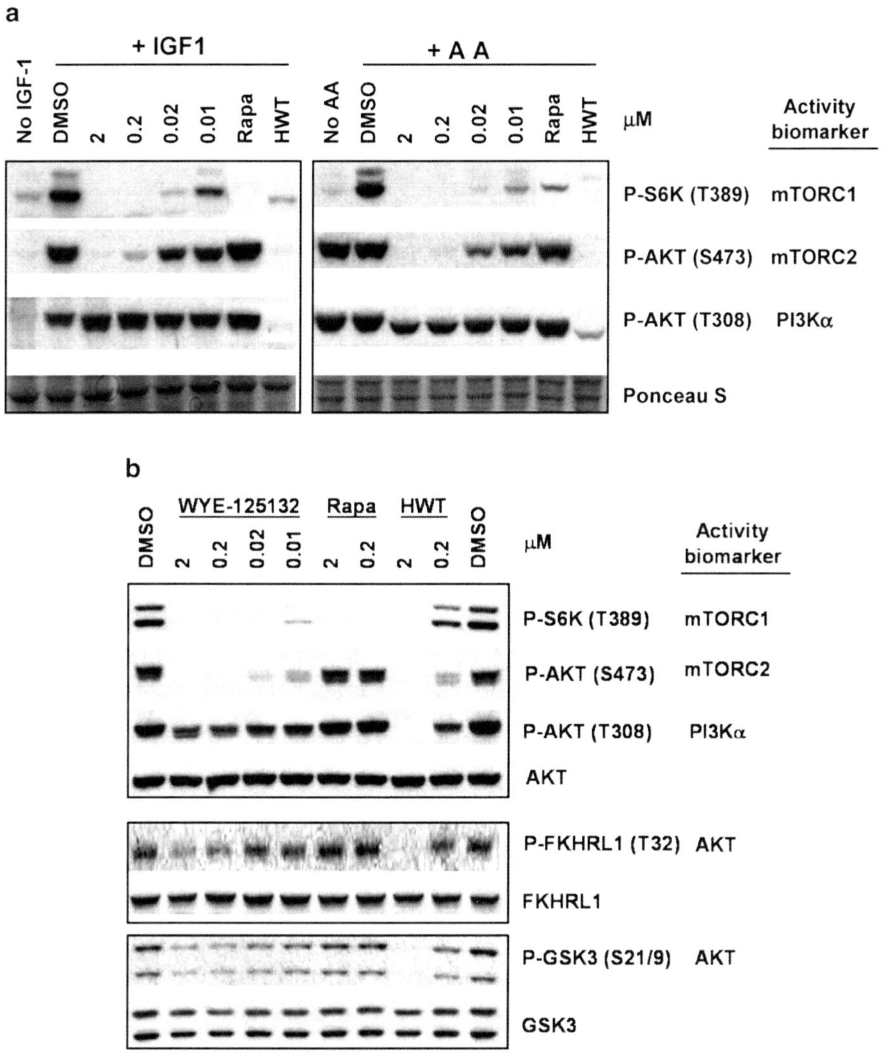

Fig. 3. Differential inhibition of mTORC1 and mTORC2 by rapamycin and WYE-125132 in cultured cells. (**a**) Inhibition of IGF1-induced mTOR activation in Rat1 cells (*left panel*) and inhibition of amino acid (AA)-induced mTOR activation in HEK293 cells (*right panel*). (**b**) Inhibition of mTOR signaling function in PI3K-hyperactive MDA361 breast cancer cells. Protein lysates were immunoblotted. Note that consistent with the targeting of mTORC1 and mTORC2, WYE-125132 inhibits cellular P-S6K (T389) and P-AKT (S473) as well as AKT downstream targets P-FKHRL1 (T32) and P-GSK3 (S21/9), whereas mTORC1-selective rapamycin only inhibits P-S6K (T389). Note also that WYE-125132 does not significantly inhibit the PI3K biomarker P-AKT (T308), whereas the PI3K inhibitor 17-hydroywortmannin (HWT) inhibits P-AKT (T308) (**a**, *left panel* reproduced from ref. 5 with permission from the American Cancer Society).

3.3.3. Inhibition of mTOR in PI3K Pathway Hyperactive MDA361 Breast Cancer Cells

1. Plate MDA361 cells in six-well culture dish in growth medium (3 mL) for 24 h.
2. The cells (50–60% confluence) are treated with 3 μL inhibitor (1,000× dilution from the prepared stocks) for 6 h in growth medium.
3. Prepare total cell lysate as described for the IGF-1 and amino acids experiments.
4. Perform Western blotting (Subheading 3.2, steps 14–24) for P-S6K1(T389), P-AKT (S473), P-AKT (T308), P-FKHRL1 (T32), FKHRL1, P-GSK3 (S21/9), and GSK3. An example of the results produced is shown in Fig. 3b.

3.4. Inhibition of mTOR in MDA361 Xenograft Tumors and Antitumor Efficacy by WYE-125132

3.4.1. Inhibition of mTOR Signaling Biomarkers in MDA361 Xenograft Tumors

1. Prepare MDA361 xenograft tumors in female nude mice as described (5).
2. Stage tumors when they reach approximately 400 mm^3 and randomize them to treatment groups.
3. Formulate WYE-125132 in the specified vehicle.
4. Dose tumor-bearing mice ($n = 3$) by a single oral injection with vehicle or 50 mg/kg WYE-125132.
5. Harvest and dissect tumors at 0, 3, 6, 12, 24, and 36 h after oral injection.
6. Quickly mince tumor tissues and lyse them in the specified tumor lysis buffer using a probe homogenizer (Ultra-Turrax T8, IKA Labortechnic Staufen).
7. Centrifuge the tumor lysate for 10 min at 14,000 rpm at 4°C.
8. Carefully transfer the clear lysate into a new tube and determine the protein concentration by the Bradford method (Bio-Rad).
9. Perform Western blotting (Subheading 3.2, steps 14–24) for P-S6K1(T389), P-AKT (S473), P-AKT (T308), AKT, cleaved PARP and β-actin. An example of the results produced is shown in Fig. 4a.

3.4.2. Antitumor Efficacy of WYE-125132

1. Stage MDA361 tumors when they reach approximately 200 mm^3 and randomize them to treatment groups.
2. Formulate WYE-125132 in the specified vehicle.
3. Dose tumor-bearing mice ($n = 10$) with vehicle, 5, 10, and 20 mg/kg WYE-125132 orally via cycle dosing (one cycle consists of 5 days on and 2 days off) for four cycles.
4. Measure tumor volume twice a week using sliding vernier calipers and analyze tumor growth data (9). An example of results produced is shown in Fig. 4b.

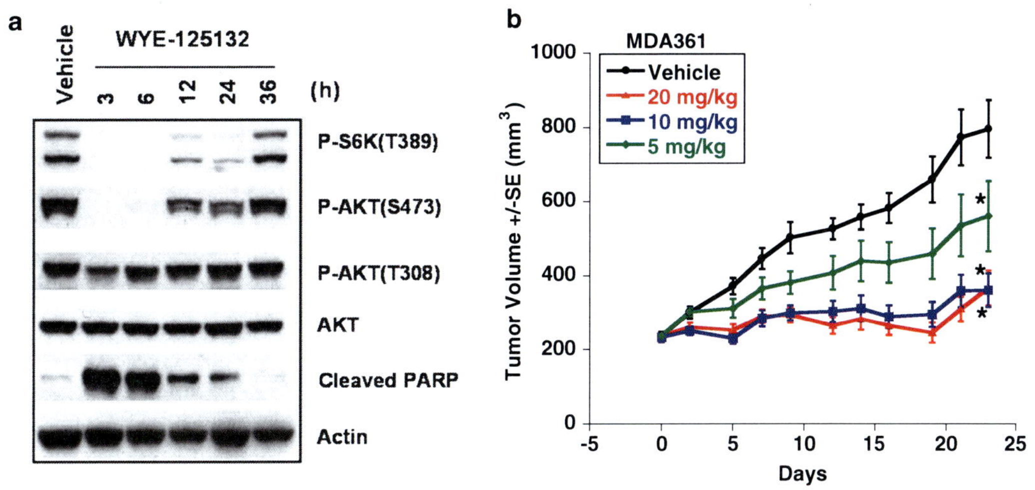

Fig. 4. WYE-125132 inhibits mTOR signaling and tumor growth in PI3K/AKT/mTOR-hyperactivated MDA361 tumors *in vivo*. (**a**) Mice bearing MDA361 tumors were dosed orally with vehicle or 50 mg/kg WYE-125132. Tumor lysates at the indicated time points after dosing are immunoblotted. (**b**) Mice bearing MDA361 tumors were dosed with vehicle, 5, 10, or 20 mg/kg WYE-125132 for cycles (one cycle consists of 5 days on and 2 days off). Tumor growth curves are shown. *$p < 0.05$, WYE-125132 versus vehicle (reproduced from ref. 5 with permission from the American Cancer Society).

4. Notes

1. 5× Basal assay buffer without $MnCl_2$, Microcystin, and BSA can be prepared and stored it in –20°C. 1× Assay buffer is made when needed by dilution of 5× basal assay buffer with molecular grade water, then supplement with $MnCl_2$, Microcystin, and BSA.
2. Although the SDS-PAGE and Western blotting protocols described here employ the Invitrogen reagents and equipment system, feel free to replace them with other compatible systems that you prefer.
3. Inhibitor dilutions can be prepared ahead of time and stored at –20°C. When working with sticky inhibitors, it is best to change pipette tips during serial dilutions.
4. A mutichannel repeater pipette is recommended for this procedure. We use the 250 μL Matrix 12-channel pipettor for delivering the enzyme and substrate mix.
5. A 1–10-μL EDP3-Plus Rainin 12-channel electronic pipette is used in adding the inhibitors. Change pipette tips as you add row by row to avoid carry over.
6. Washing is done by using a 12-channel Nunc-Immuno washer. The wells are filled to the top with washing buffer and aspirated.

7. Shaking is important to keep the beads in suspension. We use an Eppendorf Thermomixer set to mix at 700 rpm.
8. After stopping the reaction, the samples could be stored in −80°C or processed immediately for Western blotting. Since we use Invitrogen's NuPAGE precast gels and premix running buffers and blotting system, we generally follow their instructions for running SDS-PAGE and blotting. If you do not use the Invitrogen system please feel free to replace them with other compatible systems.
9. Right after the iBlotting, you may incubate the membrane in Ponceau S solution for 1 min. Rinse several times with distilled water to wash away the background. Make a scan (or photo copy) of the Ponceau S staining to document protein loading.

References

1. Manning G, Whyte DB, Martinez R, Hunter T, Sudarsanam S (2002) The protein kinase complement of the human genome. Science **298,** 1912–34
2. Abraham RT (2004) PI 3-kinase related kinases: 'big' players in stress-induced signaling pathways. DNA Repair (Amst) **3,** 883–7
3. Wullschleger S, Loewith R, Hall MN (2006) TOR signaling in growth and metabolism. Cell **124,** 471–84
4. Guertin DA, Sabatini DM (2007) Defining the role of mTOR in cancer. Cancer Cell **12,** 9–22
5. Yu K, Shi C, Toral-Barza L, Lucas J, Shor B, Kim JE et al (2010) Beyond rapalog therapy: preclinical pharmacology and antitumor activity of WYE-125132, an ATP-competitive and specific inhibitor of mTORC1 and mTORC2. Cancer Res **70,** 621–31
6. Guertin DA, Sabatini DM (2009) The pharmacology of mTOR inhibition. Sci Signal. **2,** pe24
7. Shor B, Gibbons JJ, Abraham RT, Yu K (2009) Targeting mTOR globally in cancer: Thinking beyond rapamycin. Cell Cycle **8,** 831–37
8. Toral-Barza L, Zhang WG, Lamison C, Larocque J, Gibbons J, Yu K (2005) Characterization of the cloned full-length and a truncated human target of rapamycin: Activity, specificity, and enzyme inhibition as studied by a high capacity assay. Biochem Biophys Res Commun **332,** 304–10
9. Yu K, Lucas J, Zhu T, Zask A, Gaydos C, Toral-Barza L et al (2005) PWT-458, A Novel Pegylated-17-Hydroxywortmannin, Inhibits Phosphatidylinositol 3-Kinase Signaling and Suppresses Growth of Solid Tumors. Cancer Biol Ther **4,** 538–545
10. Yu K, Toral-Barza L, Shi C, Zhang WG, Lucas J, Shor B et al (2009) Biochemical, cellular, and in vivo activity of novel ATP-competitive and selective inhibitors of the mammalian target of rapamycin. Cancer Res **69,** 6232–40

Chapter 3

Evaluation of the Nutrient-Sensing mTOR Pathway

Sungki Hong, Aristotle M. Mannan, and Ken Inoki

Abstract

mTOR, an evolutionarily conserved Ser/Thr protein kinase, belongs to the PI3K-related kinase family, which also includes DNA-PKcs, ATM, and ATR. Although other PI3K-related kinase family members have been shown to secure genomic integrity by sensing DNA damage and related stresses, mTOR is known to function as a nutrient and growth factor sensor. mTOR is the catalytic subunit of two distinct multiprotein complexes known as mTOR complex 1 (mTORC1) and mTOR complex 2 (mTORC2). In response to growth factor and nutrient availability, these complexes regulate a variety of cellular processes, such as cell growth, proliferation, and survival by modulating downstream effectors, such as S6K1, 4EBP1, and AKT. Therefore, evaluation of mTOR activity has been a clear readout in order to monitor the physiological status of cells in response to environmental cues. Here, we present the current techniques for the assessment of mTOR kinase activity in different experimental settings.

Key words: mTOR, PI3 Kinase-related kinase family, mTORC1, mTORC2, Nutrient, Growth factor, Cell survival, Phosphorylation, Kinase activity

1. Introduction

The mammalian target of rapamycin (mTOR) is a critical kinase in sensing availability of nutrients and growth factors, as well as a pivotal component in the regulation of a variety of cellular processes, such as translation, transcription, mRNA processing, autophagy, and actin cytoskeleton organization. In addition, recently mTOR has been shown to function in maturation of intracellular organelles, such as lysosomes during starvation (1, 2).

The TOR genes were originally identified in yeast as the target of rapamycin, a macrolide antibiotic of *Streptomyces hygroscopicus* that was originally isolated from soil samples in Easter Island. FDA-approved rapamycin and its derivatives are now commercially available for immunosuppression, artery stenosis, and several types of cancers. The mechanism by which rapamycin confers its inhibitory

Thomas Weichhart (ed.), *mTOR: Methods and Protocols*, Methods in Molecular Biology, vol. 821,
DOI 10.1007/978-1-61779-430-8_3, © Springer Science+Business Media, LLC 2012

effect on mTOR has been well characterized through biochemical, structural, and recent Cryo-EM studies (3, 4). Unlike other kinase inhibitors, rapamycin forms a drug-receptor complex with the cellular protein FKBP12 (immunophilin FK506-binding protein 12), which upon binding to mTOR, inhibits the kinase activity of mTOR (4, 5). mTOR forms at least two distinct multiprotein complexes in which it functions as a catalytic subunit. These complexes, termed mTOR complex 1 (mTORC1) and mTOR complex 2 (mTORC2), have unique substrates and therefore exert specific functions. mTORC1, which consists of mTOR, Raptor, PRAS40 and mLST8, phosphorylates 4EBPs, and the hydrophobic motif of S6 kinase. mTORC2, which is responsible for the phosphorylation of AKT and SGK1, is composed of mTOR, Rictor, mSin1, mLST8, and PRR5 (6, 7).

mTORC1 and mTORC2 are regulated by different mechanisms and have distinct physiological outputs. It has been well demonstrated that the activity of mTORC1 is enhanced by both growth factors and nutrients, such as amino acids, while mTORC2 is activated by growth factors but not nutrients. Furthermore, mTORC1 activity is highly sensitive to inhibition by rapamycin, in which the rapamycin-FKBP12 complex disrupts the interaction between mTOR and Raptor, an essential scaffolding protein that recruits substrates into mTORC1 (8). mTORC2 activity, however, is resistant to rapamycin, at least after a relatively short treatment, and does not respond to amino acid stimulation (9). The resistance of mTORC2 to rapamycin is consistent with findings that rapamycin effectively inhibits phosphorylation of S6K but not AKT, a substrate of mTORC2. More recently, ATP-competitive inhibitors of mTOR have been shown to block both mTORC1 and mTORC2 through direct inhibition of mTOR, the catalytic subunit of these complexes (10, 11). Moreover, these inhibitors have been shown to fully inhibit mTORC1 dependent translation, an effect that is not evident even in relatively higher doses of rapamycin treatment.

Although the mechanism by which growth factors enhance mTORC2 activity remains elusive, recent studies have revealed that mTORC1 activity is enhanced by at least two different ras-related GTPases, Rag, and Rheb, in response to growth factor and amino acid availability (12, 13). There are four Rag genes (RagA/B and RagC/D) in mammals whereas lower organisms, such as *Drosophila*, have one RagA ortholog (GTR1) and one RagC ortholog (GTR2). Each RagA or RagB forms a heterodimeric complex with either RagC or RagD, of which the RagA/B GTP and RagC/D GDP complexes are active. The activated Rag complex, which constitutively localizes on the endosomal membrane and regulated by branched amino acids, recruits mTORC1 to the endomembrane structures, thereby facilitating the activation of mTORC1 by Rheb (12). It has been shown that growth factor- and amino acid-induced activation of mTORC1 occurs through the Rheb GTPase and purified active Rheb enhances mTORC1 activity *in vitro*, thus suggesting Rheb to be a direct activator of mTORC1 (14). Recent findings have shown that upon amino acid stimulation, mTORC1 localizes to the lysosomal compartment, a process that is mediated by the Rag GTPases and the complex called Ragulator which help anchor the Rag complex on the lysosome. Following recruitment to the lysosome, Rheb is supposed to interact with mTORC1 to cause its subsequent activation (15). Given the fact that Rheb activity is regulated by growth factor signaling and that growth factors are not able to fully activate mTORC1 in the absence of amino acids (16), the aforementioned spatial and temporal mechanisms of mTORC1 activation, which occur via two distinct small GTPases, rationalizes the simultaneous requirement of both amino acids and growth factors for mTORC1 activation.

In this chapter, we summarize current methods for monitoring mTOR activity both *in vitro* and *in vivo* (cell culture and tissue). It is worth noting that a majority of the assays described here are designed to measure the crude or gross mTOR activity in cells. As the molecular mechanism of mTOR signaling reveals, we are aware that the regulation of mTOR in cells occurs in both temporal and spatial manners. For instance, mTORC1 activity is first down-regulated but up-regulated later during amino acid starvation, but this activity is most likely dependent on where or when mTORC1 is activated by extracellular nutrients or newly synthesized nutrients in the cells. Taking this into consideration, a combination of assays using biochemical and biological approaches would be necessary to accurately evaluate mTOR activity in the cells.

2. Materials

2.1. Mammalian Cell Culture and Protein Extraction

1. 10% FBS DMEM with 1 X Pen/Strep: Dulbecco's modified Eagle's medium (DMEM, Invitrogen, cat# 11995–063) supplemented with 10% FBS and 1× Pen Strep (Penicillin and Streptomycin).
2. 0.25% Trypsin–EDTA.
3. NP-40 buffer: 10 mM Tris pH 7.5, 2 mM EDTA, 100 mM NaCl, 1% NP-40, 50 mM NaF, 10 mM pryophosphate, 10 mM glycerophosphate, with freshly added protease inhibitor.
4. CHAPS buffer: 40 mM HEPES pH 7.5, 120 mM NaCl. 1 mM EDTA, 0.3% CHAPS (Sigma), 10 mM pyrophosphate, 10 mM glycerophosphoate, 50 mM NaF with freshly added protease inhibitor.
5. 4× sodium dodecyl sulfate (SDS) sample loading buffer: 200 mM Tris–HCl pH 6.8, 8% SDS, 40% glycerol, 20% 2-mercaptoethanol, 0.4% bromophenol blue.
6. Freezing media: 10% DMSO in 10% FBS DMEM.

2.2. Transfection

1. Lipofectamine.
2. Opti-MEM.
3. 2× HBS: for 500 ml, 8 g NaCl, 0.2 g Na_2HPO_4, $7H_2O$, 6.5 g HEPES in autoclaved water, pH 7, filter-sterilized and stored at −20°C.
4. 2 M Calcium chloride: $CaCl_2$ is dissolved in autoclaved water, filter-sterilized, and stored at 4°C.

2.3. Sodium Dodecyl Sulfate-Polyacrylamide Gel Electrophoresis

1. Acrylamide/bisacrylamide, liquid (37.5:1).
2. UltraPure Temed.
3. 10% SDS: 10% SDS in autoclaved water and stored at room temperature.
4. 10% Ammonium persulfate: 10% ammonium persulfate in water and stored at −20°C.
5. Separating gel buffer: 1.5 M Tris–HCl pH 8.8 and stored at either room temperature or 4°C.
6. Stacking gel buffer: 1 M Tris–HCl pH 6.8 and stored at either room temperature or 4°C.
7. 10× SDS gel running buffer: For 1 liter, 144 g of glycine, 30.25 g of Tris, and 10 g of SDS.

2.4. Antibodies

2.4.1. Antibodies for Western Blotting and Immunoprecipitation

Phospho-Akt (Ser473) monoclonal (587 F11) Antibody (Cell Signaling, cat# 4051), Akt antibody (Cell Signaling, cat# 9272), mTOR Antibody for Western blotting (Cell Signaling, cat# 2972), mTOR Antibody for IP (Sata Cruz, N-19, lot# D0507), Phospho-p70 S6 Kinase (Thr389) (108D2) Rabbit mAb (Cell Signaling, cat# 9234), p70 S6 Kinase (49D7) Rabbit mAb (Cell Signaling, cat# 2708), S6 ribosomal protein (5G10) Rabbit mAb (Cell Signaling, cat# 2217), Phospho-S6 Ribosomal Protein (Ser235/236) Antibody (Cell Signaling, cat# 2211), 4EBP1 (53H11) Rabbit mAb (Cell Signaling, cat# 9644), Phospho-4E-BP1 (Thre37/46)(236B4) Rabbit mAb (Cell Signaling, cat 2855), Rabbit anti-Rictor antibodies (Bethyl, for WB, cat# A300-459A or A300-458A, for IP, cat# A300-458A, both antibodies lot# are 1). Purified mouse antibody mono HA.11 (Covance, cat# MMS101P).

2.4.2. Antibodies for Immunoflourescence Staining

Alexa Fluor 594 goat anti-rabbit IgG (H+L), Alexa Fluor 488 goat anti-rabbit IgG (H+L), Alexa Fluor 635 anti-mouse IgG (H+L), Alexa Fluor 488 anti-mouse IgG (H+L): These secondary antibodies are from Invitrogen and diluted to 1:1,000 at use. LAMP-2 (H4B4) (Santa Cruz, cat# sc-18822, 1:100 dilution), mTOR (7C10) Rabbit mAb (Cell Signaling, cat# 2983, 1:250 dilution), and anti-Raptor, clone 1H6.2 (Millipore, cat# 05–1470, 1:500 dilution). For immunohistochemistry, HRP-conjugated anti-rabbit antibody (Dako, cat# K4002), pS6 (Cell Signaling, cat# 4857), pAKT (Cell Signaling, cat# 3787 or 9266), and p-mTOR (Cell Signaling, cat# 2976).

2.5. Western Blotting

1. Immobilon-P transfer membrane (PVDF membrane, pore size 0.45 μm).
2. 3MM CHR (Blotting paper).
3. 1× Transfer buffer: For 4 liters, 12.1 g of Tris, 57.6 g of glycine, and 800 ml of MeOH.
4. 1× TBST: 0.05 % Tween-20 in 1× TBS.
5. Blocking solution: 5% nonfat dry milk in 1× TBST.
6. Stripping buffer: For 1 liter, 1.88 g of glycine, and 10 g of SDS pH 2.

2.6. In Vitro *Kinase Assay for mTOR*

1. Kinase lysis buffer: CHAPS buffer described above.
2. Radioisotope: Adenosine 5′triphosphate (gamma-32P).
3. G Sepharose 4 Fast Flow.
4. Low salt washing buffer: 40 mM HEPES–KOH, pH 7.4, 150 NaCl, 2 mM EDTA.
5. Kinase reaction buffer: 25 mM HEPES–KOH, pH 7.4, 50 mM KCl, 10 mM $MgCl_2$, and 250 mM ATP stored at −80°C.

2.7. In Vitro *Kinase Assay for S6K1*

1. Kinase lysis buffer: 10 mM KPO_4/1 mM EDTA, pH 7.05, 5 mM EGTA pH 7.2, 10 mM $MgCl_2$, 50 mM β-glycerophosphate pH 7.2, 0.5 % NP-40, 0.1 % Brij-35, 0.1 % deoxycholic acid, pH 7.28, stored at 4°C.
2. 6× Kinase reaction buffer: 120 mM HEPES pH 7.2, 60 mM $MgCl_2$, 1 mg/ml BSA. Freeze 1 ml at −20°C until use. Add 1.26 μl β-mercaptoethanol to 1 ml when used.
3. Washing buffer A: 1 % NP-40, 0.5 % sodium deoxycholic acid, 100 mM NaCl, 10 mM Tris pH 7.2, 1 mM EDTA and keep it at 4°C.
4. Washing buffer B: 1 M NaCl, 0.1 % NP-40, 10 mM Tris pH 7.2, and stored at 4°C.
5. Washing buffer ST: 150 mM NaCl, 50 mM Tris–HCl, 5 mM Tris–base, pH 7.2, and stored at 4°C.

2.8. GST Purification

1. GST constructs: For mTORC1 kinase assays, GST-4EBP1 and GST-S6K1 are expressed. For mTORC2 kinase assay, GST-AKT is mostly used. For S6K1 kinase assays, GST-S6 is expressed.
2. PBST: PBS buffer containing 1 mM EDTA, 0.1% β-mercato-ethanol, 1% TX-100, and protease inhibitor.
3. STE buffer: 0.1 M NaCl, 10 mM Tris–HCl pH 8.0, 1 mM EDTA pH 8.0.
4. $MgCl_2$-NaCl buffer: 1.5 M NaCl with 12 mM $MgCl_2$.
5. Lysozyme: 1 mg/ml.
6. DNase I: 1 mg/ml.

7. HNTG buffer: 20 mM HEPES pH 7.5, 150 mM NaCl, 10% glycerol, 0.1% Triton-X100.
8. Immobilized Glutathione agarose beads (Thermo scientific).
9. L-glutathione reduced.
10. GST elution buffer: 10 mg of L-glutathione per ml of 100 mM Tris pH 8.0.
11. Slide-A-Lyzer Dialysis Cassette (Thermo Scientific).
12. Dialysis buffer: 10 mM Tris, pH 7.5, 50 mM NaCl, 0.05% β-mercaptoethanol, 0.5 mM EDTA, and stored at 4°C.

2.9. Immunofluorescence Staining

1. Round cover slips: Microscope cover glass (Fisherbrand).
2. 4% Paraformaldehyde.
3. Permeabilization buffer: 0.2% Triton X-100 in PBS at room temperature.
4. Blocking solution: 20% FBS in PBS with 0.02% sodium azide.
5. Antibody dilution solution: 10% FBS in PBS, which should be made freshly before dilution of antibodies.
6. Prolong Gold antifade reagent with DAPI (mounting media).
7. Hydrophobic pen: Pen for immunohistochemistry (Dako).
8. Liquid DAB + Substrate Chromogen system (Dako).
9. Antibodies for IHC and IF are mentioned above.

2.10. Other Chemicals and Reagents

1. Rapamycin.
2. Wortmannin.
3. Complete Mini, EDTA-free.
4. 50× NaF: O.5 M NaF and stored at room temperature.
5. 100× Sodium pyrophosphate: 1 M and stored at room temperature.

3. Methods

Since the mTOR pathway is extremely sensitive to levels of growth factors and nutrients, maintaining cells in fresh media is always recommended as this keeps mTOR more active, especially when compared to cells cultured in a lower energy environment. This notion also applies to the monitoring of *in vivo* mTOR activity, since the time points at which tissue samples are harvested from animals are critical for obtaining consistent results. If we intend to monitor mTOR activation by stimuli, the cells should be starved with growth factors (serum free media) ahead to suppress basal mTOR activity. This helps to maximize the difference between vehicle and agonist-induced mTOR activation. It is also important to minimize the time during which sampling occurs and therefore in cell culture, we avoid washing the cells with PBS before adding lysis buffer. It is always recommended to determine whether the basal activity of mTOR has been considerably decreased or increased in order to observe the changes in mTOR activity by either negative or positive stimuli.

There are several ways of evaluating mTOR activity in response to nutrient or growth factor stimulation. Western blotting using cell extracts examines mTOR kinase activity by monitoring any changes in the phosphorylation, a post-translational modification that affects functional activity, of known mTOR substrates, such as S6K1, 4EBP1, and AKT. In addition, the mTOR activity can be directly accessed by *in vitro* kinase assays, using immunoprecipitated mTOR from the cells. Using either purified S6K1 or AKT as a substrate, the mTORC1 or mTORC2 activity can be measured *in vitro*, respectively. It is also able to separately measure mTORC1 and mTORC2 activity by immunoprecipitated mTORC1 or mTORC2 using Raptor or Rictor antibody, respectively. Furthermore, the assessment of S6K1 activity can be recognized as a physiological way of monitoring any changes in mTORC1 activity. Recently, it has been demonstrated that in response to amino acid stimulation, mTOR is localized to the lysosomal structure, thus suggesting that given a short span of stimulation, determination of mTOR localization may be a cellular biological way of evaluating mTOR activity. Here, we describe well-established and updated methods, materials, and particularly antibodies that have been confirmed and actively used in this field.

3.1. Total Cell Lysate Preparation for Western Blotting

1. For six-well plates, remove media completely by aspiration and immediately add 200 µl cold NP-40 buffer containing protease and phosphatase inhibitors.
2. Tap the plate until cells float and incubate on ice for 15 min with occasional shaking.
3. Collect supernatant into 1.5 ml microcentrifuge tubes.
4. Spin at 15,000 × *g* for 15 min at cold room.
5. Transfer 150 µl supernatant into a new tube.
6. Add 50 µl 4× SDS sample buffer and vortex very briefly.
7. Denature proteins at a 100°C heating block for 5 min.
8. Cool down the samples at room temperature.
9. Spin the boiled samples at 15,000 × *g* for 30 s at room temperature.
10. Samples can be used immediately or can be stored at −20°C.

3.2. Assessment of mTOR Activity Using Phospho-Specific Antibodies Using Cell Lysates

1. Split cells into 6-well plates one day before harvest. On the day of harvest, cell density should reach to 80% of a well.
2. To inhibit mTOR activity of mTORC1, cells can be incubated with growth factors and nutrients depletion conditions, such as HBSS, DPBS, or leucine-deficient RPMI for 30–60 min after

the removal of growth media, serum-starved with serum free DMEM (growth factor free condition) for overnight, or directly treated with 20–50 nM rapamycin for 30–60 min. For nutrient stimulation to activate mTOR, cells are incubated with DPBS, HBSS, or leucine-deficient RPMI, and then DMEM or complete RPMI without FBS is added after the removal of amino acid-deficient media. To activate mTOR by serum or growth factors, cells should be serum-starved with DMEM without FBS for overnight to 24 h and then 10% FBS DMEM or DMEM with 600 μM insulin for 15–60 min. In order to induce complete inhibition or maximal induction of mTOR activity by growth factors, HEK293E and Hela cells have been known to respond better than HEK293T cells.

3. Prepare NP-40 buffer with freshly added protease inhibitor and phosphatase inhibitors.
4. Prepare total cell extracts as described above.

3.3. Western Blotting for the mTOR Signaling Pathway

1. Assemble sodium dodecyl sulfate-polyacrylamide gel electrophoresis (SDS-PAGE) gel and the PVDF membrane in transfer buffer.
2. For mTOR, Rictor or Raptor Western blot, use 6 or 8% SDS-PAGE for separation and transfer proteins onto membrane at the cold room for 180 min at fixed 350 mA. For S6K1, S6, or 4EBP1, use 8% or 12% SDS-PAGE with 2 h transfer. When quality of transfer buffer is good, voltage should go up to 220 V immediately and goes down to 110 at the end of the transfer.
3. Disassemble the transfer and wash the membrane with TBST briefly.
4. Incubate the membrane with blocking solution for 1 h.
5. During blocking, prepare primary antibody in 5% BSA TBST with 0.02% sodium azide. Recommended antibody dilution: mTOR (1:1,000) and mLST8 (1:1,000). For mTORC1 analysis, Raptor (1:1,000), S6K1 (1:1,000), pThr389 S6K1 (1:1,000), S6 (1:5,000), pSer235/236 S6 (1:5,000), pSer240/244 (1:5,000), 4EBP1 (1:5,000), and pThr37/46 4EBP1 (1:5,000). For mTORC2 analysis, Rictor (1:1,000), pSer473 Akt (1:1,000), and Akt (1:1,000).
6. Rinse the membrane with TBST several times until blocking solution is completely removed.
7. Add primary antibody (usually 10 ml) and incubate at cold room overnight.
8. On the next day, collect primary antibodies in 15 ml polystyrene disposable tubes to recycle these antibodies and keep these antibodies solution at −20°C for next uses (see Note 1).
9. Wash the membrane with TBST three times for 30 min.
10. Add secondary antibody (1:5,000) in blocking solution and incubate for 2 h at room temperature with rocking.

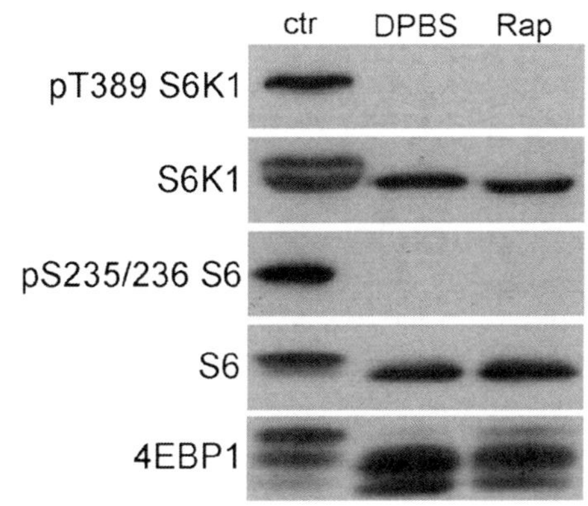

Fig. 1. mTORC1 inhibition by acute nutrient starvation and rapamycin treatment. HEK293T cells were either treated with DPBS for nutrient and growth factor starvation for 1 h or treated with 50 nM rapamycin for 30 min. Total cell lysates were analyzed by Western blotting using indicated antibodies.

11. Wash with TBST four times for 60 min.
12. Completely remove TBST and add ECL mix into the container.
13. Expose the blot on X-ray film in the dark room.
14. Usually, the signal of mTOR, Raptor, Rictor appears within 2–5 min. The signal of phospho- or total S6 and 4EBP1 on X-ray film usually appears within 20 s. S6K1, Akt, and the signal of phosphorylated proteins of these appear within 1 or 2 min. An example of the results produced is shown in Fig. 1 (see Note 2).

3.4. Preparation of Substrate for in Vitro *Kinase Assay*

In order to monitor kinase activity of mTORC1, S6K1 (ribosomal protein S6 kinase 1) and 4EBP1 (4E-binding protein 1) have been used as substrates. Functional readout of mTORC1 activity also can be confirmed through examination of changes in either S6K1 activity on phosphorylation of its substrates or levels of 4EBP1 binding to the Cap structure of mRNA (m7GTP sepharose beads). GST-Akt is mostly used for mTORC2 kinase assays and His tagged Akt is commercially available.

3.4.1. Preparation of GST-S6K1 from Mammalian Cells

1. Grow HEK293T cells in five 15 cm plates.
2. Transfect 20 μg of DNA into 50% confluent cells using calcium phosphate method.
3. 48 h post-transfection, treat cells with 20 nM rapamycin or DPBS for 1 h to induce dephosphorylation of S6K1.
4. Rinse cells with ice-cold PBS one time to remove serum.
5. Lyse cells with 1 ml of PBST buffer per plate by incubation on ice for 15 min.
6. Collect supernatant into microcentrifuge tubes.
7. Spin at cold room for 15 min at 15,000 × *g*.
8. Transfer supernatant into a 15 ml polystyrene disposable tube.

9. Add 50 μl of PBST-washed glutathione sepharose beads and rock it at cold room for 4 h.
10. Wash three times with PBST and then wash once with PBS.
11. Elute the GST-S6K1 with 50 μl of GST elution buffer for 4 h at cold room.
12. Spin and collect the eluent.
13. Dialyze the eluent with dialysis buffer for 3 h. Dialysis buffer is changed every hour.
14. Load several dilutions of samples (1, 2, and 4 μl of the eluent) with serial dilutions of BSA (0.25, 0.5, 1, 2, and 4 μg), which can be a concentration reference.
15. Make aliquots and snap-freeze GST-S6K1 solution using liquid nitrogen.
16. Stored at −80°C until use.

3.4.2. Purification of GST-S6 from Bacteria

1. Grow GST-S6 transformed bacteria in 12 ml LB with proper antibiotics for overnight.
2. Scale up overnight culture to 250 ml LB and grow bacteria for additional 1 h at 37°C.
3. Induce protein expression with 0.2 mM IPTG for 4–5 h at 37°C.
4. Pellet bacteria at 4,000 × *g* for 10 min.
5. Resuspend pellet in 10 ml cold STE buffer.
6. Transfer suspension into a 50-ml disposable tube.
7. Add 1 ml of lysozyme (10 mg/ml) and 40 μl of 1 M DTT.
8. Mix and incubate on ice for 15 min.
9. Add 250 μl of 10% NP-40 and invert tube several times.
10. Place the tube at −70°C for overnight or longer time.
11. Thaw pellet at 37°C.
12. Add 15 ml of NaCl-Mg buffer and 150 μl of Dnase I (1 mg/ml).
13. Pass suspension through a syringe with a 26-gauge needle twice.
14. Incubate on ice for 3 h with occasional mixing.
15. Pellet insoluble proteins at 15,000 × *g* for 20 min at 4°C.
16. Add 500 μl of GST beads (slurry in cold PBS) into clarified lysates.
17. Rotate at 4°C for 4 h.
18. Wash three times with cold HNTG buffer.
19. Wash twice with cold PBS.
20. Elute with 500 μl of GST elution buffer for 4 h.

21. Spin and collect the first eluent.
22. Repeat elution and collect into a new tube.
23. Dialyze and determine the amount of GST-S6 as described above (Subheading 3.4.1).

3.5. *mTOR* in Vitro *Kinase Assay*

3.5.1. In Vitro *Kinase Assay Using Endogenous mTOR*

1. Prepare two 10 cm plates of HEK293T cells. One is for mTOR IP (see Note 4) and the other plate is for control IP, to which control antibody (any goat antibody) will be added. To perform the mTORC1-specific kinase assay, Raptor should be pulled down while Rictor should be immunoprecipitated for the mTORC2-specific kinase assay.
2. Once cells reach 80% of confluency, remove media and wash once with cold PBS (see Note 3).
3. Add 1 ml CHAPS buffer containing protease inhibitors and phosphatase inhibitors into a plate.
4. Tap the plates to detach cells and incubate on ice for 20 min with occasional mixing.
5. Collect suspension into a microcentrifuge tube and spin for 15 min with 15,000 × *g* at 4°C. Transfer supernatant as a lysate into a new tube.
6. Add 10 μl of FRAP antibody (Santa Cruz, N-19) and rock it for 2–3 h at 4°C : N-19 antibody has been used only for immunoprecipitation but not for Western blotting. In the other tube, the same amount of goat antibody is added for a control.
7. Add 20 μl of G-beads into lysate and rock it for 1 h.
8. Wash beads with CHAPS buffer three times and twice with CHAPS with 500 mM NaCl.
9. Wash with low salt washing buffer once.
10. Wash beads with 1× kinase reaction buffer once.
11. Add 25 μl kinase reaction mixture containing 5 μl of 5× kinase reaction buffer, 120 ng of GST-S6K1 (for mTORC2, instead of GST-S6K1, GST-Akt is used), 200 μM ATP, and incubate for 20 min at 37°C. To determine novel phosphorylation of protein of interest, radioisotope and around 2 μg of substrate is used.
12. Terminate the reaction by addition of 10 μl of 4× SDS sample loading buffer and denature the samples at 100°C for 5 min.
13. Load 12 μl of the samples in SDS-PAGE and transfer proteins onto the membrane for 3 h at cold room.
14. To detect phosphorylated GST-S6K1, incubate membrane with pThr 389 S6K1 antibody. To monitor mTORC2, pSer473 Akt antibody is used.
15. To detect autoradiograph, expose membrane onto a phosphor screen for 2 h or overnight and develop the signal using typhoon phosphoimager (GE Healthcare).

3.5.2. In Vitro Kinase Assay Using Exogenous mTOR

1. Prepare a six-well plate.
2. Transfect 500 ng of Flag-mTOR or as a control, 1.5 μg of Flag-mTOR KD (Kinase dead) using lipofectamine. For mTORC1 kinase assays, HA-Raptor (100 ng) construct is co-transfected while HA-Rictor construct (1 μg) is co-transfected with Flag-mTOR for mTORC2 kinase assays (see Note 5).
3. 48 h post-transfection, harvest proteins using 500 μl of CHAPS buffer per well.
4. Other steps are the same as in Subheading 3.5.1, except 1 μl (1 μg) of Flag antibody is used for IP.

3.6. S6K1 In Vitro Kinase Assay

1. Prepare a 6-well plate culture of HEK293T cells.
2. Transfect 10 ng of HA-p70 construct.
3. 48 h post-transfection, lyse cells with 500 μl of S6K kinase assay lysis buffer containing protease and phosphatase inhibitors.
4. After spin, take 400 μl of supernatant for IP.
5. Add 1 μl of HA antibody and incubate at 4 °C with rocking for 1 h.
6. Add 15–20 μl of G beads and incubate another 20–30 min.
7. Wash with Buffer A.
8. Wash with Buffer B.
9. Wash with Buffer ST.
10. Wash with freshly diluted 1× kinase reaction buffer (without β-mercaptoethanol) from 6× kinase reaction buffer.
11. Add 20 μl of 1.5× kinase reaction buffer with β-mercaptoethanol, 2 μg GST-S6, 1.5 μl 1 mM cold ATP, and 5 μCi of $P^{32}\gamma$ATP.
12. Incubate 10 min at 30°C with shaking.
13. Stop the reaction by adding 10 μl of 4× SDS sample loading buffer and boil the samples for 5 min.
14. Load 20 μl of the sample a well of SDS-PAGE and transfer proteins onto PVDF membrane at cold room for 2 h.
15. Wash membrane with water shortly and wrap it with plastic wrap.
16. Expose membrane onto the phosphor screen for 2 h or overnight.

3.7. Immunofluorescence Staining of mTORC1

Samples should not be dried and must be kept in treatment conditions or PBS during the process to avoid drying samples.

1. Grow HEK293T cells in 12-well plates that have a round cover slides in each well.
2. Remove media and wash with PBS twice.
3. To fix cells, incubate cells with 1 ml of 4% paraformaldehyde and incubate 20 min at room temperature.

4. Wash with PBS three times and permeabilize cells using 1 ml permeabilizing buffer for 10 min at room temperature.
5. Wash with PBS three times.
6. Dilute mTOR antibody to 1:100 and Raptor antibody to 1:100 using antibody dilution solution.
7. Spread parafilm onto the lid of a 15-cm tissue culture plate using a handle of scissors or a bacteria spreader to attach the film evenly and firmly onto the plate.
8. Put 50 μl of diluted antibodies on the film.
9. Put the cover slips (fixed and permeabilized cells) facedown onto antibody solution and incubate at room temperature for 1 h or at 4°C for overnight. In case of overnight incubation, put folded wet kimwipes in the plate and wrap around the plate using parafilm to avoid dry.
10. Transfer cover slips into 12 wells and wash with PBS three times.
11. Dilute secondary antibodies to 1: 1,000.
12. Incubate with secondary antibodies as mentioned at step 9 under the dark but only for 30 min.
13. Wash with PBS three times and wash once with autoclaved water.
14. Put 20 μl (one big drop) of mounting media onto cover glass and put the cover slips. In this step, causing bubbles should be avoided.
15. Incubate slides for overnight under the dark.
16. Apply clear nail polish around the edge of cover slip to prevent cells from being dried.
17. Observe cells under fluorescence microscopy immediately or keep samples at 4°C in a slide box.

3.8. Evaluation of mTOR Activity in Tissues

To obtain a paraffinized sample, the animals are first perfused with 4% paraformaldehyde, fixed tissue are paraffinized, and embedded in paraffin (see Note 6).

1. Deparaffinize and hydrate sections by sequential washes in the following: Xylene 3 min (three times), 100% ethanol 1 min (twice), 95% and 80% ethanol 1 min each, and then wash in distilled water 2 min once. Rock gently in each solution.
2. Warm 300 ml of 10 mM sodium citrate, pH 6.0 to 95–100°C in a Coplin staining jar.
3. Quickly immerse slides into the hot citrate buffer and incubate for 20–40 min. The optimal incubation time needs to be determined empirically.
4. Cool the slides on the bench in citrate solution at RT for 30 min.
5. Rinse sections in PBS for 2 min twice.

6. Using a hydrophobic pen, surround the area to be stained.
7. Block sections in 2% BSA in PBS for 60 min at RT in a humidified chamber to prevent drying.
8. Remove the blocking solution.
9. Antibodies are diluted in 1% BSA, 30–50 ml per section at the appropriate dilution as indicated by the supplier and incubate for 1 h at RT or overnight at 4°C in a humidified chamber with gentle agitation.
10. Rinse with PBS for 3 min twice.
11. Block sections with peroxidase blocking solution (3% hydrogen peroxide) for 10 min.
12. Rinse in PBS for 5 min three times.
13. Add 20 μl of HRP-conjugated or diluted (200×) FITC-conjugated secondary antibody per section and incubate 1 h at RT.
14. Rinse in PBS for 10 min three times.
15. Add 20 ml of DAB solution (substrate buffer:chromogen = 50:1). DAB is oxidized and forms a stable brown end product at the site of the target antigen (Fig. 2a). For immunofluorescence, the samples are able to be monitored by fluorescence microscope (Fig. 2b).

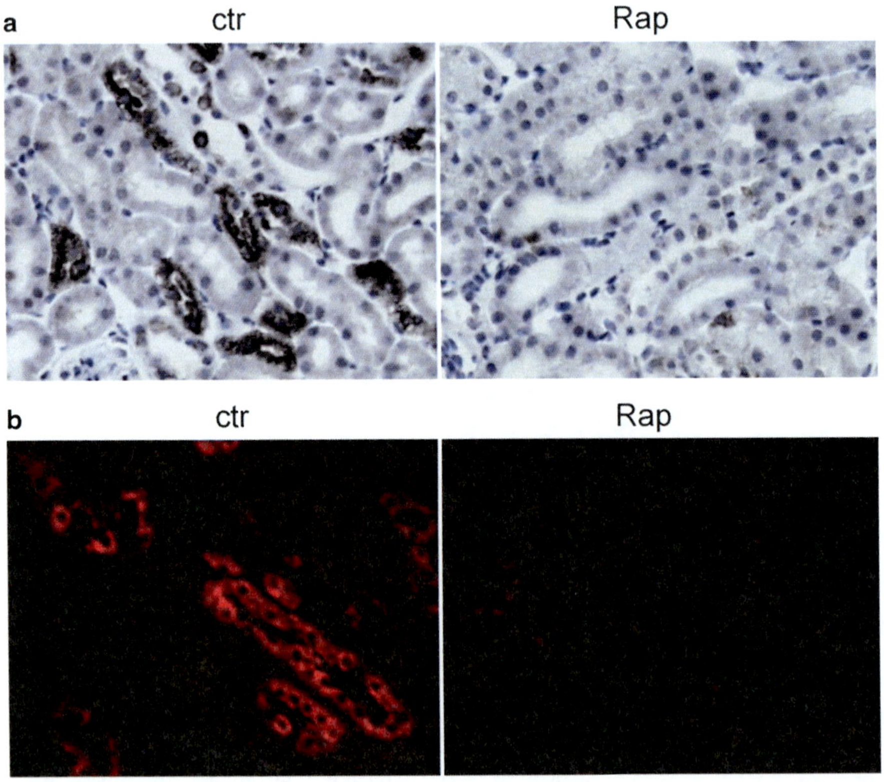

Fig. 2. mTORC1 activity in mouse tissue. Phosphorylation of S6 was monitored by immunostaining. Rapamycin treatment (1 mg/kg intra-subcutaneous injection) reduced the levels of phosphorylated S6 in renal cortex tissue. The signals were determined by the DAB method (**a**) or the immunofluorescence method (**b**), respectively.

4. Notes

1. Most of the primary antibodies can be recycled more than 5 times without any decrease in signal. Moreover, some antibodies give better results by reuse because of a decrease in nonspecificity. However, polypropylene tubes should not be used for storage of antibodies.
2. Western blotting of 4EBP1 shows different mobility shift in response to inhibition of mTOR by starvation and rapamycin treatment. Inibition of mTOR results in hypophosphorylation of 4EBP1 showing mobility shift down of 4EBP1 bands in general.
3. In case of washing cells with PBS, PBS better be applied onto plates not for the cell suspension because PBS washing can decrease mTOR activity.
4. Depending on lot numbers, the affinity of the antibody may vary such that 10 to 20 μl of antibody can be used for IP. In case of lot # D0507 of mTOR Ab for IP from Sata Cruz, 10 μl has been successful in IP.
5. To do the mTORC1 and mTORC2 kinase assay, mLST8 co-transfection is not required, since lack of mLST8 does not affect mTORC1 and 2 kinase activity *in vivo* and *in vitro*.
6. To study mTOR activity in animal tissues by immunohistochemistry (IHC), the tissue samples are prepared as either a frozen or a paraffinized section. Because the mTOR activity is very sensitive to the change in both growth factors and nutrients, the preparation for tissue samples should be done as promptly as possible.

Acknowledgments

The author would like to thank Dr Sei Yoshida for technical information on mTOR immunostaining. This work was supported by a grant from NIH (DK DK083491).

References

1. Wullschleger, S., Loewith, R., and Hall, M. N. (2006) TOR signaling in growth and metabolism, *Cell 124*, 471–484.
2. Yu, L., McPhee, C. K., Zheng, L., Mardones, G. A., Rong, Y., Peng, J., Mi, N., Zhao, Y., Liu, Z., Wan, F., Hailey, D. W., Oorschot, V., Klumperman, J., Baehrecke, E. H., and Lenardo, M. J. (2010) Termination of autophagy and reformation of lysosomes regulated by mTOR, *Nature 465*, 942–946.
3. Yip, C. K., Murata, K., Walz, T., Sabatini, D. M., and Kang, S. A. (2010) Structure of the human mTOR complex I and its implications for rapamycin inhibition, *Mol Cell 38*, 768–774.

4. Choi, J., Chen, J., Schreiber, S. L., and Clardy, J. (1996) Structure of the FKBP12-rapamycin complex interacting with the binding domain of human FRAP, *Science 273*, 239–242.
5. Schmelzle, T., and Hall, M. N. (2000) TOR, a central controller of cell growth, *Cell 103*, 253–262.
6. Sarbassov, D. D., Guertin, D. A., Ali, S. M., and Sabatini, D. M. (2005) Phosphorylation and regulation of Akt/PKB by the rictor-mTOR complex, *Science 307*, 1098–1101.
7. Sabatini, D. M. (2006) mTOR and cancer: insights into a complex relationship, *Nat Rev Cancer 6*, 729–734.
8. Kim, D. H., Sarbassov, D. D., Ali, S. M., King, J. E., Latek, R. R., Erdjument-Bromage, H., Tempst, P., and Sabatini, D. M. (2002) mTOR interacts with raptor to form a nutrient-sensitive complex that signals to the cell growth machinery, *Cell 110*, 163–175.
9. Sarbassov, D. D., Ali, S. M., Sengupta, S., Sheen, J. H., Hsu, P. P., Bagley, A. F., Markhard, A. L., and Sabatini, D. M. (2006) Prolonged rapamycin treatment inhibits mTORC2 assembly and Akt/PKB, *Mol Cell 22*, 159–168.
10. Thoreen, C. C., Kang, S. A., Chang, J. W., Liu, Q., Zhang, J., Gao, Y., Reichling, L. J., Sim, T., Sabatini, D. M., and Gray, N. S. (2009) An ATP-competitive mammalian target of rapamycin inhibitor reveals rapamycin-resistant functions of mTORC1, *J Biol Chem 284*, 8023–8032.
11. Feldman, M. E., Apsel, B., Uotila, A., Loewith, R., Knight, Z. A., Ruggero, D., and Shokat, K. M. (2009) Active-site inhibitors of mTOR target rapamycin-resistant outputs of mTORC1 and mTORC2, *PLoS Biol 7*, e38.
12. Sancak, Y., Peterson, T. R., Shaul, Y. D., Lindquist, R. A., Thoreen, C. C., Bar-Peled, L., and Sabatini, D. M. (2008) The Rag GTPases bind raptor and mediate amino acid signaling to mTORC1, *Science 320*, 1496–1501.
13. Kim, E., Goraksha-Hicks, P., Li, L., Neufeld, T. P., and Guan, K. L. (2008) Regulation of TORC1 by Rag GTPases in nutrient response, *Nat Cell Biol 10*, 935–945.
14. Sato, T., Nakashima, A., Guo, L., and Tamanoi, F. (2009) Specific activation of mTORC1 by Rheb G-protein in vitro involves enhanced recruitment of its substrate protein, *J Biol Chem 284*, 12783–12791.
15. Sancak, Y., Bar-Peled, L., Zoncu, R., Markhard, A. L., Nada, S., and Sabatini, D. M. Ragulator-Rag complex targets mTORC1 to the lysosomal surface and is necessary for its activation by amino acids, *Cell 141*, 290–303.
16. Nicklin, P., Bergman, P., Zhang, B., Triantafellow, E., Wang, H., Nyfeler, B., Yang, H., Hild, M., Kung, C., Wilson, C., Myer, V. E., MacKeigan, J. P., Porter, J. A., Wang, Y. K., Cantley, L. C., Finan, P. M., and Murphy, L. O. (2009) Bidirectional transport of amino acids regulates mTOR and autophagy, *Cell 136*, 521–534.

Chapter 4

mTOR Activity Under Hypoxia

Douangsone D. Vadysirisack and Leif W. Ellisen

Abstract

The adaptive response to hypoxia, low oxygen tension, involves inhibition of energy-intensive cellular processes including protein translation. This effect is mediated in part through a decrease in the kinase activity of mammalian target of rapamycin complex 1 (mTORC1), a master regulator of protein translation. The principle mechanism for hypoxia-induced mTORC1 inhibition, however, was not elucidated until recently. Our work has demonstrated that the stress-induced protein REDD1 is essential for hypoxia regulation of mTORC1 activity and has further defined the molecular mechanism whereby REDD1 represses mTORC1 activity under hypoxic stress. Using our studies with REDD1 as an example, we describe in detail biochemical approaches to assess mTORC1 activity in the hypoxic response. Here, we provide methodologies to monitor signaling components both downstream and upstream of the hypoxia-induced mTORC1 inhibitory pathway. These methodologies will serve as valuable tools for researchers seeking to understand mTORC1 dysregulation in the context of hypoxic stress.

Key words: mTOR, Hypoxia, Phospho-4E-BP1, Phospho-p70S6K, Phospho-S6, REDD1, TSC2, 14-3-3, Protein translation

1. Introduction

Under conditions of hypoxia, defined here as low to moderate oxygen availability, mammalian cells exhibit a reduced capacity for oxidative metabolism and as a result initiate ATP conservation by limiting energy-consuming processes, including protein synthesis (1–3). The effect of hypoxia on protein translation is mediated in part through inhibition of mammalian target of rapamycin complex 1 (mTORC1) kinase (4, 5), a central regulator of cellular growth, proliferation, and protein translation (6, 7). Regulation of cap-dependent translation initiation by mTORC1 involves the direct phosphorylation of its downstream substrates eukaryotic initiation factor 4E binding protein 1 (4E-BP1) and ribosomal

Thomas Weichhart (ed.), *mTOR: Methods and Protocols*, Methods in Molecular Biology, vol. 821,
DOI 10.1007/978-1-61779-430-8_4, © Springer Science+Business Media, LLC 2012

protein kinase S6 (p70S6K) (8–10). Several studies have indeed shown that cells exposed to hypoxic stress display a pronounced dephosphorylation of 4E-BP1 and p70S6K (4, 11). Therefore, the examination of the phosphorylation status of 4E-BP1 and p70S6K, as well as that of the p70S6K substrate ribosomal protein S6, by immunoblot analysis using commercially available phospho-site specific antibodies can serve as readouts for mTORC1 activity under hypoxic conditions.

We along with others have demonstrated that the hypoxia-inducible protein REDD1 (for *re*gulated in *d*evelopment and DNA *d*amage, also known as RTP801, DDIT4, Dig2) is necessary for hypoxia-induced inhibition of mTORC1 activity (12–14) (Fig. 1). Work in both *Drosophila* and in mammalian cells showed that REDD1 functions upstream of the tuberous sclerosis tumor suppressor complex proteins TSC1 and TSC2 to inhibit mTORC1 activity. Parallel studies showed that under normoxia, inhibitory 14-3-3 protein binds to TSC2 to suppress the function of the TSC1/2 complex, a key inhibitor of mTORC1 activity (15). We found that in response to hypoxia, REDD1 gene expression is induced, leading to REDD1-dependent dissociation of 14-3-3 and TSC2 (16) (Fig. 2). This dissociation, which appears to depend on direct, competitive binding of REDD1 to 14-3-3 within a membrane compartment, activates the TSC1/2 complex to downregulate mTORC1 activity. Thus, the assessment of REDD1/14-3-3 association and TSC2/14-3-3 dissociation by co-immunoprecipitation studies followed by immunoblot analysis provides insight into mTORC1 regulation in response to hypoxia.

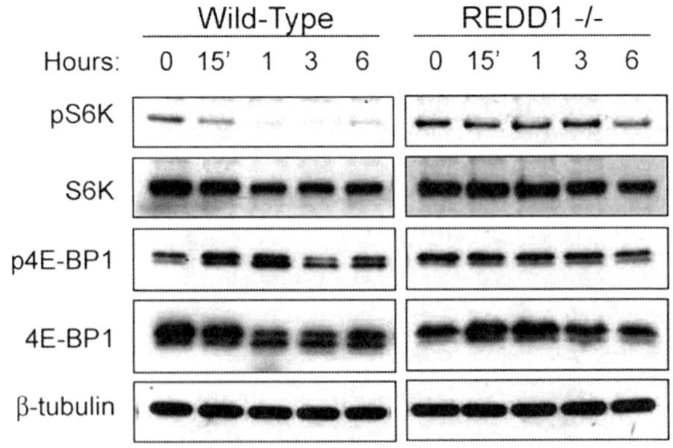

Fig. 1. REDD1 is required for inhibition of mTORC1 activity under hypoxia. Hypoxia inhibits mTORC1 activity in wild type but not in REDD1$^{-/-}$ MEFs, as evidenced by dephosphorylation of S6K (T389) and 4E-BP1 (T70). MEFs of each genotype growing in 10% serum were exposed to hypoxia (1% O_2) for the indicated times. The same blot was stripped and reprobed for the respective total proteins. Note the prominence of hypophosphorylated 4E-BP1 (*lower band*) upon hypoxic exposure of wildtype cells. Beta tubulin serves as a loading control. Adapted from Genes Dev. 22:239.

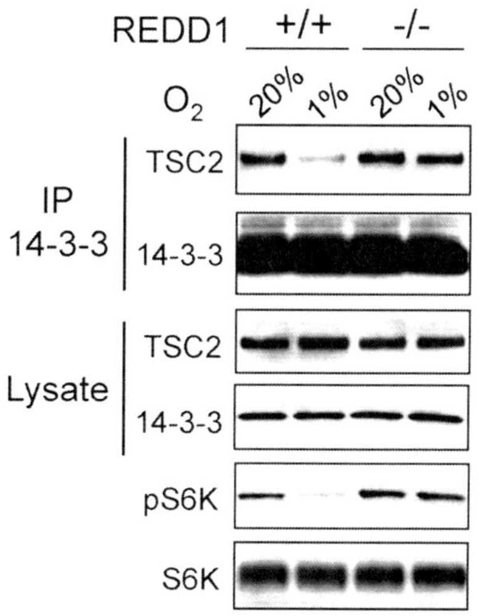

Fig. 2. REDD1 is required for hypoxia-induced TSC2/14-3-3 dissociation. MEFs of the indicated genotype were treated with hypoxia (3 h) followed by western analysis or IP for endogenous 14-3-3. Hypoxia-induced TSC2/14-3-3 dissociation and S6K (T389) dephosphorylation are both absent in REDD1$^{-/-}$ MEFs. Adapted from Genes Dev. 22:239.

Accumulating evidence suggests that the inappropriate control of mTORC1 activity in hypoxic cells confers a growth advantage and likely contributes to tumorigenesis and tumor maintenance (11, 16–18). However, the mechanism(s) by which mTORC1 activity is maintained in tumor cells under hypoxic stress remains to be fully elucidated, and further studies are warranted to clarify the interplay between aberrant mTORC1 activity, hypoxia, and tumorigenesis. The use of methodologies that provide accurate assessment of mTORC1 regulation and activity will be critical to this research effort.

2. Materials

2.1. Cell Culture

1. Primary mouse embryonic fibroblasts (MEFs) derived from 12.5–14.5 postcoitum embryos are maintained in Dulbecco's Modified Eagle's Medium containing 4.5 g/L glucose and L-glutamine, supplemented with 10% (v/v) fetal bovine serum and 1% (v/v) penicillin/streptomycin (see Note 1). Alternatively, other cell model systems of interest and their corresponding culture media can be used in lieu of primary MEFs.
2. Phosphate-buffered saline (PBS), sterilized (see Note 2).
3. Trypsin solution 0.25% in 1 mM EDTA.
4. Hypoxia cell culture incubator Heracell 150 (see Note 3).

2.2. Preparation of Cell Lysates

1. PS6 lysis buffer for phospho-4E-BP1, phospho-p70S6K, phospho-S6, and corresponding total proteins (19) contains 0.5% Nonidet P-40, 150 mM NaCl, protease, and phosphatase inhibitor cocktails. Store the buffer at 4°C.
2. Denaturing lysis buffer for co-immunoprecipitation of 14-3-3 complexes (16) contains 0.75% Nonidet P-40, 1 mM dithiothreitol (DTT) in PBS, along with protease and phosphatase inhibitor cocktails (see Note 4). Store the buffer at 4°C.

2.3. SDS-Polyacrylamide Gel Electrophoresis and Membrane Transfer

1. Bio-rad protein assay dye reagent concentrate.
2. 12% Tris–glycine polyacrylamide precast gels (see Note 5).
3. SDS-polyacrylamide gel electrophoresis (PAGE) running buffer (10×) contains 250 mM Tris–HCl, 1.92 M glycine, and 1% (w/v) sodium dodecyl sulfate (SDS) (see Note 6). Prepare 1× working solution with a 1:10 dilution of deionized distilled water. Store the 10× stock solution and the 1× working solution at room temperature.
4. 5× Laemmli sample buffer is prepared with 62.5 mM Tris–HCl pH 6.8, 20% (v/v) glycerol, 2% (w/v) SDS, 5% (v/v) 2-mercaptoethanol, and 1% (w/v) bromophenol blue (see Note 7). Store sample buffer in small aliquots at −20°C.
5. Prestained standard protein molecular weight marker.
6. Polyvinylidene fluoride (PVDF) Immobilon-FL transfer membrane cut to the dimensions of 7×8.4 cm.
7. Extra thick blot paper (mini-blot size).
8. Thick chromatography paper cut to the dimensions of 7×8.4 cm.
9. The following assumes the use of a semidry membrane transfer system. We use the Trans-blot semidry transfer cell from Bio-rad.
10. Prepare transfer solutions. Anode I buffer consists of 0.3 M Tris and 10% (v/v) methanol, pH 10.4. Anode II buffer contains 25 mM Tris and 10% (v/v) methanol, pH 10.4. Cathode buffer consists of 25 mM Tris, 192 mM glycine, and 10% (v/v) methanol, pH 9.4. Store buffers at room temperature.

2.4. Immunoblotting for Phosphorylated Downstream Targets of mTORC1

1. Prepare solution of PBS containing 0.1% (v/v) Tween-20 (PBST). Store the solution at room temperature.
2. Blocking solution is prepared either with 5% (w/v) nonfat dry milk or 5% (w/v) bovine serum albumin (BSA) fraction V in PBST (see Note 8).
3. Pertinent information regarding primary and secondary antibodies for immunoblotting, including supplier, product code, and recommended use are detailed in Table 1. It should be emphasized that these primary antibodies can be used to detect both mouse and human species. Thus, these reagents would

Table 1
Antibody reagents and suppliers

Antibody	Supplier	Product no.	Recommended use
Phospho-4E-BP1 (Thr70)	Cell Signalling Technologies	#9455	(WB) 1:1,000 1% BSA–PBST O/N
Phospho-p70S6K (Thr389)	Cell Signalling Technologies	#9205	(WB) 1:1,000 1% BSA–PBST O/N
Phospho-S6 (Ser235/Ser236)	Cell Signalling Technologies	#2211	(WB) 1:1,000 1% BSA--PBST O/N
4E-BP1	Santa Cruz Biotechnology	#6936	(WB) 1:1,000 5% milk–PBST 1 h
p70S6K	Cell Signalling Technologies	#9202	(WB) 1:1,000 5% milk–PBST 1 h
S6	Cell Signalling Technologies	#2217	(WB) 1:1,000 5% milk–PBST 1 h
β-Tubulin	Millipore	#MAB 3408	(WB) 1:10,000 5% milk–PBST 1 h
Pan-14-3-3	Thermo Scientific	#MS-1504P	(IP) 1 μg/mg protein, (WB) 1:1,000 5% milk–PBST O/N
TSC2	Santa Cruz Biotechnology	#893	(WB) 1:1,000 5% milk–PBST O/N
Goat α-mouse IgG-HRP	Biorad	#170-6516	(WB) 1:5,000 either 1% BSA or 5% PBST 1 h
Goat α-rabbit IgG-HRP	Biorad	#170-6515	(WB) 1:5,000 either 1% BSA or 5% PBST 1 h

also be useful for investigators wishing to use human cell model systems.

4. Western lightening plus-enhanced chemiluminescence (ECL) substrate reagents.
5. BioMax XAR film.
6. Autoradiography cassette.

2.5. Stripping and Reprobing Membrane for Total Protein

1. Restore western blot stripping buffer.

2.6. Co-immunoprecipitation of 14-3-3 Complexes

1. Protein G Sepharose beads.
2. Relevant information regarding antibodies for co-immunoprecipitation studies, including supplier, product code, and recommended use, are provided in Table 1.

3. Methods

The methods described below outline our established protocols for (1) treatment of cells under hypoxic conditions; (2) preparation of cell samples; (3) detection of mTORC1 downstream targets by immunoblot analysis; and (4) detection of 14-3-3 immunocomplexes by co-immunoprecipitation and immunoblot analysis. We have also highlighted some important considerations for using these techniques.

3.1. Culturing Cells Under Normoxic and Hypoxic Conditions

1. Seed cells for the described experiments when passaged cells reach 80–90% confluency. Detach cells from tissue culture plates by washing with sterile PBS, adding Trypsin–EDTA, and then placing plates back into a 37°C humidified incubator for 2–5 min.
2. Verify detachment of cells using an inverted phase contrast microscope. Once the majority of the cells have rounded up, inactivate the trypsin with the addition of prewarmed media containing serum (trypsin inhibitors are present in serum). Resuspend the cells using a serological pipette and pipette-aid (see Note 9).
3. Seed primary MEFs at a cell density of 5×10^5 cells per 10-cm plate (see Note 10).
4. Place plates in a 37°C humidified incubator at 5% CO_2 and atmospheric oxygen level or normoxia (20% O_2) for a minimum of 3–5 h to allow adherence of cells to the plates. We generally culture cells for 12–24 h postseeding.
5. Transfer plates to a 37°C hypoxia cell culture incubator and expose cells to 1% oxygen for various time points (see Note 11). Keep control plates in the normoxia incubator. Note that in our study with primary MEFs we evaluated mTORC1 activity in response to hypoxia at time points ranging from 15 min to 6 h (see Fig. 1). We observed dephosphorylation of phospho-p70S6K (Thr389) and phospho-4E-BP1 (Thr70) within 3 h in wildtype MEFs. In contrast, REDD1-null MEFs exhibited a defect in dephosphorylation of mTORC1 substrates at all time points examined.

3.2. Preparation of Cell Lysates Following Hypoxia Treatment

1. Remove plates from the hypoxia and normoxia cell culture incubators (see Note 11).
2. Quickly place the plates on ice and wash them with ice-cold, nonsterile PBS. Wash twice (see Note 12).
3. Add 400 μL of the appropriate cell lysis buffer, either PS6 or denaturing cell lysis buffer, directly to the plates.

4. Collect cells from the plates using a cell scraper then transfer cells into microcentrifuge tubes.
5. Allow the cells to lyse for 20 min on ice, inverting the microcentrifuge tubes every few minutes. Alternatively, place the microcentrifuge tubes on a 4°C rocker.
6. Centrifuge tubes at 16,000 RCF for 20 min to pellet nuclei. Use a 4°C refrigerated tabletop centrifuge.
7. Carefully transfer supernatants to new microcentrifuge tubes and place tubes on ice (see Note 13).

3.3. SDS-PAGE and Protein Transfer

1. Determine the protein concentrations of lysates using the Bio-rad protein assay dye reagent concentrate. Both the PS6 and denaturing cell lysis buffers are compatible with this Bradford protein assay. Briefly, dilute 200 μL of the protein dye concentrate with 800 μL of deionized distilled water, pipette 1 μL of lysate into the solution, wait 5 min, and then measure the absorbance at 595 nm.
2. Samples containing 25–40 μg of lysate are diluted 4:1 (v/v) with 5× Laemmli sample buffer, boiled at 95°C in a heat block for 5 min, briefly vortexed and centrifuged, and then placed on ice until use.
3. Cut open a pouch containing a 12% Tris–glycine polyacrylamide precast gel and remove the tape at the bottom of the cassette.
4. Assemble the precast gel cassette into the electrophoresis chamber. We use the Bio-rad mini-protean II apparatus.
5. Carefully lift the comb straight up to remove it from the cassette.
6. Fill the inner buffer chamber with 1× SDS-PAGE running buffer and allow the running buffer to overflow into the outer buffer chamber. This will wash and displace any air bubbles in the wells of the gel.
7. Load prestained standard protein molecular weight marker (~10 μL) and samples slowly into the wells using a P20 micropipettor (see Note 14).
8. Connect the unit to a power supply. Then run the gel at a constant voltage of 125 V until the bromophenol blue dye front has migrated to the bottom of the gel (see Note 15). This will take approximately 90 min.
9. Once electrophoresis is completed, disconnect the device from the power supply and take out the gel cassette. Open the cassette, and using a razor blade cut the upper and lower most portions of the gel and discard them. This will leave behind only the separating gel. Mark the orientation of the separating gel by cutting the top left corner. Then equilibrate the gel in Cathode buffer for 5 min.

10. The following steps detail the electrophoretic transfer of proteins from the gel to PVDF membrane using the Trans-blot semidry transfer system from Bio-rad.
11. Activate the PVDF membrane for protein transfer by wetting the membrane with 100% methanol for 1–2 min (the membrane should change from opaque to semitranslucent). Then immerse the membrane in Anode II buffer for 5 min (see Note 16).
12. Assemble the semidry transfer stack in the following manner:
    (a) Wet one extra thick blot paper in Anode I buffer and place it on the anode plate of the transfer apparatus.
    (b) Wet one piece of thick chromatography paper in Anode II buffer and place it on top of the extra thick blot paper.
    (c) Place the activated PVDF membrane on top of the stack.
    (d) Place the gel on top of the membrane, and using a serological pipette remove air bubbles between the gel and membrane by rolling it gently over the gel.
    (e) Wet one extra thick blot paper in Cathode buffer and place it on top of the gel, and remove air bubbles as before.
    (f) Place the cathode plate on top of the transfer stack.
13. Connect the transfer apparatus to a power supply and apply a constant current of 0.25 mA for 35 min.
14. After the transfer is completed, disconnect the apparatus from the power supply and remove the membrane. Mark the orientation of the membrane, as with the gel, by cutting the top left corner of the membrane.

3.4. Immunoblotting for Phosphorylated Downstream Targets of mTORC1

1. Block the membrane with 10 mL of 5% BSA–PBST. Incubate the membrane for 30 min at room temperature on a shaker.
2. Discard the blocking solution. Refer to Table 1 for the appropriate primary antibody against phospho-4E-BP1 (Thr70), phospho-p70S6K (Thr389), and phospho-S6 (Ser235/Ser236) (see Note 17). Incubate the membrane with the recommended primary antibody diluted 1:1,000 in 1% BSA–PBST overnight at 4°C on a rocker. Care should be taken to add sufficient volume of the diluted primary in order to cover the entire membrane.
3. Remove the primary antibody and rinse the membrane twice with 0.1% PBST. Next, wash the membrane twice with 10 mL 0.1% PBST for 10 min at room temperature on a shaker (see Note 18).
4. Incubate the membrane with the appropriate horseradish peroxidase-conjugated secondary antibody (diluted 1:5,000 in 1% BSA–PBST) for 40–60 min at room temperature on a shaker.

5. Discard the secondary antibody and repeat washes as before.
6. Immunoreactivity signal is visualized using ECL solutions (mixture of 1:1). The membrane is blotted on a paper towel, placed on top of a piece of plastic wrap, overlaid with ECL solution for 1 min, and after which time blotted again on a paper towel.
7. Wrap the membrane in plastic wrap, making sure to remove air bubbles. Then place the membrane inside an autoradiography cassette and expose the membrane to x-ray film for 30 s to 5 min.

3.5. Stripping and Reprobing Membrane for Total Protein

1. Once the phosphoprotein blot has been obtained, the membrane can then be stripped of primary and secondary antibodies so that it can be reprobed with an antibody that recognizes the corresponding total protein. This will allow the assessment of total protein loading. To proceed with stripping the membrane, quickly rinse the membrane of residual ECL with 0.1% PBST.
2. Incubate the membrane with 10 mL of Restore western blot stripping buffer for 15–20 min at room temperature on a shaker (see Note 19).
3. Discard the stripping buffer. Wash twice with 0.1% PBST for 10 min at room temperature on a shaker.
4. Block the membrane with 10 mL of 5% milk for 30 min at room temperature on a shaker.
5. Incubate the membrane with the appropriate primary antibody against the corresponding total protein. Refer to Table 1. Since we have found that these primary antibodies work well and provide strong detection signal, we generally incubate the membrane with the diluted primary antibody (1:1,000 in 5% milk–PBST) for 1 h at room temperature on a shaker.
6. Refer to steps 3–7 in Subheading 3.4 for immunoblotting (see Note 20).

3.6. Co-immunoprecipitation of 14-3-3 Complexes

1. Culture cells under normoxia and hypoxia as described in Subheading 3.1.
2. Prepare cell lysates in denaturing cell lysis buffer as described in Subheading 3.2.
3. Determine the protein concentration as described in Subheading 3.3.
4. Save an aliquot of 50 μL of lysate samples to be used as an input control for immunoblot analysis. These samples can be stored at −80°C until use.
5. Use equivalent amounts of cell lysates (400 μg to 1 mg protein) for co-immunoprecipitation. Transfer cell lysates into new microcentrifuge tubes and bring the volume of the lysate up to 800 μL with the denaturing cell lysis buffer.

6. Preclear lysates with the addition of 40 μL of Protein G Sepharose beads. Rotate microcentrifuge tubes for 20 min at 4°C (see Note 21). The preclearing step will help to reduce nonspecific binding of proteins to the beads and improve the signal-to-noise ratio.
7. Use a refrigerated centrifuge to pulse-spin the tubes at 6,010 × *g*. Then transfer the supernatant to new microcentrifuge tubes.
8. Add 1 μg of the 14-3-3 antibody to the lysates (refer to Table 1) and rotate the microcentrifuge tubes for 90 min (see Note 22).
9. Incubate lysates with 40 μL of prewashed Protein G Sepharose beads. Continue to rotate the microcentrifuge tubes for an additional 30 min at 4°C.
10. Pulse-spin microcentrifuge tubes at 6,010 × *g*. Remove the supernatant with a P1000 micropipettor and discard the supernatant.
11. Wash beads with 900 μL denaturing cell lysis buffer four times, and pulse-spin as described in step 10.
12. Remove the entire volume of lysis buffer after the last wash using a P200 micropipettor.
13. To the beads, add 30 μL of 2× Laemmli sample buffer (diluted in deionized distilled water). Then tap the microcentrifuge tubes to resuspend the beads.
14. Boil the microcentrifuge tubes at 95°C in a heat block for 5 min. Pulse-spin tubes and transfer equal volume of supernatant to new microcentrifuge tubes (see Note 23).
15. Load immunoprecipitated complexes as well as input lysate samples (25–40 μg) onto a SDS-PAGE gel then perform membrane transfer as described in Subheading 3.3. Subsequently, probe the membrane for detection of TSC2 in the 14-3-3 immunoprecipitated samples using the method described in Subheading 3.4 (see Note 24). In our studies, we observed that hypoxia (1% O_2) induced the dissociation of TSC2 and 14-3-3 in wildtype primary MEFs but not in REDD1-null MEFs (see Fig. 2).

4. Notes

1. Primary MEFs replicate for a limited number of population doubling, typically exhibiting rapid growth at passages 1–5, slow growth at passages 5–10, and then no growth or senescence at passages 10–25. At higher passage, a subset of primary MEFs

will escape senescence through spontaneous immortalization. Accordingly, we recommend the use of primary MEFs at an early passage of less than 6 to preclude confounding effects associated with senescence or spontaneous immortalization.

2. Unless noted otherwise, all solutions are prepared in deionized distilled water with a specific resistance of 18.2 MΩ-cm.
3. A hypoxia cell culture incubator such as the Heracell 150 has an oxygen sensor probe that allows the precise control of oxygen content. This instrument utilizes nitrogen gas to displace or flush oxygen, thereby reducing the oxygen content inside the chamber. We prefer the use of a hypoxia cell culture incubator over hypoxia mimetics such as cobalt chloride ($CoCl_2$) and desferrioxamine (DFO). There may be differences in the response of cells elicited by chemical-induced hypoxia versus ambient hypoxia as different intracellular signaling pathways might be activated or inactivated by hypoxia mimetic drugs.
4. For both the PS6 lysis buffer and denaturing lysis buffer, protease and phosphatase inhibitors are added at the time of use to maximize their efficacy. Similarly, we add DTT (1 M stock solution stored at −20°C) to the denaturing lysis buffer before use as well.
5. We have found that the use of polyacrylamide precast gels ensure a high degree of consistency and minimize the handling of toxic chemicals that are associated with preparation of self-cast gels. Note the expiration date of precast gels prior to use.
6. Caution should be used when preparing SDS-PAGE running buffer as SDS powder is hazardous. Prepare the solution in a ventilated chemical fume hood and wear a face mask when weighing SDS powder. We suggest the use of 3M Particulate Respirator N95.
7. When dispensing 2-mercaptoethanol for preparation of 5× Laemmli sample buffer, work in a ventilated chemical fume hood as it is toxic and has an unpleasant odor.
8. The choice between the use of BSA or milk as blocking agents as well as for diluting primary and secondary antibodies is dependent on whether the protein of interest is a phosphoprotein. Milk contains phosphatases which may result in the dephosphorylation of phosphoproteins. Moreover, the phosphoprotein casein is also present in milk. Primary antibodies against phosphorylated proteins may bind to casein, resulting in high background. Consequently, we commonly use BSA when probing for phosphorylated proteins.
9. It is crucial to break up any cell clumps in order to achieve at least a 95% single cell suspension for accurate cell counting with a hemocytometer and for uniform plating of cells. We suggest two methods to facilitate obtaining a single cell suspension.

One effective method to mechanically dissociate cell clumps is to place the tip of a serological pipette against the bottom of the plate while dispensing the cell suspension. As vigorous pipetting may damage the cell membrane and reduce viability of some cell types, an alternative method is to briefly rinse cells with trypsin, aspirate the solution, and then add additional trypsin to the cells followed by placing the plate back into a 37°C incubator for 2–5 min.

10. It is rarely appreciated that cells cultured in a monolayer to a high density are under some degree of hypoxic stress. This is mainly due to the high rate of oxygen consumption and limited diffusion of oxygen in a confluent cell layer. Thus, we recommend seeding cells at a moderate density such that at the time of experimental use and exposure to hypoxia, cells do not exceed 80% confluency.
11. Reoxygenation of cells is a major issue that is often encountered with the use of a hypoxia cell culture incubator. There are two precautions that we use to minimize reoxygenation of cells. Foremost, move quickly. This includes promptly closing incubator doors after placing or removing plates from the incubator. Importantly, process all samples as soon as possible once plates are removed from the hypoxia incubator. Another precaution is to increase the volume of cell culture media in the tissue culture plates.
12. For a 10-cm plate, wash with approximately 10 mL of PBS. We wash twice, and aspirate any residual PBS to ensure complete removal of serum-containing media as it can interfere with accurate determination of protein concentration.
13. At this point, lysates prepared in PS6 lysis buffer can be stored frozen at −80°C until later use. Samples prepared in denaturing lysis buffer, however, should be processed immediately to maximize detection of 14-3-3 immunocomplexes. However, if co-immunoprecipitation of 14-3-3 complexes cannot be performed immediately then add 10% glycerol (v/v) to the lysates and store the samples at −80°C. This will help to preserve protein complexes upon freezing and thawing.
14. For best results, load empty wells with 1× Laemmli sample buffer diluted in lysis buffer to ensure that adjacent sample lanes do not spread.
15. Ensure proper running of the gel by making certain that the electrodes are connected correctly.
16. Do not allow the PVDF membrane to dry as this will result in poor and blotchy transfer of proteins. Should drying of the membrane occur rewet it with methanol and rinse with Anode II buffer.

17. It should be noted that phosphorylation at Thr70 on 4E-BP1 as well as phosphorylation at Thr389 on p70S6K is rapamycin-sensitive. Furthermore, several studies have confirmed that both Thr70 on 4E-BP1 and Thr389 on p70S6K are phosphorylated directly by mTORC1. Thus, these particular phospho-site specific antibodies can be used to monitor mTORC1 activity.
18. Retain these primary antibodies as they can be reused for 10–15 subsequent experiments over several months. Exposure time may increase following extensive reuse of the diluted primary antibody.
19. It is best to use a closed container to minimize the unpleasant odor of the stripping buffer.
20. After detection of total protein by immunoblot analysis, membranes can be stored for later reprobing of other desired proteins. Rinse the membrane of ECL and allow it to dry at room temperature. The dried membrane can be stored at 4°C and is stable for up to 6 months. To reprobe the dried membrane, simply immerse it in 100% methanol for 30 s to 1 min.
21. Use a cut micropipette tip to draw up the beads in order to prevent them from being shattered or damaged. Additionally, for the preclearing step use beads that have been prewashed in the denaturing cell lysis buffer twice. Pulse-spin the microcentrifuge tube containing the beads at 6,010 × g to pellet the beads. Remove the lysis buffer with a P1000 micropipette. Then resuspend the beads in the denaturing cell lysis buffer.
22. One technical issue with the use of a pan 14-3-3 antibody is that such reagents may not allow determination of the exact 14-3-3 isoform that binds to TSC2.
23. Following the elution of captured proteins from the beads with 2× Laemmli sample buffer, the samples can be stored at −80°C until later use.
24. Refer to Table 1 for the TSC2 antibody and recommended dilutions and conditions.

Acknowledgments

The authors would like to thank Nicole Forster and Zachary M. Nash for critical reading of the manuscript and for discussion. This work was funded by RO1 CA122589 to LWE and by NRSA post-doctoral fellowship award F32 CA150633 to DDV.

References

1. Liu L, Cash TP, Jones RG, Keith B, Thompson CB, and Simon MC (2006) Hypoxia-induced energy stress regulates mRNA translation and cell growth. Mol Cell 21: 521–531
2. Wouters BG, van den Beucken T, Magagnin MG, Koritzinsky M, Fels D, and Koumenis C (2005) Control of the hypoxic response through regulation of mRNA translation. Semin Cell Dev Biol 16: 487–501
3. Hochachka PW, Buck LT, Doll CJ and Land SC (1996) Unifying theory of hypoxia tolerance: molecular/metabolic defense and rescue mechanisms for surviving oxygen lack. Proc Natl Acad Sci 93: 9493–9498
4. Arsham AM, Howell JJ, and Simon MC (2003) A novel hypoxia-inducible factor-independent hypoxic response regulating mammalian target of rapamycin and its targets. J Biol Chem 278: 29655–29660
5. Koritzinsky M, Magagnin MG, van den Beucken T, Seigneuric R, Savelkouls K, Dostie J, Pyronnet S, Kaufman RJ, Weppler SA, Voncken JW, Lambin P, Koumenis C, Sonenberg N, and Wouters BG (2006) Gene expression during acute and prolonged hypoxia is regulated by distinct mechanisms of translational control. EMBO J 25:1114–1125
6. Foster KG and Fingar DC (2010) Mammalian target of rapamycin (mTOR): conducting the cellular signaling symphony. J Biol Chem 285: 14071–14077
7. Ma XM, and Blenis J (2009) Molecular mechanisms of mTOR-mediated translational control. Nat Rev Mol Cell Biol 10: 307–318
8. Mamane Y, Petroulakis E, LeBacquer O, and Sonenberg N (2006) mTOR, translation initiation and cancer. Oncogene 25: 6416–6422
9. Burnett PE, Barrow RK, Cohen NA, Snyder SH, and Sabatini DM (1998) RAFT1 phosphorylation of the translational regulators p70 S6 kinase and 4E-BP1. Proc Natl Acad Sci 95: 1432–1437
10. Gingras AC, Raught B, Gygi SP, Niedzwiecka A, Miron M, Burley SK, Polakiewicz RD, Wyslouch-Cieszynska A, Aebersold R, and Sonenberg N (2001). Hierarchical phosphorylation of the translation inhibitor 4E-BP1. Genes Dev 15:2852–2864
11. Connolly E, Braunstein S, Formenti S, and Schneider RJ (2006) Hypoxia inhibits protein synthesis through a 4E-BP1 and elongation factor 2 kinase pathway controlled by mTOR and uncoupled in breast cancer cells. Mol Cell Biol 26: 3955–3965
12. Shoshani T, Faerman A, Mett I, Zelin E, Tenne T, Gorodin S, Moshel Y, Elbaz S, Budanova A, Chajut A, Kalinski H, Kamer I, Rozen A, Mor O, Keshet E, Leshkowitz D, Einat P, Skaliter R, and Feinstein E (2002) Identification of a novel hypoxia-inducible factor 1-responsive gene, RTP801, involved in apoptosis. Mol Cell Biol 22: 2283–2293
13. Brugarolas J, Lei K, Hurley RL, Manning BD, Reiling JH, Hafen E, Witters LA, Ellisen LW, and Kaelin WG Jr (2004) Regulation of mTOR function in response to hypoxia by REDD1 and the TSC1/TSC2 tumor suppressor complex. Genes Dev 18: 2893–2904
14. Reiling JH and Hafen E (2004) The hypoxia-induced paralogs Scylla and Charybdis inhibit growth by down-regulating S6K activity upstream of TSC in Drosophila. Genes Dev 18: 2879–2892
15. Cai SL, Tee AR, Short JD, Bergeron JM, Kim J, Shen J, Guo R, Johnson CL, Kiguchi K, and Walker CL (2006) Activity of TSC2 is inhibited by AKT-mediated phosphorylation and membrane partitioning. J Cell Biol 173: 279–289
16. DeYoung MP, Horak P, Sofer A, Sgroi D, and Ellisen LW (2008) Hypoxia regulates TSC1/2-mTOR signaling and tumor suppression through REDD1-mediated 14-3-3 shuttling. Genes Dev 22: 239–251
17. Guertin DA and Sabatini DM (2007) Defining the role of mTOR in cancer. Cancer Cell 12: 9–22
18. Schneider A, Younis RH, and Gutkind JS (2008) Hypoxia-induced energy stress inhibits the mTOR pathway by activating an AMPK/REDD1 signaling axis in head and neck squamous cell carcinoma. Neoplasia 10: 1295–1302
19. Sofer A, Lei K, Johannessen CM, and Ellisen LW (2005) Regulation of mTOR and cell growth in response to energy stress by REDD1. Mol Cell Biol 25: 5834–5845

Chapter 5

Isolation of the mTOR Complexes by Affinity Purification

Dos D. Sarbassov, Olga Bulgakova, Rakhmet I. Bersimbaev, and Tattym Shaiken

Abstract

The mammalian Target Of Rapamycin (mTOR) protein is a central component of the essential and highly conserved signaling pathway that emerged as a critical effector in regulation of cell physiology. Biochemical studies defined mTOR as the protein kinase that exists at least in two distinct complexes. The first complex has been characterized as the nutrient-sensitive mTOR complex 1 that controls cell growth and cell size by regulating protein synthesis and autophagy. The second complex of mTOR has been defined as the component of growth factor signaling that functions as a major regulatory kinase of Akt/PKB. Here, we provide the detailed methods how to purify the functional complexes of mTOR by affinity purification. In the first part, we describe the purification of the distinct mTOR complexes by immunoprecipitation. Purification of the soluble mTOR complexes is explained in the second part of this chapter.

Key words: Rapamycin, mTOR, mLST8, Raptor, mTOR complex 1, Rictor, mTOR complex 2, Cell lysis, Affinity purification, Immunoprecipitation

1. Introduction

The highly conserved mammalian target of rapamycin (mTOR) pathway plays an essential role in regulation of cell growth, proliferation, metabolism, and survival. Its central component mTOR was originally discovered as a target for the lypophilic macrolide rapamycin. The antiproliferative and growth inhibitory effects of rapamycin are dependent on its specific targeting of mTOR. Rapamycin forms a complex with its intracellular receptor FK506 binding protein (FKBP)12 (a small 12 kDa protein) and binds FKBP12 and rapamycin binding (FRB) domain of mTOR. Binding of the rapamycin/FKBP12 complex to the FRB domain interferes with the function of mTOR (1). Initially, based on the genetic screening study in yeast, the two TOR genes were identified (2)

Thomas Weichhart (ed.), *mTOR: Methods and Protocols*, Methods in Molecular Biology, vol. 821,
DOI 10.1007/978-1-61779-430-8_5,

and the biochemical purification studies revealed mTOR as the rapamycin-dependent binding protein of FKBP12 (3–5).

The biochemical studies show that mTOR exists in two distinct and large protein complexes defined as mTOR complex 1 and 2 (mTORC1 and 2) that carry different functional roles in cells (6); the diagram of the mTOR complexes is shown in Fig. 1. Within both complexes, mTOR presides with its interacting proteins mLST8 and DEPTOR. A small adaptor protein mLST8 binds tightly to the kinase domain of mTOR (7, 8) and is required for its kinase activity (9). The recently identified DEPTOR (DEP domain containing TOR interacting protein) has been shown to carry a negative role in regulation of mTOR (10).

Binding of raptor to mTOR defines its first complex mTORC1. This complex functions as a nutrient-sensitive kinase complex that regulates protein synthesis by phosphorylation of its two substrates, S6K1 and 4EBP1 (6). A role of mTORC1 in regulation of ribosomal biogenesis and autophagy has been also reported. The second complex mTORC2 is assembled by mTOR and its essential components rictor and Sin1 (7, 8, 11, 12). Rictor and Sin1 form a heterodimer that determines a mutual stability of both proteins. The second complex mTORC2 has been initially identified as a regulator of PKCa and cytoskeleton (7, 8). Later, mTORC2 has been defined as the component of growth factor signaling that functions as a major regulatory kinase of Akt/PKB by phosphorylating it on the hydrophobic Ser-473 site (13). The members of serum and glucocorticoid-inducible kinases (SGK) family related

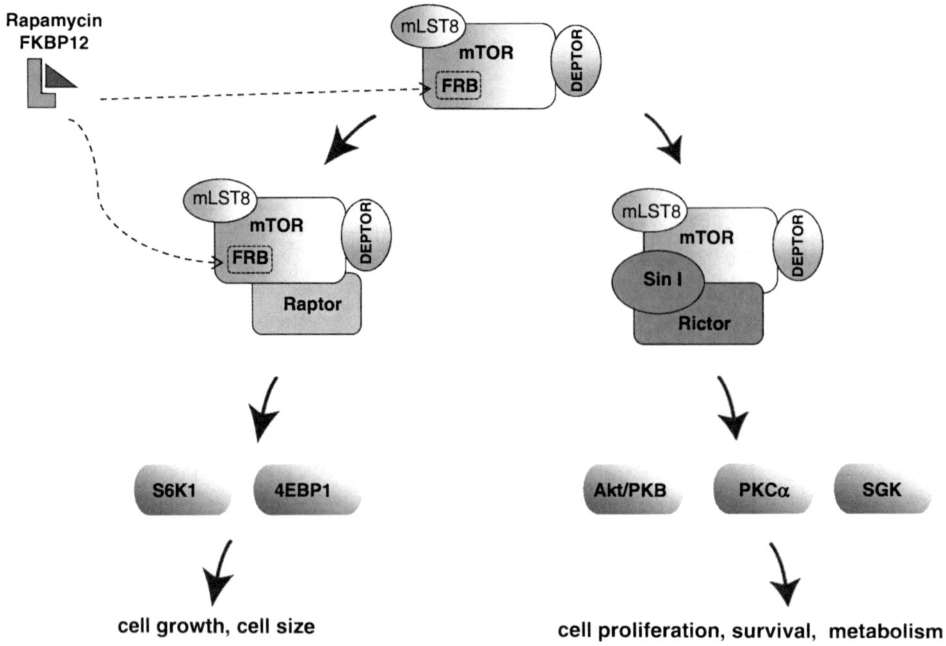

Fig. 1. A diagram of the mTOR signaling pathway.

to Akt/PKB kinases are also regulated by mTORC2 by a similar mechanism (14). In contrast to mTORC1, mTORC2 does not bind the rapamycin/FKBP12 complex, suggesting that the FRB domain is not accessible on mTORC2. A prolonged rapamycin treatment inhibits the mTORC2 assembly, where the drug is competing with the rictor/Sin1 heterodimer to bind mTOR (15).

The mTOR complex studies have been initiated by the identification and functional characterization of raptor as the mTOR interacting protein (Note 2). Replacement of the TOR genes with the tagged forms in yeast and their biochemical isolation have indicated that TOR exists in two distinct complexes and only one of them binds rapamycin/FKB12 (16). This finding has been later confirmed in mammalian cells when rictor (also named as mAVO3) has been identified as the mTOR interacting protein that forms a distinct mTOR complex and did not bind the drug complex (7, 8). These studies outlined that two distinct TOR complexes are highly conserved and their functional roles have been actively pursued in the following studies. The purpose of this chapter is to provide the detailed protocol of how to isolate the mTOR complexes based on affinity purification. We also explain the clues that led to development of this purification process. This retrospective view reflects the initial groundwork in purification of the mTOR complexes that took place in the laboratory of David M. Sabatini at the Whitehead Institute for Biomedical Research (Cambridge, MA, USA).

2. Materials

2.1. Cell Culture and Lysis

1. Dulbecco's Modified Eagle's Medium (DMEM, Invitrogen) supplemented with 10% fetal bovine serum (FBS), HyClone, Ogden, UT).
2. The human cancer cell lines (HeLa, MDA-MB-435) were obtained from the American Type Culture Collection (ATCC, Manassas, VA).
3. Cell lysis buffer stock: 40 mM HEPES, pH 7.5, 120 mM NaCl, 1 mM EDTA, 10 mM Na-pyrophosphate, 10 mM glycerophosphate, 50 mM NaF, 0.3% CHAPS (3-((3-cholamidopropyl) dimethylammonio]-1-propanesulfonate). Keep this buffer at 4°C. To complete the lysis buffer, add protease inhibitor cocktail (Roche) before lysis. All solutions made in the double-distilled water with the purity requirement shown in the Note 1.

2.2. SDS-Polyacrylamide Gel Electrophoresis/ Western Blotting

1. The 7.5% and gradient 4–15% Tris–Glycine gels, XT Sample buffer (4× Laemmli buffer), and Prestained Blue Molecular Weight Markers (all from Bio-Rad, Hercules, CA).

2. Running buffer (10×): Prepare the stock by dissolving 30 g of Tris base, 144 g of glycine, 10 g SDS in 1 L of distilled water. Dilute the stock ten times before use.
3. Mini-PROTEAN Tetra Cell (Bio-Rad, Hercules, CA).
4. Transfer buffer: To prepare 4 L of 1× CAPS transfer buffer, dissolve 8.85 g CAPS and 13 pellets of NaOH (1.2 g) in 3.6 L of distilled water and add 0.4 L ethanol.
5. Trans Blot Cell (Bio-Rad, Hercules, CA).
6. The PVDF membrane and Whatman paper were obtained from Millipore, Bedford, MA.
7. The 10× phosphate-buffered saline Tween 20 (PBST) stock solution: To prepare the stock solution, dissolve in 10 L of water 20.3 g of $NaH_2PO_4 \cdot H_2O$ (monobasic, 14.7 mM, MW = 137.99,), 115 g Na_2HPO_4 (dibasic, 81 mM, MW = 141.96), 850 g NaCl (MW = 58.44, 1.45 M), and 50 mL Tween 20 (0.5%).
8. Immobilon Western Chemiluminescent HR Substrate (Millipore, Bedford, MA).

2.3. Purification of the mTOR Complexes by Immunoprecipitation and Detection of Their Components

1. The validated, highly specific antibodies (Abs) efficient in detection or purification of the components of the mTOR complexes:
   (a) The antibodies from Santa Cruz Biotechnology (Santa Cruz, CA) – mTOR (goat FRAP antibody, #sc-1549, application: immunoprecipitation, IP); rictor antibody (goat rictor Ab, #sc-50678, application western blotting and IP); all secondary HRP-labeled antibodies (application: western blotting).
   (b) The antibody from Cell Signaling (Danver, MA) – mTOR (rabbit mTOR Ab, #2983, Cell Signaling, application: western blotting).
   (c) The antibody from SDIX (Newark, DE) – mTOR (rabbit FRAP antibody, #2813.00.02, SDI, application: western blotting and IP).
   (d) The antibodies from Bethyl (Montgomery, TX) – raptor (rabbit raptor Ab, #A300-553A, application: western blotting and IP) and rictor (rabbit rictor Ab, # A300-459A, application: western blotting and IP).
   (e) The mouse Sin1monoclonal Ab is (#05-1044) from Millipore, Bedford, MA.
2. Protein G agarose beads (#20399, Thermo Fisher Scientific/Pierce, Waltham, MA).
3. Silver staining – For convenience, we use the SilverQuest Staining kit (Invitrogen, Carlsbad, CA). Alternatively, the improved silver staining protocol is also highly sensitive and effective as described by Mortz et al. (17).

2.4. Purification of the Soluble mTOR Complexes

1. The constitutive expression of the Flag-mLST8 protein in HeLa cells is achieved by the Murine Stem Cells Virus (MSCV) retroviral expression system (Clontech, Mountain View, CA): The MSCV puromycin-resistant Flag-mLST8 mammalian expression plasmid (the plasmid submitted to Addgene by the Sabatini Laboratory, Whitehead Institute for Biomedical Research, Cambridge, MA) and the appropriate packaging cell line have been used as recommended by the manufacturer (Clontech, Mountain View, CA). The packaging cell line is designed to produces high-titer virus able to infect a broad range of actively dividing mammalian host cells.
2. One milliliter of anti-FLAG M2 agarose (Sigma, Saint Louis, MO) packed into the 20 mL column (Bio-Rad, Hercules, CA).
3. FLAG peptide (Sigma, Saint Louis, MO). The peptide sequence of the FLAG-tag is as follows: DYKDDDDK (1,012 Da).
4. The FLAG peptide elution buffer: 40 mM HEPES, pH 7.4, 500 mM NaCl, 0.1% CHAPS (Note 7). For elution step, use 10 mL of the elution buffer with 0.5 mg/mL of the Flag peptide (0.3 mg/mL peptide concentration was also effective in elution step).
5. Column regeneration buffer: 0.1 M glycine HCl, pH 3.5.
6. The column in storage buffer: 50% glycerol with PBS and 0.02% sodium azide.

3. Methods

The optimization of cell or tissue lysis conditions during purification of protein complexes is necessary to preserve the integrity of complexes. This is a particular difficult task if a novel protein complex is to be purified and characterized. To identify the appropriate lysis conditions, it is highly informative to analyze a size of the studied protein complex under different lysis conditions. Within lysis buffer, an integrity of protein complexes depends on a variety of factors, such as salt concentration, pH, and the presence or absence of calcium or magnesium ions, whereas the type and concentration of a nondenaturing detergent is the key factor that needs to be carefully assessed during purification of a protein complex (18).

In our studies to track a relative size of unknown mTOR protein complexes, we have initially applied a gel-filtration chromatography. The effects of the nonionic detergent Triton X-100 and the zwitterionic detergent CHAPS (3-[(3-cholamidopropyl) dimethylammonio]-1-propanesulfonate) on the mTOR protein were analyzed. Human cancer cells were lysed in a standard lysis buffer containing either Triton X-100 (1%) or CHAPS (0.3%).

Triton X-100 is a commonly used detergent in cell lysis and CHAPS is less-popular detergent that is mostly applied to study membrane proteins. In our study, we have selected the CHAPS detergent because it has been previously successfully applied in purification of mTOR as the rapamycin-dependent FKBP12 binding protein from rat brain tissues (4). Following fractionations of the CHAPS containing cellular lysates by a gel-filtration column, we detected mTOR in a wide range of fractions indicating no enrichment of this protein in particular fractions. The cellular lysates containing Triton X-100 were not applicable to a gel-filtration analysis because this detergent caused column pressure buildup. Instead, we have applied a simple technique of sedimentation of protein complexes by ultracentrifugation. We have assumed that in the cellular lysates

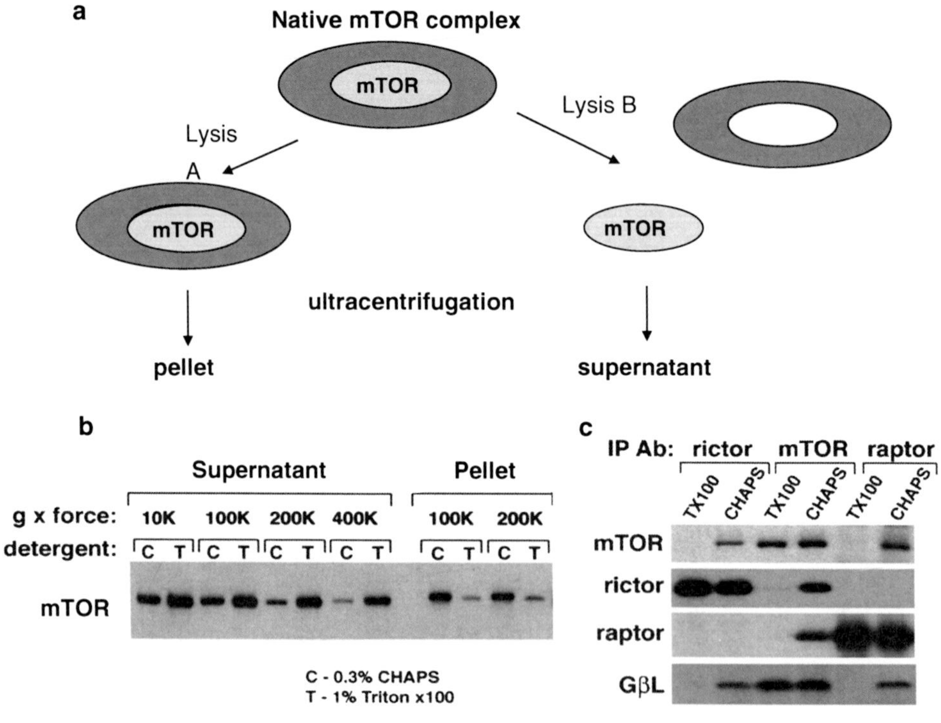

Fig. 2. Optimization of the lysis conditions to preserve the mTOR complexes. The mTOR complexes are sensitive to the Triton X-100 and preserved in the CHAPS lysis conditions. (**a**) Diagram – sedimentation of a native mTOR complex by ultracentrifugation if it is preserved in cellular lysate (lysis buffer A), disintegrated mTOR complex components are predicted to remain in supernatant (lysis buffer B). (**b**) Sedimentation of a native mTOR complex by sequential ultracentrifugation. HEK 293T cells were lysed in the standard lysis buffer containing CHAPS (0.3%) or Triton X-100 (1%). The lysates were precleared by the initial 10,000 × *g* centrifugation. The obtained supernatants of equalized protein concentration (1.5 mg in 1 mL) were applied to the sequential ultracentrifugation for 30 min at 100,000, 200,000, and 400,000 × *g*. After every ultracentrifugation step, the supernatants and pellets were analyzed by western blotting for detection of the mTOR protein. The lysis and centrifugation steps took place in cold at 4°C. (**c**) HEK 293T cells were lysed in the standard lysis buffer containing CHAPS (0.3%) or Triton X-100 (1%). The precleared lysates by the 10,000 × *g* centrifugation were applied to immunoprecipitation (IP) of rictor, raptor, and mTOR. The IP products were analyzed by immunoblotting for the presence of the indicated components of mTOR complexes. The binding of rictor and raptor to mTOR is sensitive to the Triton X-100 but not CHAPS lysis conditions. The binding of mLST8 (GβL) to mTOR is detected in both lysis conditions (**c** is reproduced from ref. 7 with permission from Elsevier Science).

an unknown multiprotein mTOR complex because of its large size is sedimented during ultracentrifugation and it does not occur if the lysis buffer caused a collapse of the complex (Fig. 2a). This simple study indicated that following ultracentrifugation of the CHAPS cellular lysate the mTOR protein was mostly present in the pellet and not in supernatant (Fig. 2b). In contrary, following ultracentrifugation, the opposite distribution of this protein was observed in the Triton X-100 cellular lysate. This was the first indication that most likely the CHAPS (Note 3) but not Triton X-100 lysis condition preserves the mTOR complex integrity. Our following studies have confirmed that the complexes of mTOR assembled with its interacting proteins raptor and rictor fall apart in the Triton X-100 lysis buffer. Importantly, both mTOR complexes are preserved when purified in the CHAPS lysis buffer as shown in Fig. 2c, and these complexes are functional as the kinase complexes as described previously (7, 9, 13, 19). Only binding of mTOR and mLST8 (GbL) is detected in both lysis conditions (Fig. 2c) indicating that the binding of this small protein to mTOR is not sensitive to the Triton X-100 detergent.

Following cellular lysis, the next important step in isolation of a protein complex by affinity purification is a specific antibody highly effective in recognition of and binding to the studied protein. The immunoprecipitation method as a low-scale affinity purification is commonly used technique in functional studies of protein complexes. In a process of IP, the antibody against studied protein is added to cellular lysate and incubated by rotation for several hours. During this incubation, the antibody recognizes and binds specifically to the protein of interest. This antibody/protein complex is pulled down by adding the protein A- or G-coated agarose beads because these bacterial proteins A and G bind with high affinity to the Fc (constant) region within the immunoglobulin (IgG) structure of the antibody (20). In most of the cases, the antigens selected to raise antibodies are based on the hydrophilicity plot analysis (a quantitative analysis of the degree of hydrophobicity or hydrophilicity of amino acids within a protein sequence) of the proteins that allows selection of the predicted hydrophilic areas within the proteins as the antibody recognition sites known as epitopes. This analysis does increase the chances that these selected peptide sequences as antibody epitopes will be located on a surface of native proteins. Development of a high-titer and good-quality polyclonal antibody from serum is usually achieved by the antigen-coupled column affinity purification, but there is no assurance that this antibody will be effective in IP. Accessibility of the peptide epitope within a native protein complex is not predictable and testing of all available antibodies is highly recommended.

In our early studies, we developed two mTOR antibodies, and only one antibody raised against peptide of the mTOR protein sequence within amino acids 221–236 has been effective in western blotting and also IP. This antibody has effectively immunoprecipitated mTOR only in the Triton X-100 but not CHAPS cellular lysates. This was an additional indication that in the CHAPS cellular lysates this particular mTOR epitope was not accessible within the mTOR complex and that following the complex disintegration by the Triton lysis buffer this antibody was effective in the pulling down of mTOR. Testing of the commercially available mTOR antibodies showed that the antibody recognizing the human N-terminal mTOR peptide sequence (Santa Cruz Biotechnology) has been highly effective in IPs in both CHAPS and Triton X-100 cell lysis conditions. In our studies, this antibody became a key tool in the purification and functional characterization of both mTOR complexes. In the functional studies of the mTOR interacting proteins, we also found that the raptor and rictor antibodies raised against their N-terminal sequences have been also effective in IPs from the CHAPS cellular lysates as shown in Fig. 2c (7).

Our recent testing of the commercially available antibodies indicated that the mTOR (SDIX), raptor (Bethyl), and rictor (Santa Cruz Biotechnology) antibodies are also effective in IPs from the CHAPS cell lysates. It was initially confirmed by a sensitive detection carried on by western blotting (Fig. 3a, b). We further validated that these antibodies are highly effective in purification of the distinct or both mTOR complexes as the resolved IP products on gels are easily detectable by the silver staining analysis (Fig. 4a, b). The mTOR complexes are stable complexes and their integrity is not sensitive to acute osmotic or oxidative stress conditions as shown in Fig. 4b, but treatment of cells with rapamycin

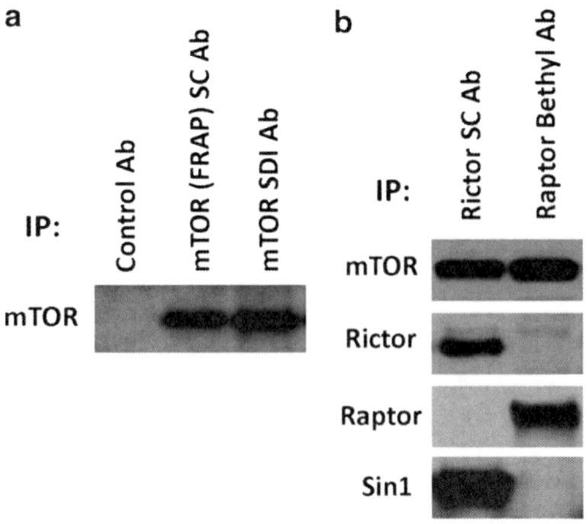

Fig. 3. Immunopurification of the mTOR complexes. MDA-MB-435 cells were lysed in the CHAPS lysis buffer and lysates were applied for immunoprecipitation by the commercially available antibodies. The mTOR SDI antibody is tested by detection of mTOR by immunoblotting (**a**). The copurification of rictor with mTOR and Sin1 in IP by the rictor SC Ab, and the raptor and mTOR proteins copurified in IP by the raptor Bethyl Ab are detected by immunoblotting (**b**).

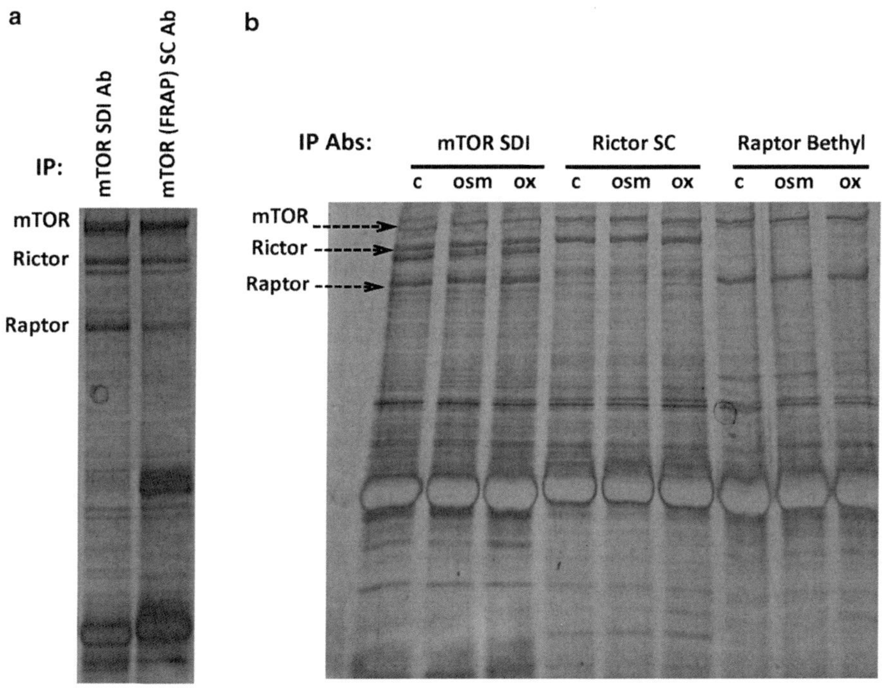

Fig. 4. Integrity of the mTOR complexes is not sensitive to the acute osmotic and oxidative stress conditions. As described in Fig. 3, MDA-MB-435 cells were lysed in the CHAPS lysis buffer and applied for IP by the validated antibodies. The IP products were run on SDS-PAGE gels and applied for silver staining. The silver staining of the IP products of two different mTOR antibodies are shown in (**a**). MDA-MB-435 cells were stressed for 30 min, lysed, and the lysates were applied for IP by the mTOR, rictor, and raptor Abs. The IP products were stained by silver staining (**b**). As indicated: *c* control cells; *osm* cells stressed by 0.5 M sorbitol (osmotic stress); *ox* cells stressed by 1 mM hydrogen peroxide (oxidative stress). In both (**a**) and (**b**) panels, as shown, the presence of mTOR, rictor, and raptor on the indicated protein bands is confirmed by the mass spectrometry analysis (data not shown).

affects both complexes by different mechanisms. As it has been previously described, the rapamycin/FKBP12 complex binds to mTORC1 by interacting with the FRB domain of mTOR that causes destabilization of mTORC1 (19). This drug complex does not bind mTORC2, but it is competing with rictor/Sin1 for binding to mTOR and thereby a prolonged rapamycin treatment of cells inhibits assembly of mTORC2 (15). Detection of the IP products as strong and distinct bands by silver staining indicates that at least 30–50 ng of the protein is purified by IP. In our previous studies, the samples for mass spectrometry analysis have been scaled up by combining 10–12 IPs in one sample and the IP products were detectable by Coomassie staining (7, 9, 19). In the present days, the sensitivity of mass spectrometry has been significantly improved and the bands detected by silver staining are sufficient for protein identification analysis.

The purification of the soluble mTOR complexes has been also recently developed (21). This purification is based on a constitutive expression and purification of the FLAG-tagged mLST8 (also known as GβL) and a specific elution of the FLAG affinity-purified

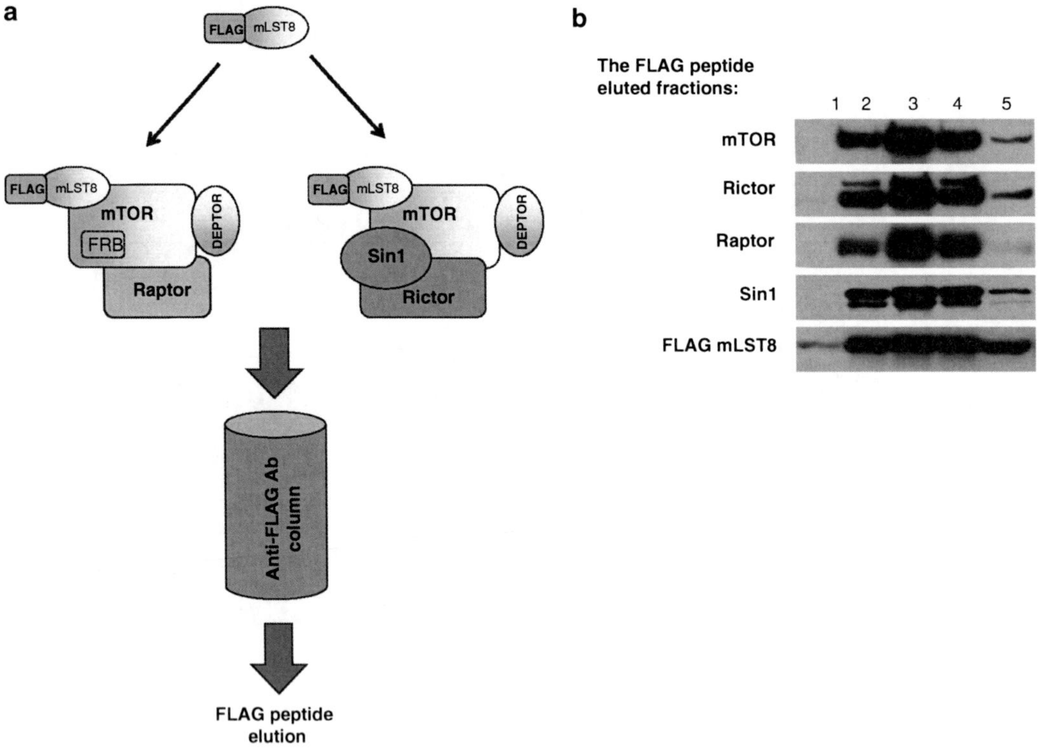

Fig. 5. Isolation of the soluble mTOR complexes. A diagram describing purification of the soluble mTOR complexes by the FLAG-tag affinity purification is shown in (**a**). Immunodetection of the components of the mTOR complexes (mTOR, rictor, raptor, Sin1, and mLST8) in the FLAG-peptide-eluted fractions is presented in (**b**).

mLST8 protein as the final purification step. The constitutively expressed mTOR interacting protein mLST8 binds endogenous mTOR protein and this heterodimer is further assembled into both mTOR complexes (mTORC1 and 2). The N-terminally tagged FLAG mLST8 protein as a component of the functional mTOR complexes is accessible for the effective FLAG M2 antibody affinity column purification. As the final step of purification, the FLAG-mLST8 protein assembled into the mTOR complexes and bound to the M2 FLAG antibody is specifically eluted by the FLAG peptide. The schematic description of this purification and detection of the components of the mTOR complexes in the eluted fractions of this purification is shown in Fig. 5. Both soluble mTOR complexes purified by the expression and purification of the recombinant Flag-mLST8 proteins are functional as the kinase complexes and in this format it is very convenient to apply the mTOR complexes for the high-throughput screening assays.

3.1. Isolation of the mTOR Complexes by Affinity Purification

1. Cell culture conditions – For immunopurification of the mTOR complexes, the human cancer cells (HeLa, HEK 293 T, MDA-MB-435) are split 1 day prior the experiment into 100-mm dishes by seeding 5 million cells per dish.

2. Cell lysis: Cells are rinsed twice with ice-cold PBS to wash away the serum proteins and medium. Keep the rinsed cells on ice, and tilt the dish to aspirate any traces of PBS. Add 1 mL of the CHAPS lysis buffer, scrape the cells, and transfer to 1.5-mL Eppendorf tube. Incubate the cell lysate by mixing on rotator (no shaking) in cold room for 20 min. Following the lysis, centrifuge the cell lysate at 12,000 × *g* for 15 min at 4°C. After centrifugation, carefully transfer the supernatant without disturbing the pellet to a new tube and determine the protein concentration of the lysate by Bradford assay. For optimal immunoprecipitation of the mTOR complexes, apply lysates containing the total amount of protein in a range of 1.5 mg protein within 600–900 mL. For every studied cell culture condition, apply the same amount of the total cell lysate protein for immunoprecipitation by adjusting protein concentration of cellular lysates by the lysis buffer.
3. Cell lysate preclearing: To minimize binding of a background (nonspecific protein binding to protein G beads) in immunoprecipitates, the preclearing step is recommended. For preclearing, add 30 μl of the protein G beads (50% slurry) and incubate on rotator for 1 h in cold room. Following incubation, spin the lysates for 1 min at 5,000 × *g* and transfer supernatant to a new tube (it is better to leave about 100 μl of the lysate on a bottom of a tube to avoid transferring the beads to a new tube; the left lysate can be used as total cell lysates for western blotting analysis).
4. To initiate immunoprecipitation, add the antibody of your interest to the lysate.
5. To purify both mTOR complexes, add 20–30 μl of mTOR antibody (FRAP antibody, Santa Cruz). To purify mTORC2, add 6 μl of rictor antibody (rictor antibody, Santa Cruz). Alternatively, Bethyl rictor antibody is also efficient in affinity purification; 2 μl of this Ab is sufficient for IP. To purify mTORC1, add 2 μl of raptor antibody (raptor antibody, Bethyl). Incubate the lysates with the antibodies on rotator in cold room for 2 h or overnight. During this incubation, the antibody binds specifically to the proteins containing the epitope recognized by the antibody. A high-quality antibody is critical to achieve an effective immunoprecipitation. The described antibodies have been validated and are effective for IP of the mTOR complexes.
6. Following the incubation to purify the antibody/protein complexes from lysate, add 20 μl of protein G (50% slurry) and incubate the samples for additional 1–2 h.
7. Wash G protein beads four times with the 1 mL of the CHAPS lysis buffer by spinning tubes for 20 s at 5,000 × *g* and aspirating the supernatant. After the third wash, it is recommended to move the beads to a new tube to prevent any contamination from proteins bound to the tube wall.

8. After the final wash, it is recommended to dry the agarose beads shortly for 2–3 seconds by aspiration with the 26-gauge needle; the needle can be washed briefly by aspiration of water between the IP samples.
9. To the dried beads, add 15–20 μl of 2× Laemli buffer and boil for 3 min to denature and release the IP products from antibodies and beads.
10. Load IP products on a polyacrylamide gel electrophoresis (PAGE) gel to separate the proteins by size. The 7.5% or gradient 4–15% acrylamide gels are recommended for a good resolution of the IP products, in this case the components of the mTOR complexes.

3.2. Detection of the Components of mTOR Complexes

1. Following resolving the IP products on the gel, transfer the proteins from a gel on PVDF membrane. Wet the PVDF membrane in ethanol for 10–20 s and rinse it in water (the usage of ethanol instead of methanol is explained in the Note 5). Transfer the proteins resolved on a gel to PVDF membrane by wet transfer apparatus filled with the CAPS buffer by running constant 450 mA for 150 min at room temperature (RT) with the buffer volume about 3.6 L. The buffer is reusable for three times; the described unit can transfer three blots simultaneously. After transferring, detect the prestained markers on PVDF membrane; the 250-kDa protein marker on the membrane is an indicator of a high-quality transfer.
2. Rinse the PVDF membrane after transfer in PBST and start blocking the membrane with PBST containing 5% milk for 1 h at RT.
3. After blocking, add the primary antibodies at 1:5,000 dilution in 5% BSA for 2 h at RT or overnight in cold (the antibodies against mTOR, raptor, and rictor from Cell Signaling Technologies are very appropriate for western applications), the antibodies are reusable as discussed in the Note 4. For high specificity, use 5% milk instead of BSA.
4. Rinse the membranes in PBST for 15–20 min by replacing PBST 3–4 times.
5. After washing the membrane, add the secondary HRP-conjugated antibody at 1:10,000 dilution in PBST containing 5% milk for 1 h at RT.
6. Rinse the membrane in PBST for 15–20 min by replacing PBST 3–4 times. To maximize detection, the final wash can be made in PBS to eliminate any traces of Tween 20.
7. Incubate a membrane in chemiluminescent HR substrate solution for 1 min and detect the protein detection signal by exposing the membrane to X-ray film (Blue film). For blotting the membrane with other antibodies we recommend to follow recommendations discussed in the Note 6.

3.3. Silver Staining

For convenience, we mostly use the Silver Staining kit (Silver Quest Staining kit) by following the protocol developed by manufacture. The gel staining with this kit is highly sensitive and does not interfere with the mass spectrometry analysis. Alternatively, the highly efficient silver staining (EMBL silver staining protocol) by Moritz et al. is described below (17).

1. Fix gel in 50% methanol and 5% acetic acid in 1 L for 20 min.
2. Wash gel in 50% methanol for 10 min in 1 L.
3. Wash gel in water for 2 h or overnight.
4. Sensitize gel in 100 mL of 0.02% $Na_2S_2O_3$ for 1 min.
5. Wash gel in water for 1 min.
6. Wash gel again with water for 1 min.
7. Incubate gel in cold 0.1% $AgNO_3$.
8. Wash gel in water for 1 min.
9. Change gel chamber (container).
10. Wash gel for 1 min in water.
11. Develop gel in 0.04% formalin, 2% Na_2CO_3.
12. Observe gel and terminate the reaction when staining is sufficient by 5% acetic acid.
13. Change solution couple of times and store gel in 1% acetic acid in cold.

3.4. Purification of the Soluble TOR Complexes

1. The constitutive expression of the Flag-mLST8 in HeLa cell line has been developed by transducing HeLa cells with the retrovirus produced by transfecting the packaging cells with the MSCVpuro-FLAG-mLST8 plasmid. The conditions for the virus production in the packaging cells by expression of the MSCV-FLAG-mLST8 plasmid and transduction of HeLa human cancer cells were applied as recommended by the manufacturer. The expression of the Flag-mLST8 in HeLa cells was confirmed by western blotting.
2. Cell culture for purification of the Flag-mLST8 protein: Prepare 12 plates of cells to apply to the 1-mL FLAG antibody column. Seven million cells were seeded to 150-mm dishes in 25 mL of the cell culture medium (10% IFS in DMEM). The cells were lysed after 42 h.
3. Cell lysis: Each plate of cells are washed with 25 mL of cold PBS, aspirated, and lysed in 1.6 mL of the CHAPS lysis buffer per plate. After scraping and collecting the cells to the 50-mL conical tube, the cells were incubated with mild agitation for 20 min on ice. The lysate (the total volume of 24 mL, about 0.4 mL is added per plate from the PBS on the surface of the plate) was transferred to the spinning tubes for centrifugation for 15 min at 15,000 × *g*. The cellular lysate concentration is

about 3 mg/mL. After centrifugation, the supernatant is used for Flag-mLST8 affinity purification.

4. Purification: Pack the 1 mL Anti-Flag M2 agarose to the 20 mL Bio-Rad column, and wash the column with PBS with a high flow by attaching the flexible plastic tubing to the column (1 m of the tubing attached to the column is sufficient to provide pressure for a high flow). The purification can take place in a lab at room temperature, but keep all your solutions including the cell lysate on ice and avoid drying the column during purification (Note 8). After washing the column with cold PBS, run the cellular lysate 5–7 times through the column on a high flow. Keep the flow through on ice to cool the cellular lysate. After running the lysates, wash the column with 40 mL of lysis buffer. Final wash is 5 mL of elution buffer. For elution step, use 10 mL of the elution buffer with 0.5 mg/mL of the Flag peptide (0.3 mg/mL peptide concentration was also effective in elution step). Collect ten fractions of 1 mL each. *Important:* Let the first mL as a first fraction to go through the column, stop the column for 10 min with the stopper, collect the second fraction; stop the column for 10 min, collect the third fraction; stop the column for 10 min and collect the fourth fraction. After fourth fraction, collect the rest of the fractions without stopping the column.
5. The eluted proteins will be mostly in second and third fractions if column was used for the first time; after regeneration of column, the binding efficiency is increased. The column is reusable for 4–5 times.
6. Storage: Freeze and store the purified Flag-mLST8 containing the mTOR complexes at –80°C. The kinase activity of the mTOR complexes is preserved after first defrosting and its activity is decreasing after freezing and defrosting.
7. Column regeneration and storage: After elution, wash the column in PBS, use 10 mL regeneration buffer (do not use this buffer for more than 20 min), and wash the column with 50 mL PBS. After regeneration, store the column in storage buffer, let 3–4 mL of the buffer to go through, stop the column, and keep at 4°C. To use the column again, run and wash the columns with PBS (50 mL).

4. Notes

1. In all the protocols described in this chapter, the double-distilled water (ddH_2O) has been used with the resistivity of 18.2 MΩ-cm.

2. Initially, the raptor/mTOR complex has been purified by stabilizing this complex in cells prior to lysis in the Triton lysis buffer by the amine-reactive chemical cross-linker dithiobis-[succinimidyl propionate (DSP, also known as a Lomant's reagent)] (19). The chemical cross-linker alters enzymatic activity of proteins and the functional activities of the raptor/mTOR complex have been studied by purifying this complex by lysing cells in CHAPS lysis buffer.
3. The lysing of cells in the standard buffer containing the zwitterionic CHAPS at 0.3% preserves both mTOR complexes. This concentration is in a lower range of the CHAPS Critical Micelle Concentration (CMC). According to the Calbiochem Detergent manual, CHAPS M.W. 614.9; CMC at 6–10 mM; aggregation number 4–14; micellar weight 6,150.
4. The high-quality antibodies are recommended to be used in 5% BSA PBST because in this solution the antibodies are reusable for 3–4 times, and for storage of the antibodies in this solution add sodium azide (0.02%).
5. Originally, methanol was recommended to wet the PVDF membrane and use it in transfer buffer. Because of high methanol toxicity, we switched to ethanol. We confirmed that ethanol works very similar to methanol in our western blot detections.
6. If the PVDF membranes are to be blotted against several antibodies that will require the antibody stripping step, it is recommended after transfer to wash the membranes in PBS and air dry for 1 h. Before using the dry membrane, wet it again in ethanol. The drying step is important to irreversibly bind proteins to the membrane and stripping of the membrane does not wash away the proteins.
7. In the FLAG elution buffer, the content of salt (0.5 M NaCl) is critical for efficient elution.
8. At any step of the anti-FLAG antibody column purification work, never let column to dry. Drying the column substantially decreases the column efficiency in binding FLAG-tagged proteins.

Acknowledgments

We gratefully acknowledge David M. Sabatini (Whitehead Institute for Biomedical Research, Cambridge, MA) for his commitment, guidance, and constructive criticism in this long-term project of purification and functional characterization of the mTOR complexes. We are also thankful to Seong A. Kang (David M. Sabatini Laboratory) for providing MSCVpuro-FLAG-mLST8 and developing purification of the soluble mTOR complexes by expression of the FLAG-mLST8 protein. This work was supported by the NIH grant CA133522 (D.D. Sarbassov).

References

1. Guertin DA, Sabatini DM: The pharmacology of mTOR inhibition. *Sci Signal* 2:pe24, 2009
2. Heitman J, Movva NR, Hall MN: Targets for cell cycle arrest by the immunosuppressant rapamycin in yeast. *Science* 253:905–909, 1991
3. Brown EJ, Albers MW, Shin TB, Ichikawa K, Keith CT, Lane WS, Schreiber SL: A mammalian protein targeted by G1-arresting rapamycin-receptor complex. *Nature* 369:756–758, 1994
4. Sabatini DM, Erdjument-Bromage H, Lui M, Tempst P, Snyder SH: RAFT1: a mammalian protein that binds to FKBP12 in a rapamycin-dependent fashion and is homologous to yeast TORs. *Cell* 78:35–43, 1994
5. Sabers CJ, Martin MM, Brunn GJ, Williams JM, Dumont FJ, Wiederrecht G, Abraham RT: Isolation of a protein target of the FKBP12-rapamycin complex in mammalian cells. *J Biol Chem* 270:815–822, 1995
6. Sarbassov DD, Ali SM, Sabatini DM: Growing roles for the mTOR pathway. *Curr Opin Cell Biol* 17:596–603, 2005
7. Sarbassov DD, Ali SM, Kim DH, Guertin DA, Latek RR, Erdjument-Bromage H, Tempst P, Sabatini DM: Rictor, a novel binding partner of mTOR, defines a rapamycin-insensitive and raptor-independent pathway that regulates the cytoskeleton. *Curr Biol* 14:1296–1302, 2004
8. Jacinto E, Loewith R, Schmidt A, Lin S, Ruegg MA, Hall A, Hall MN: Mammalian TOR complex 2 controls the actin cytoskeleton and is rapamycin insensitive. *Nat Cell Biol* 6: 1122–1128, 2004
9. Kim DH, Sarbassov DD, Ali SM, Latek RR, Guntur KV, Erdjument-Bromage H, Tempst P, Sabatini DM: GbetaL, a positive regulator of the rapamycin-sensitive pathway required for the nutrient-sensitive interaction between raptor and mTOR. *Mol Cell* 11:895–904, 2003
10. Peterson TR, Laplante M, Thoreen CC, Sancak Y, Kang SA, Kuehl WM, Gray NS, Sabatini DM: DEPTOR is an mTOR inhibitor frequently overexpressed in multiple myeloma cells and required for their survival. *Cell* 137:873–886, 2009
11. Jacinto E, Facchinetti V, Liu D, Soto N, Wei S, Jung SY, Huang Q, Qin J, Su B: SIN1/MIP1 maintains rictor-mTOR complex integrity and regulates Akt phosphorylation and substrate specificity. *Cell* 127:125–137, 2006
12. Frias MA, Thoreen CC, Jaffe JD, Schroder W, Sculley T, Carr SA, Sabatini DM: mSin1 is necessary for Akt/PKB phosphorylation, and its isoforms define three distinct mTORC2s. *Curr Biol* 16:1865–1870, 2006
13. Sarbassov DD, Guertin DA, Ali SM, Sabatini DM: Phosphorylation and regulation of Akt/PKB by the rictor-mTOR complex. *Science* 307:1098–1101, 2005
14. Garcia-Martinez JM, Alessi DR: mTOR complex 2 (mTORC2) controls hydrophobic motif phosphorylation and activation of serum- and glucocorticoid-induced protein kinase 1 (SGK1). *Biochem J* 416:375–385, 2008
15. Sarbassov DD, Ali SM, Sengupta S, Sheen JH, Hsu PP, Bagley AF, Markhard AL, Sabatini DM: Prolonged rapamycin treatment inhibits mTORC2 assembly and Akt/PKB. *Mol Cell* 22:159–168, 2006
16. Loewith R, Jacinto E, Wullschleger S, Lorberg A, Crespo JL, Bonenfant D, Oppliger W, Jenoe P, Hall MN: Two TOR complexes, only one of which is rapamycin sensitive, have distinct roles in cell growth control. *Mol Cell* 10:457–468, 2002
17. Mortz E, Krogh TN, Vorum H, Gorg A: Improved silver staining protocols for high sensitivity protein identification using matrix-assisted laser desorption/ionization-time of flight analysis. *Proteomics* 1:1359–1363, 2001
18. Simpson RJ, Adams PD, Golemis E: *Basic methods in protein purification and analysis : a laboratory manual.* Cold Spring Harbor, N.Y., Cold Spring Harbor Laboratory Press, 2009
19. Kim DH, Sarbassov DD, Ali SM, King JE, Latek RR, Erdjument-Bromage H, Tempst P, Sabatini DM: mTOR interacts with raptor to form a nutrient-sensitive complex that signals to the cell growth machinery. *Cell* 110: 163–175, 2002
20. Harlow E, Lane D: *Using antibodies : a laboratory manual.* Cold Spring Harbor, N.Y., Cold Spring Harbor Laboratory Press, 1999
21. Yip CK, Murata K, Walz T, Sabatini DM, Kang SA: Structure of the human mTOR complex I and its implications for rapamycin inhibition. *Mol Cell* 38:768–774

Chapter 6

An *In Vitro* Assay for the Kinase Activity of mTOR Complex 2

Jingxiang Huang

Abstract

The mTOR complex 2 (mTORC2) is a protein kinase complex involved in many important physiological processes through the regulation of its substrates, such as Akt, SGK1, and conventional PKC. Its activity is modulated negatively by interaction with DEPTOR and positively by the TSC1–TSC2 protein complex. To study the regulation of mTORC2 activity, it is a common practice to examine the phosphorylation of its substrates *in vivo* and verify the findings with an *in vitro* assay of kinase activity. This kinase assay measures the phosphorylation of exogenously derived Akt by mTORC2 immunoprecipitates isolated from cells and can be modified to study the effect of small molecules on mTORC2 activity.

Key words: mTOR, Rictor, Sin1, Akt, Kinase

1. Introduction

The serine/threonine kinase, mTOR, is a member of the PIKK (phosphatidylinositol kinase-related kinase) family and functions in at least two well-defined multiprotein complexes (1). In mTOR complex 1 (mTORC1), mTOR associates with RAPTOR (regulatory associated protein of mTOR) and mLST8 (mammalian lethal with sec18 protein 8). The mTOR complex 2 (mTORC2) is biochemically distinct from mTORC1. In this complex, mTOR associates with RICTOR (rapamycin-insensitive companion of mTOR), mSIN1 (stress-activated protein kinase interacting protein-1), and mLST8 and is implicated in the phosphorylation of two conserved motifs in some members of the AGC family of kinases. The most well-characterized substrate of mTORC2 is S473 on the hydrophobic motif of Akt (2). Similar motifs on protein kinase Cα (PKCα, Ser657) (3, 4) and serum- and glucocorticoid-induced protein kinase-1 (SGK1, Ser422) (5) have also been shown to be

Thomas Weichhart (ed.), *mTOR: Methods and Protocols*, Methods in Molecular Biology, vol. 821,
DOI 10.1007/978-1-61779-430-8_6, © Springer Science+Business Media, LLC 2012

phosphorylated by mTORC2. The phosphorylation of the "turn" motifs of Akt (T450) and some isoforms of PKC (e.g., T497 of PKCα) (6, 7) are also mediated by mTORC2. Through the regulation of these substrates, mTORC2 play a significant role in malignant transformation (8, 9), fat metabolism and growth (10–12).

Recent studies have demonstrated that the activity of mTORC2 is negatively regulated by DEPTOR (13), and positively regulated by the TSC1–TSC2 protein complex (14, 15); and these effects are mediated through direct interaction with mTORC2. On the contrary, phosphorylation of RICTOR by ribosomal protein S6 kinase-1 (S6K1) (16–18) and interaction of mTORC2 with PROTOR (19, 20) have no effect on mTORC2 kinase activity. Although currently known chemical inhibitors of mTOR kinase lack specificity and affect both mTORC1 and mTORC2 (21, 22), efforts to discover other regulatory mechanisms may reveal those that can be targeted by small molecules to modulate mTORC2 activity specifically. For these studies, measuring mTORC2 activity *in vitro* should be useful as a biochemical tool to complement the detection of substrate phosphorylation in the determination of the functional status of this protein kinase complex (23).

2. Materials

2.1. Cell Culture and Lysis

1. Dulbecco's Modified Eagle's Medium (DMEM) supplemented with 10% fetal bovine serum (FBS).
2. Trypsin (0.05%) and EDTA (0.53 mM) solution.
3. Insulin is dissolved at 100 μM in 0.01% HCl, filter-sterilized, and stored in single-use aliquots at −20°C.
4. Transfection: Lipofectamine 2000 (Invitrogen); Opti-MEM I medium (Invitrogen).
5. Cell lysis buffer: 40 mM HEPES pH 7.5, 120 mM NaCl, 1 mM EDTA, 10 mM tetrasodium pyrophosphate, 10 mM glycerol 2-phosphate, 50 mM NaF, 0.3% CHAPS {3-[(3-chloramidopropyl)-dimethylammonio]-1-propanesulfonate}. Store at 4°C. Add 0.5 mM sodium orthovanadate and protease inhibitors just before use.
6. Phosphate-buffered saline (10× PBS): 80 g NaCl, 2.0 g KCl, 14.4 g Na_2HPO_4, 2.4 g KH_2PO_4 in 1 l of water. 10× PBS is diluted to 1× PBS and autoclaved.
7. Sample buffer (6×): 375 mM Tris–HCl, 12% SDS, 30% β-mercaptoethanol, 30% glycerol, 0.06% bromophenol blue. The 6× sample buffer can be diluted to 1× when required.
8. Cell lifter.
9. Measurement of protein concentration: Bio-Rad Protein Assay.

2.2. Immunoprecipitation and Kinase Assay

1. Antibody for immunoprecipitation: anti-Rictor (#A300-459A, Bethyl Laboratories); preimmune rabbit immunoglobulin (Santa Cruz Biotechnology); anti-HA.11 monoclonal mouse Ab (#MMS-101P, Covance Research Products).
2. Protein A/G agarose resin beads (Thermo Scientific).
3. Kinase reaction buffer: 25 mM HEPES pH 7.4, 100 mM potassium acetate, 1 mM magnesium chloride. Prepare 10× solution, filter-sterilize, and store at 4°C.
4. Adenosine triphosphate (ATP) is dissolved in water as 10 mM stock solution, and stored in single-use aliquots at −20°C.
5. Substrate: Inactive Akt, expressed by baculovirus in Sf21 insect cells (#14-279, Millipore).
6. Alternate substrate: HA-tagged, kinase-dead mutant of Akt (K179D) expressed in HEK293E cells.
7. LY294002 is dissolved in DMSO at concentration of 15 mM, divided into single-use aliquots and stored at −20°C.

2.3. SDS-Polyacrylamide Gel Electrophoresis (SDS-PAGE) and Western Blot

1. Prestained molecular weight markers.
2. Gel reagents: 1.5 M Tris(hydroxymethyl)aminomethane (Tris) pH 8.8; 1.0 M Tris–HCl pH 6.8; 10% ammonium persulfate; 10% sodium dodecylsulfate; 30% acrylamide and bisacrylamide stabilized solution, N,N,N',N'-tetramethylethylenediamine (TEMED).
3. Running buffer: 25 mM Tris–HCl, 192 mM glycine, 0.1% (w/v) sodium dodecylsulfate. Prepare 10× solution and store at room temperature, dilute to 1× before use.
4. Transfer buffer: 25 mM Tris–HCl, 192 mM glycine, 15% (v/v) methanol. Prepare 10× solution without methanol and store at room temperature. Dilute to 1× and add methanol before use.
5. Transfer membrane: Immobilon-P PVDF membrane (Millipore).
6. Membrane wash buffer: 10 mM Tris–HCl pH 7.5, 100 mM NaCl and 0.1% Tween-20.
7. Blocking buffer: 5% (w/v) nonfat milk dissolved in membrane wash buffer. Make fresh before use.
8. Primary antibody dilution buffer: 5% (w/v) bovine serum albumin and 0.01% sodium azide in membrane wash buffer.
9. Primary antibodies: anti-mTOR (#2972, Cell Signaling Technology); anti-phospho(S473)-Akt (#4058, Cell Signaling Technology); anti-Rictor (#A300-459A, Bethyl Laboratories); mSin1 (#A300-910A, Bethyl Laboratories); mLST8 (#A300-680A, Bethyl Laboratories).
10. Secondary antibody: Goat anti-rabbit IgG, horse-radish peroxidase conjugate (Millipore).

11. Chemiluminesence detection: SuperSignal West Pico Chemiluminescent Substrate (Thermo Scientific); Blue Lite autoradiographic film (ISC Bioexpress).

2.4. Stripping and Reprobing Blots for Total Akt

1. Stripping buffer: 62.5 mM Tris–HCl pH 6.8, 2% (w/v) sodium dodecylsulfate, 1% (v/v) β-mercaptoethanol. Add β-mercaptoethanol before use.
2. Primary antibody: anti-Akt (#4691, Cell Signaling Technology).

3. Methods

Genetic studies in yeast and mouse reveal that mTOR, RICTOR, mSIN1 and mLST8 are all critical for the function of mTORC2 (3, 24). A successful mTORC2 kinase assay depends on efficient immunoprecipitation of intact mTORC2 from cell lysate. Interestingly, the integrity of this protein complex is detergent sensitive: association between the components is maintained in lysis buffers containing 0.3% CHAPS or 0.1–0.3% Tween-20, but disrupted in buffers containing Triton-X100, Nonidet P-40 or *N*-octylglucoside (4, 19, 25, 26).

3.1. Immunoprecipitation of mTORC2

1. HEK293 cells or murine embryonic fibroblasts (MEFs) are passaged when approaching confluence and seeded into maintenance and experimental culture dishes on 100-mm dishes (see Note 1). For each experimental culture, about 3×10^6 HEK293 cells should be seeded, such that the cells will be about 70% confluent the next day. MEFs are more widely divergent in terms of growth rate and should be seeded based on prior experience with the cell line.
2. 24–36 h after initial seeding, the cultures are rinsed once with DMEM (without serum) and incubated for a further 15 h in DMEM (without serum).
3. Prepare all materials for treatment of cells and cell lysis: fill shallow tray with ice; prepare cell lysis buffer and place on ice; label two microcentrifuge tubes for each sample and cool on ice; place cold PBS on ice.
4. The appropriate cultures are treated with growth factor (e.g., 100-nM insulin for 15 min).
5. Dishes are removed from the incubator and placed immediately on ice. The medium is removed by aspiration and the cells are washed once with ice-cold PBS. Aspirate the PBS and add 1 ml of cell lysis buffer to each dish (see Note 2).
6. The cells are scraped from the dish with cell lifter. Lysis buffer and cell debris (crude cell lysate) is transferred into the chilled microcentrifuge tube.

7. Rock the microcentrifuge tube containing crude cell lysate in the cold (4°C) room for 20 min and then pellet the cell debris using a cold (4°C) microcentrifuge at about 16,500 g for 10 min. Transfer supernatant (cleared lysate) into fresh microcentrifuge tube.
8. Measure protein concentration by adding 5 μl of cleared lysate to 1 ml aliquots of Bio-Rad protein assay, using 5 μl of the original lysis buffer as blank. Based on obtained values, dilute each sample such that they are of the same protein concentration and lysate volume.
9. Take out 100 μl of normalized lysate into new chilled microcentrifuge tube and add 20 μl of 6× sample buffer. Mix well and boil for 5 min, centrifuge, and cool to room temperature. This is the lysate sample that can be used subsequently for Western blots.
10. To the rest of the normalized lysate, add 5 μl of anti-Rictor antibody (200 ng/μl) or an equivalent amount of nonimmune rabbit IgG and rock the mixture in the cold room for 2 h (see Note 3).
11. During the 2-h incubation period, prepare protein A/G bead slurry: wash the protein A/G beads three times with 40-mM HEPES pH 7.5. For each wash, add 500 μl of the wash buffer, mix well, centrifuge the mixture at 6,000 g for 30 s, and aspirate to discard the supernatant. Finally, add 40-mM HEPES pH 7.5 to make 1:1 slurry.
12. Add 15 μl of washed protein A/G beads to each reaction mixture (see Note 4). Rock the mixture in the cold room for 1 h.
13. Spin down the beads in a cold centrifuge at 6,000 g for 30 s. Aspirate the supernatant, taking care not to aspirate the beads. Wash the beads four times with lysis buffer. For each wash, add 1 ml of the lysis buffer, rock the mixture in the cold room for 5 min, centrifuge the mixture at 6,000 g for 30 s, and aspirate to discard the supernatant (see Note 5). The immunoprecipitate is to be used for the kinase assay described below.

3.2. Kinase Assay of mTORC2 Activity Using Inactive Akt Derived from Insect Cells

1. Wash immunoprecipitate prepared above with kinase reaction buffer: add 1 ml of the kinase reaction buffer, rock the mixture in the cold room for 5 min, centrifuge the mixture at 6,000 g for 30 s, and aspirate to discard the supernatant (see Note 5).
2. During the wash, prepare kinase reaction mixture: for each reaction, 15 μl of kinase reaction mixture contains 500 ng inactive Akt and 500 uM ATP in kinase reaction buffer (see Note 6).
3. Add 15 μl of kinase reaction mixture to the washed immunoprecipitate. Incubate the kinase reaction in an incubating shaker set at 250 rpm and 37°C for 30 min (see Note 7).

Tap the microcentrifuge tube every 5 min to resuspend the beads within the kinase reaction mixture (see Note 8).

4. Stop the kinase reaction by adding 45 μl of 1× sample buffer. Mix well and boil for 5 min, centrifuge (6,000 g), and cool to room temperature.

3.3. Kinase Assay of mTORC2 Activity Using Kinase-Dead Akt Derived from Mammalian Cells

1. HEK293 cells are passaged when approaching confluence and seeded into maintenance and experimental culture dishes. For each experimental culture, about 3×10^6 HEK293 cells should be seeded in 100-mm dishes such that the cells will be about 70% confluent the next day. One dish of HEK293 cells is to be prepared for each kinase assay planned (see Note 9).
2. On the next day, transfect the cells with HA-tagged kinase-dead Akt (K179D)-expressing vector (described in (27)), using Lipofectamine 2000: for each dish to be transfected, dilute 10 μg of vector and 25 μl of Lipofectamine 2000 separately in 1.5 ml of Opti-MEM I; incubate at room temperature for 5 min and combine the diluted vector with diluted Lipofectamine 2000; incubate this at room temperature for 20 min and add the mixture gently to the experimental culture.
3. Twenty-four hours after transfection, the transfected cultures are rinsed once with DMEM (without serum) and incubated for a further 15 h in DMEM (without serum). The above three steps are to be done in parallel with steps 1–2 of Subheading 3.1 such that both sets of cells are ready for cell lysis on the same day.
4. Prepare all materials for treatment of cells and cell lysis: fill shallow tray with ice; prepare cell lysis buffer and place on ice; label two microcentrifuge tubes for each sample and cool on ice; place cold PBS on ice.
5. Perform cell lysis according to steps 5–9 of Subheading 3.1.
6. To the rest of the normalized lysate, add 1 μl of anti-HA.11 antibody (1 μg/μl) and rock the mixture in the cold room for about 4 h. During this time, perform steps 3–11 of Subheading 3.1.
7. Add 15 μl of washed protein A/G beads to the lysate and antibody solution from step 6 (see Note 4). Rock the mixture in the cold room for 1 h.
8. Spin down the beads in a cold centrifuge at 6,000 g for 30 s. Aspirate the supernatant, taking care not to aspirate the beads.
9. Add lysate and antibody solution from step 10 of Subheading 3.1 and rock the mixture in the cold room for 1 h.
10. Wash the beads four times with lysis buffer. For each wash, add 1 ml of the lysis buffer, rock the mixture in the cold room for

5 min, centrifuge the mixture at 6,000 g for 30 s, and aspirate to discard the supernatant (see Note 5). The immunoprecipitate is to be used for the kinase assay.

11. Wash immunoprecipitate prepared above with kinase reaction buffer: add 1 ml of the kinase reaction buffer, rock the mixture in the cold room for 5 min, centrifuge the mixture at 6,000 g for 30 s, and aspirate to discard the supernatant (see Note 5).
12. During the wash, prepare kinase reaction mixture: for each reaction, 15 μl of kinase reaction mixture contains 500 μM ATP in kinase reaction buffer.
13. Add 15 μl of kinase reaction mixture to the washed immunoprecipitate and complete the kinase assay as described in steps 3–4 of Subheading 3.2.

3.4. SDS-PAGE and Western Blot

1. Prepare at least two 8% gels: for 20 ml of separating gel, mix 9.3 ml of water, with 5.0 ml of 1.5 M Tris–HCl pH 8.8, 5.3 ml of ProtoGel, 200 μl of 10% sodium dodecylsulfate, 200 μl of 10% ammonium persulfate, and 12 μl of TEMED. Pour the gel, leaving space for a stacking gel, and overlay with water (volume required may vary depending on gel electrophoresis system used). The gel should polymerize in about 30 min.
2. Pour off the water and rinse the top of the gel twice with water.
3. Prepare the stacking gel: for 6 ml of stacking gel, mix 4.0 ml of water, 750 μl of 1.0 M Tris–HCl pH 6.8, 1.0 ml of ProtoGel, 60 μl of 10% sodium dodecylsulfate, 60 μl of 10% ammonium persulfate, and 6 μl of TEMED. Pour the stack and insert the comb. Each well formed by the comb should have a volume of at least 30 μl.
4. When the stacking gel is polymerized, wash the wells with running buffer and assemble the electrophoresis unit and load 20 μl of kinase reaction into each well and prestained molecular weight markers in a separate well.
5. Connect the electrophoresis unit to a power supply. The gel can be run at 80–100 V, until the dye front just runs off the gel.
6. After the samples have been separated by SDS-PAGE, they are transferred to a PVDF membrane electrophoretically. We use the Bio-Rad mini-PROTEAN3 electrophoresis system: soak a PVDF membrane just larger than the separating gel into methanol and prepare a tray of transfer buffer that is large enough to lay out a transfer cassette; in the transfer buffer, lay the separating gel on the presoaked PVDF membrane and place them between four sheets of similarly sized Grade 3MM Chr chromatography paper, to make a "sandwich"; ensure that there is no air bubble between the layers of the "sandwich" before placing it between two fiber pads within the transfer cassette; close the transfer cassette.

7. Place the cassette into the transfer tank containing transfer buffer such that the separating gel lies between the PVDF membrane and the cathode. Perform the transfer in a cold room or in an ice water bath at 350 mA for 75 min. After the transfer, disassemble the "sandwich" and soak the PVDF membrane in membrane wash buffer for several minutes. When transfer is complete, the prestained molecular weight markers should be visible on the membrane.
8. Place the PVDF membrane on a glass plate and cut the membrane horizontally at the level of about 80 kDa. The part of the membrane with higher molecular weight proteins will be used for immunoblots to detect the immunoprecipitated mTORC2 components, mTOR and RICTOR; the rest of the membrane will be used for detection of phosphorylated Akt and subsequently total Akt.
9. Incubate the membrane in blocking buffer for 1 h at room temperature on a rocking platform.
10. Discard the blocking buffer and rinse the membrane with membrane wash buffer prior to addition of the appropriate primary antibodies. To detect mTOR and RICTOR in the kinase reaction, anti-mTOR and anti-RICTOR antibodies are mixed together in primary antibody dilution buffer at 1:2,000 dilutions each. To detect phosphorylated Akt in the kinase reaction, use the anti-phospho-Akt antibody at 1:5,000 dilution. The membrane is incubated in primary antibody overnight on a rocking platform at 4°C.
11. The primary antibody is then removed (see Note 10) and the membrane is washed with membrane wash buffer three times for 10 min each.
12. The secondary antibody solution is a 1:5,000 dilution of the secondary antibody in blocking buffer. Add to the membrane for 1 h at room temperature on a rocking platform.
13. The secondary antibody is discarded and membrane washed three times for 10 min each with membrane wash buffer. Finally, wash the membrane with membrane wash buffer without Tween-20.
14. Once the final wash is removed from the blot, apply the chemiluminescent substrate for 5 min.
15. The blot is removed from the chemiluminescent substrate, blotted with Kim-wipes, placed between a sheet protector, and then securely taped into an X-ray film cassette.
16. Place X-ray film into the cassette to obtain several different exposures between 5 s and 4 min (see Note 11).

3.5. Stripping and Reprobing Blots for Total Akt

1. The membrane is stripped of phospho-Akt signal and reprobed with anti-Akt antibody to confirm that equal amounts of Akt substrate were added to kinase reactions and loaded into the gel. The membrane is placed in stripping buffer and incubated at 60°C for 30 min.
2. Rinse the blot thoroughly with water (until soapy bubbles formed by sodium dodecylsulfate are not present).
3. Complete the immunoblot process (steps 9–16 of Subheading 3.4).

4. Notes

1. mTORC2 immunopreciptation and kinase assay can be performed using various mammalian cell lines, but for the purposes of illustration, we use HEK293 (see Fig. 1).
2. This and subsequent steps should be performed as quickly as possible with samples placed on ice or in a 4°C cold room.
3. Anti-RICTOR antibodies are used to immunoprecipitate mTORC2 because RICTOR is a specific component in mTORC2 that is not present in mTORC1. Alternatively, one could transfect cells with epitope-tagged, Rictor-expressing construct (e.g., Addgene #11367) and immunoprecipitate the epitope-tagged protein using an epitope-specific antibody. Although the other mTORC2 components will be co-immunoprecipitated, the kinase assay is not as robust as that from immunoprecipitation of endogenous RICTOR.
4. Mix protein A/G bead slurry well each time before pipetting into the immunoprecipitation mixture. Cut the end of a 200 μl pipette tip to ensure that the beads can be aspirated together with the buffer.
5. In between washes, leave about 15 μl of supernatant to avoid aspirating the beads. After the last wash, carefully aspirate as much supernatant as possible by gently aspirating the residual 15 μl of supernatant with a pipetteman and narrow-bore pipette tip.
6. The kinase activity of RICTOR immunoprecipitates can be inhibited *in vitro* by the addition of mTOR kinase inhibitors into the kinase reaction mixture (e.g., 15 μM LY294002 in Fig. 1). This serves to demonstrate that the kinase activity of mTOR is responsible for the observed *in vitro* phosphorylation of the Akt substrate. When this control experiment is performed, an equivalent volume of DMSO is added to the other kinase reactions.

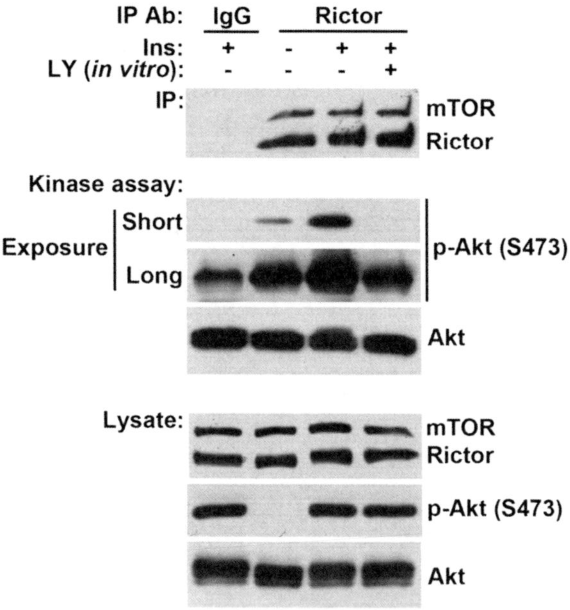

Fig. 1. Kinase activity of mTORC2 is stimulated by insulin and inhibited *in vitro* by LY294002. HEK293 cells were serum starved overnight and stimulated with insulin (Ins, 100 nM) for 15 min before lysis and immunoprecipitation (IP) with control IgG or anti-RICTOR antibodies (Ab). Where indicated, LY294002 (LY; 15 μM) was added to the kinase reaction. Kinase activity of the RICTOR immunoprecipitates was assessed by determining the increase of Akt (S473) phosphorylation in the inactive Akt substrate used. Some basal Akt phosphorylation can be detected in the Akt substrate when the film is exposed for a longer duration.

7. To determine the optimum duration for the kinase reaction, and to appreciate the dynamic range of the assay, it advised to perform a time course.
8. Ensure that the beads are well mixed with the kinase reaction mixture by gently tapping the microcentrifuge tube, taking care that the beads are not splashed onto the walls of the tube.
9. Depending on transfection efficiency, it may be necessary to use a 150 mm dish of HEK293 cells.
10. Primary antibodies can be reused several times over 2 months if stored at 4°C.
11. Determining the appropriate exposure time is important. As some level of phospho-Akt (S473) can be detected in the inactive Akt substrate under longer exposures, short exposures allow greater contrast between different experimental conditions (Fig. 1). Alternatively, a different Akt substrate with a lower level of basal Akt phosphorylation can be used (described in Subheading 3.3). The use of either substrate does not affect the results obtained (14). Kinase assays looking at the phosphorylation of mTORC2-mediated sites other than Akt (S473) have also been attempted (5, 7).

Acknowledgments

I would like to thank Christian Dibble for developing the kinase assay using kinase-dead Akt and both Dr Brendan Manning and Christian Dibble for invaluable discussion that contributed to the successful execution and optimization of the assays.

References

1. Wullschleger S, Loewith R, and Hall MN. (2006) TOR signaling in growth and metabolism, *Cell* **124**, 471–484.
2. Sarbassov DD, Guertin DA, Ali SM, and Sabatini DM. (2005) Phosphorylation and regulation of Akt/PKB by the rictor-mTOR complex, *Science* **307**, 1098–1101.
3. Guertin DA, Stevens DM, Thoreen CC, Burds AA, Kalaany NY, Moffat J, Brown M, Fitzgerald KJ, and Sabatini DM. (2006) Ablation in mice of the mTORC components raptor, rictor, or mLST8 reveals that mTORC2 is required for signaling to Akt-FOXO and PKCalpha, but not S6K1, *Dev Cell* **11**, 859–871.
4. Sarbassov DD, Ali SM, Kim DH, Guertin DA, Latek RR, Erdjument-Bromage H, Tempst P, and Sabatini DM. (2004) Rictor, a novel binding partner of mTOR, defines a rapamycin-insensitive and raptor-independent pathway that regulates the cytoskeleton, *Curr Biol* **14**, 1296–1302.
5. Garcia-Martinez JM, and Alessi DR. (2008) mTOR complex 2 (mTORC2) controls hydrophobic motif phosphorylation and activation of serum- and glucocorticoid-induced protein kinase 1 (SGK1), *Biochem J* **416**, 375–385.
6. Facchinetti V, Ouyang W, Wei H, Soto N, Lazorchak A, Gould C, Lowry C, Newton AC, Mao Y, Miao RQ, Sessa WC, Qin J, Zhang P, Su B, and Jacinto E. (2008) The mammalian target of rapamycin complex 2 controls folding and stability of Akt and protein kinase C, *EMBO J* 27, 1932–1943.
7. Ikenoue T, Inoki K, Yang Q, Zhou X, and Guan KL. (2008) Essential function of TORC2 in PKC and Akt turn motif phosphorylation, maturation and signalling, *EMBO J* **27**, 1919–1931.
8. Masri J, Bernath A, Martin J, Jo OD, Vartanian R, Funk A, and Gera J. (2007) mTORC2 activity is elevated in gliomas and promotes growth and cell motility via overexpression of rictor, *Cancer Res* 67, 11712–11720.
9. Guertin DA, Stevens DM, Saitoh M, Kinkel S, Crosby K, Sheen JH, Mullholland DJ, Magnuson MA, Wu H, and Sabatini DM. (2009) mTOR complex 2 is required for the development of prostate cancer induced by Pten loss in mice, *Cancer Cell* **15**, 148–159.
10. Soukas AA, Kane EA, Carr CE, Melo JA, and Ruvkun G. (2009) Rictor/TORC2 regulates fat metabolism, feeding, growth, and life span in Caenorhabditis elegans, *Genes Dev* **23**, 496–511.
11. Jones KT, Greer ER, Pearce D, and Ashrafi K. (2009) Rictor/TORC2 regulates Caenorhabditis elegans fat storage, body size, and development through sgk-1, *PLoS Biol* 7, e60.
12. Cybulski N, Polak P, Auwerx J, Ruegg MA, and Hall MN. (2009) mTOR complex 2 in adipose tissue negatively controls whole-body growth, *Proc Natl Acad Sci USA* **106**, 9902–9907.
13. Peterson TR, Laplante M, Thoreen CC, Sancak Y, Kang SA, Kuehl WM, Gray NS, and Sabatini DM. (2009) DEPTOR is an mTOR inhibitor frequently overexpressed in multiple myeloma cells and required for their survival, *Cell* **137**, 873–886.
14. Huang J, Dibble CC, Matsuzaki M, and Manning BD. (2008) The TSC1-TSC2 complex is required for proper activation of mTOR complex 2, *Mol Cell Biol* **28**, 4104–4115.
15. Huang J, Wu S, Wu CL, and Manning BD. (2009) Signaling events downstream of mammalian target of rapamycin complex 2 are attenuated in cells and tumors deficient for the tuberous sclerosis complex tumor suppressors, *Cancer Res* 69, 6107–6114.
16. Dibble CC, Asara JM, and Manning BD. (2009) Characterization of Rictor phosphorylation sites reveals direct regulation of mTOR complex 2 by S6K1, *Mol Cell Biol* **29**, 5657–5670.
17. Julien LA, Carriere A, Moreau J, and Roux PP. (2009) mTORC1-Activated S6K1 Phosphorylates Rictor on Threonine 1135 and Regulates mTORC2 Signaling, *Mol Cell Biol* 2010; **30**, 908–921.
18. Treins C, Warne PH, Magnuson MA, Pende M, and Downward J. (2009) Rictor is a novel target of p70 S6 kinase-1, *Oncogene* 2010; **29**, 1003–1016.

19. Pearce LR, Huang X, Boudeau J, Pawlowski R, Wullschleger S, Deak M, Ibrahim AF, Gourlay R, Magnuson MA, and Alessi DR. (2007) Identification of Protor as a novel Rictor-binding component of mTOR complex-2, *Biochem J* **405**, 513–522.
20. Woo SY, Kim DH, Jun CB, Kim YM, Haar EV, Lee SI, Hegg JW, Bandhakavi S, Griffin TJ, and Kim DH. (2007) PRR5, a novel component of mTOR complex 2, regulates platelet-derived growth factor receptor beta expression and signaling, *J Biol Chem* **282**, 25604–25612.
21. Nowak P, Cole DC, Brooijmans N, Bursavich MG, Curran KJ, Ellingboe JW, Gibbons JJ, Hollander I, Hu Y, Kaplan J, Malwitz DJ, Toral-Barza L, Verheijen JC, Zask A, Zhang WG, and Yu K. (2009) Discovery of potent and selective inhibitors of the mammalian target of rapamycin (mTOR) kinase, *J Med Chem* **52**, 7081–7089.
22. Yu K, Toral-Barza L, Shi C, Zhang WG, Lucas J, Shor B, Kim J, Verheijen J, Curran K, Malwitz DJ, Cole DC, Ellingboe J, Ayral-Kaloustian S, Mansour TS, Gibbons JJ, Abraham RT, Nowak P, and Zask A. (2009) Biochemical, cellular, and in vivo activity of novel ATP-competitive and selective inhibitors of the mammalian target of rapamycin, *Cancer Res* **69**, 6232–6240.
23. Huang J, and Manning BD. (2009) A complex interplay between Akt, TSC2 and the two mTOR complexes, *Biochem Soc Trans* **37**, 217–222.
24. Wullschleger S, Loewith R, Oppliger W, and Hall MN. (2005) Molecular organization of target of rapamycin complex 2, *J Biol Chem* **280**, 30697–30704.
25. Loewith R, Jacinto E, Wullschleger S, Lorberg A, Crespo JL, Bonenfant D, Oppliger W, Jenoe P, and Hall MN. (2002) Two TOR complexes, only one of which is rapamycin sensitive, have distinct roles in cell growth control, *Mol Cell* **10**, 457–468.
26. Yang Q, Inoki K, Ikenoue T, and Guan KL. (2006) Identification of Sin1 as an essential TORC2 component required for complex formation and kinase activity, *Genes Dev* **20**, 2820–2832.
27. Manning BD, Tee AR, Logsdon MN, Blenis J, and Cantley LC. (2002) Identification of the tuberous sclerosis complex-2 tumor suppressor gene product tuberin as a target of the phosphoinositide 3-kinase/akt pathway, *Mol Cell* **10**, 151–162.

Chapter 7

Overexpression or Downregulation of mTOR in Mammalian Cells

Mahmoud Khalil and Ivan Gout

Abstract

This chapter presents an overview of the methods that have been used to overexpress or downregulate the level of mTOR isoforms in mammalian cells. The techniques of transient overexpression, generation of stable cell lines, retroviral- and lentiviral-mediated overexpression or downregulation are discussed.

Key words: Mammalian target of rapamycin, Methods, Transient and stable overexpression, Downregulation

1. Introduction

The mammalian target of rapamycin (mTOR) was identified at the same time by several independent groups as the cellular target for the immunosuppressant drug rapamycin, a naturally occurring antifungal macrolide (1–5). These studies were preceded by the pioneering discovery of two TOR genes in yeast, TOR1 and TOR2, by M. Hall and colleagues (6). In mammals, there is one mTOR gene which encodes two splicing isoforms, mTORα and mTORβ. The domain organization and regulatory sites of phosphorylation of mTOR isoforms are shown in Fig. 1a. Since mTORβ was identified only a year ago, most studies on defining the role of mTOR in signal transduction and the regulation of cellular functions have solely involved originally cloned mTORα isoform. Northern and Western blot analysis indicate that both isoforms are ubiquitously expressed in mammalian tissues, while mTORα is the predominantly expressed isoform at both RNA and protein levels (Fig. 1c). Since Northern blot analysis has revealed at least four mRNA transcripts

Thomas Weichhart (ed.), *mTOR: Methods and Protocols*, Methods in Molecular Biology, vol. 821,
DOI 10.1007/978-1-61779-430-8_7, © Springer Science+Business Media, LLC 2012

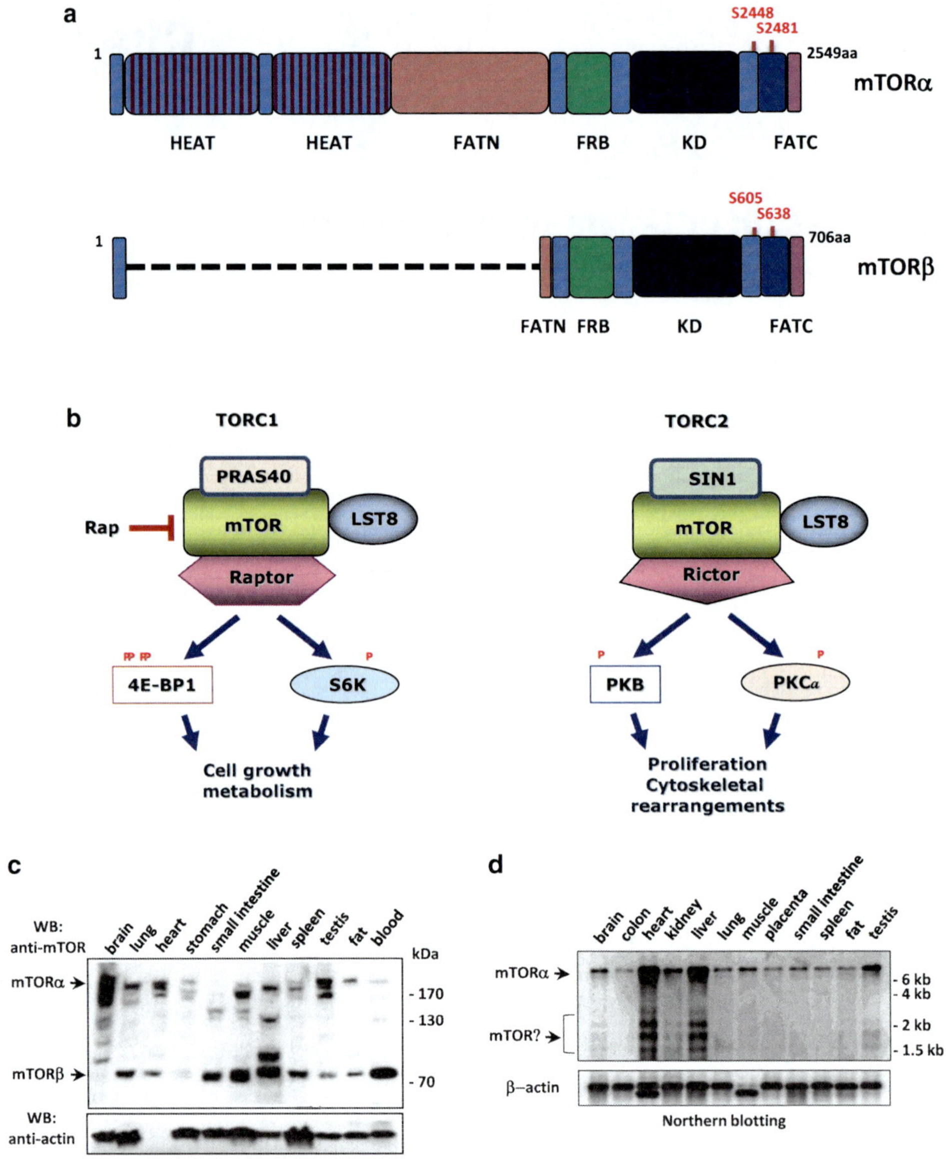

Fig. 1. Domain organization, signaling complexes, and expression pattern of mTOR splicing isoforms. (**a**) Structural features of mTORα and mTORβ splicing isoforms. HEAT (*H*untingtin, *E*F3, PP2*A*, and *T*OR1); FAT (*F*RAP/TOR-*A*TM-*T*RRAP); FRB (FKBP12-Rapamycin-Binding); KD (Kinase Domain). (**b**) Schematic presentation of TORC1 and TORC2, and key downstream signaling effectors. (**c**) Western blot analysis of the mTOR expression in rat tissues. 30 μg of total protein extracts were immunoblotted with the C-terminal mTOR polyclonal antibodies from Cell Signaling (mTOR-CS) (*top panel*). The membrane was reprobed with antiactin antibodies (*lower panel*). (**d**) Northern blot analysis of mTOR transcripts in human tissues. The 3′ mTOR probe was used to probe the blot (*top panel*) and then the expression β-actin analyzed by reprobing the membrane (*lower panel*). Expression analysis of both mTOR isoforms at mRNA and protein levels was published by Panasyuk et al. (7).

with the mTOR specific probe, the existence of yet undiscovered mTOR splicing isoforms is feasible.

Interestingly, the overexpression of mTORα in mammalian cells does not result in the upregulation of TOR signaling, nor the induction of cell growth, proliferation, or survival. By contrast, the overexpression of mTORβ isoform significantly shortens the G1/S transition of the cell cycle, induces cell proliferation and survival. As a consequence, mTORβ, but not mTORα, has the potential to induce oncogenic transformation when stably overexpressed in immortalized cell lines and to promote the growth of tumors in a xenograft model (7). The inability of overexpressed mTORα to induce mTOR-mediated signaling and cellular functions could be explained by the complexity of its activation and substrate presentation in two mutually exclusive complexes, mTOR complex 1 (mTORC1) and mTORC2, which coordinate cellular functions in response to mitogenic stimulation, nutrient and energy sufficiency. The main components of TORC1 and TORC2 signaling complexes and their key downstream signaling effectors are shown in Fig. 1b. As mTORβ has the ability to signal via both TORC1 and TORC2, and is also sensitive to rapamycin, the existence of yet unidentified regulators/downstream effectors of mTORβ signaling might explain the observed differences in overexpression studies (7).

Interaction between rapamycin and the immunophilin FKBP12 generates a highly potent and specific inhibitor of mTORC1. The mTORC1 complex signals to eIF4E-binding protein 1 (4E-BP) and S6 kinases (S6Ks), which are key regulatory components of protein synthesis (8, 9). It has been demonstrated that the activity of mTOR in the mTORC2 complex can also be inhibited by prolong (24 h) treatment of cells with rapamycin. It appears that the rapamycin–FKBP12 complex does not bind directly to mTORC2, but long-term treatment of cells with rapamycin disrupts the assembly of functional mTORC2 through an unknown mechanism. It has been proposed that the FKBP12–rapamycin interaction with mTOR might block subsequent binding of other components of TORC2, such as Rictor and SIN1 (10). However, it is not clear why the rapamycin-mediated inhibition of mTORC2 assembly only occurs in certain cell types.

Genetic studies in model organisms targeting different components of TOR regulatory complexes and the use of rapamycin/rapalogs uncovered a critical role of mTOR in integrating signaling information from growth factors, nutrient and energy sufficiency to coordinate cell growth, proliferation and survival via the regulation of cellular biosynthetic processes and autophagy (11–13). Deregulation of the mTOR signaling pathway has been implicated in a diverse range of human pathologies, including cancer, autoimmunity, cardiovascular, and neurodegenerative diseases and metabolic disorders such as diabetes (14–17). This has prompted researchers from academia and pharmaceutical companies to develop novel mTOR inhibitors, ranging from derivatives of rapamycin to ATP-competitive compounds (18). At present, sirolimus (a rapamycin homolog) is used in the clinic for treating patients with coronary stenosis (sirolimus-eluting stents), while temsirolimus is indicated as the first line therapy for patients with renal cell carcinoma.

As mTOR signaling has been the subject of intense drug discovery activities, there are numerous clinical trials in progress for testing the efficacy of novel mTOR inhibitors.

Here, we will describe the protocols which have been used successfully for overexpression as well as downregulation of mTOR in mammalian cells. These include the following: (a) different approaches for transient transfection of cells with expression plasmids containing mTOR splicing isoforms or mTOR specific shRNAs, (b) generation of stable cells overexpressing mTOR under inducible or noninducible promoters, (c) retroviral, adenoviral, and lentiviral-mediated overexpression/downregulation of target genes in mammalian cells.

1.1. Transient Overexpression of mTOR in Mammalian Cells

To study cellular functions of mTOR, cDNA sequences corresponding to both isoforms of mTOR can be cloned into mammalian expression vectors with or without tagged sequences. The most common tags that have been used for mTOR expression in mammalian cells include Myc (EQKLISEEDL), FLAG (DYKDDDDK), HA (YPYDVPDYA), and EE (EFMPME). In addition, GST or GFP fusion constructs of mTOR have also been used in various experimental setups. Some of these plasmids are commercially available. For example, Addgene offers a panel of mTOR expression constructs which have been generated and tested in research laboratories. These include pRK5/myc-mTORwt, pRK5/myc-mTOR kinase-dead, pcDNA3.1/FLAG-mTORwt, pcDNA3.1/AU1-mTORE2419K, and others (http://www.addgene.org/). The above plasmids attain a high copy number during bacterial growth, thus ensuring a good yield from plasmid purifications. The use of high quality and endotoxin-free plasmid DNA is critical for achieving high level and reproducible expression of mTOR by transient transfection in mammalian cells. Transient transfection offers a short-lived expression of mTOR and therefore transfected cells should be examined within a few days after transfection. The efficiency of transfection varies between different cell lines and therefore it should be determined by reporter gene assays or by cotransfection of GFP-expressing plasmids. Three protocols are presented below for transient transfection of mammalian cells: calcium phosphate coprecipitation, electroporation, and lipid-based transfection.

1.1.1. Calcium Phosphate-Based Transfection

This transfection approach is based on the ability of calcium phosphate to form insoluble precipitates with DNA under certain conditions. When calcium–DNA precipitates are added to cultured cells, they stick to the cell surface and then are taken into the cells by endocytosis. Calcium–DNA precipitates are clearly visible on the cell surface the day after transfection when examined by phase contrast microscopy. Most calcium phosphate-based transfection protocols are based on a method developed by Jordan et al. (19) who rigorously optimized the conditions for efficient transfection of mammalian cells. This protocol is optimized for transfection of HEK 293 cells, but could be easily adapted for both adherent and nonadherent cell lines. In contrast to other transfection approaches, the calcium phosphate-based technique is inexpensive and does not require specialized equipment, such as a Gene Pulser electroporator. We have used this approach extensively to study the role of overexpressed S6Ks and other signaling molecules in a panel of cell lines.

1.1.2. Transfection of Cells by Electroporation

Electroporation uses high-voltage electric shocks to open up pores in the cell, allowing the diffusion of large highly charged molecules such as DNA into the cytoplasmic compartment. This technique is based on the use of an electroporation device, which can subject cells to high-voltage electrical pulses of defined magnitude and length (20). Electroporation has been used extensively for the introduction of plasmid DNA into prokaryotic and eukaryotic cells. The protocols have been established for both transient and stable transfection of mammalian cells. The optimal conditions for efficient transfection vary between different cell types, so this procedure requires fine-tuning. In brief, exponentially growing cells are mixed with plasmid DNA in an appropriate buffer and then transferred into an electroporation cuvette. The cells in the cuvette are subjected to electrical pulses of defined magnitude and length: this generally is in the range of 1.5 kV at 25 μF. If excessive cell death is observed (more than 60%) the length of the pulse can be lowered by reducing the capacitance. After a short recovery on ice following the pulse, the cells are cultured in normal growth medium to allow the expression of the protein of interest.

1.1.3. Lipid-Based Transfection of Mammalian Cells (Lipofection)

The method of lipid-mediated DNA transfection into mammalian cells was originally developed by Felgner (21). It is based on an ionic interaction between negatively charged DNA and liposomes containing cationic lipids with positively charged head groups and one or two hydrocarbon chains. In this procedure, DNA is not encapsulated within unilamellar liposomes (600–1,200 nm in size) but forms tight complexes with the positively charged surface of the liposomes. When added to cells, these "transfection complexes" interact with the negatively charged cell membrane and enter the cell through endocytosis.

There are several advantages of cationic liposome-mediated transfection when compared to other methods, including the following: (a) high efficiency of transfection of a wide variety of mammalian cell lines, (b) relatively low cell toxicity, (c) in addition to DNA, the procedure is also adapted for transfection of cells with siRNAs, synthetic oligonucleotides, proteins, and viruses. A relatively high cost of transfection reagents is the main disadvantage of lipofection.

A comprehensive range of transfection reagents with efficient and reproducible delivery of plasmid DNA, siRNA, DNA oligonucleotides into mammalian cell lines has been developed and offered to researchers by different companies, including Invitrogen, MBI Fermentas, Roche Diagnostics, etc.

1.2. Generation of Stable Cell Lines for the Expression of mTOR

In contrast to transient expression, stable expression of a protein of interest in mammalian cell is more time consuming and requires careful selection and maintenance of generated cell lines. To date, the number of selectable markers available for selection of stable cells is limited. The most commonly used selection markers in expression plasmids include G418, neomycin, zeocin, hygromycin, and puromycin. Taking into account that some of these antibiotics are very expensive, the cost of generating (weeks/months) and maintaining (months/years) stable cell lines may be a factor in experimental design. A number of expression vectors/systems for stable overexpression have been developed. Here, we will describe the use of the tetracycline-regulated expression system which has been used for stable expression of mTORβ in HEK 293 cells (Fig. 2b).

1.2.1. Tetracycline-Inducible Expression System

The tetracycline-regulated expression system (T-Rex System) in mammalian cells uses regulatory elements from the *E. coli* Tn10-encoded tetracycline (Tet) resistance operon (22). Tetracycline regulation in the T-Rex System is based on the binding of tetracycline to the tetracycline repressor (Tet repressor), and subsequent derepression of the promoter which controls expression of the gene of interest (23). To create stable cell lines, the gene of interest is cloned into the inducible expression vector, pcDNA4/TO, and the resulting construct is cotransfected with the regulatory plasmid pcDNA6/TR into mammalian cells. The production of stable cell lines depends on a random homologous recombination event to integrate the plasmid into the host genome, so that it replicates in tandem with the chromosomal DNA. The pcDNA4/TO plasmid contains a zeocin resistance gene that can be used to select for clones containing the recombinant construct. A number of cell lines with stable integration of the regulatory plasmid, pcDNA6/TR, are commercially available, including T-REx HEK 293, Jurkat, CHO, HeLa.

1.3. The Use of Viral Expression Systems

DNA can also be introduced into eukaryotic cells by using viral expression systems. Distinct types of viral vectors have been developed for specific and efficient gene delivery into mammalian cells. The most commonly used viral expression systems employ modified retroviruses, adenoviruses and lentiviruses. It has been demonstrated that viral vectors can drive the expression of target genes in both dividing and nondividing cells (24). Viral-mediated expression has proved to be more efficient than other gene transfer strategies described above. However, constructing a recombinant virus is time consuming and often requires specialized tissue culture facilities. It is therefore important to carefully plan the production and the use of recombinant viruses to ensure that the optimal virus is constructed for the experimental design and that safety is guaranteed.

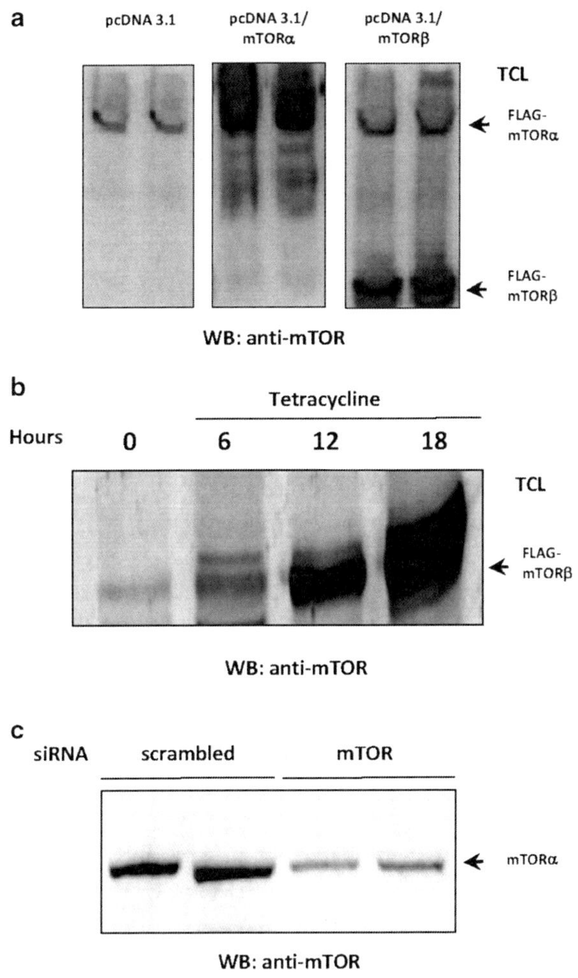

Fig. 2. Overexpression and downregulation of mTOR splicing isoforms in mammalian cells. (**a**) Transient overexpression of mTORα and mTORβ in HEK 293 cells. ExGen 500 transfection reagent was used to transfect HEK 293 cells with pcDNA3.1, pcDNA3.1/FLAG-mTORα and pcDNA3.1/FLAG-mTORβ constructs. Two days later, total cell lysates were immunoblotted with the mTOR-CS antibodies. (**b**) Tetracycline-induced expression of mTORβ in a HEK 293 stable cell line. The expression of FLAG-tagged mTORβ in HEK 293/FLAG-mTORβ stable cells was induced by the addition of tetracycline (final concentration 1 μg/ml). Total cell lysates from noninduced and induced (6, 12 and 18 h) cells were probed with the mTOR-CS antibodies for the expression of FLAG-mTORβ. (**c**) siRNA-mediated downregulation of mTORα expression in HEK 293 cells. Scrabbled and mTOR specific siRNAs were transfected into HEK 293 cells using Lipofectamine 2000. Two days later, the expression of mTORα was examined by immunoblotting of total cell lysates with mTOR-CS antibodies.

1.3.1. Retroviral Expression System

Retrovirus expression system is based on replication-competent viruses isolated from rodents or chickens. Retrovirus vectors are modified in different ways to introduce a gene of interest into mitotic cells *in vivo* or *in vitro*. The integration machinery of naturally occurring retroviruses is utilized to stably integrate a single copy or a few copies of the viral genome into a host chromosome.

At the first stage of retrovirus production, a retroviral vector with the gene of interest is introduced into the retroviral packaging cells. Transfection is the method of choice, as it allows many copies of the retroviral vector plasmid DNA to be taken up by each cell. It is important to achieve a high level of expression with consequently high retroviral titer.

Although it is possible to scale up, the highest retroviral titers are obtained using 60 mm dishes. The Protocol described here can be used to cotransfect retroviral structural genes together with the retroviral vector into HEK 293 cells or to transfect a retroviral vector into HEK 293-derived packaging cell lines.

1.3.2. Lentiviral Expression System

In recent years, lentivirus vectors have become very popular for target gene delivery and expression. It is mainly due to their genetic stability and safety. Lentiviral vectors can deliver and express target genes in a wide variety of proliferating and nonproliferating cells (25). At present, a panel of lentiviral vectors/systems has been developed for noninducible or inducible expression of the transgene (26). Lentiviral particles are usually produced in a packaging cell line HEK 293T transfected with three plasmids: (a) the HIV-1 based lentiviral transfer plasmid (e.g., pSIN), (b) the packaging plasmid (e.g., pCMVΔR8.91), and (c) the VSV-G protein envelope plasmid (e.g., pMD.G).

1.4. Molecular Approaches for Downregulation of mTOR Expression

RNA interference is a naturally occurring regulatory process which mediates sequence-specific posttranscriptional gene silencing by short double-stranded RNA during the development and in response to growth factors, cellular metabolites, and extracellular stresses. This phenomenon has been exploited for targeted downregulation of gene expression by transduction of cells with synthesized double-stranded small interfering RNA (siRNA), consisting of 21–25 nucleotides. siRNAs have two main characteristics that make them unique: (a) they do not encode proteins and (b) they interfere posttranscriptionally with mRNA. Different methodologies have been developed for the induction of siRNA-mediated RNA interference in mammalian cells, including microinjection of siRNA, transfection of siRNA using cationic lipids, electroporation, transfection of plasmids containing siRNA expression cassettes, or infection of cells by recombinant viruses that express siRNA. Specialized expression libraries have been made with short hairpin RNA (shRNA), which has the potential to stably express small interfering RNA in target cells.

Here, we present a liposome-mediated method for siRNA transfection. The liposome transfection of mammalian cells with siRNA shares some similarity with DNA transfection protocols. Similar considerations apply for siRNA and DNA transfection protocols, although special transfection reagents were developed to achieve optimal transfection efficiencies with siRNA. For some cell lines, specific DNA transfection reagents such as Lipofectamine 2000 (Invitrogen) appear to work the best.

2. Materials

2.1. Transfection protocols

2.1.1. Calcium Phosphate-Based Transfection

1. All reagents should be sterile or aseptically prepared.
2. Mammalian cell lines (HEK 293, NIH 3T3, MCF7, U937, etc.).
3. Complete culture medium (depending on cell lines used).
4. Plasmid DNA (10–50 μg per transfection).
5. 2.5 M $CaCl_2$.
6. 2× HEPES-buffered saline, pH 7.1. For efficient transfection, the optimal pH of this buffer should be between 7.05 and 7.12. Note that the pH of the 2× HEPES solution can change, even when stored at –20 C. Therefore, check pH before transfection.
7. 0.1× TE (pH 7.6).
8. Optional: 100 mM Chloroquine, 500 mM Sodium butyrate, 15% v/v Glycerol in 1× HEPES-buffered saline.

2.1.2. Transfection of Cells by Electroporation

1. Exponentially growing mammalian cells (HEK 293, NIH3T3, MCF7, U937, etc.).
2. Complete culture medium (depending on cell lines used).
3. Plasmid DNA (10–50 μg per transfection).
4. Electroporation buffer: 20 mM HEPES (pH 7.0), 137 mM NaCl, 5 mM KCl, 0.7 mM Na_2HPO_4, 6 mM glucose, store at 4°C.
5. Electroporator (e.g., Gene Pulser MXcell).
6. Electroporation cuvettes.

2.1.3. Transfection of NIH3T3 Cells Using Lipofectamine

1. Exponentially growing mammalian cells (HEK 293, NIH3T3, MCF7, U937, etc.).
2. Complete culture medium (depending on cell lines used).
3. Reduced serum medium: Opti-MEM® I (contains 2% FBS).
4. Serum-free medium: DMEM without serum or antibiotics.
5. Plasmid DNA, siRNA, etc.
6. Transfection reagents: Lipofectamine 2000, Lipofectin, ExGen 500, FuGENE.

2.2. Production of stable cell lines

2.2.1. Generation of Stable Cell Lines for the Expression of mTOR

1. T-Rex HEK 293 cell line stably expressing the tetracycline (Tet) repressor.
2. pcDNA4/TO-mTORβ expression vector.
3. Complete culture medium.
4. Antibiotics: zeocin and blasticidin.
5. 1 mg/ml tetracycline.
6. Transfection reagents: Lipofectamine 2000, Lipofectin, ExGen 500, FuGENE.

2.3. Retroviral and lentiviral expression

2.3.1. Retroviral Expression System

1. FuGENE 6 transfection reagent.
2. HEK 293T cells *or* appropriate packaging cell line.
3. 293T cell growth medium with.

2.3.2. Lentiviral Expression System

1. 293T human embryonic kidney cell line.
2. 293T cell growth medium (Opti-MEM®I) with and without 10 mM HEPES.
3. Lentiviral transfer plasmid (e.g., pSIN).
4. Lentiviral packaging construct DNA (e.g., pCMVΔR8.91; (25)).
5. VSV-G expressing plasmid DNA (e.g., pMD.G).
6. FuGENE 6 transfection reagent.

2.4. Downregulation of mTOR Expression

1. Mammalian cells to be transfected (e.g., HEK 293).
2. Opti-MEM® I reduced serum medium.
3. Annealed siRNA or shRNA plasmids.
4. Lipofectamine 2000.

3. Methods

3.1. Transfection protocols

3.1.1. Calcium Phosphate-Based Transfection

1. Replate exponentially growing HEK 293 cells by trypsinization at a density of 2×10^5 cells/cm^2 in 100 mm tissue culture dishes. Culture cells for 24 h at 37°C in an incubator with 5% CO_2. Change the medium 1 h before transfection. For efficient transfection, the cells should spread well and cover approximately 60–80% of the dish, as the capacity of cells to take up DNA correlates to the surface area of the cell exposed to the medium.
2. To prepare the calcium phosphate–DNA mixture, combine 100 μl of 2.5 M $CaCl_2$ with 25 μg of plasmid DNA in a sterile 15 ml Falcon tube and use 0.1× TE (pH 7.6) to bring the final volume to 1 ml. After mixing, add equal volume of 2× HEPES buffered saline and quickly tap the side of the tube to mix the ingredients. Leave the mixture at room temperature (RT) for 10 min.
3. Aspirate the culture medium and carefully overlay the cells with the suspension of calcium phosphate–DNA precipitate. After mixing, the medium will appear yellow-orange and slightly cloudy. Incubate the transfected cells at 37°C with 5% CO_2 for 6 h and then remove the medium and the DNA precipitate by aspiration. Wash the cells twice with tissue culture PBS, before adding complete DMEM medium. Culture cells for 24–72 h, depending on the experimental setup.
4. The expression of the protein of interest could be examined by various approaches, including Western blot analysis of total cell

lysates, immunoprecipitation followed by Western blotting or enzymatic assay, immunofluorescence analysis, etc.

5. Several additional steps have been proposed to increase the uptake of DNA and therefore increase the efficiency of transfection. These include the following: (a) treatment of cells with 1 mM chloroquine in the presence of the calcium phosphate–DNA coprecipitate, (b) exposure of cells to 15% glycerol 1× HEPES-buffered saline, (c) treatment of cells with 2 mM sodium butyrate after the removal of the calcium phosphate–DNA coprecipitate. The length of these optional treatments should be determined experimentally for each cell line.

3.1.2. Transfection of Cells by Electroporation

1. Harvest exponentially growing HEK 293 cells by trypsinization and pellet them by centrifugation at 200 × *g* for 5 min at 4°C.
2. Wash collected cells in tissue culture PBS, centrifuge (200 × *g* for 5 min at 4°C) and then resuspend in 0.5× volume of the complete DMEM medium. Count cells using a hemocytometer (optimal 1×10^7 cell per electroporation).
3. After centrifugation, suck off the medium and resuspend the pelleted cells in the ice-cold electroporation buffer at 2.5×10^7 cells/ml. From this point work quickly but carefully.
4. Add 0.4 ml of cells into prelabeled electroporation cuvettes first and later 30 μg of prepared plasmid DNA. Pipette the mixture up and down, but do not introduce air bubbles into the suspension of cells by vigorous pipetting. If you do several transfections, mix DNA with cells at the interval of 1.5 min. Incubate at RT for 10 min.
5. Set the parameters on the electroporator, which have been determined for a particular cell line. For HEK 293 cells, we used the following conditions: capacitance value – 1,050 μF; voltage – 260 V; internal resistance – infinite value. Test the settings on a blank cuvette containing phosphate-buffered saline before beginning electroporation of cells.
6. Place a cuvette in the holder of the electroporator and pulse the cells only once with preset conditions (record the settings and the duration of the electroporation).
7. After electroporation, place the cuvette on ice for 10 min and then transfer electroporated cells into a Falcon tube containing 10 ml of complete DMEM medium. Mix cells gently by pipetting to break the clumps and plate out the cells onto a 100 mm culture dish. Culture electroporated cells at 37°C in an incubator with 5% CO_2 for 24–96 h.
8. Analyze the cells for transient expression of the protein of interest.

3.1.3. Transfection of NIH3T3 Cells Using Lipofectamine

1. NIH3T3 cells were transfected using Lipofectamine 2000™ reagent (Invitrogen) as instructed by the manufacturer with small modifications. One day prior to transfection, cells were seeded at around 1×10^6 cells per 60 mm dish in medium lacking antibiotics. This number of cells is sufficient to give a confluence of at least 90% at the time of transfection.
2. For each transfection, prepare the DNA and lipid solutions in separate sterile tubes: (a) dilute plasmid DNA (3 μg per 60 mm dish) in 250 μl of Opti-MEM® I reduced serum medium; (b) add 10 μl of Lipofectamine 2000 to 250 μl of Opti-MEM® I. Incubate both mixtures for 5 min at RT and then mix gently by pipetting up and down several times. DNA–Lipofectamine 2000 complexes were allowed to form by incubation at RT for 20 min.
3. While the transfection complexes are forming, wash the cells to be transfected three times with serum-free medium. After the third wash, add 0.5 ml of serum-free medium to each 60 mm dish and incubate at 37°C until the transfection mixture is added. Note, that the presence of serum and antibiotics strongly inhibits the efficiency of transfection.
4. Overlay the DNA–Lipofectamine 2000 mixture onto the cells and incubate dishes at 37°C in a CO_2 incubator for 6 h.
5. Wash cells twice with serum-free medium. Feed the cells with complete medium, containing 10% FCS and antibiotics. Keep the cells under standard culturing conditions and analyze the expression of the protein of interest 24–72 h after the start of transfection.

3.1.4. Transfection of HEK 293 Cells Using ExGen 500

1. This protocol is adapted for transfection of HEK293 cells using ExGen 500 reagent. Unlike several alternative transfection reagents, the use of ExGen 500 allows transfection of mammalian cells in normal culture medium, containing serum and antibiotics. It also exhibits a relatively low toxicity, so the cells can tolerate long exposure to a transfection solution (up to 24 h) without adverse effects.
2. The day before transfection, exponentially growing HEK 293 cells were harvested and plated at 5×10^5 cells per 60 mm dish to give a confluence of approximately 60% at the time of transfection.
3. Next day, DNA–ExGen 500 complexes were prepared for transfection by adding 3 μg of plasmid DNA and 9 μl of ExGen (per construct per 60 mm dish) into 200 μl of sterile filtered 150 mM NaCl. The mixture was vortexed immediately for 10 s and incubated at RT for 10 min to allow the formation of the DNA–ExGen complexes.
4. During this incubation, the medium was removed from cultured cells and replaced with 2 ml of fresh DMEM containing 10%

serum. The transfection mixture was added to the cells dropwise and dishes were incubated for 24 h at 37°C in a CO_2 incubator.

5. The transfection medium was replaced with a complete DMEM medium the following day. Cells were typically harvested 48 h after transfection.
6. This protocol was used extensively for transient overexpression of mTORα and mTORβ in various cell lines. Transient overexpression of both mTOR isoforms in HEK 293 cells is shown in Fig. 2a.

3.2. Generation of Stable Cell Lines for the Expression of mTOR

1. Transfect T-Rex HEK 293 cell with pcDNA4/TO-mTORβ expression vector using ExGen 500, as described above.
2. 48 h after transfection, split the cells 1/20 into 100 mm dishes and grow them in complete DMEM medium, containing 5 μg/ml of blasticidin and 100 μg/ml of zeocin.
3. Change selection medium every 3–4 days and monitor cell survival over the period of two/three weeks to identify colonies of cells with stable integration of the expression construct.
4. Discrete colonies were selected and isolated in individual plastic chambers to allow them to be trypsinized and removed. In this study, ten clones were selected and screened for tetracycline-regulated expression of mTORβ.
5. To induce the expression, tetracycline was added to exponentially growing cells at a final concentration of 1 μg/ml. The level of expression was monitored by Western blotting at various time-points after induction.
6. Those cell clones that tested positive for expression were frozen in DMEM medium containing 50% FBS and 10% DMSO and stored in liquid nitrogen.

3.3. Retroviral and lentiviral expression

3.3.1. Retroviral Expression System

1. 24 h before transfection, plate HEK 293 cells at density 1.0×10^6 in a 100 mm tissue culture dish. Cell clumping should be minimized.
2. Cells are transfected the following day (at approximately 40% confluence) with a transfection cocktail, containing 10 μg of retroviral vector DNA and 12 μl of FuGENE 6 transfection reagent, according to the manufacturer's recommendations.
3. Next day, remove the transfection medium and culture the cells in fresh complete medium for an additional 24 h.
4. Harvest the retroviral supernatant 48 h following transfection by gently removing the supernatant from the cells. Filter the supernatant through a 0.45 μm filter to remove live cells.
5. After harvesting, the efficiency of transfection can be tested by immunoblot or immunofluorescence analysis of transfected cells. Optimal transfection efficiencies should range from 30%

to 60%. If the cells are sparse at 24 h postinfection, it may be necessary to wait until 72 h to obtain high-titer supernatant.

6. If the retroviral supernatant is to be used within 1–2 hr, store it on ice. For longer intervals, freeze the supernatant on dry ice and transfer it to –80°C. Freezing does not appear to cause >2-fold drop in titer as long as the supernatant does not undergo more than one freeze/thaw cycle.
7. Infectious titer can be quantitated based on marker staining, e.g., using *lacZ*, or green fluorescent protein (GFP). Calculate viral titer using a standard protocol.

3.3.2. Lentiviral Expression System

1. One day before the transfection, split confluent HEK 293T cells into 100 mm dishes at 1×10^6 cells per dish.
2. On the day of the transfection, wash the cells (80% confluence is expected) with PBS and add 8 ml Opti-MEM® I medium per plate.
3. For each plate to be used mix three plasmids as follows: 1.5 μg pSIN/mTORβ, 1 μg of the packaging plasmid pCMVΔR8.91, and 1 μg of the VSV-G protein envelope plasmid pMD.G. Bring the volume up to 50 μl with Opti-MEM® I medium.
4. Mix 10 μl FuGENE 6 transfection reagent with 40 μl Opti-MEM® I medium and incubated for 5 min at RT.
5. Mix both FuGENE and DNA solutions, and incubate at RT for 15 min. Add the mixture dropwise to prewashed HEK 293T cells, gently swirl the plate, and incubate at 37°C for 5 h.
6. After the incubation, the transfection medium is removed and the cells are washed with PBS and cultured in complete DMEM medium. 48 h posttransfection, the medium containing the lentivirions is collected, passed through a 0.45 μm filter to remove any cell debris, and either used immediately or quickly frozen at –80°C.
7. Each time lentivirus is prepared its quality (infectivity) is tested on HEK293T cells before using it for the infection of other cells. To test this, 1 ml of GFP-lentivirus is thawed on ice and added to a 60 mm dish of 70% confluent HEK293T cells. The infectivity is measured 48–72 h postinfection by assessing the amount of green fluorescent cells (infected cells) compared to non fluorescent cells (uninfected cells). Virus preparations are considered as suitable for further studies only when 80% of cells are infected.
8. The efficiency of infection depends to a large extent on the ratio of lentivirions to target cells (multiplicity of infection, MOI). To achieve efficient gene delivery of the transgene into cells the value of MOI should range between 300 and 1000 (27). To attain these MOI values, highly concentrated vector particle preparations are required. Several methods have been developed to concentrate lentiviral vector particles (28). The

stability of the VSV-G envelope protein allows generation of high-titer lentiviral vector stocks by ultracentrifugation. Generated viral stocks are centrifuged at 50,000 × *g* for 90 min, 4°C. A pale-yellow pellet (approx. 2–5 mm in diameter) should be visible at the bottom of the tube after centrifugation.

9. For the infection of target cells, 5×10^5 MCF7 cells were seeded into a T75 cm^2 flask. The following day (at approximately 30% confluence), the medium was aspirated and the cells were infected with lentiviral or retroviral stocks (the value of MOI is determined for each experimental setup) in the presence of 8 μg/ml polybrene. The cells were then incubated at 37°C for 24 h. The following day, the medium is replaced with 10 ml of fresh culture medium. 72 h postinfection cells are harvested for testing target gene expression. For stable expression antibiotic selection medium is added 4 days postinfection (6 μg/ml puromycin). The selection medium is changed every 3–4 days of culturing.

4. Molecular Approaches for Downregulation of mTOR Expression

1. 24 h before lipofection of siRNA, harvest exponentially growing HEK 293 cells by trypsinization and replate them onto 60 mm tissue culture dishes at a density of 5×10^5 cells/dish. Add 5 ml of growth medium, and incubate the cultures for 20–24 h at 37°C in a humidified incubator with 5% CO_2 until the cells are 50–75% confluent.
2. Before transfection, prepare the DNA and lipid solutions in sterile tubes:
   (a) For the siRNA solution, 200–400 pmol of siRNA was diluted in 250 μl of Opti-MEM®I reduced serum medium.
   (b) For the lipid solution, 10 μl of Lipofectamine 2000 was diluted in a separate aliquot of Opti-MEM®I.
3. Incubate both mixtures for 5 min at RT. Combine both solutions and mix gently by pipetting up and down several times. Incubate at room temperature for 20 min to allow the formation of DNA–lipid complexes.
4. While the DNA–lipid solution is incubating, aspirate the medium and 2 ml of fresh medium (Antibacterial agents should be omitted during transfection) is added to the 60 mm dish and return the washed cells to a 37°C humidified incubator with an atmosphere of 5 % CO_2.
5. The siRNA–Lipofectamine 2000 complexes were then added to each plate.

6. Incubate the cells with the complexes at 37°C in a CO_2 incubator for 48–72 h until ready to assay for gene knockdown.
7. The siRNA were manufactured and purified by HPLC by MWG biotech. They were dissolved in sterile siMAX Universal buffer to give a concentration of 1 pmol/μl. siRNA-mediated downregulation of mTORα is shown in Fig. 2c.

Acknowledgments

The authors would like to thank Prof. D. Saggerson for critical reading of the manuscript. This work was supported in part by grants from the Association for International Cancer Research (AICR) and Atlantic Philanthropies/LICR Clinical Discovery Program.

References

1. Brown, E.J., Albers, M.W., Shin, T.B., Ichikawa, K., Keith, C.T., Lane, W.S., and Schreiber, S.L. (1994). A mammalian protein targeted by G1-arresting rapamycin-receptor complex. Nature *369*, 756–758.
2. Chen, J., Zheng, X.F., Brown, E.J., and Schreiber, S.L. (1995). Identification of an 11-kDa FKBP12-rapamycin-binding domain within the 289-kDa FKBP12-rapamycin-associated protein and characterization of a critical serine residue. Proc. Natl. Acad. Sci USA *92*, 4947–4951.
3. Chiu, M.I., Katz, H., and Berlin, V. (1994). RAPT1, a mammalian homolog of yeast Tor, interacts with the FKBP12/rapamycin complex. Proc. Natl. Acad. Sci. USA *91*, 12574–12578.
4. Sabatini, D.M., Erdjument-Bromage, H., Lui, M., Tempst, P., and Snyder, S.H. (1994). RAFT1: a mammalian protein that binds to FKBP12 in a rapamycin-dependent fashion and is homologous to yeast TORs. Cell *78*, 35–43.
5. Sabers, C.J., Martin, M.M., Brunn, G.J., Williams, J.M., Dumont, F.J., Wiederrecht, G., and Abraham, R.T. (1995). Isolation of a protein target of the FKBP12-rapamycin complex in mammalian cells. J Biol Chem. *270*, 815–822.
6. Heitman, J., Movva, N.R., and Hall, M.N. (1991). Targets for cell cycle arrest by the immunosuppressant rapamycin in yeast. Science *253*, 905–909.
7. Panasyuk, G., Nemazanyy, I., Zhyvoloup, A., Filonenko, V., Davies, D., Robson, M., Pedley, R.B., Waterfield, M., and Gout, I. (2009). mTORbeta splicing isoform promotes cell proliferation and tumorigenesis. J Biol Chem. *284*, 30807–30814.
8. Brunn, G.J., Hudson, C.C., Sekulic, A., Williams, J.M., Hosoi, H., Houghton, P.J., Lawrence, J.C., Jr., and Abraham, R.T. (1997). Phosphorylation of the translational repressor PHAS-I by the mammalian target of rapamycin. Science *277*, 99–101.
9. Burnett, P.E., Barrow, R.K., Cohen, N.A., Snyder, S.H., and Sabatini, D.M. (1998). RAFT1 phosphorylation of the translational regulators p70 S6 kinase and 4E-BP1. Proc. Natl. Acad. Sci USA *95*, 1432–1437.
10. Sarbassov, D.D., Ali, S.M., Sengupta, S., Sheen, J.H., Hsu, P.P., Bagley, A.F., Markhard, A.L., and Sabatini, D.M. (2006). Prolonged rapamycin treatment inhibits mTORC2 assembly and Akt/PKB. Mol Cell *22*, 159–168.
11. Laplante, M. and Sabatini, D.M. (2009). An emerging role of mTOR in lipid biosynthesis. Curr Biol. *19*, 1046–1052.
12. Wullschleger, S., Loewith, R., and Hall, M.N. (2006). TOR signaling in growth and metabolism. Cell *124*, 471–484.
13. Yang, Q., Guan, KL. (2007). Expanding mTOR signaling. Cell Res. *17*, 666–681.
14. Easton, J.B. and Houghton, P.J. (2006). mTOR and cancer therapy. Oncogene *25*, 6436–6446.

15. Goberdhan, D.C., Ogmundsdottir, M.H., Kazi, S., Reynolds, B., Visvalingam, S.M., Wilson, C., and Boyd, C.A. (2009). Amino acid sensing and mTOR regulation: inside or out? Biochem Soc. Trans. *37*, 248–252.
16. Hoeffer, C.A., and Klann, E. (2010). mTOR signaling: at the crossroads of plasticity, memory and disease. Trends Neurosci. *33*, 67–75.
17. Sabatini, D.M. (2006). mTOR and cancer: insights into a complex relationship. Nat. Rev. Cancer *6*, 729–734.
18. Sudarsanam, S., and Johnson, D.E. (2010). Functional consequences of mTOR inhibition. Curr Opin Drug Discov Devel. *13*, 31–40.
19. Jordan, M., Schallhorn, A., and Wurm, F.M. (1996). Transfecting mammalian cells: optimization of critical parameters affecting calcium-phosphate precipitate formation. Nucleic Acids Res. *24*, 596–601.
20. Shigekawa, K., and Dower, W.J. (1988). Electroporation of eukaryotes and prokaryotes: a general approach to the introduction of macromolecules into cells. Biotechniques *6*, 742–751.
21. Felgner, P.L., Gadek, T.R., Holm, M., Roman,R.,Chan,H.W.,Wenz,M.,Northrop,J.P., Ringold, G.M., and Danielsen, M. (1987). Lipofection: a highly efficient, lipid-mediated DNA-transfection procedure. Proc. Natl. Acad. Sci USA *84*, 7413–7417.
22. Hillen, W., and Berens, C. (1994). Mechanisms underlying expression of Tn10 encoded tetracycline resistance. Annu. Rev. Microbiol. *48*, 345–369.
23. Yao, F., Svensjo, T., Winkler, T., Lu, M., Eriksson, C., and Eriksson, E. (1998). Tetracycline repressor, tetR, rather than the tetR-mammalian cell transcription factor fusion derivatives, regulates inducible gene expression in mammalian cells. Hum. Gene Ther. *9*, 1939–1950.
24. Hendrie, P.C., and Russell, D.W. (2005). Gene targeting with viral vectors. Mol Ther. *12*, 9–17.
25. Zufferey, R., Nagy, D., Mandel, R.J., Naldini, L., and Trono, D. (1997). Multiply attenuated lentiviral vector achieves efficient gene delivery in vivo. Nat. Biotechnol. *15*, 871–875.
26. Cockrell, A.S. and Kafri, T. (2007). Gene delivery by lentivirus vectors. Mol Biotechnol. *36*, 184–204.
27. Case, S.S., Price, M.A., Jordan, C.T., Yu, X.J., Wang, L., Bauer, G., Haas, D.L., Xu, D., Stripecke, R., Naldini, L., Kohn, D.B., and Crooks, G.M. (1999). Stable transduction of quiescent CD34(+)CD38(−) human hematopoietic cells by HIV-1-based lentiviral vectors. Proc. Natl. Acad. Sci USA *96*, 2988–2993.
28. Salmon, P., and Trono, D. (2006). Production and titration of lentiviral vectors. Curr Protoc Neurosci. Chapter *4*:Unit 4.21.

Chapter 8

Detection of Cytoplasmic and Nuclear Functions of mTOR by Fractionation

Margit Rosner and Markus Hengstschläger

Abstract

Subcellular localization constitutes the environment in which proteins act. It tightly controls access to and availability of different types of molecular interacting partners and is therefore a major determinant of protein function and regulation. Originally thought to be a mere cytoplasmic kinase the mammalian target of rapamycin (mTOR) has recently been localized to various intracellular compartments including the nucleus and specific components of the endomembrane system such as lysosomes. The identification of essential binding partners and the structural and functional partitioning of mTOR into two distinct multiprotein complexes warrant the detailed investigation of the subcellular localization of mTOR as part of mTORC1 and mTORC2. Upon establishment of experimental conditions allowing cytoplasmic/nuclear fractionation at high purity and maximum mTOR complex recovery we have previously shown that the mTOR/raptor complex (mTORC1) is predominantly cytoplasmic whereas the mTOR/rictor complex (mTORC2) is abundant in both compartments. Moreover, the mTORC2 complex components rictor and sin1 are dephosphorylated and dynamically distributed between the cytoplasm and the nucleus upon long-term treatment with the mTOR-inhibitor rapamycin. These findings further demonstrate that the here presented and detailly described fractionation procedure is a valuable tool to study protein localization and cytoplasmic/nuclear protein shuttling in the context of expanding mTOR signalling.

Key words: mTORC1, mTORC2, CHAPS, Fractionation, Cytoplasm, Nucleus, Cytoplasmic-nuclear translocation

1. Introduction

The serine/threonine kinase mTOR (mammalian target of rapamycin) has attracted much attention as a major regulator of most fundamental biological processes such as cell growth and operates in two distinct multiprotein complexes termed mTORC1 containing mTOR, raptor, and mLST8, and mTORC2 containing mTOR,

Thomas Weichhart (ed.), *mTOR: Methods and Protocols*, Methods in Molecular Biology, vol. 821,
DOI 10.1007/978-1-61779-430-8_8, © Springer Science+Business Media, LLC 2012

rictor, sin1, and mLST8. Both complexes have distinct upstream regulators and downstream effectors providing evidence for a signalling network that is tightly controlled at multiple levels (1, 2). One such level is the spatial organization of proteins in specific intracellular compartments restricting access to and availability of molecular interactors. Whole cell analysis is a valuable and indispensible approach to characterize a protein's basic cellular function. However, localization of a protein to a specific intracellular compartment as well as subcellular redistribution assays to monitor protein shuttling can significantly enhance our understanding of known functions and even help to assign new functions.

Consistent with its early defined role in controlling translation mTOR has originally been considered as a mere cytoplasmic kinase operating in endomembrane-containing compartments of the cytosol (3, 4). Early reports on mTOR subcellular distribution provided evidence for mTOR nuclear localization but failed to demonstrate functional relevance of the observed regulations. Whereas mTOR was previously shown to be indirectly involved in transcriptional regulation by controlling the cytoplasmic/nuclear shuttling of transcription factors, recent reports have demonstrated direct interaction of mTOR with promoters of various genes providing functional relevance of mTOR nuclear localization (5–7). The number of reports suggesting distinct sites of mTOR intracellular localization and its dynamic shuttling between compartments upon various stimuli is constantly increasing. Reported sites of mTOR subcellular localization include various types of endomembrane-containing compartments such as mitochondria, endosomes, and the endoplasmic reticulum as well as specific sites in the nucleus such as PML nuclear bodies (Table 1 and refs. 8–22 therein).

Notably, the majority of *in vitro* studies on mTOR function including biochemical fractionation have been performed under lysis conditions known to interfere with mTOR complex formation. The integrity of different kinds of mTOR complexes is well known to be sensitive to disruption by various detergents including Triton X-100 and NP40, both commonly used for the preparation of whole cell lysates as well as subcellular fractions. Instead, the detergent CHAPS was proven to efficiently recover mTOR complexes (23). These findings clearly illustrate the need for particular fractionation conditions. For studies on localization of mTOR, mTOR complex components and mTOR complexes we therefore aimed at establishing appropriate fractionation conditions to maintain the integrity of mTOR complexes at concomitant minimal cross-contamination of cytoplasmic and nuclear fractions. The here described protocol is based on a previous publication and combines the usage of the detergent CHAPS for preparation of cytoplasmic fractions and the extraction of soluble, nuclear proteins by repeated

Table 1
Specific sites of mTOR subcellular localization

Compartment/organelle	Cells	Experimental approach	References
Mitochondria	3T3, HEK293T, Jurkat	Fractionation, immunofluorescence	(8, 9)
Endoplasmic reticulum	C2C12, HEK293T, HeLa, MRC-5, REF, Rh30	Fractionation, immunofluorescence	(10, 11)
Golgi apparatus	C2C12, HEK293T, HeLa, MRC-5, REF, Rh30	Fractionation, immunofluorescence	(10, 11)
Late endosmes/ lysosomes	HeLa, HEK293T, MEF	Immunofluorescence	(12, 13)
Plasma membrane/lipid rafts (detergent-insoluble)	MEF	Fractionation	(14)
Nucleus	C2C12, CV-1, HCT8, HCT29, HCT116, HEK293, HeLa, IMR-90, MCF7, MEF, MG63, Rh1, Rh4, Rh30	Fractionation, immuno-fluorescence, chromatin immunoprecipitation	(15–22)

freeze and thaw cycles in high salt extraction buffer, a condition tested to preserve mTOR complex integrity. Following this approach and using primary human fibroblasts, we have previously demonstrated that the mTOR complex components raptor, rictor, sin1, and mLST8 can be detected in both, the cytoplasm and the nucleus although at different abundance. Unexpectedly, although mTOR and raptor are expressed at high abundance in the nuclear fraction, the complex of mTOR and raptor is hardly detectable in the nucleus but is highly abundant in the cytoplasm. In contrast, the mTOR/rictor complex can be detected at reasonable levels in both compartments. Moreover and importantly, we could further show that mTORC2 complex components rictor and sin1 undergo phosphorylation and that rapamycin-mediated abrogation of phosphorylation is accompanied by specific cytoplasmic translocation of nuclear rictor and sin1 (24). These findings provide evidence for the suitability of this protocol to study cytoplasmic/nuclear protein shuttling. Further detailed investigations are warranted to detect cytoplasmic and nuclear functions of mTOR by identifying compartment-specific binding partners, upstream regulators, and downstream effectors.

2. Materials

2.1. IMR-90 Cell Culture

1. IMR-90 primary fibroblasts, CCL-186 (ATCC, American Type Culture Collection, Manassas, VA, USA).
2. Dulbecco's Modified Eagle's Medium (DMEM) at 4.5 g/l glucose, supplemented with 10% foetal bovine serum, 2 mM L-Glutamine, 60 mg/l penicillin, and 100 mg/l streptomycin sulphate.
3. Dulbecco's PBS (DPBS) (1×) without Ca and Mg.
4. Trypsin/EDTA solution at 0.4% (wt/vol) trypsin and 0.02% (vol/vol) Na-EDTA in 1× PBS. Store in aliquots at −20°C.
5. Standard cell culture disposables including sterile 100 mm cell culture dishes and sterile serological pipettes.

2.2. Cytoplasmic/Nuclear Fractionation

1. Trypsin/EDTA solution at 37°C, DMEM complete growth medium, and 1× PBS at 4°C.
2. Extraction buffer 1 (1×) for the extraction of cytoplasmic proteins: 20 mM Tris, pH 7.6, 0.1 mM EDTA, 2 mM $MgCl_2$ in water. Store at 4°C. Shortly before use, extraction buffer 1 is supplemented with protease inhibitors as described in the Subheading 3.2.2 (point 2) to yield ready-to-use extraction buffer.
3. Extraction buffer 2 (1×) for the extraction of nuclear proteins: 20 mM HEPES, pH 7.9, 400 mM NaCl, 25% glycerol, 1 mM EDTA, 0.5 mM NaF, 0.5 mM Na_3VO_4, 0.5 mM DTT in water. Store in aliquots at −20°C. Shortly before use, extraction buffer 2 is supplemented with protease inhibitors as described in the Subheading 3.2.4 (point 1) to yield ready-to-use extraction buffer.
4. PIM (*P*rotease *I*nhibitor *M*ix) stock solution (100×): 200 μg/ml aprotinin, 200 μg/ml leupeptin, 30 μg/ml benzamidine chloride, 1 mg/ml trypsin inhibitor in water. Store in aliquots at −20°C. PIM is used at a final 1× concentration.
5. PMSF (phenylmethylsulfonyl fluoride) stock solution (100×): 100 mM in isopropanol. Store in aliquots at −20°C. PMSF is used at a final 1× concentration.
6. CHAPS (3-[(3-cholamidopropyl)dimethylammonio]-1-propanesulfonate) stock solution (10×): 6% (wt/vol) in water. Store at room temperature.
7. Protein assay dye reagent.
8. Protein sample buffer (4×): 200 mM Tris, pH 6.8, 400 mM DTT, 8% (wt/vol) SDS, 0.4% (wt/vol) bromophenol blue, 40% glycerol in water. Store in aliquots at −20°C.
9. Standard molecular biology equipment including single-channel pipettes, centrifuge, heating block, plastic-, and glassware.

2.3. SDS-Polyacrylamide Gel Electrophoresis

1. 30% Acrylamide/bisacrylamide solution, 37.5:1.
2. Separating buffer (4×): 1.5 M Tris, pH 8.8, 0.4% (vol/vol) SDS. Store at 4°C.
3. Stacking buffer (4×): 0.5 M Tris, pH 6.8, 0.4% (vol/vol) SDS. Store at 4°C.
4. APS (ammoniumpersulfate) stock solution (100×): 10% (wt/vol) in water. Store at 4°C.
5. TEMED (N,N,N',N'-tetramethylethylenediamine).
6. Isopropanol.
7. Running buffer (10×): 250 mM Tris, 1.94 M glycine, 1% (wt/vol) SDS in water. Store at room temperature.
8. Prestained molecular weight marker (170–10 kDa).
9. Standard equipment for protein electrophoresis such as the Mini Protean 3 electrophoretic cell for vertical mini gel electrophoresis (Bio-Rad, Hercules, CA, USA).

2.4. Western Blotting

1. Transfer buffer (1×): 48 mM Tris, 386 mM glycine, 0.1% (wt/vol) SDS, 20% (vol/vol) methanol in water. Store at 4°C.
2. Nitrocellulose blotting membrane.
3. 3MM Chr chromatography paper.
4. Standard equipment for Western blotting such as the Mini Trans-Blot electrophoretic transfer cell (Bio-Rad).

2.5. Immunodetection

1. TBS (Tris-buffered saline) stock solution (10×): 1.5 M NaCl, 0.5 M Tris, pH 7.4 in water. Store at room temperature.
2. TBS-T (1×): 1× TBS supplemented with 0.1% (vol/vol) Triton X-100. Store at room temperature.
3. Blocking buffer (1×): 5% (wt/vol) non-fat, dry milk in 1× TBS-T. Can be stored for up to 3 days at 4°C but is recommended to be freshly prepared.
4. Primary antibody dilution buffer (1×): 5% (wt/vol) BSA (bovine serum albumin) in 1× TBS-T. Recommended to be freshly prepared.
5. Primary antibodies include anti-mTOR (for WB, *Western blot* and IP, *immunoprecipitation*), #2972 (Cell Signaling, Danvers, MA, USA), anti-raptor (IP), #A300-553A (Bethyl Laboratories, Montgomery, TX, USA), anti-raptor (WB), #A300-506A (Bethyl Laboratories), anti-rictor (IP), #A300-458A (Bethyl Laboratories), anti-rictor (WB), #A300-459A (Bethyl Laboratories) and anti-sin 1 (IP, WB), #A300-910A (Bethyl Laboratories). For a detailed list of antibodies recommended to detect subcellular marker proteins to verify purity of obtained fractions see Table 2.

Table 2
Recommended subcellular marker proteins

Designation	Full name	Organelle	Mol. Wt. (kDa)	Antibody selling company	Clone#	Cat#	WB dilution
Cytoplasm							
α-Tubulin		Cytoskeleton	60	Calbiochem	DM1A	CP06	1:2000
β-Tubulin		Cytoskeleton	55	Cell Signaling	9F3	2128	1:1000
HSP90	Heat shock protein 90 kDa	Cytosol	92	Cell Signaling		4874	1:1000
Calnexin		Endoplasmic reticulum	90	BD Transduction Laboratories	37	610523	1:500
COX IV	Cytochrome c oxidase IV	Mitochondria	17	Cell Signaling	3E11	4850	1:2000
SOD-2	Superoxid dismutase 2	Mitochondria	25	Stressgen		SOD-110	1:2000
Rab5	Ras-related protein Rab-5A	Early endosomes	25	Cell Signaling		2143	1:1000
ATG12	Autophagy-related protein 12	Autophagosomes	53, 16	Cell Signaling		2010	1:1000
Nucleus							
NUP98	Nucleoporin 98 kDa	Nuclear envelope	98	Cell Signaling	C39A3	2598	1:1000
Lamin A/C		Nuclear lamina	65, 74	BD Transduction Laboratories	14	612162	1:500
TopoIIβ	DNA topoisomerase 2-beta	Nucleoplasm	180	BD Transduction Laboratories	40	611492	1:500
Jun							
Histon H3		Nucleoplasm	39	BD Transduction Laboratories	3	610326	1:500
CENP-A							
		Nucleosome	17	Cell Signaling		9715	1:2000
Fibrillarin	Centromere protein A	Centromere	17	Cell Signaling		2186	1:1000
		Nucleolus	37	Cell Signaling	C13C3	2639	1:2000
Shuttling							
Importinα	*alt.* Karyopherinα	Cytoplasm/Nucleus	58	BD Transduction Laboratories	2	610485	1:1000

6. The secondary antibody dilution buffer is equivalent to the blocking buffer.
7. Secondary antibodies include HRP-linked anti-rabbit IgG from goat, #A120-101P (Bethyl Laboratories) and HRP-linked anti-mouse IgG from goat, #A90-116P (Bethyl Laboratories).
8. ECL chemiluminescent Western blotting substrate.
9. Films for autoradiography such as ECL Hyperfilm (high sensitivity) (GE healthcare, Piscataway, NJ, USA) or CL-XPosure film (low sensitivity) (Pierce, Thermo Fisher Scientific, Rockford, IL, USA).

3. Methods

Numerous cytoplasmic/nuclear fractionation protocols are available which differ in the way how disruption of the cellular organization, the so-called homogenization, is achieved and how the homogenate is further separated to yield different fractions. Different detergents in varying concentrations and diverse buffers are used making it sometimes very difficult to compare results obtained with different methods. Care should be taken when using commercially available kits as for some of these insufficient information on the underlying method, composition of buffers, or type of detergent can significantly impede interpretation of results.

The usage of detergents in the preparation of cytoplasmic fractions is very popular as they are highly efficient in solubilizing endomembrane-containing compartments thereby releasing membrane-bound or integral membrane proteins such as the ER (endoplasmic reticulum) protein calnexin. NP40 is widely used for preparation of whole cell lysates as well as for biochemical fractionation as it gives good results in terms of yield and purity. Although NP40-containing lysis buffer can be used in cytoplasmic/nuclear fractionation studies on mTOR localization (Fig. 1a), one has to be aware that the highest rate of mTOR complex recovery is obtained when using CHAPS-containing lysis buffer (Fig. 1b). NP40 and the related detergent Triton X-100 do not allow the investigation of mTOR complexes due to disruption of mTOR complex integrity, mainly influencing the mTOR/raptor interaction. Notably, extraction of proteins in high salt buffer and repeated freezing and thawing in liquid nitrogen, a condition completely free from the usage of any detergent (*w/o*), gives comparable results to CHAPS (Fig. 1c).

The protocol described here combines advantages of both extraction conditions enabling the investigation of the cytoplasmic/nuclear distribution of mTOR complexes as well as the analysis of mTOR subcellular localization under conditions of preserved mTOR complex integrity. CHAPS-containing hypotonic buffer is

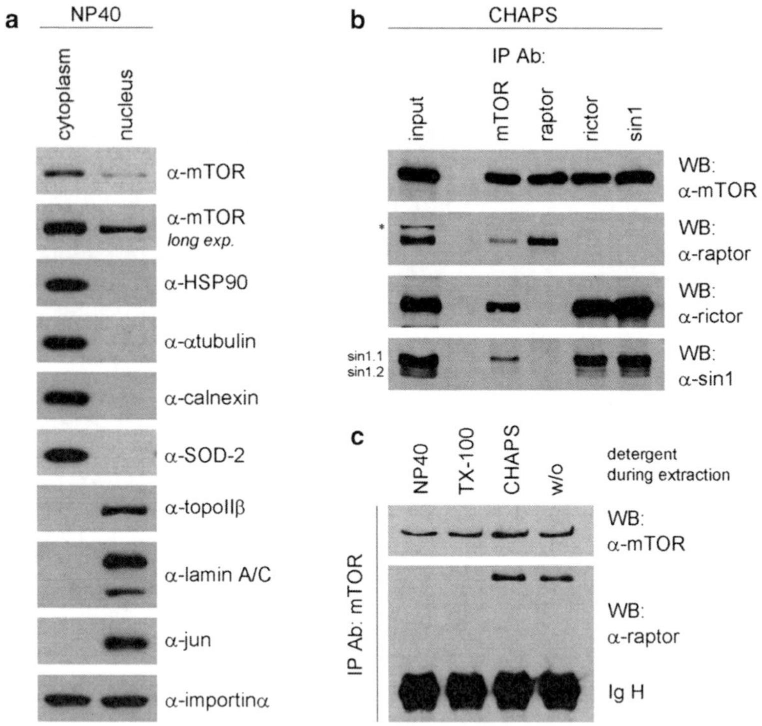

Fig. 1. Utility of various detergents for different types of mTOR studies. (**a**) IMR-90 fibroblasts were separated into cytoplasmic and nuclear fractions using NP40-containing extraction buffer and analyzed for the subcellular distribution of mTOR protein via immunoblotting. Purity of fractions was proven by co-analyzing a panel of specific marker proteins as indicated (*for detailed information on recommended subcellular marker proteins see Table* 2). (**b**) IMR-90 whole cell lysates prepared in CHAPS-containing lysis buffer were used for immunoprecipitation of endogenous mTOR, raptor, rictor, and sin1. mTOR assembly into mTORC1 and mTORC2 was confirmed via immunoblotting using indicated antibodies. The *asterisk* indicates a non-specific band. (Reproduced from ref. 24 by permission of Oxford University Press.) (**c**) IMR-90 total protein was either prepared by cell lysis with indicated detergents or extracted by repeated freeze and thaw cycles in high salt extraction buffer (*w/o*). mTOR immunoprecipitates were analyzed for co-immunoprecipitated raptor via immunoblotting.

used to yield cytoplasmic fractions, whereas high salt extraction buffer in combination with repeated freeze and thaw cycles is used to obtain fractions of soluble nuclear protein. An experimental flow chart of the procedure is presented in Fig. 2.

It is essential to stress that with traditional subcellular fractionation procedures like the one described here complete purification is almost impossible but rather favours the enrichment of particular fractions. Therefore, it is indispensible to characterize obtained fractions and to assess their quality and purity, both done by analyzing a defined set of subcellular marker proteins. This substantially helps to identify limitations and advantages of the procedure and provides the only appropriate means for correct interpretation of obtained results and comparison of different preparations. This

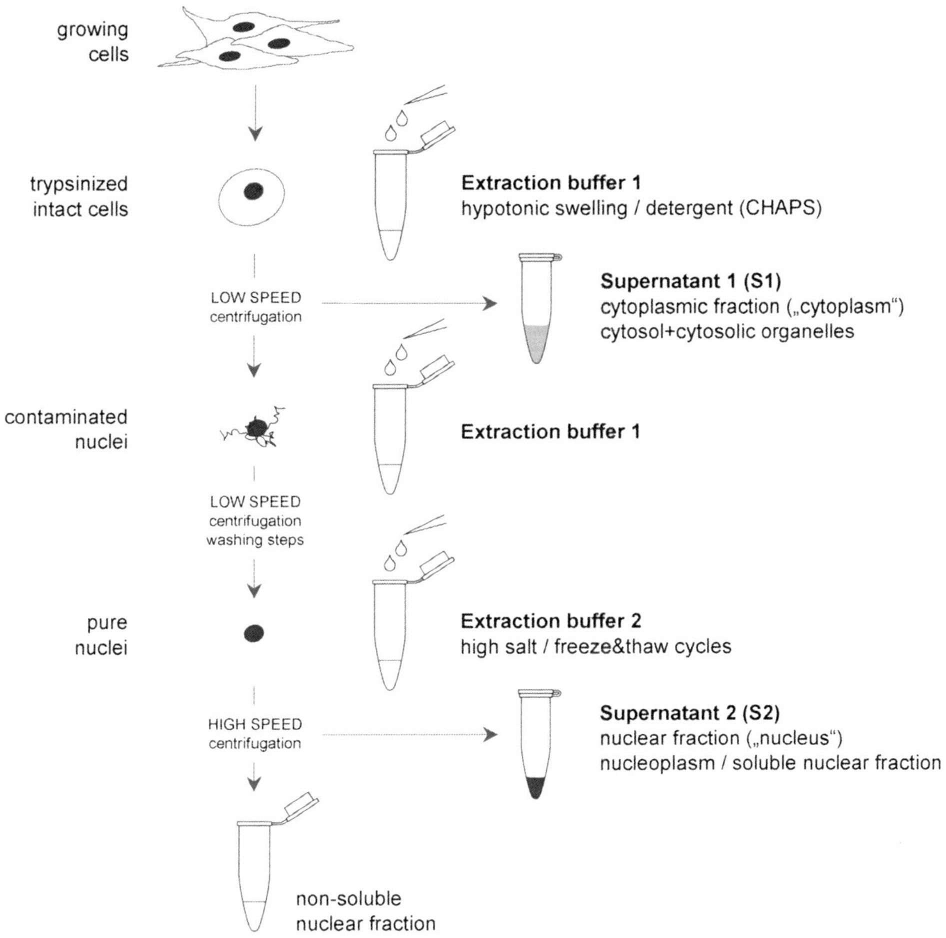

Fig. 2. Experimental flow chart of the fractionation procedure. Trypsinized, intact cells are lysed in hypotonic, CHAPS-containing lysis buffer (Extraction buffer 1) to yield a cytoplasmic fraction including the cytosol and cytosolic organelles (Supernatant 1 "cytoplasm"). Intact nuclei contaminated with unbroken or partly lysed cells are separated by low-speed centrifugation and washed in cytoplasmic extraction buffer to obtain a fraction of pure nuclei which is extracted by repeated freeze and thaw cycles in high salt extraction buffer (Extraction buffer 2). Supernatants are collected via high-speed centrifugation and constitute the nuclear fraction which consists of soluble nuclear proteins (Supernatant 2 "nucleus").

is a straightforward protocol which allows convenient and reproducible separation of fractions enriched in specific subcellular organelles. As proven by immunoblot analyses of a subset of different marker proteins the cytoplasmic fraction (*Supernatant 1, S1*, Fig. 2) contains cytoskeletal proteins, cytosolic proteins, and proteins representative for endomembrane-containing organelles such as the endoplasmic reticulum, endosomes, mitochondria, and autophagosomes. Extraction of purified nuclei (*Supernatant 2, S2*, Fig. 2) yields proteins localized to the nuclear envelope, the nuclear lamina, chromatin-associated proteins, proteins diffusely distributed in the nucleoplasm as well as proteins localized to the nucleolus (see data figures later in the protocol and Table 2). In addition, proteins known to shuttle between the cytoplasm and the nucleus such as the transport protein importinα (also known as karyopherinα) can be efficiently analyzed.

As far as purity is concerned, this protocol gives good results for both fractions as proven by the analysis of the same set of marker proteins described above. Neither increased cytoplasmic contamination of nuclear fractions (mainly due to the extraction of contaminated nuclei) nor excessive contamination of cytoplasmic extracts with nuclear proteins (most likely as a consequence of premature lysis of nuclei during cytoplasmic extraction or due to slightly damaged, leaky nuclei) can be observed (see data figures later in the protocol). In summary, this procedure yields fractions of high purity which can be used for subsequent immunoprecipitation of mTOR and mTOR complex components to study the subcellular distribution of mTOR complexes.

It is important to note that this protocol has been established using primary IMR-90 fibroblasts which are commercially available at the American type Culture Collection (ATCC). All data presented have been obtained from experiments with these cells. Unlike other cells preferentially used in mTOR studies such as HEK293 or 3T3, which are highly transfectable, rapidly dividing and yield high amounts of protein but are either transformed or immortalized, IMR-90 fibroblasts are non-transformed, non-immortalized human cells derived from foetal lung and harbour a normal diploid karyotype and finite lifetime. In the course of experiments, these cells turned out to be a valuable tool for studies on mTOR function as they are not prone to difficulties usually encountered with transformed, tumour-derived cell lines like chromosomal instability due to extended culture periods. Moreover, IMR-90 fibroblasts can be transfected with either plasmid DNA or siRNA at high efficiencies, can be grown with reasonable proliferation rates and are suitable to perform biochemical analyses known to be essential in mTOR studies such as co-immunoprecipitation of proteins at the endogenous level. In addition, they are suitable to investigate cell biological phenomenons including the regulation of cell cycle and cell size. In contrast to HeLa or HEK293 cells which cannot be arrested via serum deprivation IMR-90 cells can be efficiently synchronized in G0/G1 (an average rate of 90% G0/G1 synchronized cells can be obtained) and can be used to study serum-mediated re-entry into the cell cycle (25, 26).

3.1. IMR-90 Cell Culture

1. Since day-to-day consistency is crucial in maintaining IMR-90 fibroblasts, it is recommended to stick to standardized cultivation procedures descending from healthy cells at a limited number of 35–40 population doubling levels (PDLs). IMR-90 cells are grown in standard DMEM at 4.5 g/l glucose supplemented with 10% foetal bovine serum and favour culture conditions where cells are never too sparse. Frequent medium

changes without subculturing cells or too frequent subculturing should be avoided. Under favourable growth conditions IMR-90 fibroblasts have an average doubling time of 48–72 h and are subcultured at 100% confluency at an average splitting ratio of about 1:2.5 to 1:4. Cells typically reach confluency within 3–5 days, can be grown for up to 5–7 days without changing growth medium and can be kept confluent for more than 2 days (see Notes 1 and 2).

2. The day before fractionation confluent cells are harvested, seeded at 5×10^5 to 8×10^5 cells per 100 mm dish and are typically incubated overnight or for up to 24 h resulting in about 40–50% confluent cultures (see Notes 3–5).

3.2. Cytoplasmic/ Nuclear Fractionation

3.2.1. Step 1: Cell Harvesting

1. Cells are washed once with cold (4°C) 1× PBS and incubated with 0.5 ml trypsin–EDTA solution (prewarmed to 37°C) per dish at room temperature. Within 2–4 min of incubation cells are completely detached from the culture surface. Detaching cells can be easily identified without microscopic screening, as they come off as white clusters when culture dishes are tilted.
2. Immediately after cells have detached, trypsin is inactivated by the addition of 2 ml prechilled serum-containing DMEM growth medium. Cells are gently resuspended four to five times and are collected in a 15 ml centrifuge tube. Tubes are centrifuged at 4°C and $200 \times g$ for 3 min. If handling large-scale samples such as necessary when harvesting cells for immunoprecipitation of endogenous mTOR complexes it is recommended not to trypsinize all dishes at once but to handle stacks of around five dishes one after another (see Note 6).
3. The supernatant is discarded and the cell pellet is carefully resuspended in 1 ml cold 1× PBS. This represents a washing step to eliminate remaining traces of trypsin as well as BSA which originates from the serum-containing growth medium used to inactivate the trypsin (see Note 7). Cells are centrifuged at 4°C and $200 \times g$ for another 3 min.
4. The supernatant is discarded, the cells are resuspended in 1 ml cold 1× PBS, and transferred to a 1.5 ml centrifuge tube. During this last washing step the cell suspension usually contains 80–90% intact cells that can be verified under the phase contrast microscope. Representative pictures of growing cells and trypsinized intact cells are presented in Fig. 3a, b. Cell harvesting is completed by a final centrifugation step at 4°C and $200 \times g$ for 10 min.

3.2.2. Step 2: Extraction of Cytoplasmic Proteins

1. After centrifugation, the supernatant is thoroughly discarded. Remaining PBS should be removed by using a low volume single-channel pipette. The remaining cell pellet consists of intact cells purified from trypsin and growth medium.

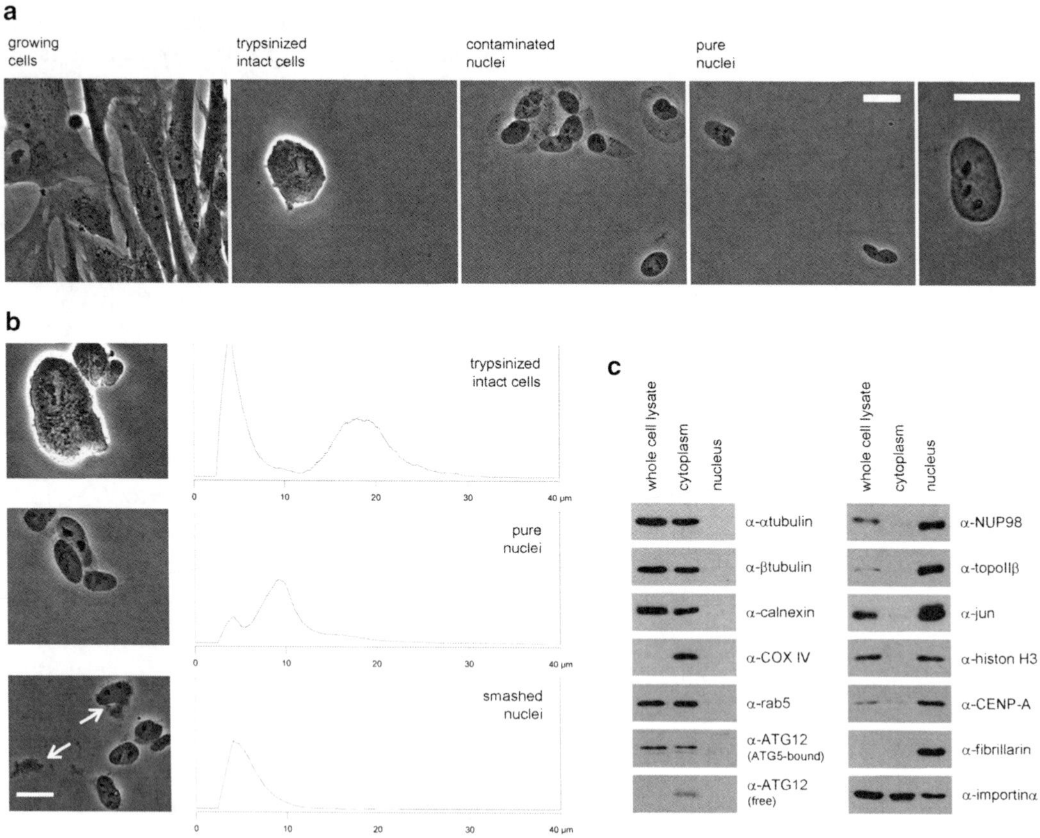

Fig. 3. Monitoring quality and purity of obtained fractions. (**a**) Stepwise separation of IMR-90 fibroblasts into cytoplasmic and nuclear fractions was monitored via phase contrast microscopy. Representative pictures of the keysteps of the procedure are shown including pictures of growing cells, trypsinized intact cells, contaminated nuclei, and pure nuclei. The outermost picture shows a single pure nucleus at higher magnification. Scale bars represent 10 μm. (**b**) Fractionated IMR-90 cells were analyzed for morphology as described in (**a**) and additionally examined for qualitative and quantitative enrichment of nuclear fractions via investigation of cell size distribution on a CASY cell counter and analyzer. Representative analyses of trypsinized intact cells (*upper panel*), pure nuclei (*middle panel*), and smashed nuclei derived from an experiment performed under suboptimal conditions (*lower panel*) are presented. *Arrows* indicate heavily damaged nuclei. Scale bar represents 10 μm. (**c**) IMR-90 cytoplasmic and nuclear fractions derived from a single fractionation experiment were analyzed for purity via immunoblotting using antibodies specific to different subcellular marker proteins. The *left panel* of blots shows analyses of cytoplasmic marker proteins, whereas the *right panel* represents analyses of nuclear marker proteins and of a shuttling protein. *Note that the apparent absence of signals for COX IV, free ATG12 and fibrillarin in whole cell lysates is due to different detection limits resulting from different lysis conditions used for whole cell lysates versus cytoplasmic and nuclear fractions. (for detailed information on recommended subcellular marker proteins see Table* 2).

2. Ready-to-use cytoplasmic extraction buffer (extraction buffer 1, Fig. 2) is supplemented with protease inhibitors by adding 10 μl 100× PIM and 10 μl 100× PMSF to 980 μl 1× extraction buffer 1. The cell pellet is resuspended in five packed cell volumes extraction buffer 1 and incubated for 2 min at room temperature and another 10 min on ice. This represents a preparative step for cell lysis as cell swelling under hypotonic

conditions is induced. Cells at 40–50% confluency harvested from one 100 mm dish usually give a final pellet volume of about 20–30 μl, five packed cell volumes therefore correspond to 100–150 μl extraction buffer 1.

3. Disruption of the cell membrane is achieved by the addition of CHAPS at a final concentration of 0.6%. For a sample volume of 100 μl add 10 μl of a 6% CHAPS stock solution (see Note 8). Gently mix by snapping the tube with your fingers. Broken cells are then homogenized by passing them through a 20-g needle for three times. Alternatively, a 200 μl tip can be used. The introduction of air bubbles during this step should be avoided as this leads to loss of material.
4. Cytoplasmic extracts are separated by low speed centrifugation at 4°C and 500 × *g* for 5 min. After centrifugation the majority of supernatant (about 80%) constituting the cytoplasmic fraction is carefully removed and transferred to a new microcentrifuge tube (Fig. 2). The remaining 20% are discarded. This is done to avoid carryover of intact, slightly damaged or leaky nuclei to the cytoplasmic fraction (see Note 9). The remaining pellet contains nuclei with low purity contaminated with unbroken or partly lysed cells and non-soluble parts of the plasma membrane.

3.2.3. Step 3: Purification of Cell Nuclei

1. For further purification nuclei are washed in extraction buffer 1 supplemented with 0.6% CHAPS (Fig. 2). This is done by adding 10–15 packed cell volumes of detergent-containing extraction buffer 1 to the nuclear pellet which has an average pellet volume of about 15 μl. Nuclei are gently resuspended by pipetting up and down for two to three times using a 200- or 1,000 μl tip. A small fraction of these contaminated nuclei can be monitored under the phase contrast microscope. For a representative brightfield picture of this early nuclear fraction compare Fig. 3a. Nuclei are then centrifuged at 4°C and low speed (500 × *g*) for 5 min.
2. This washing step is repeated twice and results in a pure fraction of nuclei suitable for extraction (see Note 10). The quality of the nuclear preparation can be roughly estimated either microscopically (Fig. 3a, b) or by means of cell size analyses using a CASY cell counter and analyzer or similar equipment. Compared to trypsinized, intact cells pure nuclei come up as an enriched fraction of significantly smaller size (Fig. 3b, *upper and middle panels*). An example of heavily damaged nuclei is included in Fig. 3b, *lower panel* (also compare Note 8). Substantially damaged or smashed nuclei can be identified as a small fraction merging with the cellular debris fraction of less than 4 μm diameter on average. So purified nuclei can likewise be analyzed on a flow cytometer (data not shown).

3.2.4. Step 4: Extraction of Nuclear Proteins

1. Ready-to-use nuclear extraction buffer (extraction buffer 2, Fig. 2) is supplemented with protease inhibitors by adding 5 µl 100× PIM and 5 µl 100× PMSF to 490 µl 1× extraction buffer 2. The nuclear pellet is resuspended in 1–2 packed cell volumes extraction buffer 2 and heavily vortexed (see Note 11).
2. Nuclei resuspended in extraction buffer 2 are quickly frozen by placing the tube in a bath of liquid nitrogen for a few seconds up to 1 min. The frozen tube is quickly thawed by clasping it with hands.
3. This freeze and thaw cycle is repeated once. Thereafter, samples are incubated on ice for a minimum of 20 min. High-speed centrifugation at 4°C and 20,000 × *g* for 20 min is performed to separate soluble nuclear proteins from the non-soluble nuclear fraction. The supernatant constituting the obtained nuclear fraction is removed and transferred to a new microcentrifuge tube (see Note 12) (Fig. 2).
4. Obtained fractions are assayed for protein content using the Protein assay dye reagent from Bio-Rad according to the manufacturer's instructions. Equal amounts of cytoplasmic and nuclear fractions can be analyzed in different downstream applications such as immunoblotting and immunoprecipitation. Conditions to immunoprecipitate endogenous mTOR complexes have been adopted from previous publications and are applied to cytoplasmic and nuclear fractions without further modification (23, 24).

3.3. SDS-PAGE and Western Blotting

Denaturing sodium dodecylsulfate-polyacrylamide gel electrophoresis (SDS-PAGE) and Western Blotting are performed according to standard techniques using the materials outlined in the Materials section 2 (24–26).

1. Briefly, equal amounts (10–30 µg) of fractions are mixed with 4× protein sample buffer to yield 1× concentration and are boiled at 95°C for 5 min.
2. Depending on the molecular weight of the analyzed protein(s) denatured samples are then separated on 7% (mTOR, raptor, rictor and sin1, subcellular marker proteins ≥50 kDa), 11% or 15% (subcellular marker proteins ≤50 kDa) acrylamide gels in 1× running buffer at 100 V for 90 min. For the preparation of 6 ml of a 7% separating gel (sufficient for one mini gel) 1.4 ml acrylamide, 1.5 ml 4× separating buffer, and 3.1 ml water are supplemented with 60 µl 100× APS and 5 µl TEMED. Isopropanol is used to gently overlay the separating gel and is

discarded after the gel has polymerized, typically after 10–15 min. For the preparation of 1.5 ml of a 5% stacking gel (sufficient for one mini gel) 250 μl acrylamide, 375 μl 4× stacking buffer, and 875 μl water are supplemented with 15 μl 100× APS and 1.25 μl TEMED.

3. Using transfer cassettes, 3MM whatman filter paper and foam sheets separating gels are transferred to nitrocellulose membranes in 1× transfer buffer at 350 mA and 4°C for 60 min.

3.4. Immunodetection of mTOR and Organelle-Specific Marker Proteins

Immunodetection of mTOR, mTOR complex components, and subcellular marker proteins is performed according to standard techniques using the materials outlined in the Materials section 2 (24–26).

1. Briefly, nitrocellulose membranes are soaked in water to remove traces of transfer buffer and are incubated in blocking solution for 1 h at room temperature.
2. Membranes are rinsed in 1× TBST-T to remove remaining blocking solution and are incubated with primary antibodies diluted in primary antibody dilution buffer at 4°C for 12–16 h (overnight). All antibodies mentioned in the Materials section 2 were diluted according to the manufacturer's instructions. For a detailed list of subcellular marker proteins recommended to confirm identity and verify purity of obtained fractions see Table 2.
3. After washing the membranes three times in 1× TBS-T for 10 min, blots are incubated with HRP-linked secondary antibodies diluted in blocking buffer (1:10,000) for 1 h at room temperature.
4. Washing steps as described above are repeated and signals are detected via chemiluminescence and autoradiography using ECL chemiluminescent Western blotting substrate and films of different sensitivity. A representative immunoblot analysis of subcellular marker proteins is presented in Fig. 3c, demonstrating enrichment of various cytoplasmic marker proteins in the cytoplasmic fraction and the accumulation of nuclear marker proteins in the nuclear fraction without significant cross-contamination between fractions (see Note 13). Results obtained with the presented fractionation protocol are shown in Fig. 4 and include the analysis of the cytoplasmic/nuclear distribution of mTORC1 and mTORC2 (Fig. 4a) and the cytoplasmic translocation of nuclear rictor upon long-term treatment with the mTOR-inhibitor rapamycin (Fig. 4b) (24).

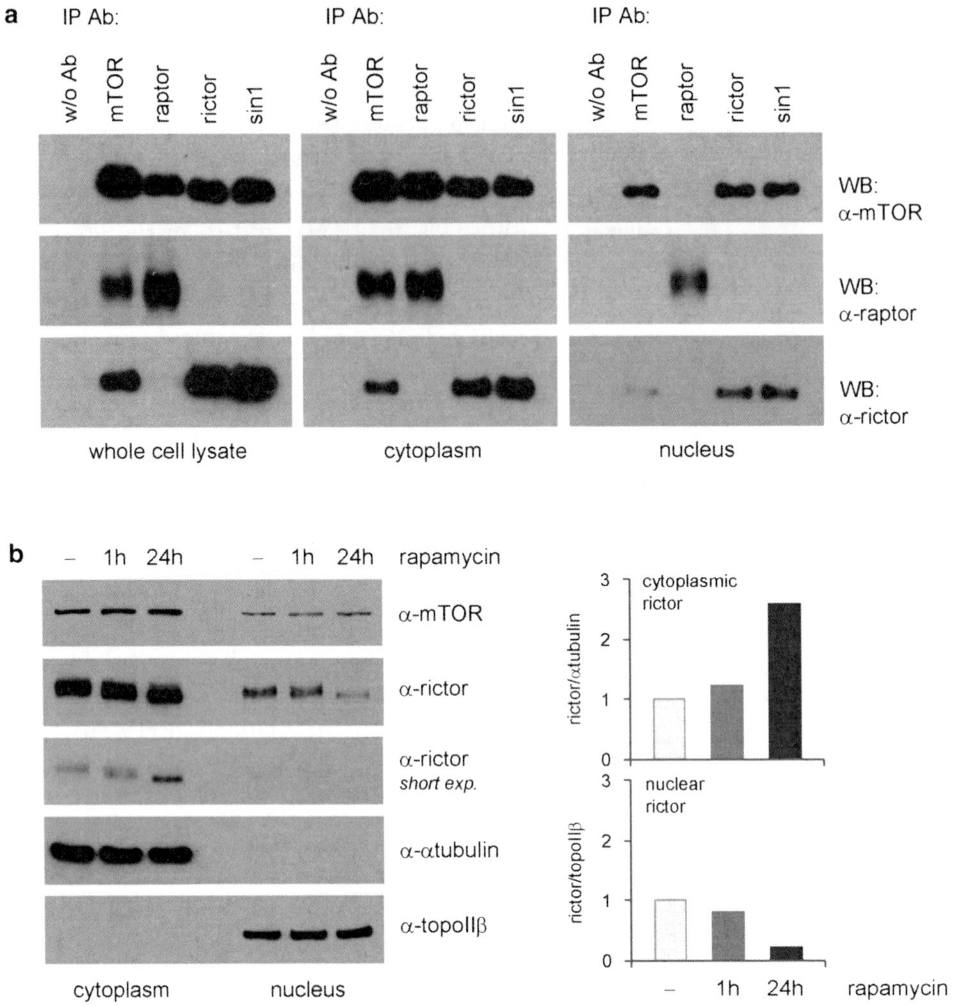

Fig. 4. Subcellular distribution of mTOR complexes and rapamycin-mediated cytoplasmic translocation of nuclear rictor. (**a**) IMR-90 whole cell lysates and subcellular fractions were analyzed for cytoplasmic/nuclear distribution of mTOR complexes via immunoprecipitation and immunoblotting using indicated antibodies. (**b**) Logarithmically growing IMR-90 fibroblasts were treated with 100 nM rapamycin for indicated time points and separated into cytoplasmic and nuclear fractions. Protein levels of mTOR and rictor were examined via immunoblotting. α-Tubulin and topoIIβ were co-analyzed to prove purity of fractions (*left panel*). In addition, cytoplasmic and nuclear levels of rictor were densitometrically analyzed (*right panel*) (reproduced from ref. 24 with permission from Oxford University Press).

4. Notes

1. IMR-90 fibroblasts are reported to be capable of attaining 58 PDLs before the onset of senescence. Under favourable growth conditions one passage equates to 2.5 PDLs. Cells are obtained from ATCC at passage number 10 (PDL 25). Accordingly, they can theoretically be subcultured for about 13 additional passages before becoming senescent. Since cellular senescence is known to be associated with intracellular

redistribution of proteins, cells are recommended to be grown for no more than eight additional passages (≤45 total PDLs) to avoid putative effects. Cells grown for more than ten additional passages (≥50 total PDLs, ≥ passage number 20) display reduced proliferation rates and unhealthy morphology. In addition, IMR-90 fibroblasts are recommended to be regularly analyzed by standard karyotyping to confirm a normal diploid karyotype.

2. This protocol has been proven to work in several cell lines including those preferentially used for mTOR studies such as HEK293, NIH3T3 and the recently established clonal lines of human amniotic fluid stem cells (hAFSCs) (26). In terms of purity, it seems as it gives slightly better results for fibroblast lines. We assume that this is partly due to the fact that fibroblasts usually grow in a single cell manner and can be easily detached from the culture surface yielding a single cell suspension. Single cells are then efficiently lysed as they are evenly dispersed in the lysis buffer without a significant number of cells remaining unbroken. Epithelial HEK293 cells which tend to grow in clusters can hardly be dispersed into a single cell suspension and stick together in clusters of about 3–8 cells even when harshly trypsinized. Limited access of lysis buffer to cells in the centre of clusters may lead to insufficient cell lysis with a high amount of cells remaining unbroken which causes contamination of nuclear pellets and carryover of cytoplasmic proteins into the nuclear fraction. For HEK293 cells, it is recommended to scale up the amount of cytoplasmic extraction buffer by multiplying the volume of extraction buffer 1 by 1.5 or 2.0.
3. One 100 mm dish of 40–50% confluent cells usually yields enough cytoplasmic and nuclear protein to run multiple polyacrylamide gels for immunoblot analyses. The limiting factor is the amount of nuclear protein which is typically five to six times less than the one of the cytoplasmic fraction. One 100 mm dish of 40–50% confluent cells typically yields 250 μg cytoplasmic protein extract and 50 μg nuclear protein extract. Dependent on the abundance of the analyzed protein in cytoplasmic and nuclear fractions polyacrylamide gels are run at 10–25 μg protein on average. If the obtained fractions are used for subsequent immunoprecipitations of endogenous proteins at 350–500 μg per sample the number of plates is recommended to be scaled by ten, means ten dishes for one immunoprecipiation.
4. Subculturing and replating cells before fractionation favours the outcome of the procedure as cells are evenly distributed in a single cell manner which facilitates trypsinization to yield a single cell suspension which then can be efficiently lysed.
5. The average cell cycle distribution of freshly plated cells is about 50% G0/G1 cells, 40% cells in S phase and 10% G2/M

cells representing a cell population under highly proliferative conditions. The cytoplasmic/nuclear ratio of mTOR heavily depends on growth conditions. In proliferating cells, the amount of nuclear mTOR in relation to cytoplasmic mTOR is higher compared to resting cells (data not shown). Thus, for fractionation studies on the subcellular localization of mTOR it is essential that cell seeding and harvesting is performed under constant conditions with minimal variation between single experiments to be able to compare results from different experiments.

6. In contrast to many researchers who lyse cells directly on the culture dish we use trypsin for cell harvest which has the advantage that the protein concentration can be thoroughly adjusted by the addition of a defined volume of lysis buffer. In addition, cells can be efficiently detached from the culture dish and are dispersed in a single cell suspension. When properly employed this facilitates cell lysis. Prolonged incubation with trypsin or overtypsinization with high volumes of trypsin results in cell clumping, inefficient cell lysis, or premature cell breakage.
7. Inefficient removal of BSA will influence immunoblot results as it will be non-specifically detected as prominent band of around 60–70 kDa resulting in high background staining.
8. The CHAPS concentration demonstrated to preserve the mTOR/raptor complex in whole cell analyses at highest efficiency is 0.3% (23). This concentration was tested in cytoplasmic/nuclear fractionation but turned out to be less effective in that significant cross-contamination of nuclear extracts with cytoplasmic proteins was observed, most likely due to incomplete disruption of cell membranes. We also tested a final concentration of 1.0% CHAPS in our experiments but this did not only cause the expected disruption of the cell membrane but rather resulted in premature breakage of nuclei and nuclear leakage. After the final washing step the remaining nuclei appeared heavily damaged and partly smashed. A representative brightfield picture of such an unfavourable condition is shown in Fig. 3b, *lower panel*. A final CHAPS concentration of 0.6% was proven to yield both, efficient disruption of cell membranes at preserved integrity of cell nuclei (24).
9. High speed centrifugation at $100{,}000 \times g$ or more is usually performed to further separate the cytoplasmic fraction into cytosolic and endomembrane-containing parts. Since this step is omitted from this protocol repeated freezing and thawing of cytoplasmic extracts upon storage at –80°C should be avoided as this tends to cause precipitation of specific protein fractions within the cytoplasmic fraction. A protein's cytoplasmic/nuclear expression ratio analyzed by immunoblotting can therefore significantly differ when freshly prepared or repeatedly frozen and thawed samples are examined.

10. During washing steps the nuclear pellet is usually getting smaller (about 10–20% of the nuclear pellet volume is lost), most likely due to the fact that unbroken cells that have not been efficiently lysed during the cytoplasmic extraction step get broken. Increased loss of nuclear pellet volume during washing steps is indicative for premature nuclear breakage. This is usually accompanied by nuclei getting sticky due to the release of genomic DNA. This can happen if the amount of detergent in extraction buffer 1 is too high or if too many washing steps are performed. The sensitivity of nuclear integrity towards the amount of detergent can significantly vary between cell lines and has to be thoroughly tested if using this protocol for cells other than described here.
11. It is hardly avoidable that some nuclei get prematurely broken during purification what makes the pellet of cell nuclei slightly viscous. Enforced pipetting is necessary to resuspend nuclei in extraction buffer 2 to get a homogenous solution.
12. Unlike the cytoplasmic fraction, the nuclear fraction obtained through high-speed centrifugation can be repeatedly frozen and thawed upon storage.
13. When assessing purity of preparations via immunoblotting of specific subcellular marker proteins prolonged exposure times of autoradiography films are recommended to visualize potential cross-contamination. This is especially true if the studied protein of interest is low abundant in a particular fraction, e.g. it is possible that without correct adjustment of exposure times a putative nuclear localization turns out to be mere cytoplasmic contamination of nuclear fractions.

Acknowledgments

The authors want to thank all laboratory members for their helpful discussion and assistance in developing this protocol. This work was supported by the Herzfelder'sche Familienstiftung and the Österreichische Nationalbank.

References

1. Laplante M, Sabatini DM (2009) mTOR signaling at a glance. J Cell Sci 122:3589–3594.
2. Foster KG, Fingar DC (2010) Mammalian target of rapamycin (mTOR): conducting the cellular signaling symphony. J Biol Chem 19:14071–14077.
3. Averous J, Proud CG (2006) When translation meets transformation: the mTOR story. Oncogene 25:6423–6435.
4. Proud CG (2007) Signalling to translation: how signal transduction pathways control the protein synthetic machinery. Biochem J 403: 217–234.
5. Yang TT, Yu RY, Agadir A et al (2008) Integration of protein kinases mTOR and extracellular signal-regulated kinase 5 in regulating nucleocytoplasmic localiztaion of NFATc4. Mol Cell Biol 28:3489–3501.

6. Cunningham JT, Rodgers JT, Arlow DH et al (2007) mTOR controls mitochondrial oxidative function through a YY1-PGC-1alpha transcriptional complex. Nature 450:736–740.
7. Tsang CK, Liu H, Zheng XF (2010) mTOR binds to the promoters of RNA polymerase I- and III-transcribed genes. Cell Cycle 9: 953–957.
8. Desai BN, Myers BR, Schreiber SL (2002) FKBP12-rapamycin-associated protein associates with mitochondria and senses osmotic stress via mitochondrial dysfunction. Proc Natl Acad Sci USA 99:4319–4324.
9. Schieke SM, Phillips D, McCoy JP Jr et al (2006) The mammalian target of rapamycin (mTOR) pathway regulates mitochondrial oxygen consumption and oxidative capacity. J Biol Chem 281:27643–27652.
10. Drenan RM, Liu X, Bertram PG et al (2004) FKBP12-rapamycin-associated protein or mammalian target of rapamycin (FRAP/mTOR) localization in the endoplasmic reticulum and Golgi apparatus. J Biol Chem 279:772–778.
11. Liu X, Zheng XF (2007) Endoplasmic reticulum and Golgi localization sequences for mammalian target of rapamycin. Mol Biol Cell 18:1073–1082.
12. Flinn RJ, Yan Y, Goswami S et al (2010) The late endosome is essential for mTORC1 signaling. Mol Biol Cell 21:833–841.
13. Sancak Y, Bar-Peled L, Zoncu R et al (2010) Ragulator-Rag complex targets mTORC1 to the lysosomal surface and is necessary for its activation by amino acids. Cell 14:290–303.
14. Partovian C, Ju R, Zhuang ZW et al (2008) Syndecan-4 regulates subcellular localization of mTOR Complex2 and Akt activation in a PKCalpha-dependent manner in endothelial cells. Mol Cell 32:140–149.
15. Kim JE, Chen J (2000) Cytoplasmic-nuclear shuttling of FKBP12-rapamycin-associated protein is involved in rapamycin-sensitive signaling and translation initiation. Proc Natl Acad Sci USA 97:14340–14345.
16. Park IH, Bachmann R, Shirazi H et al (2002) Regulation of ribosomal S6 kinase by mammalian target of rapamycin. J Biol Chem 277: 31423–31429.
17. Zhang X, Shu L, Hosoi H et al (2002) Predominant nuclear localization of mammalian target of rapamycin in normal and malignant cells in culture. J Biol Chem 277:28127–28134.
18. Bernardi R, Guernah I, Jin D et al (2006) PML inhibits HIF-1 alpha translation and neoangiogenesis through repression of mTOR. Nature 442:779–785.
19. Panasyuk G, Nemazanyy I, Zhyvoloup A et al (2009) mTORbeta splicing isoform promotes cell proliferation and tumorigenesis. J Biol Chem 284:30807–30814.
20. Goh ET, Pardo OE, Michael N et al (2010) Involvement of heterogeneous ribonucleoprotein F in the regulation of cell proliferation via the mammalian target of rapamycin/S6 kinase 2 pathway. J Biol Chem 285:17065–17076.
21. Kantidakis T, Ramsbottom BA, Birch JL et al (2010) mTOR associates with TFIIIC, is found at tRNA and 5S rRNA genes, and targets their repressor Maf-1. Proc Natl Acad Sci USA 107:11823–11828.
22. Shor B, Wu J, Shakey Q et al (2010) Requirement of the mTOR kinase for the regulation of Mafl phosphorylation and control of RNA polymerase III-dependent transcription in cancer cells. J Biol Chem 285: 15380–15392.
23. Kim DH, Sarbassov DD, Ali SM et al (2002) mTOR interacts with raptor to form a nutrient-sensitive complex that signals to the cell growth machinery. Cell 110:163–175.
24. Rosner M, Hengstschläger M (2008) Cytoplasmic and nuclear distribution of the protein complexes mTORC1 and mTORC2: rapamycin triggers dephosphorylation and delocalization of the mTORC2 components rictor and sin1. Hum Mol Genet 17: 2934–2948.
25. Rosner M, Fuchs C, Siegel N et al (2009) Functional interaction of mammalian target of rapamycin complexes in regulating mammalian cell size and cell cycle. Hum Mol Genet 18:3298–3310.
26. Rosner M, Siegel N, Fuchs C et al (2010) Efficient siRNA-mediated prolonged gene silencing in human amniotic fluid stem cells. Nat Protoc 5:1081–1095.

Chapter 9

Evaluation of Rapamycin-Induced Cell Death

Lorenzo Galluzzi, Eugenia Morselli, Oliver Kepp, Ilio Vitale, Aména Ben Younes, Maria Chiara Maiuri, and Guido Kroemer

Abstract

Mammalian target of rapamycin (mTOR) is an evolutionarily conserved kinase that integrates signals from nutrients and growth factors for the coordinate regulation of many cellular processes, including proliferation and cell death. Constitutive mTOR signaling characterizes multiple human malignancies, and pharmacological inhibitors of mTOR such as the immunosuppressant rapamycin and some of its nonimmunosuppressive derivatives not only have been ascribed with promising anticancer properties *in vitro* and *in vivo* but are also being extensively evaluated in clinical trials. mTOR inhibition rapidly leads to the activation of autophagy, which most often exerts prosurvival effects, although in some cases it accompanies cell death. Thus, depending on the specific experimental setting (cell type, concentration, stimulation time, and presence of concurrent stimuli), rapamycin can activate/favor a wide spectrum of cellular responses/phenotypes, ranging from adaptation to stress and survival to cell death. The (at least partial) overlap among the biochemical and morphological responses triggered by rapamycin considerably complicates the study of cell death-associated variables. Moreover, rapamycin presumably triggers acute cell death mainly via off-target mechanisms. Here, we describe a set of assays that can be employed for the routine quantification of rapamycin-induced cell death *in vitro*, as well as a set of guidelines that should be applied for their correct interpretation.

Key words: Apoptosis, Autophagosomes, Clonogenic assays, Immunoblotting, LC3, Mitochondrial membrane permeabilization, Phosphatidylserine exposure

Abbreviations

AnnV	Annexin V
ATCC	American type culture collection
BSA	Bovine serum albumin
CRTC1	CREB-regulated transcription coactivator 1
CV	Crystal violet
DAPI	4′,6-Diamidino-2-phenylindole dihydrochloride
dH_2O	Deionized water
$DiOC_6(3)$	3,3′-Dihexiloxalocarbocyanine iodide
DTT	Dithiothreitol
EDTA	Ethylenediaminetetraacetic acid

Thomas Weichhart (ed.), *mTOR: Methods and Protocols*, Methods in Molecular Biology, vol. 821,
DOI 10.1007/978-1-61779-430-8_9, © Springer Science+Business Media, LLC 2012

EGFR	Epidermal growth factor receptor
FITC	Fluorescein isothiocyanate
FSC	Forward scatter
HBSS	Hank's balanced salt solution
IARC	International Agency for Research on Cancer
IC_{50}	Half maximal inhibitory concentration
IM	Inner mitochondrial membrane
IMS	Mitochondrial intermembrane space
$\Delta\psi_m$	Mitochondrial transmembrane potential
MMP	Mitochondrial membrane permeabilization
mTOR	Mammalian target of rapamycin
NTP	National Toxicology Program
OM	Outer mitochondrial membrane
OSHA	Occupational Safety and Health Administration
PBS	Phosphate-buffered saline
PFA	Paraformaldehyde
PI	Propidium iodide
PI3K	Phosphoinositide 3-kinase
PLSCR1	Phospholipid scramblase 1
PMSF	Phenylmethylsulfonyl fluoride
PS	Phosphatidylserine
ROS	Reactive oxygen species
RT	Room temperature
SDS	Sodium dodecylsulfate
SF	Surviving fraction
siRNA	Small interfering RNA
SSC	Side scatter
TBS	Tris-buffered saline
TMRM	Tetramethylrhodamine methyl ester
TORC	TOR complex

1. Introduction

The coordinate regulation of cell growth (increase in cell volume) and duplication (increase in cell number) is assured by multiple molecular mechanisms, most of which are controlled by the enzymatic activity of the kinase mammalian target of rapamycin (mTOR) (1). Numerous signals emanating from the extracellular milieu (e.g., glucose, amino acid, and growth factor availability) converge on the activation of mTOR, which in turn transduces a wide array of positive signals for cell growth and proliferation (1). Constitutive signaling through mTOR, most often due to activating mutations in growth factor receptors (e.g., epidermal growth factor receptor, EGFR) or in intracellular signal transducers that operate upstream of mTOR (e.g., phosphoinositide 3-kinase, PI3K; AKT), characterizes numerous human malignancies (2). Consistent with these observations, pharmacological inhibition of mTOR with the macrolide rapamycin or some analogues has been associated with antitumor effects, *in vitro* and *in vivo* (3). Moreover, rapamycin (which has been used for decades as an immunosuppressant to avoid graft rejection) as well as some of its nonimmunosuppressive derivatives (e.g., CCI-779 and RAD001) are being tested in phase II/III clinical trials, as single agents or within combination therapies, for the treatment of various hematological and solid tumors (4–6).

In all eukaryotes, at least two distinct TOR-containing multiprotein complexes exist. These complexes are referred to as TORC1 (i.e., TOR complex 1, also known as CRTC1, i.e., CREB-regulated transcription coactivator 1) and TORC2 (CRTC2), respectively (7), and exert partially overlapping functions (8). As compared to TORC1, TORC2 exhibits reduced sensitivity to acute pharmacological inhibition with rapamycin (7, 8), which can be restored (at least partially, in some cell types) by prolonged exposure or by the use of agents that directly target the kinase domain of mTOR (9, 10).

The global response to rapamycin may be considerably influenced by cell intrinsic (e.g., genotype and cell cycle phase) and extrinsic factors (e.g., concentration, stimulation time, and nutrient availability). For this reason, rapamycin can induce a continuum of biological responses that range from strong prosurvival mechanisms such as autophagy to cell death. Most often, rapamycin simultaneously triggers multiple biochemical cascades that not only manifest in a partially overlapping fashion but also exhibit considerable levels of cross talk (11). This may render the analysis of rapamycin-induced phenomena, and in particular that of rapamycin-induced cell death, relatively complex.

Cell death can manifest with distinct biochemical and morphological features, some of which are particularly useful for its quantitative and qualitative assessment. Until recently, the existence of three major cell death modes was commonly accepted by the scientific community: (1) apoptosis manifesting with cell shrinkage (pyknosis), chromatin condensation, nuclear breakdown (karyorrhexis), and the generation of apoptotic bodies, (2) autophagic cell death, characterized by the massive accumulation of autophagosomes and limited chromatin condensation, and (3) necrosis, exhibiting cellular and organellar swelling in the absence of features of apoptosis or autophagy (12, 13). Apoptosis has classically been associated with the early permeabilization of mitochondrial membranes and with the massive activation of caspases (14, 15), autophagic cell death with the maturation of the lysosomal protein LC3 (whose yeast ortholog is known as Atg8) (16), and necrosis with a premature plasma membrane permeabilization followed by a rather disorganized dismantling of the cytoplasmic content (17, 18).

Now, it is becoming increasingly clear that such a clear-cut categorization does not properly reflect the complexity of the cellular demise and that mixed cell death phenotypes (for instance characterized by the coexistence of apoptotic and necrotic features) are common (19). Moreover, the appropriateness of the term "autophagic cell death" has been the subject of an intense debate (20, 21). Most often, indeed, the activation of autophagy by lethal stimuli represents an ultimate attempt of cells to cope with stress, and the pharmacological/genetic inactivation of autophagy accelerates, rather than preventing, cell death (22, 23). Finally, the concept of necrosis as a merely accidental cell death mechanism has been discarded, and it is now established that necrosis can be regulated both in its occurrence and in its course (18, 24, 25).

Recently, prominent investigators in the fields of cell death and autophagy have discussed and agreed on sets of guidelines for the use and interpretation of assays to monitor these two processes (16, 26). In line with these indications, here we propose some techniques for the qualitative and quantitative evaluation of rapamycin-induced cell death on a routine basis. For practical reasons, we describe the application of these methods to human cervical carcinoma HeLa cells, the most widely employed *in vitro* model of human malignancies. With minor changes, however, our protocols can be promptly adapted to other adherent immortalized and primary cells, as well as to cells growing in suspension.

2. Materials

2.1. Common Materials

Disposables

1. 1.5 mL microcentrifuge tubes.
2. 15 and 50 mL conical centrifuge tubes.
3. 5 mL, 12 × 75 mm FACS tubes (BD Falcon, San Jose, USA).
4. 6-, 12-, 24-well plates for cell culture.
5. 75 or 175-cm^2 flasks for cell culture.

Equipment

1. Bright-field inverted microscope: XDS-1R (Optika, Ponteranica, Italy), equipped with 10, 25, and 40× long working distance planachromatic objectives.
2. Cytofluorometer: FACScan or FACSVantage (BD, San Jose, USA), equipped with an argon ion laser emitting at 488 nm and controlled by the operational/analytical software CellQuest™ Pro (BD).

Reagents

1. Complete growth medium for HeLa cells: DMEM containing 4.5 g/L glucose, 4 mM L-glutamine, and 110 mg/L sodium pyruvate supplemented with 100 mM HEPES buffer and 10% fetal bovine serum (FBS) (PAA Laboratories GmbH, Pasching, Austria) (see Note 1).
2. Hank's balanced salt solution (HBSS).

3. Phosphate-buffered saline (PBS, 1×): 137 mM NaCl, 2.7 mM KCl, 4.3 mM Na_2HPO_4, and 1.4 mM KH_2PO_4 in deionized water (dH_2O), adjust pH to 7.4 with 2 N NaOH.
4. Propidium iodide (PI) (Molecular Probes-Invitrogen, Carlsbad, USA), stock solution in dH_2O, 1 mg/mL, stored at 4°C under protection from light (see Notes 2 and 3).
5. Rapamycin (Tocris Biosciences, Ellisville, USA), stock solution in DMSO, 10 mM, stored at −20°C (see Note 4).
6. Trypsin–EDTA: 0.25% trypsin – 0.38 g/L (1 mM) EDTA × $4Na^+$ in HBSS.

2.2. Preliminary Assessment of Short-Term Cell Death: Exclusion of Vital Dyes

1. Counting chamber: improved Neubauer or Bürker hemocytometer.
2. Trypan blue (TB), stock solution in PBS, 0.4% (w/v), stored at room temperature (RT) (see Notes 5 and 6).

2.3. Preliminary Assessment of Long-Term Cell Death: Clonogenic Assays

1. Fixative/staining solution: 0.25% (w/v) crystal violet (CV), 3% (v/v) formaldehyde, and 70% (v/v) methanol in water, stored at RT (see Notes 7–12).
2. Rinsing buffer: sodium chloride (NaCl) in PBS, 0.9% (w/v), stored at RT.

2.4. Quantification of Apoptotic Features by Cytofluorometry: Mitochondrial Membrane Permeabilization

1. 3,3′-Dihexiloxalocarbocyanine iodide ($DiOC_6(3)$) (Molecular Probes-Invitrogen), stock solution in 100% ethanol, 40 μM, stored at −20°C under protection from light (see Notes 13 and 14).
2. Carbonyl cyanide *m*-chlorophenylhydrazone (CCCP) (Sigma-Aldrich, St. Louis, USA), stock solution in 100% ethanol, 10 mM, stored at −20°C (see Note 15).

2.5. Quantification of Apoptotic Features by Cytofluorometry: Phosphatidylserine Exposure

1. Binding buffer (20×) (Miltenyi Biotec, Bergisch Gladbach, Germany), stored at 4°C (see Note 16).
2. Fluorescein isothiocyanate-conjugated Annexin V (FITC-AnnV) (Miltenyi Biotec), stored at 4°C (see Note 17).

2.6. Quantification of Autophagic Markers by Immuno-fluorescence Microscopy: LC3 Aggregation

1. 12 mm Ø coverslips.
2. 76 mm × 26 mm slides for fluorescence microscopy.
3. Fixative solution: 4% paraformaldehyde (PFA). For 10 mL: dissolve 0.4 g of PFA in 500 μl dH_2O, add one drop of 2 N NaOH, heat to 65–70°C until solution clears, add 9.5 mL PBS, let cool to RT, and move the fixative solution to an ice bath (see Notes 18 and 19).
4. Fluorescence microscope: IRE2 microscope equipped with a DC300F camera (Leica Microsystems GmbH, Wetzlar, Germany).

5. Fluoromount-G™ mounting medium (Southern Biotech, Birmingham, USA).
6. Hoechst 33342 (Molecular Probes-Invitrogen), stock solution in dH_20, 10 mg/mL, stored at 4°C under protection from light (see Notes 20 and 21).
7. Lipofectamine™ 2000 transfection reagent (Gibco-Invitrogen), stored at 4°C.
8. Opti-MEM® reduced serum medium supplemented with Glutamax™ and phenol red (Gibco-Invitrogen), stored at 4°C.

2.7. Quantification of Autophagic Markers by Immunoblotting: LC3 Maturation

1. 3 MM® Whatmann filter paper.
2. Binding buffer: 0.1% Tween 20 in TBS supplemented with 3% (w/v) BSA, stored at 4°C (see Note 22).
3. Blocking buffer: 0.1% Tween 20 in TBS supplemented with 5% (w/v) nonfat powdered milk (commonly found in food stores), stored at 4°C (see Note 22).
4. Bovine serum albumin (BSA), stock solution 10 mg/mL in dH_2O, stored at 4°C (see Note 22).
5. Antibodies employed in immunoblotting techniques are listed in Table 1 (see Note 23).
6. Chemiluminescent substrate: SuperSignal West Pico Chemiluminescent Substrate (Pierce Biotechnology, Rockford, USA), stored at RT (see Note 24).
7. *DC* Protein Assay (Bio-Rad, Hercules, USA), stored at RT (see Notes 25 and 26).

Table 1
Antibodies for immunoblotting

Primary antibody	Specificity	Source organism	Company [a]	Ref.
Anti-GAPDH (6C5)	monoclonal IgG_1	Mouse	Chemicon International	mAB374
Anti-LC3 (LC3B)	polyclonal IgG	Rabbit	Cell Signaling Technology	#2775
Anti-p62	polyclonal IgG	Rabbit	Santa Cruz Biotechnology	Sc-25575
Secondary antibody	**Label**	**Source organism**	**Company [a]**	**Ref.**
Anti-mouse	Horseradish peroxidase	Goat	Southern Biotech	1010-05
Anti-rabbit	Horseradish peroxidase	Goat	Southern Biotech	4010-05

[a]Cell Signaling Technology, Danvers, USA; Chemicon International, Temecula, USA; Santa Cruz Biotechnology, Santa Cruz, USA; and Southern Biotech, Birmingham, USA

8. Dehybridization buffer: Restore Western Blot Stripping Buffer (Pierce Biotechnology), stored at 4°C (see Note 27).
9. Electrophoresis apparatus: XCell4 SureLock™ Mini-Cell (Invitrogen).
10. Electrotransfer module: XCell II™ Blot Module (Invitrogen).
11. Electrotransfer sponges: XCell II™ blotting pads (Invitrogen).
12. Film processor: Curix 60 tabletop film processor (Agfa, Mortsel, Belgium).
13. Methanol (see Notes 11 and 12).
14. Nitrocellulose membrane.
15. NP40 lysis buffer: 1% NP40, 20 mM HEPES, 10 mM potassium chloride (KCl), 1 mM ethylenediaminetetraacetic acid (EDTA) (Sigma-Aldrich), 10% glycerol, 1 mM orthovanadate (Sigma-Aldrich), 1 mM phenylmethylsulfonyl fluoride (PMSF) (Sigma-Aldrich), 1 mM dithiothreitol (DTT) (Sigma-Aldrich), and complete EDTA-free protease inhibitor cocktail (Roche Diagnostics GmbH, Mannheim, Germany, one tablet/10 mL), stored at −20°C (see Notes 28–31).
16. NuPAGE® antioxidant (1×) (Invitrogen), stored at 4°C (see Note 32).
17. NuPAGE® LDS sample buffer (4×) (Invitrogen), stored at 4°C.
18. NuPAGE® MES SDS running buffer (20×) (Invitrogen), stored at 4°C or RT (see Notes 33 and 34).
19. NuPAGE® Novex 4–12% Bis-Tris precast gels (Invitrogen), stored at 4°C or RT (see Notes 34 and 35).
20. NuPAGE® Novex gel knife (Invitrogen).
21. NuPAGE® sample reducing agent (10×) (Invitrogen), stored at 4°C or RT (see Notes 34 and 36).
22. NuPAGE® transfer buffer (20×) (Invitrogen), stored at 4°C or RT (see Notes 34 and 35).
23. Photographic films: 18 cm×24 cm Amersham Hyperfilm™ ECL™ (GE Healthcare Life Sciences, Uppsala, Sweden).
24. Ponceau S staining solution: 0.1% (w/v) Ponceau S and 5.0% (w/v) acetic acid in dH_2O (or ready-made, commercially available) (see Note 37).
25. Protein molecular weight (MW) markers: Precision Plus Protein™ Standard (Bio-Rad).
26. Rinsing buffer: 0.1% Tween 20 in TBS, stored at 4°C (see Note 38).
27. Tris-buffered saline (TBS, 1×): 150 mM NaCl, 50 mM Tris in dH_2O, adjust pH to 7.4 with 1 N HCl.

3. Methods

3.1. Cell Culture and Rapamycin Administration

1. Upon thawing, HeLa cells are routinely cultured in complete growth medium within 175 cm^2 flasks (37°C, 5% CO_2) (see Notes 39 and 40).
2. When confluence approaches 70–90%, cells are detached with trypsin–EDTA (see Notes 41 and 42) to constitute fresh maintenance cultures (see Note 43) and to provide cells for experimental determinations.
3. The latter are carried out in 6-, 12-, or 24-well plates, depending on the specific experimental setting (see below).
4. 12–24 h after seeding (to allow cells to readapt and recover normal growth rates), rapamycin is administered in fresh, complete growth medium (see Note 44). The final concentration of rapamycin and the stimulation time required to induce detectable levels of cell death may exhibit consistent variation depending on cell type (see Note 45), as well as on cell-extrinsic variables (e.g., medium composition, cell density, and medium consumption). As a guideline, HeLa cultures treated with 50 μM rapamycin for 48 h in complete growth medium usually contain 70–90% cells characterized by plasma membrane breakdown.
5. Cells treated with an equivalent volume of DMSO diluted in fresh, complete medium will provide appropriate negative control conditions (see Note 46). The convenient positive controls depend on the specific experimental setting (see below).

3.2. Preliminary Assessment of Short-Term Cell Death: Exclusion of Vital Dyes

As recently outlined by a large group of prominent investigators in the field of cell death (26), the study of the cellular demise should entail a primary estimation aimed at addressing the following fundamental question: "are cells truly dead?" Although this may seem a trivial point, multiple cell-death-related phenomena also occur in cell-death-unrelated settings, which represents a nonnegligible source of overestimation of the cellular demise (27). Thus, it is critical to determine whether bona fide cell death occurs (and roughly to which extent) before starting with more precise quantitative/qualitative estimations of its morphological/biochemical manifestations. One of the easiest and most commonly employed techniques to quantify short-term cell death relies on the use of exclusion dyes. As all vital dyes, exclusion dyes differentially stain live and dead cells, and in particular they only penetrate into the latter, i.e., cells that have lost the integrity of the plasma membrane. This phenomenon can be easily quantified in light microscopy with the help of common hemocytometric chambers.

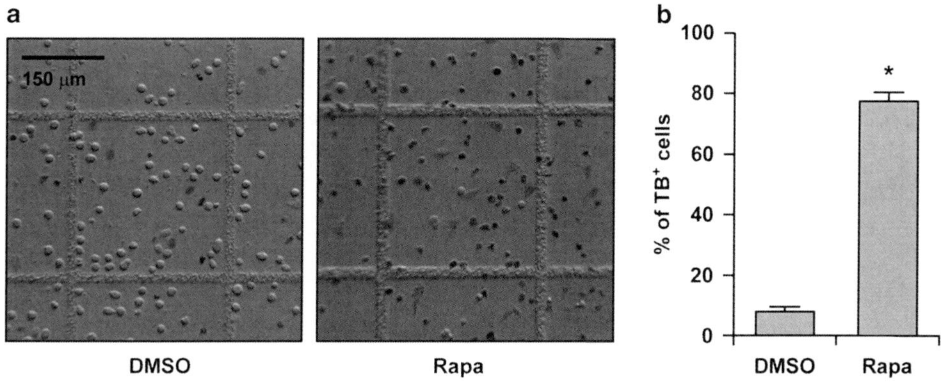

Fig. 1. Trypan blue exclusion assay. Human cervical carcinoma HeLa cells were seeded in 12-well plates, let adhere for 12 h, and then treated with 50 μM rapamycin (Rapa) or with an equivalent volume of DMSO (negative control) for additional 48 h. Finally, the cells were processed for a Trypan blue (TB) exclusion assay as detailed in Subheading 3.2. (**a**) Representative counting chambers filled with DMSO-treated (*left panel*) and rapamycin-treated (*right panel*) HeLa cells. (**b**) Quantitative data. *Columns* report the percentage of TB positive (TB^+) cells (mean ± SEM, $n = 3$, $^*p < 0.05$ as determined by unpaired Student's *t*-test).

Trypan Blue Exclusion

1. HeLa cells are seeded in 12- or 24-well plates (see Note 47), let adhere for 12–24 h (see Note 44), and treated with rapamycin (see Subheading 3.1, steps 4 and 5 and Notes 45 and 46).
2. At the end of the stimulation period, culture supernatants are collected in 15 mL centrifuge tubes (see Note 48).
3. Adherent cells are detached with trypsin–EDTA (see Notes 49–52) and gathered with the corresponding supernatants (see Note 53).
4. Samples are centrifuged at 300 × *g* for 5 min (RT) (see Note 54), supernatants are discarded and cell pellets are resuspended in an appropriate volume of PBS (see Note 55).
5. Small aliquots of these suspensions are carefully diluted (1:5–1:20) in 0.04% TB (w/v in PBS) (see Notes 55 and 56) and loaded onto the counting chamber of a classical (e.g., improved Neubauer, Bürker) hemocytometer.
6. Bright-field microscopy (×100–400 magnification) allows for the rapid discrimination between living and dead cells. Indeed, while the former exhibit a bright appearance and are surrounded by a light blue halo, the latter assume an easily recognizable dark blue cytoplasmic staining (see Notes 57 and 58) (Fig. 1).

3.3. Preliminary Assessment of Long-Term Cell Death: Clonogenic Assays

At reduced concentrations, several cytotoxic chemicals fail to induce detectable levels of acute cell death. In this case, cells may die with a relevantly delayed kinetics, especially when strong pro-survival responses such as autophagy are simultaneously activated by the lethal trigger (as in the case of rapamycin). Alternatively, cells exposed to low amounts of rapamycin can lose their proliferative potential and enter a quiescent state known as senescence, which cannot be considered as a bona fide cell death instance (12) but may result in delayed cell death (26). Thus, assays that monitor the cellular demise occurring in an acute fashion need to be complemented by techniques for the detection of delayed cell death, as well as for the discrimination of cell death from senescence. Clonogenic assays constitute the gold standard approach to address these experimental questions.

Clonogenic Assays

1. 0.1–20 × 10^3 HeLa cells are seeded in 6-well plates (in 2–3 mL complete growth medium), let adhere for 12–24 h (see Note 44), and treated with rapamycin (see Subheading 3.1, steps 4 and 5 and Notes 45 and 46).
2. After the desired period of time, stimulation medium is discarded, cells are washed twice with prewarmed PBS and standard culture conditions are restored for up to 14 days, to allow for the generation of colonies (see Note 59).
3. Colonies are washed 2–3 times with rinsing buffer and stained by incubation in 2–3 mL fixative/staining solution for 10 min at RT (see Note 60).
4. Fixative/staining solution is removed, stained colonies are rapidly rinsed twice with dH_2O, and quantified by light microscopy (see Notes 60–65) (Fig. 2).

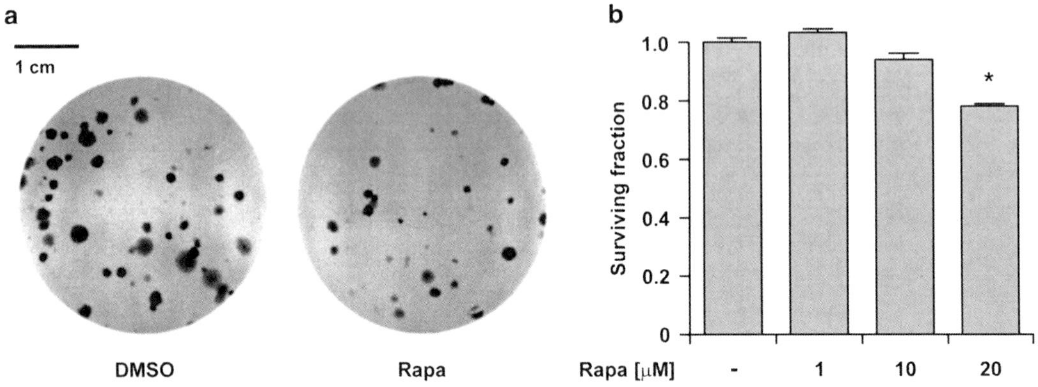

Fig. 2. Clonogenic assay. Human cervical carcinoma HeLa cells were seeded in 6-well plates, let adhere for 12 h, and then treated with the indicated concentration of rapamycin (Rapa) or with an equivalent volume of DMSO (negative control) for additional 48 h. Thereafter, the cells were allowed to form colonies for 14 days and eventually processed for a clonogenic assay, as detailed in Subheading 3.3. (**a**) Representative plates containing colonies from DMSO-treated (*left panel*) and rapamycin-treated (*right panel*) HeLa cells. (**b**) Quantitative data. *Columns* illustrate the surviving fraction (see also Note 64) normalized to that of DMSO-treated cells (mean ± SEM, $n = 3$, $^*p < 0.05$ as determined by unpaired Student's *t*-test).

3.4. Quantification of Apoptotic Features by Cytofluorometry: Mitochondrial Membrane Permeabilization

Mitochondria are critical regulators of programmed cell death, in particular when this is triggered by intracellular events and executed via the intrinsic pathway of apoptosis. In response to a wide array of intracellular stress conditions, both pro- and anti-apoptotic pathways are activated and their signals converge to and are opposed to each other at the level of mitochondria. If lethal stimuli predominate, permanent mitochondrial membrane permeabilization (MMP) occurs in most mitochondria, entailing a series of catastrophic consequences for the cell, which becomes irremediably committed to death (15, 28). The proapoptotic consequences of MMP include but are not limited to the following: (1) dissipation of the mitochondrial transmembrane potential ($\Delta\psi_m$), which results in respiratory chain uncoupling, overgeneration of reactive oxygen species (ROS) and the abrupt arrest of most mitochondrial metabolic functions and (2) the release into the cytosol of proteins that are normally confined and exert vital functions within the mitochondrial intermembrane space (IMS) (27, 29). MMP can be provoked by several distinct (yet partially overlapping) molecular cascades, which are initiated by specific factors either at the inner or the outer mitochondrial membrane (IM and OM, respectively) (15, 30). Irrespective of these mechanistic details, permanent MMP is commonly considered as the first irreversible phenomenon in the cascade of events lining up intrinsic apoptosis, a point-of-no-return that has been exploited for the development of several assays for the detection of cell death (26, 31, 32). One of the most widely employed technique for the detection of MMP relies on the $\Delta\psi_m$-sensitive dye 3,3′-dihexiloxalocarbocyanine iodide ($DiOC_6(3)$), mostly due to its rapid mitochondrial equilibration and negligible quenching effects, at least at low concentrations (10–50 nM) (33–36).

MMP Detection by $DiOC_6(3)$

1. HeLa cells previously seeded in 12- or 24-well plates (see Note 47) and allowed to adhere for 12–24 h (see Note 44) are treated with rapamycin (see Subheading 3.1, steps 4 and 5 and Notes 45 and 46) and collected as described above (see Subheading 3.2, steps 2–4 and Notes 48–54).
2. Cell pellets are then resuspended in 200–400 μL of prewarmed growth medium containing 10–50 nM $DiOC_6(3)$ (see Notes 66–68).
3. For staining, samples are kept at 37°C (5% CO_2) for 20–40 min (see Notes 69–71), under protection from light (see Note 72).
4. To concomitantly assess the integrity of the plasma membrane, a vital dye of choice (i.e., a probe that can discriminate between live and dead cells based on the plasma membrane permeability) can be integrated in the staining protocol (see Notes 73–75).

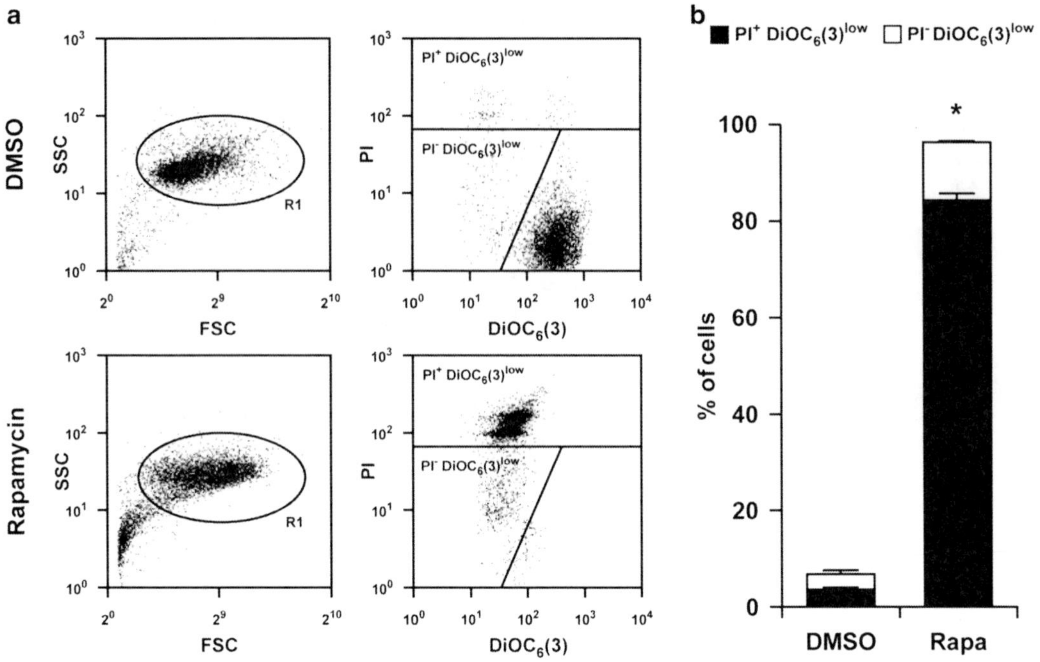

Fig. 3. Cytofluorometric assessment of mitochondrial membrane permeabilization. Human cervical carcinoma HeLa cells were seeded in 12-well plates, let adapt for 12 h, and then treated with 50 μM rapamycin (Rapa) or with an equivalent volume of DMSO (negative control) for additional 48 h. The cells were then processed for the cytofluorometric quantification of mitochondrial transmembrane potential ($\Delta\psi_m$) dissipation and plasma membrane breakdown, as detailed in Subheading 3.4. (**a**) Representative *dot plots* of DMSO-treated (*upper panels*) and rapamycin-treated (*lower panels*) HeLa cells. *Left panels* depict emblematic forward scatter (FSC) versus side scatter (SSC) *dot plots*, which allow for gating (*see* also Notes 77 and 80). *Right panels* illustrate typical $DiOC_6(3)$ fluorescence versus propidium iodide (PI) fluorescence *dot plots*, on which the actual quantification is performed (limited to region R1). (**b**) Quantitative data. *Black* and *white columns* depict the percentage of dead ($PI^+ DiOC_6(3)^{low}$) and dying ($PI^- DiOC_6(3)^{low}$) cells, respectively (mean ± SEM, $n=3$, $^*p<0.05$ as determined by unpaired Student's *t*-test).

5. Samples are analyzed by means of a classic cytofluorometer allowing for the acquisition of light scattering data and fluorescence in (at least) two separate channels (e.g., green and red) (see Note 76).
6. Statistical analyses are performed by first selecting events that exhibit normal light scattering parameters (forward and side scatter, i.e., FSC and SSC, respectively) (see Notes 77–80) (Fig. 3).

3.5. Quantification of Apoptotic Features by Cytofluorometry: Phosphatidylserine Exposure

Once the point-of-no-return represented by MMP has been passed, apoptotic cells activate an ensemble of degradative phenomena that also account for the morphological manifestations of apoptosis. These processes are driven by the massive and self-amplificatory activation of caspases (which is known as the caspase cascade) (37) as well as by numerous other caspase-dependent and-independent enzymes including nucleases and noncaspase proteases (e.g., calpains and cathepsins) (38, 39). One of the earliest biochemical manifestations of the execution phase of apoptosis is the translocation of phosphatidylserine (PS) on the outer leaflet of the plasma membrane (40), from where it provides an "eat-me" signal to phagocytic cells for the immunologically silent internalization of apoptotic bodies (41). Initially, the exposure of PS (which in healthy cells is confined to the cytosolic leaflet of the plasma membrane) was believed to occur through a caspase-dependent mechanism catalyzed by the PS-translocating enzyme phospholipid scramblase 1 (PLSCR1) (42, 43). Now, it has become clear not only that the activation of PLSCR1 can occur independently of caspases (44) and in cell death-unrelated settings (45), but also that PS externalization can be mediated by PLSCR1-independent mechanisms driven by sustained increases in cytosolic Ca^{2+} (46, 47). Still, the cytofluorometric quantification of externalized PS by labeled Annexin V (AnnV, a Ca^{2+}-dependent phospholipid-binding protein with high affinity for PS) remains one of the most common techniques for the detection of apoptotic cells (48).

Detection of PS Exposure by FITC-AnnV

1. HeLa cells are seeded in 12- or 24-well plates (see Note 47), allowed to adhere for 12–24 h (see Note 44), treated with rapamycin (see Subheading 3.1, steps 4 and 5 and Notes 45 and 46), and collected as described above (see Subheading 3.2, steps 2–4 and Notes 48–54).
2. Cells are gently washed 1–2 times (see Notes 81 and 82) in 250–500 μL cold (4°C) 1× binding buffer, followed by centrifugation at 300×*g* for 5 min (4°C) and removal of supernatants.
3. For staining, cells are resuspended in 50–100 μL 1× cold (4°C) binding buffer previously complemented with 2–4 μL FITC-AnnV (see Note 83), and kept for 30–45 min on ice (see Notes 84 and 85) under protection from light (see Note 86).
4. To remove unbound FITC-AnnV, samples are diluted with 500–1,000 μL cold (4°C) 1× binding buffer and washed 1–2 times (see Subheading 3.5, step 2 and Note 82).
5. Finally, cells are resuspended (and kept for 2–5 min) in 250–500 μL cold (4°C) 1× binding buffer previously supplemented with 0.5–1 μg/mL PI, followed by cytofluorometric analysis (see Subheading 3.4, steps 5 and 6 and Notes 76–78, 80 and 87) (Fig. 4).

3.6. Quantification of Autophagic Markers by Immunofluorescence Microscopy: LC3 Aggregation

Rapamycin-mediated inhibition of mTOR is one of the most effective experimental perturbations to induce autophagy (49). A large number of techniques have been developed for the detection of autophagy in eukaryotic cells, based on distinct technological platforms and characterized by specific advantages and drawbacks (16, 50). In this context, one of the most important issues is the

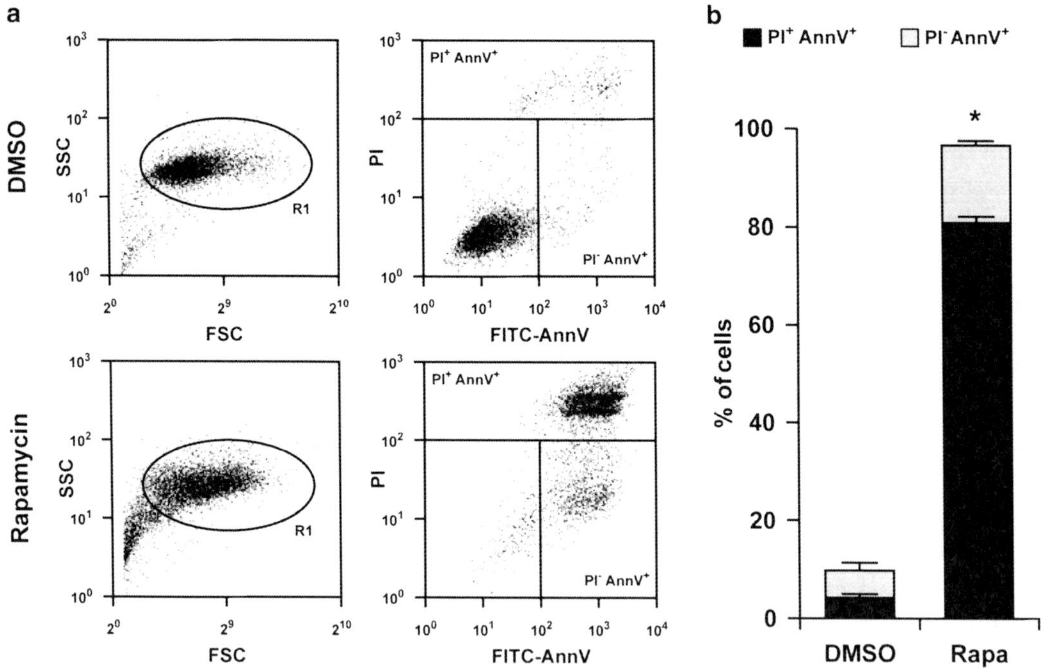

Fig. 4. Cytofluorometric assessment of phosphatidylserine exposure. Human cervical carcinoma HeLa cells were seeded in 12-well plates, let adhere for 12 h, and then treated with 50 μM rapamycin (Rapa) or with an equivalent volume of DMSO (negative control) for additional 48 h. The cells were then processed for the cytofluorometric quantification of phosphatidylserine exposure, as described in Subheading 3.5. (**a**) Typical *dot plots* of DMSO-treated (*upper panels*) and rapamycin-treated (*lower panels*) HeLa cells. *Left panels* illustrate representative forward scatter (FSC) versus side scatter (SSC) *dot plots*, which are required for gating (see also Notes 77 and 80). *Right panels* depict typical fluorescein isothiocyanate-conjugated Annexin V (FITC-AnnV) fluorescence versus propidium iodide (PI) fluorescence *dot plots*, which allow for quantification (limited to region R1). (**b**) Quantitative data. *Black* and *light gray columns* depict the percentage of dead (PI^+ $AnnV^+$) and dying (PI^- $AnnV^+$) cells, respectively (mean ± SEM, $n = 3$, $^*p < 0.05$ as determined by unpaired Student's *t*-test).

need for differentiating between the accumulation of autophagosomes (which also can derive from the inhibition of lysosomal degradation) and a bona fide functional enhancement of the autophagic flow (which entails the accumulation of autophagosomes, but also an increased degradation of autophagic substrates) (16). During autophagy, the cytosolic protein LC3 (which in its immature variant is called LC3-I) matures by lipidation (it is then referred to as LC3-II) and translocates to forming autophagosomes (51). The generation of a chimeric protein constituted by LC3 fused to the C-terminus of the green fluorescent protein (GFP-LC3) has allowed for the development of a rapid method to quantify autophagosomes by fluorescence microscopy (16, 50).

Fluorescence Microscopy-Assisted Quantification of GFP-LC3 Aggregation.

1. Sterile 12 mm Ø coverslips are deposited in 24-well plates (see Notes 88 and 89).

2. 50–100 × 10^3 HeLa cells are seeded in 24-well plates (in 0.5 mL growth medium per well) and let adhere for 12–24 h (see Note 44).
3. The cells are transiently transfected with a GFP-LC3-encoding plasmid by means of Lipofectamine™ 2000, according to the manufacturer's instructions (see Notes 90–94), and cultured for 24 h (or until they start to express GFP-LC3, see Note 95) prior to the administration of rapamycin (see Subheading 3.1, steps 4 and 5 and Notes 45 and 46).
4. An appropriate positive control for the accumulation of autophagosomes can be obtained by incubating otherwise untreated (but GFP-LC3-transfected) cells in HBSS (upon the removal of growth medium and 1–2 washes in prewarmed PBS or HBSS) during the last 6 h of the assay (see Note 96).
5. After the desired period of time, growth medium is removed, the cells are washed twice with PBS and fixed in 400 μL fixative solution for 20–30 min at RT (see Note 97).
6. Fixative solution is discarded, the cells are washed 2–3 times with PBS, and nuclear counterstaining is performed by the addition of 2 μM Hoechst 33342 in PBS (200 μL per well) for 10–15 min (see Note 98).
7. After two additional washes in PBS (see Note 99), coverslips are mounted onto 76 mm × 26 mm slides for fluorescence microscopy by means of the Fluoromount-G™ mounting medium (see Note 100).
8. Finally, slides are stored (see Note 101) or examined by fluorescence microscopy. For the visualization of GFP-LC3, an oil immersion objective (50–65× magnification) should be employed. Suitable excitation and emission filters should be chosen according to the following absorption and emission peaks: 488/507 nm for LC3-GFP and 352/461 nm for Hoechst 33342 (see Notes 102–104) (Fig. 5).

3.7. Quantification of Autophagic Markers by Immunoblotting: LC3 Maturation and p62 Degradation

The proautophagic processing of LC3-I involves at least two distinct steps, namely, cleavage at Met 121 and conjugation with phosphatidylethanolamine, which are catalyzed by the essential autophagic modulators ATG4 and ATG7/ATG3, respectively (52, 53). As discussed above, LC3-I is a cytosolic protein, whereas LC3-II relocalizes to forming autophagosomes and plays an important function in membrane dynamics by mediating tethering and hemifusion (54). LC3-II also differs from LC3-I relative to its electrophoretic mobility. In nonautophagic conditions, indeed, the bulk of LC3 migrates as a single band with an apparent MW of ~18 kDa (corresponding to LC3-I). Conversely, when autophagy is induced, LC3-specific antibodies allow for the visualization of a second band that is characterized by an apparent MW of approximately

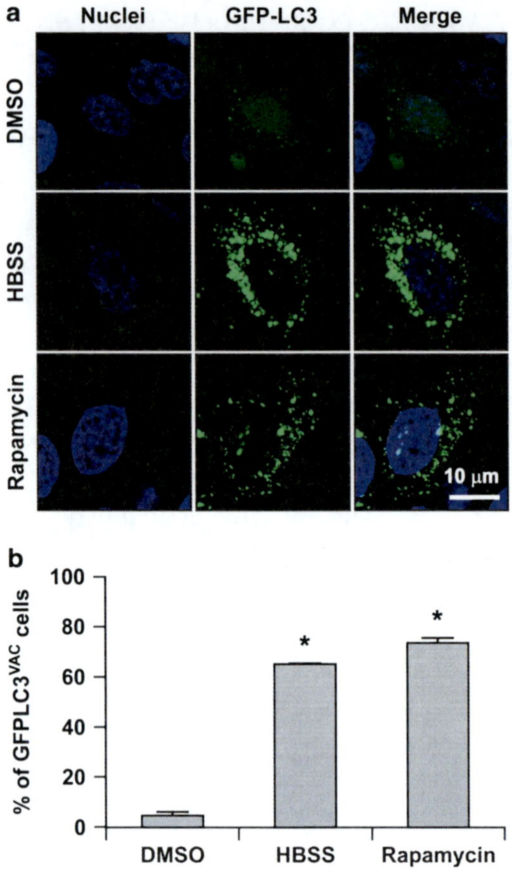

Fig. 5. Immunofluorescence microscopy for the assessment of GFP-LC3 aggregation. Human cervical carcinoma HeLa cells were seeded in 12-well plates, let adapt for 12 h, and then transfected with a plasmid coding for a green fluorescent protein (GFP)-LC3 chimera. Twenty-four hours later, the cells were administered with 10 μM rapamycin, with an equivalent volume of DMSO (negative control) or kept in HBSS (positive control) for additional 6 h. Finally, the cells were processed for the assessment of GFP-LC3 aggregation by immunofluorescence microscopy, as described in Subheading 3.6. (**a**) Representative immunofluorescence micrographs of DMSO-treated (*upper panel*), HBSS-treated (*mid panel*), and rapamycin-treated (*lower panel*) HeLa cells. (**b**) Quantitative data. *Columns* illustrate the percentage of cells exhibiting the cytoplasmic aggregation of GFP-LC3 (GFP-$LC3^{vac}$ cells, mean ± SEM, $n = 3$, $*p < 0.05$ as determined by unpaired Student's *t*-test).

~15–16 kDa (corresponding to LC3-II). Immunoblotting aimed at estimating the relative amount of LC3-I versus LC3-II provides an additional means to determine the accumulation of autophagosomes (16, 50). This can be conveniently combined with an immunoblotting-based technique for the detection of functional autophagy based on the protein p62 (also known as SQSTM1) (16, 50). p62 serves as a link between LC3 and ubiquitinated substrates, becomes incorporated into formed autophagosomes and is eventually degraded within auto(phago)lysosomes (55). Inhibition of autophagy correlates with increased levels of p62

(56), further strengthening the notion that the steady state levels of this protein can be used (with some limitations) as an indicator of the autophagic status (16, 50).

3.7.1. Preparation of Samples

1. HeLa cells are seeded in 6-well plates (see Note 47), let adhere for 12–24 h (see Note 44), treated with rapamycin (see Subheading 3.1, steps 4 and 5 and Notes 45 and 46), and collected as described above (see Subheading 3.2, steps 2–4 and Notes 48–54).
2. Positive control for LC3-I→LC3-II conversion can be obtained by incubating untreated cells in HBSS (upon the removal of growth medium and 1–2 washes in prewarmed PBS or HBSS) during the last 3–4 h of the assay (see Note 105).
3. The cells are washed in 500–1,000 μL cold (4°C) PBS, followed by centrifugation at 300–500 × *g* for 5 min (4°C) and removal of supernatants (see Note 106).
4. Pellets are carefully lysed in 30–50 μL of cold (4°C) NP40 lysis buffer (see Notes 107 and 108).
5. The protein concentration of lysates is determined by means of the *DC* Protein Assay, following the manufacturer's instructions (see Note 109).
6. 20–80 μg of proteins from each lysate (see Note 110) are mixed with 5–7.5 μL NuPAGE® LDS sample buffer (4×) and the final volume is adjusted to 20–30 μL with lysis buffer (see Notes 111 and 112).
7. Samples are thermally denatured by incubation at 100°C for 5 min. Upon cooling to RT, the samples are ready to be loaded onto a sodium dodecylsulfate (SDS)-polyacrylamide gel (see Notes 113–115).

3.7.2. Electrophoresis

1. These instructions refer to an Invitrogen XCell4 SureLock™ Mini-Cell system for gel electrophoresis and NuPAGE® Novex 4–12% Bis-Tris precast gels. Detailed instructions for the use of a BioRad mini-PROTEAN 3 system can be found in ref. 50.
2. For a single unit, which can accommodate up to two precast gels, use dH_2O to prepare 500 mL 1× NuPAGE® MES SDS running buffer (see Notes 116 and 117).
3. Unwrap the gel(s), peel off the tape covering the slot on the back of the gel cassette (see Note 118), and gently remove (by one smooth motion) the disposable comb(s) while paying particular attention at preserving the integrity of wells (see Note 119).
4. Gently wash wells with dH_2O (or with 1× running buffer) 1–2 times, to remove possible polyacrylamide leftovers, and assemble the electrophoresis unit by following the manufacturer's instructions (see Notes 120 and 121).

5. Pour 1× running buffer in the chamber of the unit that is delimited by gels until wells are entirely submerged (see Note 122), add 300 μL 1× NuPAGE® antioxidant (see Note 123), and gently remove bubbles that may have formed (see Note 124).
6. Pour the rest of the running buffer in the outer chamber of the unit and proceed to load each well with one sample prepared as described above (see Subheading 3.7.1 and Notes 44–54 and 105–115). 1–2 Wells are loaded with Precision Plus Protein™ Standard MW markers (see Note 125).
7. Immediately after loading (see Note 126), connect the unit to the power supply and separate proteins in constant-field mode (see Note 127). Importantly, the run should be protracted until the 10 kDa band of the MW marker reaches (but not trespass) the horizontal line that delineates the zone of the gel that will be cut for the electrotransfer (gel "foot") (see Subheading 3.7.3, step 5 and Notes 128 and 132).

3.7.3. Electrotransfer

1. For a single unit, which can accommodate up to two precast gels, prepare 1 L 1× NuPAGE® transfer buffer by mixing 50 mL 20× NuPAGE® transfer buffer, 200 mL methanol, 1 mL 1× NuPAGE® antioxidant, and 749 mL dH_2O (see Notes 116 and 117).
2. Five XCell II™ blotting pads and four sheets of 3 MM® Whatmann filter paper of approximately the same size than the running gel (9 cm × 9 cm) are equilibrated in 1× transfer buffer for 1–2 min at RT.
3. Two nitrocellulose membranes of approximately the same size of the running gel (9 cm × 9 cm) are activated by incubation in dH_2O for 2–5 min at RT, then equilibrated in transfer buffer (see Note 129).
4. The gel unit is disconnected from the power supply and disassembled. With the help of a NuPAGE® Novex gel knife, precast gels are opened and the upper plastic shield is discarded (see Note 130).
5. The gel "foot" is pushed out of its slot by means of the gel knife and each gel is deposited on a clean and flat surface previously wetted with transfer buffer. Wells and gel "feet" are cut away and discarded (see Notes 131 and 132).
6. The transfer "sandwich" is prepared by overlaying the following components (presoaked in transfer buffer) on top of the cathodic core (−) of the XCell II™ Blot Module (conditions for two gels): two blotting pads, one filter paper, the first gel, the first nitrocellulose membrane, one filter paper, one blotting pad, one filter paper, the second gel, the second nitrocellulose membrane, one filter paper, and two blotting pads (as indicated by the manufacturer) (see Notes 133 and 134).

7. The anodic core (+) of the blot module is then used to close the transfer sandwich, which is subsequently inserted into the XCell4 SureLock™ Mini-Cell system.
8. Fill the blot module with 1× transfer buffer until the sandwich is entirely covered (see Note 135).
9. Fill the outer chamber with ~650 mL dH_2O until the water level reaches ~2 cm from the top of the unit (see Note 136).
10. Place the lid on top of the unit, connect it to the power supply and electrotransfer proteins in constant-field mode (30 V) for 1 h, as recommended by the manufacturer (see Notes 137–139).
11. Once the transfer is complete, disconnect the transfer unit from the power supply, disassemble the blot module, and withdraw nitrocellulose membranes (see Note 140), on which colored MW markers should be clearly visible (see Notes 141 and 142).
12. At this stage, bound proteins can be visualized (as a preliminary indicator of an adequate electrotransfer as well as of the equal loading of lanes) by incubating membranes in Ponceau S staining solution for 2–5 min (see Notes 143 and 144).
13. To allow for the parallel detection of LC3-I→LC3-II conversion (MW < 20 kDa), p62 degradation (MW = ~62 kDa), and glyceraldehyde-3-phosphate dehydrogenase (GAPDH, MW = ~37 kDa) abundance (as a definitive control of the equal loading of lanes), each nitrocellulose membrane should be horizontally cut with scissors approximately at the level of the 25 and 50 kDa MW marker bands (see Notes 145–147).
14. Upon cutting, Ponceau S should be removed by quickly rinsing membranes in PBS, dH_2O, or rinsing buffer (see Note 148).

3.7.4. Immunoblotting for the Detection of LC3 and p62

1. To block unspecific binding sites, nitrocellulose membrane fragments are incubated on a rocking platform in 10 mL of blocking buffer (30–60 min, 4°C) (see Note 149).
2. Blocking buffer is discarded and membrane fragments are rapidly washed in rinsing buffer and incubated with the corresponding primary antibodies. The following conditions should be employed: anti-LC3, 1/500–1/1,000 in either binding or blocking buffer (overnight at 4°C); anti-p62, 1/500–1/1,000 in either binding or blocking buffer (overnight at 4°C); anti-GAPDH, 1/10,000 in either binding, blocking, or rinsing buffer (45–60 min RT) (see Notes 150–152).
3. Primary antibodies are removed (see Note 153) and membrane fragments are washed three times in rinsing buffer (10 min, RT).
4. Upon washing, membranes are incubated for 45–60 min at RT (on the rocking platform) with freshly prepared 1/5,000–1/10,000

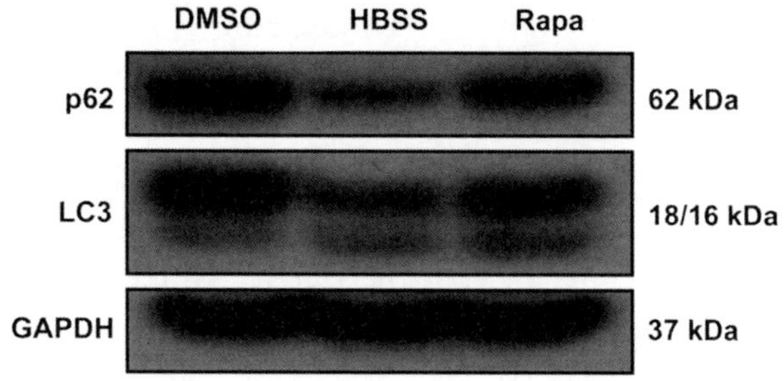

Fig. 6. Immunoblotting for the assessment of LC3-I→LC3-II conversion and p62 degradation. Human cervical carcinoma HeLa cells were seeded in 12-well plates, let adhere for 12 h, and then treated with 10 μM rapamycin (Rapa), with an equivalent volume of DMSO (negative control) or kept in HBSS (positive control, see also Notes 105 and 163) for additional 3 h. Finally, the cells were subjected to immunoblotting to monitor LC3-I→LC3-II conversion and p62 degradation, as described in Subheading 3.7. GAPDH was employed as a loading control (see also Note 147).

dilutions (in blocking buffer) of goat anti-rabbit (for the detection of LC3 and p62) or goat anti-mouse (for the detection of GAPDH) horseradish peroxidase (HRP)-coupled secondary antisera.

5. Secondary antibodies are discarded (see Note 154), and membrane fragments are again washed three times in rinsing buffer (10 min, RT).
6. Each membrane fragment is overlaid for 2–5 min with 500–1,000 μL SuperSignal West Pico Maximum Sensitivity chemiluminescent substrate, prepared shortly beforehand as indicated by the manufacturer (see Note 155).
7. Finally, a photographic film is exposed to the membrane fragments for 3–5 min in the dark room (see Notes 156–159), prior to development and fixation in a common tabletop film processor (see Notes 160–163) (Fig. 6).

3.8. Guidelines for the Correct Evaluation of Rapamycin-Induced Cell Death

Depending on both cell-intrinsic and cell-extrinsic variables, the administration of rapamycin can elicit a number of biochemical cascades whose outcomes are as diverse as adaptation to stress (survival) and cell death. In some cases, these signaling pathways are activated by rapamycin in a near-to-simultaneous fashion, which largely complicates the study of rapamycin-induced events. It should also be noted that rapamycin concentrations higher than 10 μM are likely to promote cell death mainly through off-target mechanisms. Indeed, while in the low micromolar range (1–10 μM) rapamycin robustly and rapidly (3–6 h) triggers autophagy (implying that it efficiently inhibits mTOR) (Fig. 5 and refs. 50, 57), similar doses neither induce cell death (even upon prolonged exposure, 24–48 h) (Fig. 7) nor affect clonogenicity (Fig. 2).

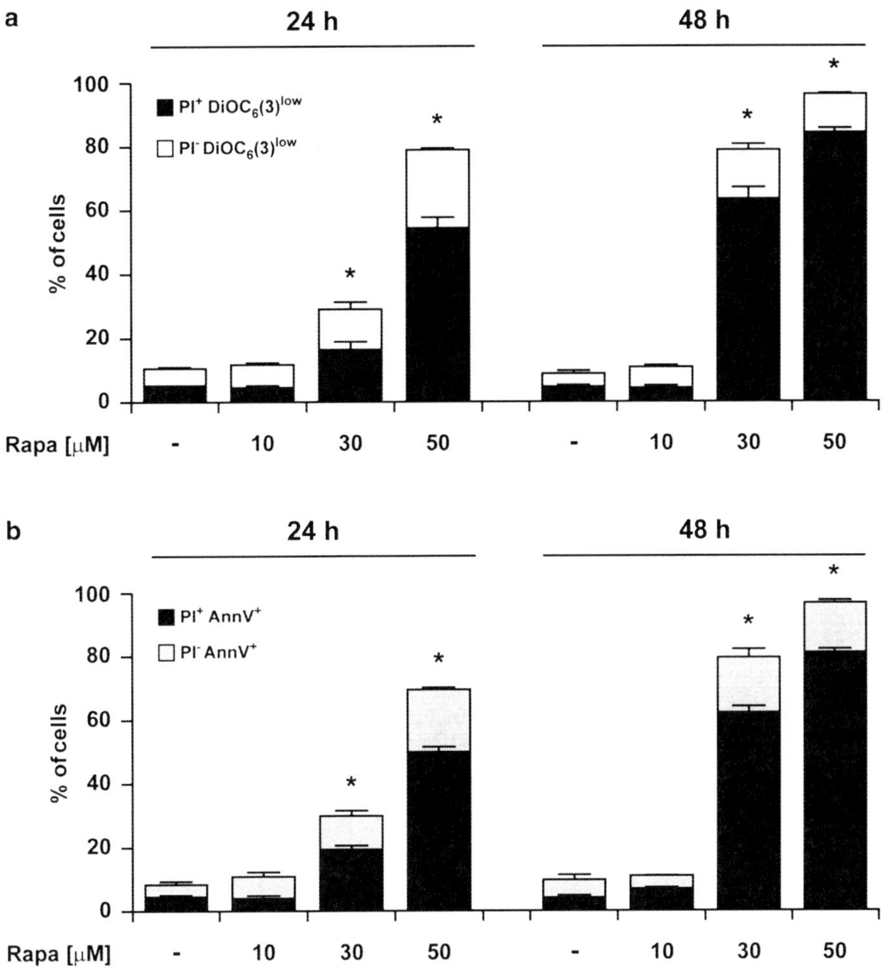

Fig. 7. Kinetics of cell death induction by rapamycin. Human cervical carcinoma HeLa cells were seeded in 12-well plates, let adhere for 12 h, and then treated with the indicated concentration of rapamycin (Rapa) or with an equivalent volume of DMSO (negative control) for further 24–48 h. The cells were then processed for the cytofluorometric quantification of mitochondrial transmembrane potential ($\Delta\psi_m$) dissipation and plasma membrane breakdown (as detailed in Subheading 3.4) or for the assessment of phosphatidylserine exposure (as described in Subheading 3.5) (**a**) *Black* and *white columns* depict the percentage of dead (PI^+ $DiOC_6(3)^{low}$) and dying (PI^- $DiOC_6(3)^{low}$) cells, respectively (mean ± SEM, $n = 3$, $^*p < 0.05$ as determined by unpaired Student's *t*-test). (**b**) *Black* and *light gray columns* illustrate the percentage of dead (PI^+ $AnnV^+$) and dying (PI^- $AnnV^+$) cells, respectively (mean ± SEM, $n = 3$, $^*p < 0.05$ as determined by unpaired Student's *t*-test).

With these considerations in mind, we propose a short set of guidelines aimed at providing researchers with a practical tool to approach the study of rapamycin-induced cell death.

1. In the experimental setting of choice, the occurrence of bona fide cell death in response to rapamycin should be assessed first, followed by a rough quantitative estimation as well as by an approximate determination of the timeline of the process.
2. When an idea has been obtained about these basic parameters, a more precise quantitative and qualitative evaluation of the

molecular pathways activated by rapamycin in the experimental scenario under consideration should be attempted. In particular, the investigator should aim at answering the following questions:

- By which cell death subroutine are cells succumbing upon the administration of rapamycin? Biochemical and kinetic determinations should be combined for the investigation of the most common cell-death-related parameters (26, 31, 32).
- Can rapamycin-induced cell death be totally ascribed to mTOR inhibition? If so, other mTOR inhibiting agents (e.g., CCI-779, RAD001, and AP23576) (58) as well as small interfering RNA (siRNA)-mediated depletion of mTOR should entirely reproduce the biochemical/morphological manifestations triggered by rapamycin. If not, which targets of rapamycin other than mTOR may be involved? The existence of putative additional rapamycin-binding proteins that might mediate (or at least participate into) the lethal effects of this macrolide should be elucidated.
- Is there an important cell-intrinsic component that affects rapamycin-induced cell death? The possibility that signaling cascades converging on mTOR are constitutively active or blocked due to genetic/epigenetic features of the cellular model under investigation should be explored.
- Is there an important cell-extrinsic component that affects rapamycin-induced cell death? Nutrient, amino acid and growth factor availability should be taken into account as factors that may considerably influence the effects of rapamycin.
- Is there an autophagic component in the observed phenomena? The accumulation of autophagosomes and the functional induction of autophagy should be estimated (16, 50).
- How does this autophagic component (if any) relate to rapamycin-induced cell death? If the autophagic pathway contributed to cell death execution, pharmacological/genetic inhibition of autophagy would block (or at least inhibit/retard) cell death. Conversely, if autophagy served a mere prosurvival purpose, its suppression would accelerate cell death. This should be determined by combining the administration of rapamycin with pharmacological agents that inhibit autophagy (e.g., 3-methyladenine) or with siRNAs for the depletion of essential autophagic modulators (16, 50).

The elucidation of these issues will undoubtedly help researchers with the challenging task of finely characterizing rapamycin-induced cell death.

4. Notes

1. Optimal growth conditions (e.g., medium composition, supplements) vary according to the cell line of choice, and may quite dramatically influence not only growth rates but also the response to rapamycin. In particular, the use of suboptimal growth conditions can per se trigger autophagy via mTOR inhibition, thereby relevantly aggravating the effects of rapamycin. Unless changes in the composition of the growth medium are part of the experimental protocol, we suggest using the medium that is most suitable for the cell line of choice, as recommended by the American Type Culture Collection (ATCC, Manassas, USA).
2. PI is not currently listed as a carcinogen by the National Toxicology Program (NTP), the International Agency for Research on Cancer (IARC) or the Occupational Safety and Health Administration (OSHA) but should be treated as such. Symptoms of exposure are undefined but this reagent should always be handled using protective gloves, clothing, and eyewear.
3. When stored under protection from light at 2–8°C, PI is stable for 18 months.
4. When stored sealed at −20°C, undissolved rapamycin (powder) is stable for at least 2 years. It is recommendable to prepare the stock solution of rapamycin in small aliquots (10–50 μL), which should be used within 6–8 months from preparation. To prevent unwarranted physicochemical modifications (potentially resulting in reduced levels of active rapamycin in the stock solution), it is also advisable to avoid each aliquot to undergo more than 1–2 freezing–thawing cycles.
5. TB is not currently listed as a carcinogen by the NTP, IARC, or OSHA. Information on the human health effects from exposure to this substance is limited. Upon direct contact, TB may behave as an irritant for the eyes, skin, gastrointestinal tract, and respiratory system.
6. Undissolved TB appears as a dark purple-brown powder, which is stable under ordinary conditions of use and storage (RT, sealed vials). TB solutions in aqueous solvents (e.g., dH_2O and PBS) tend to form precipitates with time, which are quickly dissolved by agitation unless solvent has evaporated to a significant extent (and TB concentration has trespassed the maximal level of solubility). In this case, the stock solution of TB should be prepared de novo.
7. CV is not currently listed as a carcinogen by the NTP, IARC, or OSHA. Information on the human health effects from exposure to this substance is limited. Upon direct contact,

CV may behave as an irritant for the eyes, skin, gastrointestinal tract, and respiratory system.

8. Undissolved CV appears as a dark purple powder, which is stable under ordinary conditions of use and storage (RT, sealed vials). CV solutions in methanol/ethanol are highly inflammable and should therefore be kept away from heat, fire, sparkling sources, and oxidizers.
9. CV solutions in methanol/ethanol are stable for at least 1 year under ordinary conditions of use and storage (RT, sealed vials), although they tend to become overconcentrated due to solvent evaporation. Overconcentrated solutions can be easily recognized by the presence of a purple precipitate that does not readily dissolve upon agitation. In this case, the CV solution should be prepared de novo.
10. Formaldehyde causes burns and is very toxic by inhalation, ingestion, and through skin absorption (which occurs very readily). Moreover, formaldehyde is a mutagen and is considered as a probable human carcinogen, and should hence always be handled using protective gloves, clothing, and eyewear.
11. Methanol is volatile and highly toxic due to the formation of formaldehyde and formic acid (an inhibitor of mitochondrial respiration causing hypoxia and lactate accumulation) by hepatic alcohol dehydrogenases.
12. Methanol is highly inflammable and should therefore be kept away from heat, fire, sparkling sources, and oxidizers. As a note, methanol flames are almost invisible in normal light conditions, but may be detected by the heat generated or the burning of other materials.
13. $DiOC_6(3)$ is not currently listed as a carcinogen by NTP, IARC, or OSHA, yet may behave as an irritant for the eyes, skin, and respiratory system.
14. Prolonged exposure to light of the stock solution should be avoided to prevent photobleaching.
15. CCCP may behave as an irritant upon inhalation, ingestion, or contact with the skin. Its thermal decomposition may yield toxic gases including carbon monoxide (CO), carbon dioxide (CO_2), hydrogen chloride (HCl), and hydrogen cyanide (HCN).
16. This reagent contains sodium azide (NaN_3), which is toxic by ingestion and under acidic conditions may release the highly toxic gas hydrazoic acid. Thus, it should always be manipulated with maximal care and preferably under an appropriate fume hood.
17. Under appropriate storage conditions (4°C, under protection from light) FITC-AnnV is stable for at least 12 months.

Exposure to light of the FITC-AnnV reagent should be carefully avoided to prevent photobleaching.

18. PFA is very toxic by inhalation and should be handled under an appropriate fume hood.
19. As the stability of PFA in solution is limited, it should be prepared shortly before use and kept on ice throughout the entire experiment.
20. Hoechst 33342 is currently included among substances that may cause concern to man owing to possible mutagenic effects but for which the available information is inadequate for making satisfactory assessments. It should be handled anyway with the maximal care, by using protective gloves, clothing, and eyewear.
21. As compared to other chromatinophilic dyes (e.g., 4′,6-diamidino-2-phenylindole dihydrochloride, DAPI) (Molecular Probes-Invitrogen), Hoechst is particularly prone to photobleaching. Exposure to light of the stock solution should, therefore, be carefully prevented.
22. Storage at 4°C of binding and blocking buffers should not exceed 2 weeks, unless 0.02% NaN_3 is added as a preservative to prevent microbial contamination (see also Note 16). The same applies to rehydrated BSA (lyophilized BSA has a shelf life of 1 year at 4°C).
23. Each antibody should be handled and stored as indicated following the manufacturer's instructions. Moreover, antibodies should never be subjected to steep temperature changes and should always be kept on ice when solutions for immunoblotting are prepared.
24. This material is considered hazardous by the OSHA Hazard Communication Standard and may behave as an irritant for the eyes, skin, and respiratory system.
25. The *DC* Protein Assay kit contains sodium hydroxide (NaOH), which is corrosive and may cause severe burns of the skin, eyes, and gastrointestinal system upon contact/ingestion.
26. The *DC* Protein Assay kit has a shelf life of at least 6 months from the date of purchase.
27. This material is considered hazardous by the OSHA Hazard Communication Standard, is corrosive and may cause severe burns of the skin, eyes, and gastrointestinal system upon contact/ingestion.
28. EDTA is not currently listed as a carcinogen by NTP, IARC, or OSHA, yet may behave as a mild irritant for the eyes, skin, and respiratory system.
29. PMSF is toxic by inhalation, upon contact with skin, and if swallowed, and should therefore always be manipulated with maximal care and preferably under an appropriate fume hood.

30. DTT is considered hazardous by the OSHA Hazard Communication Standard, and can cause skin irritation and severe ocular damage.
31. Before the addition of the protease inhibitor cocktail, the lysis buffer can be stored at 4°C for up to 6 months. Once the protease inhibitor tablets have been added, it is recommendable either to use the lysis buffer within 24 h (storage at 4°C) or to store small aliquots at –20°C. These are stable for at least 6 months, but should not be refrozen upon the first use.
32. This reagent contains sodium bisulfate and *N*,*N*-dimethylformamide (DMF), which is readily absorbed by inhalation/through the skin and can act as an irritant for the eyes, skin, gastrointestinal tract, and respiratory system.
33. This reagent contains SDS, which can irritate the eyes and the skin, and hence should be handled using protective gloves, clothing, and eyewear.
34. These products are stable for at least 12 months under ordinary conditions of use and storage (4°C or RT, close bottles or sealed packages).
35. This product contains no substances which at their given concentration, are considered to be hazardous to health.
36. This reagent contains 500 mM DTT (see also Note 30).
37. Information on the human health effects deriving from exposure to Ponceau S is limited. It should therefore be handled with the maximal care, by using protective gloves, clothing, and eyewear.
38. Tween 20 is not currently listed as a carcinogen by NTP, IARC, or OSHA, yet may behave as a mild irritant for the eyes, skin, and respiratory system.
39. Depending on specific requirements (e.g., physical room within incubators, need for very large amounts of cells), 175-cm^2 flasks can be substituted for with smaller or larger supports for cell culture.
40. While HeLa cells promptly grow on untreated supports, pretreated (e.g., collagen-coated, gelatin-coated) flasks may be required for other cell types.
41. Indicatively, in these conditions, a 175 cm^2 flask contains $15–20 \times 10^6$ HeLa cells. This may considerably vary for other cell lines, mainly depending on cell size.
42. The subculture of both highly underconfluent and excessively confluent flasks should be avoided as it may result in a genetic drift and in the overcrowding-dependent sufferance of the cell population, respectively, both of which negatively affect experimental determinations.

43. It is recommendable to allow freshly thawed cells to readapt for 2–3 passages before using them in experimental assessments. Similarly, it is good practice to define a maximal number of passages and to discard maintenance cultures once this limit has been reached, to avoid genetic population drifts. This strategy requires an appropriate liquid nitrogen stock of cells, which should be generated from a homogenous and healthy population before the beginning of the experiments.
44. Quickly adhering cells (e.g., HeLa) can be treated after 12 h, whereas at least 24–36 h should be allowed for cells with longer adaptation times (e.g., HCT 116). Premature treatment may result in excessive toxicity due to suboptimal culture conditions. Light microscopy-assisted observation of the cell morphology normally suffices to determine the adherence/adaptation status.
45. Whenever possible, we recommend to treat cells with (at least) three different concentrations of rapamycin (ideally $IC_{50} \times 10^{-1}$, $IC_{50} \times 10^{0}$, and $IC_{50} \times 10^{+1}$), to avoid insufficient or excessive stimulation.
46. With regard to this, it should be remembered that, also at very low concentrations, organic solvents such as DMSO may be toxic per se or may promote specific cellular responses (e.g., differentiation). Such effects and the concentration of DMSO at which they appear should be carefully determined at a preliminary stage, and should be taken into careful consideration for the definition of the experimental plan.
47. Cell density at seeding depends on cell size, proliferation rate, well surface, and total assay duration. To avoid the appearance of overcrowding-dependent toxicity, control cultures at the end of the assay should exhibit confluence levels that never exceed 85%. For assays lasting no more than 72 h, $100–200 \times 10^3$ ($50–100 \times 10^3$) HeLa cells in 1 mL (0.5) complete growth medium can indicatively be seeded in 12- (24-)well plates.
48. Supernatants from treated cultures should not be discarded since they contain the fraction of cells which underwent apoptosis-dependent detachment from the substrate during the stimulation period.
49. Trypsinization time is a function of cell type, viability, and culture density. For HeLa cells (and most cell lines), 5 min at 37°C are amply sufficient to fully detach a totally confluent, healthy population. It is, however, recommendable to check for complete cell detachment by visual inspection or by rapid observation in light microscopy.
50. The detachment of some cell types may require slightly prolonged trypsinization. This may also be necessary if traces of medium (containing FBS, which inactivates trypsin) remain

in the wells after the collection of supernatants. The latter occurrence can be circumvented by rapidly washing wells with prewarmed PBS immediately before the addition of trypsin–EDTA solution.

51. Excessive trypsinization (>10 min) should be prevented, since it may result per se in some extent of cell sufferance. At least initially, this is limited to a minimal level of treatment-unrelated $\Delta\psi_m$ dissipation and hence does not affect assays that monitor phenomena associated with end-stage cell death such as TB exclusion and PI incorporation. Conversely, this can lead to the (slight) overestimation of both $\Delta\psi_m$ loss and PS exposure. Trypsinization can be replaced by mechanical scraping, although this may also induce $\Delta\psi_m$ loss in sensible cells (see also Note 70).

52. As dead cells float (or are very loosely attached to the substrate) and stressed/dying cells detach faster than their healthy counterparts, it is important to ensure that all cells have been collected by trypsinization. This is critical to avoid the overestimation of cell-death-related parameters due to an unbalanced loss of healthy cells from samples.

53. FBS contained in supernatants definitely inactivates trypsin at this stage. Should supernatants be scarce (<500 μL) or should derive from FBS-free culture conditions, complete trypsin inactivation may be warranted by adding 3–5 mL prewarmed complete growth medium to each sample.

54. While apoptotic cells are more dense, and hence precipitate faster than their healthy counterparts, cells undergoing necrosis are swollen and exhibit reduced precipitation rates. Insufficient centrifugation should therefore be avoided as it may lead to over-/underestimation of cell death.

55. PBS can be substituted for by complete growth medium or another balanced salt solution, but not by dH_20, as this will lead to osmotic lysis of the vast majority of cells.

56. The number of cells may exhibit consistent variations among samples, depending on cell density at seeding, well surface, rapamycin concentration, and duration of the assay. To avoid excessively or insufficiently dense counting fields, at this stage it is recommended to perform multiple dilutions of each sample (ideally three, i.e., 1:5, 1:10, and 1:20), of which only the best 1–2 will be used for quantification.

57. For each sample, the percentage of cells exhibiting plasma membrane breakdown (TB^+) should be assessed in a statistically relevant population (at least 200 cells).

58. It is important to remember that the TB exclusion assay tends to underestimate cell death, as it is intrinsically unsuitable for the detection of cells in the early stage of the process (which

still have intact plasma membranes), as well as cells that have already undergone disintegration (whose particulate debris (1) are too small to be visible in light microscopy and (2) have been partially lost at centrifugation).

59. Alternatively, the challenge with rapamycin can be prolonged until the end of the assay. Nevertheless, complete growth medium (either supplemented with rapamycin or not) has to be replaced with fresh one every 2–3 days, to avoid the reduction of clonogenic potential that may derive from glucose and/or amino acid deprivation.

60. Washing should be gentle to avoid the detachment of colonies.

61. Upon use, the fixative/staining solution can be collected, filtered (with a 0.22 μm-pore filter, to remove detached cells and particulate debris) and subsequently reemployed for at least 1–2 times.

62. Most often, only colonies composed of at least 50 cells are scored.

63. Quantification can be performed manually (ideally by at least three independent operators) or upon image acquisition and analysis with specific plug-ins of the Java®-based, open-source software ImageJ (freely available at http://rsb.info.nih.gov/ij/).

64. Results are normally expressed in terms of surviving fraction (SF), which is calculated as the ratio between the number of colonies observed at the end of the assay and the number of cells seeded at its beginning. In control conditions (untreated/solvent-treated cells or wild type cells), SF represents plating efficiency. In clonogenic assays, the IC_{50} is defined as the concentration of a chemical that is able to reduce the SF to half of the plating efficiency.

65. Ideally, colonies should be quantified shortly (1–2 days) after fixation/staining. Still, CV labeling is stable and plates can be stored for very long periods (>1 year) at 4°C or RT. Before quantification, plates stored for more than 1 month should be rinsed with dH_2O to minimize background (which may have accumulated with time, see also Note 60).

66. In these experimental conditions, $DiOC_6(3)$ quickly accumulates within healthy cells' mitochondria in a virtually nonsaturatable fashion. Thus, to avoid intrasample staining variations due to different absolute quantities of this fluorochrome, it is crucial to stain all samples with an identical volume of (the same) $DiOC_6(3)$-containing medium.

67. While at these concentrations $DiOC_6(3)$ exhibits negligible self-quenching effects (see also Note 70) (33–36), higher $DiOC_6(3)$ concentrations result in relevant quenching of the

fluorescent signal and in the staining of extra-mitochondrial compartments such as the endoplasmic reticulum (32).

68. To avoid the persistence of cell clumps that may have formed during centrifugation, which lead to an uneven staining of the cell population (see also Note 70), resuspension should be quite vigorous. Vortexing samples for 10–15 s is appropriate for most cell lines.

69. To avoid $DiOC_6(3)$-dependent toxicity, staining should not be prolonged for more than 45 min. Very sensible cell lines may exhibit some extents of $DiOC_6(3)$-dependent sufferance even with minimal staining times.

70. Excessively short staining, insufficient $DiOC_6(3)$ concentrations (<10 nM, see also Note 67) as well as the presence of cell clumps (see also Note 68) may lead to the uneven staining of the cell population, which favors the overestimation of dying ($DiOC_6(3)^{low}$) cells. This can be quite easily recognized by comparing the microscopic appearance of control cultures with the percentage of $DiOC_6(3)^{low}$ events recorded therein. If control cultures look normal, no more than 10–15% $DiOC_6(3)^{low}$ cells should be observed. If this percentage is higher, one of the abovementioned circumstances may have occurred, or cells might have suffered at detachment (see also Note 51).

71. When numerous samples (>24) are analyzed, it is recommended to limit the variation of the time intervening between labeling and the actual cytofluorometric acquisition. To this aim, samples should be divided in groups of 20–24, and subsequent series should be stained at 10–15 min intervals from each other.

72. Although in these conditions $DiOC_6(3)$ is minimally affected by photobleaching, as a good cytofluorometric practice, samples should be protected shielded from light from this step onward (see also Note 86).

73. The vital dye should be chosen by carefully taking into account its spectral properties in relationship to those of the employed mitochondrial probe ($DiOC_6(3)$, in this case) so that the fluorescent emission from the two molecules does not overlap and can be subsequently discriminated.

74. As a guideline, $DiOC_6(3)$ (which emits in green, absorption/emission peaks at 484/501 nm) can be associated with the vital dye propidium iodide (PI, emitting in red, absorption/emission peaks at 535/617 nm) (31). PI staining should be performed shortly before cytofluorometric analysis, by adding the dye directly to samples at a final concentration of 0.5–1 μg/mL.

75. As an alternative to $DiOC_6(3)$, 150 nM tetramethylrhodamine methyl ester (TMRM, fluorescing in red, absorption/emission

peaks at 543/573 nm) can be used to quantify $\Delta\psi_m$ together with 10 μM DAPI (that emits in blue, absorption/emission peaks at 358/461 nm) for the assessment of viability.

76. For instance, we most often employ a FACScan or a FACSVantage cytofluorometer (BD, San José, USA) equipped with an argon ion laser (emitting at 488 nm) for excitation. In this case, the following channels are appropriate for fluorescence detection: FL1 for $DiOC_6(3)$ and FITC (see also Note 87); FL2 for TMRM; FL3 for PI. As a note, DAPI requires UV excitation and hence is intrinsically incompatible with these technical platforms.

77. As a reminder, while FSC depends on cell size, SSC reflects the refractive index, which in turn is influenced by multiple parameters including cellular shape and intracellular complexity (i.e., presence of cytoplasmic organelles and granules).

78. Cytofluorometers are normally provided with dedicated software for instrumental control and analysis. Unfortunately, the execution of nearly all these packages is limited to the computer that controls the cytofluorometer and, very often, to a single operating system (i.e., Mac). A number of analytical software packages (for both Mac and PC-based systems) are available online free of charge. For PC users, we recommend WinMDI 2.8 (© 1993–2000, Joe Trotter), originally developed for Windows® 3.1 but compatible with new versions of the operating system (freely available at http://www.cyto.purdue.edu/flowcyt/software.htm).

79. At the analytical stage, it should be kept in mind that some $\Delta\psi_m$-sensitive probes including $DiOC_6(3)$ and TMRM are influenced by mitochondrial $\Delta\psi_m$-unrelated (e.g., mitochondrial mass) as well as by mitochondria-independent parameters (e.g., cell size, plasma membrane potential, and activity of multidrug resistance pumps) (33, 35, 36). To adequately evaluate these possibilities, each sample should be divided in two aliquots before centrifugation (Subheading 3.2, steps 3 and 4), and only one of these two sets should be preincubated with 50–100 μM CCCP (a ionophore causing complete and irreversible $\Delta\psi_m$ dissipation) for 5–10 min before staining (37°C, 5% CO_2).

80. If $DiOC_6(3)$ (or FITC-AnnV) has been used in association with PI, evaluation of $\Delta\psi_m$ (or PS exposure) can be limited to cells with intact plasma membranes. This may be especially useful to exclude necrotic cells (be them the result of primary or secondary necrosis) from the analysis. To this aim, a double gating, first on events characterized by normal FSC and SSC (see also Note 77) and second on PI^- events, can be performed.

81. Two washes at a reduced centrifugation speed (150 × *g*) are recommended if samples contain or may be contaminated with platelets. The presence of platelets should be carefully avoided since activated platelets not only expose PS (and hence fix AnnV) but also stick to cells (59), thereby consistently increasing the possibility of false positive results.
82. To avoid plasma membrane damage, which also results in treatment-unrelated (and hence false) AnnV positivity, washes should be performed as gently as possible.
83. The manufacturer recommends the use of 10 μL FITC-AnnV/100 μL binding buffer for the staining of up to 10^6 cells. In our experience, to adequately label samples derived from 24- and 12-well plates, respectively (even in the case of highly confluent cell cultures), the volume of FITC-AnnV can be safely scaled down to 2–4 μL. However, the amount of FITC-AnnV that is required to achieve a correct staining may vary depending on cell type (as distinct cells may differ from each other relative to plasma membrane surface or intracellular levels of PS). Thus, a few preliminary experiments aimed at optimizing staining conditions are recommended.
84. As an alternative, the entire protocol can be performed at room temperature. In this case, the staining time should not exceed 15–30 min (see also Note 85).
85. Please note that excessive staining times as well as temperatures >25°C may result in unspecific cell labeling, thereby leading to false positive results (see also Note 84).
86. Protection from light is critical as it prevents photobleaching of the FITC fluorochrome, which is much more sensitive to this phenomenon than $DiOC_6(3)$ (see also Note 72).
87. FITC emits in green (absorption/emission peaks at 494/521 nm) (see also Note 76).
88. Coverslips are presterilized by incubation for 15–30 min in 100% ethanol. As an alternative, nonsterile coverslips can be deposited in plates, followed by short (2–5 min) exposure to UV light (under the safety cabinet as if this was undergoing the daily sterilization cycle). Please note that UV sterilization requires plates to be open (as covers may adsorb a nonnegligible fraction of the UV light) and placed as close as possible to the UV source. At cabinet restart, the laminar flow will intervene to preserve sterility until covers are put back in place.
89. When scarcely adhering cells are employed, coverslips can be coated with poly-L-lysine directly upon their delivery to the plate. This is achieved by incubating coverslips in 0.01% poly-L-lysine solution (Sigma-Aldrich) for 10 min at 37°C, followed by rinsing in PBS. It should be noted that some cell types are

particularly sensitive to poly-L-lysine and fail to grow on poly-L-lysine-coated coverslips.

90. Although manufacturer's instructions recommend to transfect cells at 50% of confluence, we found that transfection rates are optimal when cells are slightly more confluent (70–90%). Still, excessive confluence (resulting in restrained proliferation) should be carefully avoided, as it leads to a significant drop in transfection efficiency.
91. When diluted Lipofectamine™ 2000 and the plasmid DNA are mixed, the solution may become cloudy.
92. Liposome-mediated plasmid transfection is carried out entirely at RT under a common safety cabinet. However, it is recommendable to keep the tubes containing plasmid stocks and the Lipofectamine™ 2000 reagent on ice (and to return them to storage conditions immediately after use), to minimize solvent evaporation and avoid liposome degradation.
93. To avoid intrawell variations of transfection efficiency as well as local toxicity, transfection complexes (which exhibit a very high affinity for plasma membranes) should be added to cells very slowly and dropwise, while trying to cover the whole surface of the growth medium.
94. HeLa cells are easily transfectable, and the use of Lipofectamine™ 2000 rarely results in transfection rates lower than 70%. For hardly transfectable cells, other liposome-based reagents including Attractene® (Qiagen, Hilden, Germany) can be tested. Electroporation might also be considered as an ultimate alternative to liposome-mediated transfection. As a note, a minimum transfection efficiency of ~50% is required to quantify GFP-LC3 aggregation by fluorescence microscopy.
95. We recorded the expression of the GFP-LC3 fusion protein as soon as 12 h after transfection. Faster expression is not excluded, but should be controlled in each specific experimental setting.
96. In HeLa cells, this reportedly leads to 40–50% cells displaying cytoplasmic GFP-LC3 aggregation (57). Other cell lines may respond to glucose and amino acid deprivation by activating autophagy to different extents or with different kinetics.
97. Washes should be performed by gentle aspiration and pipetting to avoid excessive detachment of cells from coverslips.
98. For nuclear counterstaining, Hoechst 33342 can be replaced with 300 nM DAPI. In this case, coverslips should be incubated at RT for no longer than 5–10 min.
99. To reduce the intensity of the background signal from nuclear counterstaining, the number washes can be increased up to 5.

100. Insufficient drying of the mounting medium may provoke the annoying displacement of coverslips during microscopic observation.

101. Generally, stained and mounted coverslips can be stored at 4°C under protection from light for several weeks. However, a time-dependent decrease in fluorescence (whose rapidity depends on the probe/fluorochrome in use as well as on storage conditions and notably is faster for Hoechst 33342 and DAPI than for GFP) will eventually result in the impossibility to analyze the samples.

102. To avoid cross-channel interference, we recommend performing a separate photographic acquisition for each channel (and eventually merge the images, if necessary).

103. Commercial software packages allowing for single-channel and merged editing such as Adobe® Photoshop® (Adobe Systems Inc., San Jose, USA) or ImageJ (see also Note 63) can be employed to facilitate the analysis.

104. For each experimental condition, the frequency of cells exhibiting the cytoplasmic aggregation of GFP-LC3 should be determined within a statistically relevant population (at least 200 cells). Moreover, the number of autophagosomes/cell should also be quantified in a representative sample of (at least) 50 cells.

105. LC3-I → LC3-II conversion is an early manifestation of autophagy, and as the autophagic flow proceeds LC3-II is degraded within auto(phago)lysosomes, thereby becoming barely detectable. For this reason, cells that will constitute positive controls should not be incubated in HBSS for more than 4 h. This notion should also be kept in mind when immunoblotting results are analyzed (see also Note 163).

106. To minimize proteolytic degradation, from this step onward samples should always be kept on ice (until the addition of loading buffer and denaturation or storage).

107. Lysis should be performed in the minimal volume of lysis buffer that allows for the complete dissolution of cell pellets (30 μL normally suffice for samples derived from 6-well plates). Lysis may be facilitated by thorough pipetting, vortexing, or sonication. If some particulate debris remains undissolved, 5–20 μL supplemental lysis buffer can be added, followed by another round of pipetting, vortexing, or sonication. Keeping the lysate volume to a minimum is important to avoid an excessive dilution of the protein content of samples, which may turn out to be problematic at gel loading (see also Notes 111 and 112).

108. Lysates can be stored at −20°C for prolonged periods (several months) without any significant degradation of their protein content.

109. Protein quantification is performed by interpolating a calibration curve built from a serial dilution of BSA. BSA standards are rarely (if ever) included in protein quantification kits and should be prepared shortly before use. Either dH_2O or (more properly) lysis buffer can be used as solvent.

110. The amount of total proteins that should be loaded onto the polyacrylamide gel for separation depends on multiple variables, including size and expression levels of the protein of interest as well as the performance of the antibodies used at detection. Although LC3 is small (15–18 kDa), in our experimental conditions 30–50 μg total proteins normally suffice to visualize the corresponding band upon immunoblotting. The same amount is also appropriate for the detection of p62 (62 kDa).

111. When the amount of total proteins for loading has been determined, lysates that will be separated on the same polyacrylamide gel are diluted in the minimal volume that allows for the loading of the same quantity of proteins and the addition of 4× loading buffer (to a final 1× concentration). Keeping the loading volume to a minimum facilitates the loading procedure (see also Notes 107 and 112).

112. According to the manufacturer, these are the maximal loading volumes for 1 mm-thick NuPAGE® Novex 4–12% Bis-Tris precast gels (Invitrogen): 1-well gel: 700 μL/well; 9-well gel: 28 μL/well; 10-well gel: 25 μL/well; 15-well gel: 18 μL/well; and 17-well gel: 15 μL/well. The volume that actually fits into such wells is 10–15% higher if loading is performed very slowly while keeping the pipette tip at the well upper limit (see also Notes 107 and 111).

113. Boiling is intended to integrate the denaturing activity of lysis and loading buffers (which contain SDS, reducing agents and calcium chelators), thereby fully resolving inter- and intramolecular bonds that are responsible for tertiary and quaternary protein structures. This step ensures that the subsequent electrophoretic separation is based on the MW of protein subunits with no (or little) influence from protein-to-protein interactions and native structural features (see also Note 114).

114. Incomplete denaturation may have either of the following consequences at electrophoresis. (Partially) folded polypeptides often migrate with an apparently lower MW (as compared to the predicted value), as they retain (at least in part) some structural compactness that favors their progress through the molecular sieve provided by the gel. Conversely, proteins that are engaged in unresolved protein–protein interactions migrate with an apparently higher MW, which (at least theoretically) reflects the sum of the MW of all interacting polypeptides (see also Note 113).

115. Alternatively, samples can be stored at −20°C for prolonged periods (several months) without significant degradation of their protein content.

116. 1× NuPAGE® MES SDS running buffer and 1× NuPAGE® transfer buffer are stable under ordinary conditions of storage (4°C, close bottles).

117. Used 1× NuPAGE® MES SDS running buffer and 1× NuPAGE® transfer buffer can be recycled for at least 2–3 additional electrophoretic runs and electrotransfers, respectively.

118. Failure to do so will result in low or no current during the electrophoretic run, due to failing electric circuit within the electrophoresis unit.

119. If wells get seriously damaged (their walls are broken), adjacent samples may mix up at loading. In this case, the use of a new gel should be envisioned. Very frequently, the damage is limited (walls are distorted but intact), and repair can be attempted with the help of a pipette tip.

120. Polyacrylamide leftovers may impede correct gel loading, result in the mix up of adjacent samples and/or affect the entry of proteins in polyacrylamide, and hence should be carefully avoided.

121. If only one polyacrylamide gel is used, replace the second gel cassette with the plastic buffer dam provided with the XCell4 SureLock™ Mini-Cell unit.

122. Importantly, the running buffer should completely cover the wells but not create a communication between the upper and lower chambers, as this would compromise the correct electrical flow within the unit.

123. The reducing agents that are contained in the NuPAGE® LDS sample buffer do not comigrate with proteins through the gel in a neutral pH environment. Although disulfide bonds are poorly reactive at neutral pH and hence poorly prone to reoxidize, some reoxidation may occur during electrophoresis in the absence of an antioxidant. The NuPAGE® Antioxidant migrates with the proteins during electrophoresis, thereby fully preventing reoxidization.

124. We recommend to remove bubbles to eliminate any potential source of mix up between adjacent samples at loading.

125. Prestained markers that cover various MW ranges and allow for the visual follow-up of migration are available from distinct commercial provider. For practical reasons, we routinely use the large-range Precision Plus Protein™ Standard MW markers (Bio-Rad), covering from 10 to 250 kDa. An appropriate alternative for the detection of LC3 and p62 is provided by the low-range Natural Prestained SDS-PAGE Standards (Bio-Rad), covering from 14.4 to 97.4 kDa.

126. As a good practice, to avoid the diffusion of samples within the gel (in particular of those that were loaded first) and hence maximize resolution, it is recommended to perform loading as quickly as possible (though very carefully) and to start the electrophoretic run immediately thereafter.

127. According to the manufacturer, protein samples loaded on NuPAGE® Novex 4–12% Bis-Tris precast gels in NuPAGE® MES SDS running buffer should be separated at 200 V, resulting in a runtime of ~35 min and in expected currents at the start and at the end of the run of 110–125 and 70–80 mA/gel, respectively. Although it results in increased runtimes, we normally perform separation at 150–180 V, as this relevantly ameliorates resolution (especially for low MW proteins such as LC3-I and LC3-II).

128. In most cases in which the 10 kDa MW marker trespasses this limit, LC3-II (and sometimes also LC3-I) cannot be detected. This occurs either due to the fact that the corresponding band(s) has (have) run out of the gel at electrophoresis and hence has (have) been lost in the running buffer, or because the band(s) is (are) too close to the gel "foot" and therefore is (are) cut away for the electrotransfer (see also Note 132).

129. As an alternative, polyvinylidene fluoride (PVDF) membranes can be employed. In this case, activation should be performed by incubating membranes for 30–60 s in methanol, ethanol or isopropanol, followed by brief rinsing in dH_2O and equilibration in transfer buffer.

130. Use caution while inserting the gel knife between the two plastic shields to avoid excessive pressure toward the gel. While removing the upper plastic shield, to minimize the risk of gel rupture, proceeds slowly and ensure that the gel remains entirely adherent to the lower plastic plate.

131. Alternatively, wells and gel "feet" can be cut away when gels still lay on the lower plastic shield, followed by assembly of the transfer "sandwich" (Subheading 3.7.3, step 6).

132. Cutting the gel "feet" is obligatory, as the correct assembly of the transfer "sandwich" will otherwise be prevented. For this reason, any protein that may have migrated down to this level of the gel is irremediably lost (see also Note 128).

133. During assembly, the formation of bubbles between the various layers of the "sandwich" should be carefully avoided, as this will locally interfere with protein transfer and result in the generation of protein-free zones on membranes (which are easily visualized by subsequent Ponceau S staining, Subheading 3.7.3, step 12, see also Note 143). To remove bubbles, a common 5 mL pipette can be gently rolled on top of each component before the addition of the next one.

134. During assembly, great attention should be made at respecting the reciprocal order among the anodic core (+) of the blot module (which attracts proteins that are negatively charged by SDS), membranes, gels, and the cathodic core (–) of the blot module. Although this might seem trivial, the loss of proteins due to a "reverse" electrotransfer occurs more frequently than one would predict.

135. Do not fill all the way to the top of the blot module as this will only generate extra conductivity and heat.

136. This serves to dissipate part of the heat produced during the run.

137. Expected currents at the start and at the end of the run are 170 and 110 mA, respectively.

138. These settings are appropriate for both nitrocellulose and PVDF membranes (see also Note 129).

139. The efficient electrotransfer of high MW proteins (>100 kDa) may require more than 1 h. In this case, the temperature of the electrotransfer unit might increase excessively, if not controlled, due to the intense current. To avoid this problem, the entire electrotransfer procedure can be performed in a cold room (4°C).

140. Upon careful rinsing in PBS or dH_2O, sponges can be reused. Filter papers are discarded.

141. If MW markers are still (partially) visible on the gels (and poorly on membranes), the electrotransfer has occurred with suboptimal efficiency. When an obvious reason for this can be identified and resolved, and if "sandwiches" have not been completely dismantled (see also Note 142), the transfer module can be reassembled and another electrotransfer run can be started.

142. "Sandwiches" should be removed from the transfer module with the maximal care and the correct transfer of MW markers should be checked prior to their dismantling (by slightly opening the layers of the "sandwich" at one corner). This is very important as it allows for the proper reconstitution of "sandwiches" and for another electrotransfer run if the efficiency of the first one has been suboptimal (see also Note 141).

143. Ponceau S staining allows for the identification of gross transfer problems (e.g., bubbles, see also Note 133) as well as for highly unequal lane loading. Stained membranes can be wrapped in transparent plastic film and photocopied or scanned, to keep track of the correct transfer.

144. As it contains 5.0% acetic acid, the Ponceau S staining solution favors the fixation of electrotransferred proteins onto nitrocellulose membranes. For this reason, it is advisable to incubate nitrocellulose membranes in Ponceau S staining

solution for at least 5 min, whereas 1–2 min normally suffice for PDVF membranes (for which the staining only serves visualization purposes).

145. Cutting should be avoided if – for any reason – the electrophoretic run has not proceeded to completion (as witnessed by the incomplete resolution of MW marker bands along the membrane). In this case, intact membranes should be subjected to one round of immunoblotting for each antibody, separated from each other by multiple washes and dehybridization (see also Note 161).

146. It is advisable to cut membranes before washing away the Ponceau S stain (Subheading 3.7.3, step 14). In this case, colored protein bands provide indeed a valuable guide for cutting membranes along an approximately straight line that involves a MW range as narrow as possible.

147. Loading control should be performed with an antibody that recognizes a highly expressed protein, whose amount is stable in most experimental settings. As an alternative to GAPDH, α-tubulin (MW = ~52 kDa) or β-actin (MW = ~45 kDa) levels can be monitored to ensure the equal loading of lanes.

148. Used Ponceau S solutions retain nearly all their staining power, and hence can be reused several times over prolonged periods (several months). Precipitates and polyacrylamide leftovers that may accumulate should be removed every now and then by filtering the solution with a 0.22 μm filter.

149. Alternatively, binding buffer can be used. Binding buffer is recommended for both blocking and hybridization (Subheading 3.7.4, step 2 and Note 151) by the majority of antibody suppliers, as it is characterized by a strictly defined molecular composition (and hence is less prone to interfere with the antibody/antigen interaction). Nevertheless, the use of blocking buffer should be preferred (unless specifically indicated in the antibody datasheet) as it generally results in reduced unspecific background signal at revelation (Subheading 3.7.4, steps 6–7 and Notes 155–160).

150. To avoid degradation, primary antibody stocks should always be kept in near-to-storage conditions, in particular concerning temperature. This can be achieved by preparing hybridization solutions as rapidly as possible, and with the help of ice baths (for antibodies stored at 4°C) or precooled ice packs (for antibodies stored at −20°C). It is a good laboratory practice to split antibodies at dispatch into aliquots that are sufficient for 1–2 membranes.

151. The anti-GAPDH, anti-LC3 and anti-p62 antibodies that we routinely employ (see also Table 1) are perfectly compatible

with blocking buffer (see also Note 149). This may not apply to other antibodies that recognize the same proteins.

152. As an alternative, hybridization with anti-LC3 and anti-p62 antibodies can be achieved by incubating membranes in the corresponding solutions for 2 h at RT.

153. Used hybridization solutions can be stored and reemployed as follows: anti-GAPDH (stored at –20°C) at least ten times; anti-LC3 (stored at –20°C) 4–5 times; anti-p62 (stored at 4°C) 1–2 times. However, please note that the storage of binding and blocking buffers at 4°C should not exceed 2 weeks (see also Note 22).

154. As secondary antibodies are coupled to an enzymatic activity, it is not recommendable to store them for long periods upon first usage. In exceptional cases, however, they can be stored at 4°C for no more than 24 h and reused once (usually with suboptimal, yet acceptable, results).

155. Attention should be paid to ensure that membranes are entirely covered by the chemiluminescent substrate.

156. Photographic films are highly sensitive to light and should be handled only in the dark room. Films accidentally exposed to light are useless and can be discarded. Accidental exposure of a box can be quickly checked by developing one of its films. Upon development and fixation, properly handled films that have never been exposed to light are virtually transparent. The presence of large black zone indicates a quite intense exposure. Haloed films result from prolonged exposure to low intensity light.

157. Exposure time for the correct visualization of protein bands varies according to the intensity of the HRP-catalyzed chemiluminescent reaction. In turn, this is influenced by several factors including the amount of protein in the band, the affinity of primary and secondary antibodies, the presence of contaminants that inhibit HRP, and the time passed since the addition of the chemiluminescent substrate. As a general rule, exposure times between 1 and 5 min provide good results. When proteins are small or scarcely expressed (or when the antibodies display reduced affinity), longer exposures (up to 10–12 h) should be attempted. On the contrary, when proteins are large or well represented in the sample (or when the antibodies exhibit high affinity), a few seconds may be sufficient. Some trials may be required to obtain the correct exposure time. Given the time-dependent decay of the luminescent signal, short exposures should be performed before longer ones. Indicatively, exposures of 1–5 min are appropriate for visualizing both LC3I/II and p62. On the contrary, the signal from the anti-GAPDH antibody is usually very strong and an exposure of 1 s is largely sufficient (see also Notes 158 and 161).

158. Revelation of the anti-GAPDH antibody is often associated with overexposure and/or with the visualization of "fuzzy" bands (due to the improper displacement of the film relative to the membrane, see also Notes 157 and 161). To circumvent these issues, exposure of the anti-GAPDH antibody can be performed 30–120 min after the addition of the chemiluminescent substrate, when the efficiency of the enzymatic reaction has decreased enough to allow for standard exposure times.

159. If the LC3I/II and p62 bands cannot be detected upon overnight exposure, membranes can be collected and washed three times in rinsing buffer (10 min, RT), followed by rehybridization with freshly prepared secondary antibodies, washing and revelation with the SuperSignal West Femto Chemiluminescent Substrate (Pierce Biotechnology), as recommended by the manufacturer (Subheading 3.7.4, steps 4–7 and Notes 154 and 155).

160. Chemiluminescence can also be detected with dedicated imaging devices, such as the G:BOX Chemi XT16 gel analysis system (Syngene, Frederick, USA). Although punctual sensitivity might be suboptimal, these imaging devices allow for signal integration, which represents a considerable advantage when the chemiluminescence reaction is weak and optimal film exposure may become hard to achieve.

161. If cutting could not be performed (and hence hybridization involved a single primary antibody, see also Note 145), after revelation membranes are washed three times in rinsing buffer (10 min, RT) and antibodies are removed by incubating them in dehybridization buffer for 30–60 min (RT or 37°C), as recommended by the manufacturer, followed by extensive washing with dH_2O. Membranes processed as such can be reprobed with another primary antibody starting from the saturation of unspecific binding sites onward (Subheading 3.7.4, steps 1–7 and Notes 154 and 155). In this case, as the anti-GAPDH antibody has a very strong affinity for GAPDH and provides a very high signal (which may be hardly dehybridized, see also Notes 157 and 158), immunoblotting to monitor the equal loading of lanes should be performed as the final round.

162. Upon revelation, membranes are washed three times in rinsing buffer (10 min, RT) and can be stored at 4°C (in rinsing buffer) for up to 6 months. If additional immunoblotting rounds become required within this period, stored membranes can be normally processed starting from the saturation of unspecific binding sites onward (Subheading 3.7.4, steps 1–7 and Notes 154 and 155).

163. Immunoblotting data should be interpreted by keeping in mind the following profiles: healthy cells are characterized by

high levels of p62 and of LC3-I, but barely detectable LC3-II; cells in the initial stages of autophagy exhibit diminished levels of p62 and the progressive LC3-I→LC3-II conversion (which continues until LC3-I becomes nearly undetectable and LC3-II levels are maximal); as autophagy proceeds, the amount of LC3-I remains unchanged while that of LC3-II progressively decreases (as the levels of p62 keep on doing), and finally LC3-I/II and p62 all become virtually undetectable (see also Note 105).

Acknowledgments

GK is supported by the Ligue Nationale contre le Cancer (Equipe labellisée), Agence Nationale de la Recherche, Cancéropôle Ile-de-France, Fondation pour la Recherche Médicale, Institut National du Cancer, European Commission (Active p53, Apo-Sys, RIGHT, TransDeath, ChemoRes, ApopTrain), and Fondation pour la Recherche Médicale. LG, EM, and OK are supported by Apo-Sys, the ApopTrain Marie Curie training network of the European Union, and INSERM, respectively.

References

1. Sarbassov, D. D., Ali, S. M., and Sabatini, D. M. (2005) Growing roles for the mTOR pathway, *Curr Opin Cell Biol* **17**, 596–603.
2. Guertin, D. A., and Sabatini, D. M. (2005) An expanding role for mTOR in cancer, *Trends Mol Med* **11**, 353–361.
3. Law, B. K. (2005) Rapamycin: an anti-cancer immunosuppressant? *Crit Rev Oncol Hematol* **56**, 47–60.
4. Faivre, S., Kroemer, G., and Raymond, E. (2006) Current development of mTOR inhibitors as anticancer agents, *Nat Rev Drug Discov* **5**, 671–688.
5. Decker, T., Sandherr, M., Goetze, K., Oelsner, M., Ringshausen, I., and Peschel, C. (2009) A pilot trial of the mTOR (mammalian target of rapamycin) inhibitor RAD001 in patients with advanced B-CLL, *Ann Hematol* **88**, 221–227.
6. Kapoor, A., and Figlin, R. A. (2009) Targeted inhibition of mammalian target of rapamycin for the treatment of advanced renal cell carcinoma, *Cancer* **115**, 3618–3630.
7. Loewith, R., Jacinto, E., Wullschleger, S., Lorberg, A., Crespo, J. L., Bonenfant, D., Oppliger, W., Jenoe, P., and Hall, M. N. (2002) Two TOR complexes, only one of which is rapamycin sensitive, have distinct roles in cell growth control, *Mol Cell* **10**, 457–468.
8. Bhaskar, P. T., and Hay, N. (2007) The two TORCs and Akt, *Dev Cell* **12**, 487–502.
9. Sarbassov, D. D., Ali, S. M., Sengupta, S., Sheen, J. H., Hsu, P. P., Bagley, A. F., Markhard, A. L., and Sabatini, D. M. (2006) Prolonged rapamycin treatment inhibits mTORC2 assembly and Akt/PKB, *Mol Cell* **22**, 159–168.
10. Zeng, Z., Sarbassov dos, D., Samudio, I. J., Yee, K. W., Munsell, M. F., Ellen Jackson, C., Giles, F. J., Sabatini, D. M., Andreeff, M., and Konopleva, M. (2007) Rapamycin derivatives reduce mTORC2 signaling and inhibit AKT activation in AML, *Blood* **109**, 3509–3512.
11. Castedo, M., Ferri, K. F., and Kroemer, G. (2002) Mammalian target of rapamycin (mTOR): pro- and anti-apoptotic, *Cell Death Differ* **9**, 99–100.
12. Galluzzi, L., Maiuri, M. C., Vitale, I., Zischka, H., Castedo, M., Zitvogel, L., and Kroemer, G. (2007) Cell death modalities: classification and pathophysiological implications, *Cell Death Differ* **14**, 1237–1243.
13. Kroemer, G., Galluzzi, L., Vandenabeele, P., Abrams, J., Alnemri, E. S., Baehrecke, E. H., Blagosklonny, M. V., El-Deiry, W. S., Golstein, P.,

Green, D. R., Hengartner, M., Knight, R. A., Kumar, S., Lipton, S. A., Malorni, W., Nunez, G., Peter, M. E., Tschopp, J., Yuan, J., Piacentini, M., Zhivotovsky, B., and Melino, G. (2009) Classification of cell death: recommendations of the Nomenclature Committee on Cell Death 2009, *Cell Death Differ* **16**, 3–11.

14. Garrido, C., Galluzzi, L., Brunet, M., Puig, P. E., Didelot, C., and Kroemer, G. (2006) Mechanisms of cytochrome c release from mitochondria, *Cell Death Differ* **13**, 1423–1433.
15. Kroemer, G., Galluzzi, L., and Brenner, C. (2007) Mitochondrial membrane permeabilization in cell death, *Physiol Rev* **87**, 99–163.
16. Klionsky, D. J., Abeliovich, H., Agostinis, P., Agrawal, D. K., Aliev, G., Askew, D. S., Baba, M., Baehrecke, E. H., Bahr, B. A., Ballabio, A., Bamber, B. A., Bassham, D. C., Bergamini, E., Bi, X., Biard-Piechaczyk, M., Blum, J. S., Bredesen, D. E., Brodsky, J. L., Brumell, J. H., Brunk, U. T., Bursch, W., Camougrand, N., Cebollero, E., Cecconi, F., Chen, Y., Chin, L. S., Choi, A., Chu, C. T., Chung, J., Clarke, P. G., Clark, R. S., Clarke, S. G., Clave, C., Cleveland, J. L., Codogno, P., Colombo, M. I., Coto-Montes, A., Cregg, J. M., Cuervo, A. M., Debnath, J., Demarchi, F., Dennis, P. B., Dennis, P. A., Deretic, V., Devenish, R. J., Di Sano, F., Dice, J. F., Difiglia, M., Dinesh-Kumar, S., Distelhorst, C. W., Djavaheri-Mergny, M., Dorsey, F. C., Droge, W., Dron, M., Dunn, W. A., Jr., Duszenko, M., Eissa, N. T., Elazar, Z., Esclatine, A., Eskelinen, E. L., Fesus, L., Finley, K. D., Fuentes, J. M., Fueyo, J., Fujisaki, K., Galliot, B., Gao, F. B., Gewirtz, D. A., Gibson, S. B., Gohla, A., Goldberg, A. L., Gonzalez, R., Gonzalez-Estevez, C., Gorski, S., Gottlieb, R. A., Haussinger, D., He, Y. W., Heidenreich, K., Hill, J. A., Hoyer-Hansen, M., Hu, X., Huang, W. P., Iwasaki, A., Jaattela, M., Jackson, W. T., Jiang, X., Jin, S., Johansen, T., Jung, J. U., Kadowaki, M., Kang, C., Kelekar, A., Kessel, D. H., Kiel, J. A., Kim, H. P., Kimchi, A., Kinsella, T. J., Kiselyov, K., Kitamoto, K., Knecht, E., Komatsu, M., Kominami, E., Kondo, S., Kovacs, A. L., Kroemer, G., Kuan, C. Y., Kumar, R., Kundu, M., Landry, J., Laporte, M., Le, W., Lei, H. Y., Lenardo, M. J., Levine, B., Lieberman, A., Lim, K. L., Lin, F. C., Liou, W., Liu, L. F., Lopez-Berestein, G., Lopez-Otin, C., Lu, B., Macleod, K. F., Malorni, W., Martinet, W., Matsuoka, K., Mautner, J., Meijer, A. J., Melendez, A., Michels, P., Miotto, G., Mistiaen, W. P., Mizushima, N., Mograbi, B., Monastyrska, I., Moore, M. N., Moreira, P. I., Moriyasu, Y., Motyl, T., Munz, C., Murphy, L. O., Naqvi, N. I., Neufeld, T. P., Nishino, I., Nixon, R. A., Noda, T., Nurnberg, B., Ogawa, M., Oleinick, N. L., Olsen, L. J., Ozpolat, B., Paglin, S., Palmer, G. E., Papassideri, I., Parkes, M., Perlmutter, D. H., Perry, G., Piacentini, M., Pinkas-Kramarski, R., Prescott, M., Proikas-Cezanne, T., Raben, N., Rami, A., Reggiori, F., Rohrer, B., Rubinsztein, D. C., Ryan, K. M., Sadoshima, J., Sakagami, H., Sakai, Y., Sandri, M., Sasakawa, C., Sass, M., Schneider, C., Seglen, P. O., Seleverstov, O., Settleman, J., Shacka, J. J., Shapiro, I. M., Sibirny, A., Silva-Zacarin, E. C., Simon, H. U., Simone, C., Simonsen, A., Smith, M. A., Spanel-Borowski, K., Srinivas, V., Steeves, M., Stenmark, H., Stromhaug, P. E., Subauste, C. S., Sugimoto, S., Sulzer, D., Suzuki, T., Swanson, M. S., Tabas, I., Takeshita, F., Talbot, N. J., Talloczy, Z., Tanaka, K., Tanida, I., Taylor, G. S., Taylor, J. P., Terman, A., Tettamanti, G., Thompson, C. B., Thumm, M., Tolkovsky, A. M., Tooze, S. A., Truant, R., Tumanovska, L. V., Uchiyama, Y., Ueno, T., Uzcategui, N. L., van der Klei, I., Vaquero, E. C., Vellai, T., Vogel, M. W., Wang, H. G., Webster, P., Wiley, J. W., Xi, Z., Xiao, G., Yahalom, J., Yang, J. M., Yap, G., Yin, X. M., Yoshimori, T., Yu, L., Yue, Z., Yuzaki, M., Zabirnyk, O., Zheng, X., Zhu, X., and Deter, R. L. (2008) Guidelines for the use and interpretation of assays for monitoring autophagy in higher eukaryotes, *Autophagy* **4**, 151–175.
17. Golstein, P., and Kroemer, G. (2005) Redundant cell death mechanisms as relics and backups, *Cell Death Differ* **12 Suppl 2**, 1490–1496.
18. Golstein, P., and Kroemer, G. (2007) Cell death by necrosis: towards a molecular definition, *Trends Biochem Sci* **32**, 37–43.
19. Galluzzi, L., Blomgren, K., and Kroemer, G. (2009) Mitochondrial membrane permeabilization in neuronal injury, *Nat Rev Neurosci* **10**, 481–494.
20. Galluzzi, L., Vicencio, J. M., Kepp, O., Tasdemir, E., Maiuri, M. C., and Kroemer, G. (2008) To die or not to die: that is the autophagic question, *Curr Mol Med* **8**, 78–91.
21. Kroemer, G., and Levine, B. (2008) Autophagic cell death: the story of a misnomer, *Nat Rev Mol Cell Biol* **9**, 1004–1010.
22. Boya, P., Gonzalez-Polo, R. A., Casares, N., Perfettini, J. L., Dessen, P., Larochette, N., Metivier, D., Meley, D., Souquere, S., Yoshimori, T., Pierron, G., Codogno, P., and Kroemer, G. (2005) Inhibition of macroautophagy triggers apoptosis, *Mol Cell Biol* **25**, 1025–1040.
23. Gonzalez-Polo, R. A., Boya, P., Pauleau, A. L., Jalil, A., Larochette, N., Souquere, S., Eskelinen, E. L., Pierron, G., Saftig, P., and Kroemer, G. (2005) The apoptosis/autophagy

paradox: autophagic vacuolization before apoptotic death, *J Cell Sci* **118**, 3091–3102.

24. Galluzzi, L., Kepp, O., and Kroemer, G. (2009) RIP kinases initiate programmed necrosis, *J Mol Cell Biol*
25. Galluzzi, L., and Kroemer, G. (2008) Necroptosis: a specialized pathway of programmed necrosis, *Cell* **135**, 1161–1163.
26. Galluzzi, L., Aaronson, S. A., Abrams, J., Alnemri, E. S., Andrews, D. W., Baehrecke, E. H., Bazan, N. G., Blagosklonny, M. V., Blomgren, K., Borner, C., Bredesen, D. E., Brenner, C., Castedo, M., Cidlowski, J. A., Ciechanover, A., Cohen, G. M., De Laurenzi, V., De Maria, R., Deshmukh, M., Dynlacht, B. D., El-Deiry, W. S., Flavell, R. A., Fulda, S., Garrido, C., Golstein, P., Gougeon, M. L., Green, D. R., Gronemeyer, H., Hajnoczky, G., Hardwick, J. M., Hengartner, M. O., Ichijo, H., Jaattela, M., Kepp, O., Kimchi, A., Klionsky, D. J., Knight, R. A., Kornbluth, S., Kumar, S., Levine, B., Lipton, S. A., Lugli, E., Madeo, F., Malomi, W., Marine, J. C., Martin, S. J., Medema, J. P., Mehlen, P., Melino, G., Moll, U. M., Morselli, E., Nagata, S., Nicholson, D. W., Nicotera, P., Nunez, G., Oren, M., Penninger, J., Pervaiz, S., Peter, M. E., Piacentini, M., Prehn, J. H., Puthalakath, H., Rabinovich, G. A., Rizzuto, R., Rodrigues, C. M., Rubinsztein, D. C., Rudel, T., Scorrano, L., Simon, H. U., Steller, H., Tschopp, J., Tsujimoto, Y., Vandenabeele, P., Vitale, I., Vousden, K. H., Youle, R. J., Yuan, J., Zhivotovsky, B., and Kroemer, G. (2009) Guidelines for the use and interpretation of assays for monitoring cell death in higher eukaryotes, *Cell Death Differ* **16**, 1093–1107.
27. Galluzzi, L., Joza, N., Tasdemir, E., Maiuri, M. C., Hengartner, M., Abrams, J. M., Tavernarakis, N., Penninger, J., Madeo, F., and Kroemer, G. (2008) No death without life: vital functions of apoptotic effectors, *Cell Death Differ* **15**, 1113–1123.
28. Ferri, K. F., and Kroemer, G. (2001) Mitochondria – the suicide organelles, *Bioessays* **23**, 111–115.
29. Ravagnan, L., Roumier, T., and Kroemer, G. (2002) Mitochondria, the killer organelles and their weapons, *J Cell Physiol* **192**, 131–137.
30. Brenner, C., and Grimm, S. (2006) The permeability transition pore complex in cancer cell death, *Oncogene* **25**, 4744–4756.
31. Galluzzi, L., Vitale, I., Kepp, O., Seror, C., Hangen, E., Perfettini, J. L., Modjtahedi, N., and Kroemer, G. (2008) Methods to dissect mitochondrial membrane permeabilization in the course of apoptosis, *Methods Enzymol* **442**, 355–374.
32. Galluzzi, L., Zamzami, N., de La Motte Rouge, T., Lemaire, C., Brenner, C., and Kroemer, G. (2007) Methods for the assessment of mitochondrial membrane permeabilization in apoptosis, *Apoptosis* **12**, 803–813.
33. Metivier, D., Dallaporta, B., Zamzami, N., Larochette, N., Susin, S. A., Marzo, I., and Kroemer, G. (1998) Cytofluorometric detection of mitochondrial alterations in early CD95/Fas/APO-1-triggered apoptosis of Jurkat T lymphoma cells. Comparison of seven mitochondrion-specific fluorochromes, *Immunol Lett* **61**, 157–163.
34. Petit, P. X., Lecoeur, H., Zorn, E., Dauguet, C., Mignotte, B., and Gougeon, M. L. (1995) Alterations in mitochondrial structure and function are early events of dexamethasone-induced thymocyte apoptosis, *J Cell Biol* **130**, 157–167.
35. Zamzami, N., Marchetti, P., Castedo, M., Decaudin, D., Macho, A., Hirsch, T., Susin, S. A., Petit, P. X., Mignotte, B., and Kroemer, G. (1995) Sequential reduction of mitochondrial transmembrane potential and generation of reactive oxygen species in early programmed cell death, *J Exp Med* **182**, 367–377.
36. Zamzami, N., Marchetti, P., Castedo, M., Zanin, C., Vayssiere, J. L., Petit, P. X., and Kroemer, G. (1995) Reduction in mitochondrial potential constitutes an early irreversible step of programmed lymphocyte death in vivo, *J Exp Med* **181**, 1661–1672.
37. Inoue, S., Browne, G., Melino, G., and Cohen, G. M. (2009) Ordering of caspases in cells undergoing apoptosis by the intrinsic pathway, *Cell Death Differ* **16**, 1053–1061.
38. Kroemer, G., and Martin, S. J. (2005) Caspase-independent cell death, *Nat Med* **11**, 725–730.
39. Kroemer, G., and Jaattela, M. (2005) Lysosomes and autophagy in cell death control, *Nat Rev Cancer* **5**, 886–897.
40. Martin, S. J., Reutelingsperger, C. P., McGahon, A. J., Rader, J. A., van Schie, R. C., LaFace, D. M., and Green, D. R. (1995) Early redistribution of plasma membrane phosphatidylserine is a general feature of apoptosis regardless of the initiating stimulus: inhibition by overexpression of Bcl-2 and Abl, *J Exp Med* **182**, 1545–1556.
41. Wu, Y., Tibrewal, N., and Birge, R. B. (2006) Phosphatidylserine recognition by phagocytes: a view to a kill, *Trends Cell Biol* **16**, 189–197.
42. Naito, M., Nagashima, K., Mashima, T., and Tsuruo, T. (1997) Phosphatidylserine externalization is a downstream event of interleukin-1 beta-converting enzyme family protease activation during apoptosis, *Blood* **89**, 2060–2066.

43. Zhuang, J., and Cohen, G. M. (1998) Release of mitochondrial cytochrome c is upstream of caspase activation in chemical-induced apoptosis in human monocytic tumour cells, *Toxicol Lett* **102–103**, 121–129.
44. Bailey, K., Cook, H. W., and McMaster, C. R. (2005) The phospholipid scramblase PLSCRl increases UV induced apoptosis primarily through the augmentation of the intrinsic apoptotic pathway and independent of direct phosphorylation by protein kinase C delta, *Biochim Biophys Acta* **1733**, 199–209.
45. Smrz, D., Lebduska, P., Draberova, L., Korb, J., and Draber, P. (2008) Engagement of phospholipid scramblase 1 in activated cells: implication for phosphatidylserine externalization and exocytosis, *J Biol Chem* **283**, 10904–10918.
46. Balasubramanian, K., Mirnikjoo, B., and Schroit, A. J. (2007) Regulated externalization of phosphatidylserine at the cell surface: implications for apoptosis, *J Biol Chem* **282**, 18357–18364.
47. Williamson, P., Christie, A., Kohlin, T., Schlegel, R. A., Comfurius, P., Harmsma, M., Zwaal, R. F., and Bevers, E. M. (2001) Phospholipid scramblase activation pathways in lymphocytes, *Biochemistry* **40**, 8065–8072.
48. Vermes, I., Haanen, C., Steffens-Nakken, H., and Reutelingsperger, C. (1995) A novel assay for apoptosis. Flow cytometric detection of phosphatidylserine expression on early apoptotic cells using fluorescein labelled Annexin V, *J Immunol Methods* **184**, 39–51.
49. Foster, D. A., and Toschi, A. (2009) Targeting mTOR with rapamycin: one dose does not fit all, *Cell Cycle* **8**, 1026–1029.
50. Tasdemir, E., Galluzzi, L., Maiuri, M. C., Criollo, A., Vitale, I., Hangen, E., Modjtahedi, N., and Kroemer, G. (2008) Methods for assessing autophagy and autophagic cell death, *Methods Mol Biol* **445**, 29–76.
51. Kabeya, Y., Mizushima, N., Ueno, T., Yamamoto, A., Kirisako, T., Noda, T., Kominami, E., Ohsumi, Y., and Yoshimori, T. (2000) LC3, a mammalian homologue of yeast Apg8p, is localized in autophagosome membranes after processing, *EMBO J* **19**, 5720–5728.
52. Tanida, I., Ueno, T., and Kominami, E. (2004) Human light chain 3/MAP1LC3B is cleaved at its carboxyl-terminal Met121 to expose Gly120 for lipidation and targeting to autophagosomal membranes, *J Biol Chem* **279**, 47704–47710.
53. Ichimura, Y., Kirisako, T., Takao, T., Satomi, Y., Shimonishi, Y., Ishihara, N., Mizushima, N., Tanida, I., Kominami, E., Ohsumi, M., Noda, T., and Ohsumi, Y. (2000) A ubiquitin-like system mediates protein lipidation, *Nature* **408**, 488–492.
54. Nakatogawa, H., Ichimura, Y., and Ohsumi, Y. (2007) Atg8, a ubiquitin-like protein required for autophagosome formation, mediates membrane tethering and hemifusion, *Cell* **130**, 165–178.
55. Bjorkoy, G., Lamark, T., Brech, A., Outzen, H., Perander, M., Overvatn, A., Stenmark, H., and Johansen, T. (2005) p62/SQSTM1 forms protein aggregates degraded by autophagy and has a protective effect on huntingtin-induced cell death, *J Cell Biol* **171**, 603–614.
56. Komatsu, M., Wang, Q. J., Holstein, G. R., Friedrich, V. L., Jr., Iwata, J., Kominami, E., Chait, B. T., Tanaka, K., and Yue, Z. (2007) Essential role for autophagy protein Atg7 in the maintenance of axonal homeostasis and the prevention of axonal degeneration, *Proc Natl Acad Sci USA* **104**, 14489–14494.
57. Tasdemir, E., Maiuri, M. C., Galluzzi, L., Vitale, I., Djavaheri-Mergny, M., D'Amelio, M., Criollo, A., Morselli, E., Zhu, C., Harper, F., Nannmark, U., Samara, C., Pinton, P., Vicencio, J. M., Carnuccio, R., Moll, U. M., Madeo, F., Paterlini-Brechot, P., Rizzuto, R., Szabadkai, G., Pierron, G., Blomgren, K., Tavernarakis, N., Codogno, P., Cecconi, F., and Kroemer, G. (2008) Regulation of autophagy by cytoplasmic p53, *Nat Cell Biol* **10**, 676–687.
58. Vignot, S., Faivre, S., Aguirre, D., and Raymond, E. (2005) mTOR-targeted therapy of cancer with rapamycin derivatives, *Ann Oncol* **16**, 525–537.
59. Schroit, A. J., and Zwaal, R. F. (1991) Transbilayer movement of phospholipids in red cell and platelet membranes, *Biochim Biophys Acta* **1071**, 313–329.

Chapter 10

Evaluation of mTOR-Regulated mRNA Translation

Valentina Iadevaia, Xuemin Wang, Zhong Yao, Leonard J. Foster, and Christopher G. Proud

Abstract

mTOR, the mammalian target of rapamycin, regulates protein synthesis (mRNA translation) by affecting the phosphorylation or activity of several translation factors. Here, we describe methods for studying the impact of mTOR signalling on protein synthesis, using inhibitors of mTOR such as rapamycin (which impairs some of its functions) or mTOR kinase inhibitors (which probably block all functions).To assess effects of mTOR inhibition on general protein synthesis in cells, the incorporation of radiolabelled amino acids into protein is measured. This does not yield information on the effects of mTOR on the synthesis of specific proteins. To do this, two methods are described. In one, stable-isotope labelled amino acids are used, and their incorporation into new proteins is determined using mass spectrometric methods. The proportions of labelled vs. unlabeled versions of each peptide from a given protein provide quantitative information about the rate of that protein's synthesis under different conditions. Actively translated mRNAs are associated with ribosomes in polyribosomes (polysomes); thus, examining which mRNAs are found in polysomes under different conditions provides information on the translation of specific mRNAs under different conditions. A method for the separation of polysomes from non-polysomal mRNAs is described.

Key words: Target of rapamycin, Protein synthesis, Polyribosome analysis, Pulsed SILAC, Stable isotope labelling

1. Introduction

The mammalian target of rapamycin (mTOR) is a protein kinase that forms signalling complexes with other proteins in the cell. One such complex is mTOR complex 1 (mTORC1); in addition to mTOR, it contains the proteins raptor and mLst8 (also termed GβL) (1). Raptor interacts with proteins which are substrates for phosphorylation by mTOR. Phosphorylation of several of these

Thomas Weichhart (ed.), *mTOR: Methods and Protocols*, Methods in Molecular Biology, vol. 821,
DOI 10.1007/978-1-61779-430-8_10, © Springer Science+Business Media, LLC 2012

substrates is inhibited by the macrolide antibiotic rapamycin, which binds to mTOR as a complex with the immunophilin FKBP12. The use of rapamycin has revealed that mTORC1 regulates many key cellular processes, including protein synthesis, ribosome biogenesis, the cell cycle, and cell division (1–5). Indeed, mTORC1 signalling appears to play an important role in tumorigenesis; this has stimulated interest is using rapamycin (or related compounds) in anti-cancer therapy.

The mechanism by which rapamycin impairs mTORC1 function remains incompletely understood; importantly, it does not directly inhibit its kinase activity. More recently, ATP-competitive inhibitors of the kinase activity of mTOR have been developed. Their use has confirmed what had been suspected earlier, i.e. that some functions of mTORC1 are not affected by rapamycin. One example of an mTOR kinase inhibitor is PP242 (6).

The best-understood substrates for mTORC1 are proteins involved in modulating protein synthesis. These include the ribosomal protein S6 kinases, S6K1 and S6K2 (7), which phosphorylate several proteins involved in the controlling the initiation or elongation stages of mRNA translation (protein synthesis) (2, 8). mTORC1 also phosphorylates the eukaryotic initiation factor-4E-binding proteins (4E-BPs); these proteins can bind to and inhibit eIF4E, the protein that binds the "cap" structure of the mRNA, preventing it from forming active initiation complexes with other partners. 4E-BPs are phosphorylated at several sites by mTORC1; this causes their release from eIF4E, which is now available for translation initiation. eIF4E and some targets for control by S6 kinases (such as eukaryotic elongation factor 2, whose cognate kinase is controlled by S6 kinases) are important for general protein synthesis; other proteins that form complexes with eIF4E, such as eIF4B (which is a substrate for the S6 kinases (9)), are likely important for translation of specific mRNAs, such as those whose 5′-untranslated regions containing inhibitor secondary structure. mTORC1 signalling may affect both overall protein synthesis and the translation of certain specific mRNAs.

Furthermore, rapamycin is known to impair the translation of mRNAs that contain a special feature known as a 5′-TOP (5′-terminal tract of oligopyrimidines; (10)). This feature is a negative regulator of translation, but its effect is overcome by treatment of cells with agents that activate mTORC1, such as fresh serum, and this activating effect is impaired by rapamycin (10), which causes such mRNAs to shift out of actively translating polyribosomes (10).

This chapter describes methods for studying the control of general protein synthesis by mTORC1 signalling, and procedures for studying the influence of mTORC1 signalling on the translation of specific mRNAs (i.e. on the synthesis of specific proteins).

2. Materials

2.1. Cell Culture and Lysis

1. This method can be applied to many types of cells. Dulbecco's Modified Eagle's Medium (DMEM) is used for many cell lines, supplemented with 10% (v/v) foetal bovine serum (FBS) and 1:100 dilution of penicillin/streptomycin (to give final concentrations of 50 units/ml penicillin and 50 mg/ml streptomycin; antibiotics; Invitrogen).
2. Solution of trypsin (0.25% w/v) and ethylenediamine tetraacetic acid (EDTA) (1 mM) from Gibco/Invitrogen.
3. Stimuli are dissolved at the appropriate concentration (usually 100× the required final concentration) in Milli-Q water or dimethylsulphoxide (for example), stored in aliquots at −20°C and then added to the tissue culture medium as required. Concentrations will vary with the stimulus and perhaps with the cell type, and may have to be determined experimentally.
4. Inhibitors such as rapamycin or PP242 are dissolved at a stock concentration of 100 μM or 10 mM respectively in DMSO, stored in aliquots at −20°C and then added to the cell culture as required (1:1,000 for rapamycin).
5. Cell lysis buffer: 50 mM Tris–HCl, pH 7.5, 50 mM β-glycerophosphate, 100 mM NaCl, 0.5 mM EDTA, 1% (v/v) Triton. Kept at 4°C; appropriate amount is taken as required and supplemented with protease inhibitor cocktail (Roche), 0.5 mM sodium orthovanadate, 0.1% (v/v) 2-mercaptoethanol. Kept on ice.
6. Cell lifters (also called cell scrapers).

2.2. Measurement of Protein Synthesis by Incorporation of Radiolabelled Amino Acids

1. Radioactive materials: L-(^{35}S)methionine or L-(^{3}H)phenylalanine are stored at −20°C, and are added to the cells at final concentration of, typically, 5 μCi/ml. If the incorporated radioactivity is low, more radiolabelled precursor can be used.
2. Whatman 3MM papers are cut into 2×2 cm squares with one corner folded and labelled with *pencil* (not pen).
3. 5% (w/v) trichloroacetic acid (TCA), see Note 1.
4. 100% ethanol.
5. A large glass beaker (e.g. 1 l) with a fitted metal mesh holder.
6. Scintillation counter, 4 ml vials and scintillation fluid (e.g. Ultima gold, Perkin Elmer).

2.3. pSILAC Approach

1. PBS (phosphate-buffered saline): 4.3 mM Na_2HPO_4, 137 mM NaCl, 2.7 mM KCl, 1.4 mM KH_2PO_4. Sterilize by autoclaving and store at room temperature.

2. Heavy amino acids stock solutions: Prepare 1,000× stock solution for each amino acid by dissolving in PBS. $^{13}C_6$,$^{15}N_4$ lysine-8 (154 mg/ml; Silantes: 356112); $^{13}C_6$,$^{15}N_4$ arginine-10 (89 ml/ml; Silantes: 356102). Mix until thoroughly dissolved, then aliquot into small volumes and store at −20°C until needed.
3. Dialyzed serum: dialysis membrane from ethanol storage solution and rinse with distilled water. Secure clamp to one end of the membrane (see Note 2). Fill dialyze membrane with serum and close other end with another clamp. Immerse dialysis membrane in a beaker containing a large volume of PBS at 4°C; change after 4 h and incubate with fresh PBS overnight at 4°C with gentle stirring of the buffer. Remove the serum from dialysis membrane with a Pasteur pipette and store at −20°C.
4. Prepare pSilac media, mix well together the amino acid (1×), the SILAC DMEM minus L-lysine and L-arginine (Thermo Scientific Pierce: PN89985), and 10% (v/v) dialysed foetal calf serum. The medium must then be filter-sterilized in aseptic condition and incubated at 37°C for 15 min to warm up the medium.
5. pSILAC Lysis buffer: 50 mM ammonium bicarbonate and 1% sodium deoxycholate.

2.4. For Polysome Fractionation

1. "Polysome Cell Lysis buffer" (TNM; this is a very mild lysis buffer): 50 ml: 10 mM Tris–HCl pH 7.5, 10 mM NaCl, 10 mM $MgCl_2$. Store at room temperature.
2. 10% mix of Triton X-100 and deoxycholate (DOC).
3. 1 M DTT: dissolve 3.1 g of DTT in distilled water and store at −20°C (see Note 3).
4. Recombinant RNAsin Ribonuclease Inhibitor (Promega N2515) at −20°C.
5. Proteinase K.
6. SDS, 20% sodium dodecyl sulphate.
7. Carrier tRNA, ribonucleic acid transfer can be prepared by dissolving yeast tRNA at 10 mg/ml in 0.1 M of NaCl. Extract the RNA twice with phenol (pH 7.6) and twice with chloroform, then precipitate the RNA with 2.5 volumes of 100% ethanol at room temperature. Recover the RNA by centrifugation in a microfuge at 15,700 × *g* for 15 min at 4°C, dissolve at concentration of 10 mg/ml in sterile water and store at −20°C.
8. 2X Polysome Gradient buffer: 60 mM Tris–HCl pH7.5, 200 mM NaCl, 20 mM $MgCl_2$.
9. 20% Sucrose: 50 ml 2× Polysome Gradient buffer and 20 g Ultra pure sucrose; dissolve and make up to 100 ml with RNAse-free distilled water; store at −20°C.

10. Similarly for 26, 32, 38, 44, or 50% sucrose, dissolve required amount of sucrose in 2× Polysome Gradient Buffer, make to 100 ml, and store at −20°C (see Note 4).
11. 60% (w/v) sucrose dissolved in distilled water (200 ml).
12. Phenol pH 4.3, chloroform, isopropanol, and ethanol.
13. Trichloroacetic acid (TCA) and acetone.

3. Methods

The classical method for measuring protein synthesis uses a radiolabelled amino acid, e.g. L-[^{35}S]methionine; L-[^{3}H]phenylalanine, to label the newly synthesized proteins. Label that has been incorporated into new proteins must then be separated from free (unincorporated) label. This is commonly achieved by precipitating the protein with trichloroacetic acid (TCA; see Note 1), in which amino acids, but not proteins, are soluble. Measurement of incorporated radiolabel by scintillation counting provides the data for global protein synthesis in living cells.

When working with radioisotopes, the proper safety rules for work with radioactive materials must be followed during the entire experimental procedure and waste must be disposed of correctly.

Protein synthesis is under very tight control by many parameters. To obtain reliable and reproducible results, it is very important to maintain the cells in the proper growth state, i.e. usually when the cells are about 80% confluent. This may need to be determined empirically when using a "new" cell type. High confluence, or serum starvation for too long, may impair the responsiveness of the cells; serum starvation for too short a time may mean that some signalling pathways are still active, again impairing responsiveness.

3.1. Assessment of the Impact of mTORC1 Signalling on Overall Rates of Protein Synthesis

1. This protocol applies to cells that grow in adherent monolayer culture – for cells that grow in suspension, see Note 5.
2. The cells are passaged when approaching confluence to provide experimental cultures on 60 mm dishes or in six-well plates. One dish/well is required for each experimental data point. Duplicates or triplicates are needed for each data point. The cells are cultured to reach 80–90% confluence.
3. To serum starve the cells, cultures are rinsed twice with DMEM (without serum) and incubated for overnight or a few hours with MEFs in DMEM (without serum). This depends upon the cell type and appropriate conditions need to be established.
4. For the treatment, the stimulus and inhibitors are added into the culture dishes at 1:1,000 dilution. To make sure that the

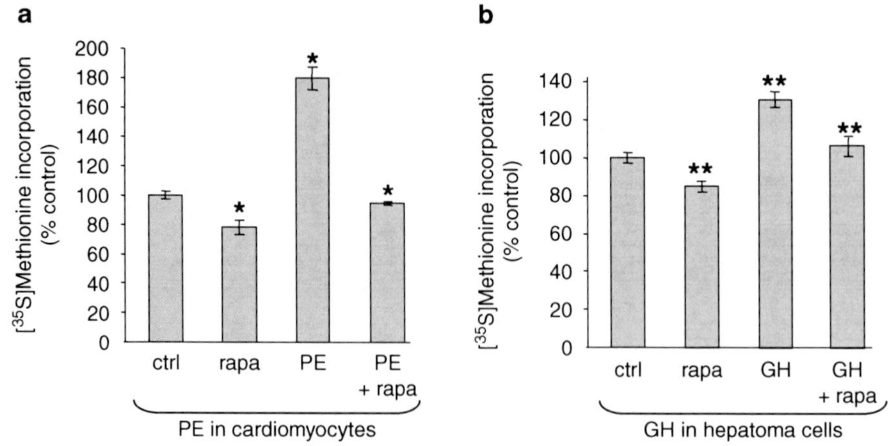

Fig. 1. Protein synthesis data. (**a**) Cardiomyocytes were pretreated with rapamycin (100 nM, 30 min, where indicated) and then, where shown, phenylephrine (PE, 10 μM) was added. 90 min later, [^{35}S]methionine was added and the cells were incubated for a further 45 min. *$p < 0.05$ vs. control. Data are given as means ± SE, $n = 3$. Data are taken from ref. 16. (**b**) H4IIE hepatoma cells were starved of serum for 30 h and then pretreated with rapamycin (50 nM, 30 min) prior to addition of GH . After 30 min, (^{35}S)methionine was added, and 1 h later, the cells were extracted. Data are given as means ± SD, $n = 4$. Data are taken from ref. 13. In each case, samples were processed for measurement of incorporation of label as described in the text. Rates of incorporation in untreated cells are set at 100. **$p < 0.01$ vs. control.

reagents (often dissolved in DMSO) are mixed well with the culture media, take 500 ml of media out and add appropriate volume of stock reagents to it (i.e. 2 μl of stock for 2 ml of cells growing in six-well plates), mix well by pipetting up and down a few times, and then the mixer is added back to the dish. Inhibitors are usually added 30–45 min before the stimulus (i.e. insulin). Stimuli are then added for further times, i.e. insulin for 25 min; see also legend to Fig. 1. Then add labelled amino acid.

5. Incubation of the cells with L-[^{35}S]methionine; L-[^{3}H]phenylalanine: from this step on, all procedures are performed according the radioactive safety rules in an appropriate location and with protective gloves, labcoat and safety glasses.
6. L-[^{35}S]methionine; (5 μCi/ml) or L-[^{3}H]phenylalanine is added and the cells are incubated for a further 30 min or 1 h in the incubator.
7. At the end of incubation time, the medium is removed carefully to a proper tube for radioactive waste disposal (see Note 6). Cells are washed quickly twice with ice-cold PBS. A small volume of lysis buffer is added to dishes on ice (see Note 7). The cells are scraped and transferred into an appropriately labelled microfuge tube.
8. The tubes are closed and spun at 15,700 × g at 4°C for 10 min. The supernatant is transferred to a fresh labelled tube (in an

ice bucket) and the protein concentration is determined by performing Bradford (11) assays (caution: radioactive).

9. 3MM paper filters are lined up in a radio-isotope safety tray.
10. Equal amounts of lysate are loaded carefully onto each filter (minimal amount of 25 μg (protein) of lysate is usually needed to obtain enough radio-labelled protein). The volume of each sample should be kept between 10 and 30 μl. Avoid loading large volumes of lysate as excess lysate leaks through the filters.
11. The filters are left on the bench to dry for 30 min.
12. The filters are immerged in 50 to 100 ml of 5% (w/v) TCA in a glass beaker inside a metal mesh holder to avoid damage to the filters. The TCA solution is boiled for 2 min in a ventilated hood (see Note 8). This step is to wash off the unbound radio-labelled material from the filters; boiling ensures that radiolabelled amino acyl-tRNAs are hydrolysed. The TCA solution contains most of the radiolabel and is discarded safely; 50–100 ml of fresh TCA is added to repeat once more the boiling/washing step. Care must be taken when holding the hot glass beaker and discarding the boiled TCA to the radioactive disposal sink. To avoid the filters slipping out of the glass beaker, it is better to lift up the metal holder before discarding the TCA.
13. Filters are rinsed briefly in 100% ethanol (to remove residual TCA) and laid out in the right order before being put into the oven to dry at 100°C for 10 min (or until completely dry).
14. The dried filters are folded carefully to fit into the 4 ml vials that already contain 3–4 ml of appropriate scintillation fluid and counted for 10 min or until appropriate accuracy has been attained (to achieve an error of ±1%, 10^4 disintegration events have to be counted).
15. Typical data are shown in Fig. 1. When assessing the effects of signalling inhibitors such as rapamycin on the activation of protein synthesis, it is essential to test their effects on unstimulated cells too, as they may impair the basal rate of protein synthesis as well as the stimulation by the agonist, and it is crucial to correct for this. The data show that phenylephrine (PE) and growth hormone (GH) stimulate the rate of amino acid incorporation (i.e. protein synthesis) in cardiomyocytes or hepatoma cells respectively. This stimulation is largely, but not completely, blocked by rapamycin. Thus, signalling through mTORC1 plays an important role in the activation of protein synthesis in different cell types by diverse stimuli (PE works through a G-protein coupled receptor and activates mTORC1 through Ras and MAP kinase (ERK) signalling (12), while GH works through a quite different type of receptor and activates mTORC1 signalling via PI 3-kinase and protein kinase B (13)).

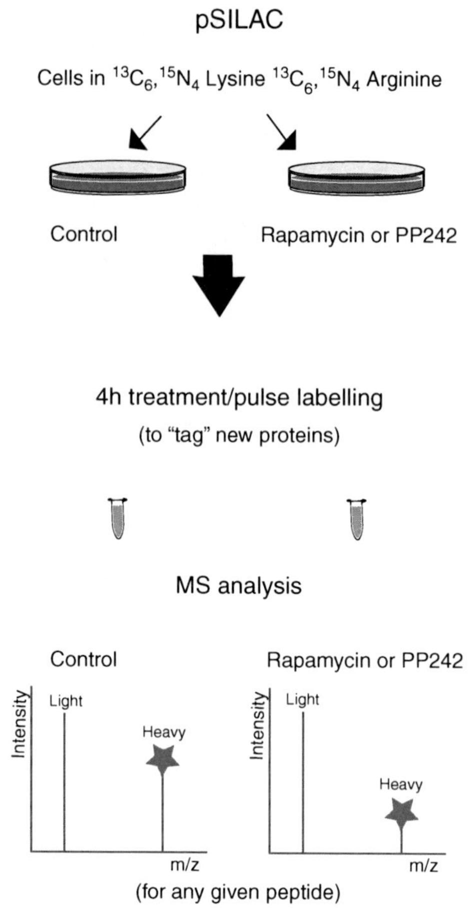

Fig. 2. Scheme of pSILAC method. The scheme depicts the work flow described in the text, *m*/*z*, mass–charge ratio of peptides.

3.2. pSILAC Method

1. The day before the experiment: split (HeLa) cells (50% confluent) into two 100-mm Petri plate with 10 ml of fresh DMEM media with 10% of foetal calf serum (see Fig. 2 for a schematic outline of the method).
2. On the day of the experiment: first aspirate the medium from the flask and then rinse the cells twice with 5–10 ml of pre-warmed PBS.
3. Add the SILAC medium to the cells (which should now be about 70% confluent) with or without PP242 (1 μM) or rapamycin 100 nM).
4. Incubate the cells at 37°C in 5% CO_2 for 4–6 h.
5. Wash twice with cold PBS and then add 300 μl of pSILAC lysis buffer to the plate.
6. Scrape the cells and transfer into clean tube (1.5 ml microfuge tube).

7. Immediately incubate at 95°C for 5–10 min (see Note 9).
8. Spin at 15,700 × *g* in a microfuge for 10 min at room temperature.
9. Transfer the supernatant to another tube.
10. Measure the protein concentration with Bradford or BCA method.
11. Aliquot 25 μg of total protein and continue with reduction (0.5 μg dithiothreitol for 30 min at 37°C), alkylation (2.5 μg iodoacetamide for 30 min at 37°C), and trypsinization (0.5 μg trypsin overnight at 37°C).
12. Dilute the sample two-fold with a solution of 1% trifluoroacetic acid, 3% (v/v) acetonitrile, 0.5% (v/v) acetic acid.
13. Centrifuge at 15,700 × *g* for 5 min.
14. Concentrate the supernatant containing the peptides, with filtered on C_{18} Stop And Go Extraction tips (14), and then elute directly into a 96-well plate.
15. Analyze the peptide mixtures on a 1100 Series nanoflow high performance liquid chromatograph (Agilent) on-line coupled via a nanoelectrospray ion source (Proxeon, Odense, Denmark) to a linear trapping quadrupole-Orbitrap (LTQ-Orbitrap) tandem mass spectrometer. For this purpose, inject the peptides directly onto a reversed phase (3 μm diameter ReproSil-Pur C18, Dr. Maisch, Ammerbuch-Entringen, Germany) column manually packed into a 15-cm long, 75 μm inner diameter fused silica emitter.
16. Elute the peptides directly into the LTQ-Orbitrap using a linear gradient from 4.8% acetonitrile in 0.5% acetic acid to 24% acetonitrile in 0.5% acetic acid over 60 min at a flow rate of 200 nl/min.
17. Acquire a full-range scan in the Orbitrap.
18. Sample data are shown in Fig. 3. They reveal that the mTOR kinase inhibitor PP242 and rapamycin each strongly inhibit the

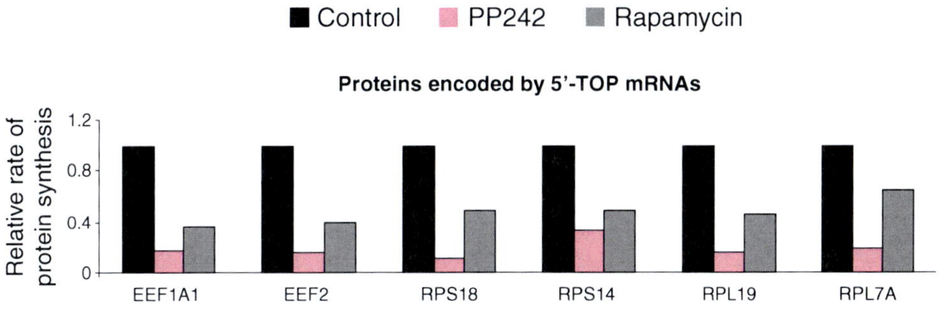

Fig. 3. Sample pSILAC data. Data are shown for the relative rates of synthesis of selected proteins, whose synthesis is strongly inhibited by PP242 or rapamycin. Control rates are all normalized to unity.

synthesis of the proteins for which data is displayed here. These proteins, which are encoded by members of the 5′-TOP (tract of pyrimidines) family of mRNAs are known to be translationally controlled by mTOR (10).

3.3. Preparation of Cell Lysates for Polysome Analysis to Study the Effects of mTORC1 Signalling on the Translation of Specific mRNAs

1. Place the dish of cells on ice and remove the medium by aspiration (Fig. 4 shows an outline of this approach).
2. Rinse the dish 3× with ice-cold PBS.
3. Prepare the lysis buffer mix : 268 μl of TNM lysis buffer and 30 μl of Triton/DOC, 40 U of recombinant RNAsin inhibitor, 1 μl 1 M DTT.
4. Add the lysis buffer mix directly to the cell plate.
5. Scrape the cells and transfer into clean tube (1.5 ml microfuge tube).
6. Incubate on ice for 2 min.
7. Spin at 15,700 × *g* in Eppendorf centrifuge at 4°C for 5 min.
8. Transfer the supernatant from the nuclei pellet to another tube and add 3 μl of Heparin (10 μg/μl).
9. Freeze in liquid nitrogen and store at −80°C.

3.4. Preparation of Density Gradients

The methods assumes the use of a Beckman SW41 rotor. Other rotors, from this and other manufacturers, are also suitable but the volumes and centrifugation conditions would have to be modified accordingly (see schematic diagram, Fig. 4).

1. Place the (e.g. Beckman) polyallomer centrifuge tube (14 × 89 mm; Beckman catalogue number 358120) in dry ice. Use a long Pasteur pipette to place 1.5 ml of 50% sucrose solution in the bottom of the tube.
2. When the first layer is frozen, layer on the next (lighter) sucrose solution (i.e. 44%), and so on until all six solutions have been layered. Cover the tube with aluminium foil; gradients may be stored at −20°C until required.
3. The night before they are needed, place at 4°C overnight for to allow gradient to thaw and for gradient equilibration (a linear gradient will form in 14–15 h). Once thawed, gradients must be handled with care to avoid mixing.

3.5. Density Gradient Centrifugation

1. Turn on Beckman ultracentrifuge, make sure the vacuum is on and the temperature is set at 4°C. Cool down the buckets for the rotor in advance.
2. Remove the gradient from the cold room taking great care to not disturb the gradient.

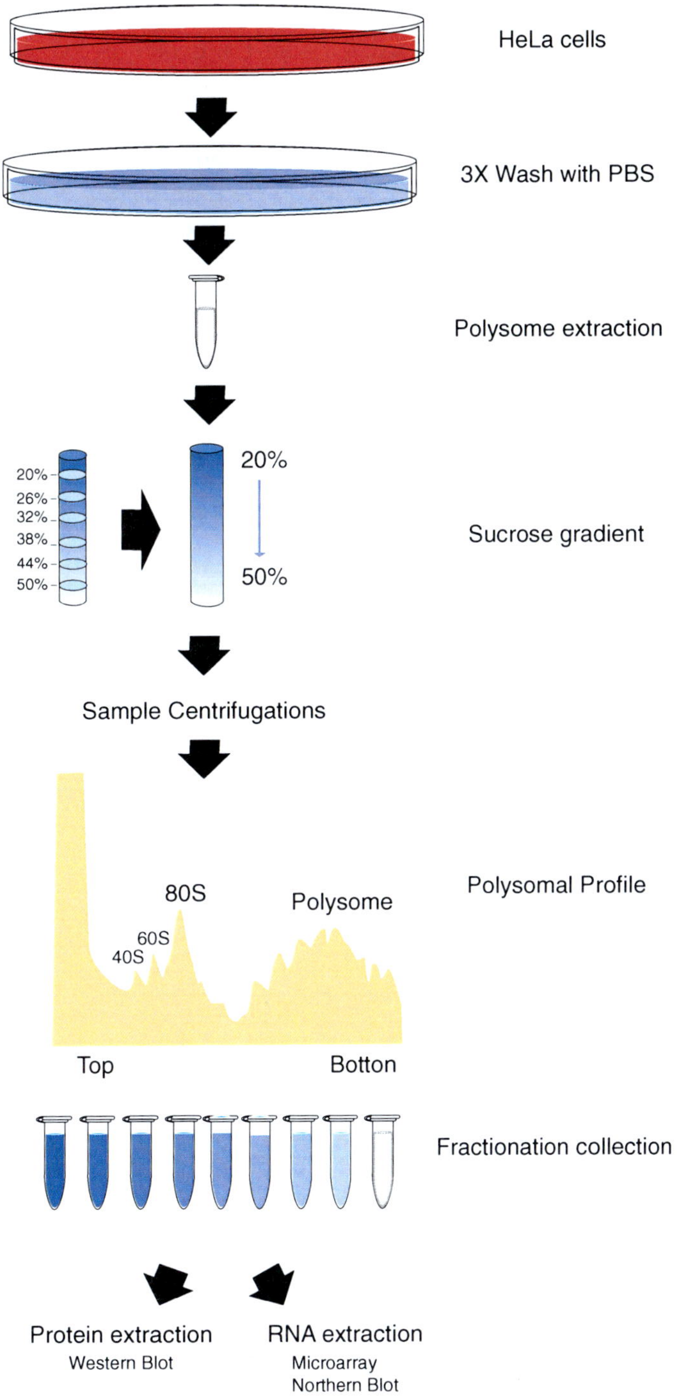

Fig. 4. Outline of sucrose gradient/polysome method. The scheme depicts the procedure described in the text, including the layering of sucrose solutions of different densities to create the gradients.

3. Carefully apply the lysate on top of the gradient. Remember to record which bucket of the rotor contains which sample! (Writing on the tube will probably be lost.)
4. Carefully insert the gradient into the pre-chilled bucket. Screw the metal cup onto the bucket and place in the swing-out (e.g. Beckman SW41) rotor (see Note 10).
5. Centrifuge (for this rotor) at 234,000 × *g* for 110 min at 4°C (see Note 11). Do not use the brake.

3.6. Fraction Collection

1. Immediately after the run, the tube should be removed carefully from the bucket (important: take great care not to disturb the gradient) and carefully place on ice while waiting for the analysis.
2. Connect the pump, fractionator, flow cell/monitor (e.g. ISCO UA-6 UV-VIS, set to record absorbance at 254 nm), and fraction collector, with a dummy tube in the fractionator.
3. Make sure that the gradient fractionator tubing contains 60% sucrose up to the bottom of the dummy tube. Remove the dummy tube, by loosening the collar sufficiently.
4. Insert the metal needle into one gradient, (be sure to not create any waves in the gradient) and ensure that the collar is tight enough to avoid any leak (but do not overtighten). Switch on the pump to push 60% (heavy) sucrose into the bottom of the gradient, thus displacing the gradient through the flow cell and then into the fraction collector. The top of the gradient emerges first. Monitor with ISCO UA-6 UV-VIS detector the absorbance at 260 nm.
5. Put the microfuge tube rack on the fraction collectors (Gilson FC 203B/FC 204) with nine microfuge tubes 2 ml with screw cups off.
6. Collect the fractions from the sucrose gradient with a constant *volume* (nine fractions of 1 ml each).
7. Label the gradient fractions.
8. Wash the optical units and the tubing every time with RNAse/DNAse-free water and make sure that all water is out of the tube before processing the next sample.

3.7. RNA Extraction (see Note 12)

1. To each fraction add 30 μl of SDS 20% and 0.25 μg/μl carrier tRNA and mix inverting 3–4 times.
2. To each fraction add 20 μl of 5 mg/ml Proteinase K and incubate the samples at 37°C for 2 h (15).
3. Add 250 μl of phenol/chloroform pH 4.5 to each fraction and vortex very well for 1 min creating an emulsion.
4. Spin at maximum speed (13,400 × *g*) in a microfuge for 10 min at 4°C.

5. Transfer the aqueous phase (top layer) into a new microcentrifuge 2 ml tube.
6. Add 250 μl of chloroform to each fraction and vortex well (this is an optional step just to remove all residue of phenol from the sample).
7. Spin at maximum speed (13,400 × *g*) in a benchtop centrifuge for 10 min at 4°C.
8. Transfer the aqueous phase to a fresh microfuge 2 ml tube.
9. Add 0.7 ml of isopropanol to each sample.
10. Mix well but gently by inverting tubes five times.
11. Incubate at –20°C for 2 h or longer.
12. Centrifuge at 13,400 × *g* for 30 min at 4°C.
13. When the spin is complete, place the sample on ice and remove all isopropanol, taking care not to touch the RNA pellet at the bottom of the tube.
14. Add 1 ml of cold 80% ethanol to each fraction.
15. Centrifuge 13,400 × *g* for 30 min at 4°C (during the spin, leave the RNAse/DNAse-free water solution on ice).
16. Place the sample on ice and remove the ethanol taking case not to touch the RNA pellet on the bottom of the tube.
17. Remove also residual 80% ethanol (any drops) with P20 pipette.
18. Air-dry the pellet for 10 min on ice (all ethanol should be removed before to resuspend the pellet in water).
19. Dissolve pellets in RNAse/DNAse-free water.

3.8. Protein Precipitation (see Note 13)

1. Add to each fraction 10% TCA before (the colour of the solution will change to yellow).
2. Incubate immediately on ice for 45 min.
3. Centrifuge at maximum speed, 13,400 × *g* for 30 min at 4°C.
4. Remove the supernatant taking care to leave the protein pellet intact.
5. Add to the pellet 1 ml of 5% TCA.
6. Vortex each tube for 10 s.
7. Centrifuge at maximum speed, 13,400 × *g* for 10 min at 4°C.
8. Remove the supernatant.
9. Wash with 1 ml acetone.
10. Vortex each tube for 10 s.
11. Spin the tubes in microcentrifuge at maximum speed, 15,700 × *g* for 10 min at 4°C.
12. Repeat steps 9–11 four to five times (see Note 14).

13. After the last wash, leave the tubes uncapped at room temperature to allow the acetone to evaporate (do not over-dry pellet; otherwise, it may not dissolve properly).
14. SDS-PAGE loading buffer and vortex thoroughly to dissolve protein pellet and leave overnight in shaking at room temperature (see Note 15).

4. Notes

1. TCA is highly corrosive and must be handled with care: wear gloves, laboratory coats, and eye protection.
2. Always use gloves when handling dialysis membrane.
3. Do not autoclave DTT.
4. One will need to add some water to the buffer to allow larger quantities of sucrose to dissolve.
5. For cells that grow in suspension, cells must be collected by centrifugation and can easily be counted after the serum starvation and then aliquotted for the treatment protocol.
6. Note that the medium contains almost all the radioactivity and must be handled and disposed of accordingly.
7. A small volume of lysis buffer is used to increase the protein concentration to avoid having to load too much lysate onto the filters.
8. Be very careful when heating the TCA not to heat it so strongly that it boils too vigorously and spills out of the beaker.
9. Do not allow the heating to exceed the time mentioned because it increases the precipitation.
10. Take care to balance the samples, make sure that they are paired correctly when putting on the rotor. At least two gradients must be prepared every time; empty buckets must be used is running less than the maximal number of gradients.
11. Centrifugation must start as soon as possible.
12. *Extraction of RNA* from the gradient fractions allows one to study which mRNAs are associated with polysomes, and with which are in non-polysomal material, under different conditions (e.g. ±mTOR inhibitors). The levels of different mRNAs in the resolved fractions can be determined by Northern blotting using appropriate probes or, e.g. following reverse-transcription by quantitative (real-time) PCR.
13. *Precipitation of protein* allows one to study the distribution across the gradient of ribosomal proteins, to confirm which fractions contain polysomes plus 80S subunits (will have proteins

from the 60S and 40S subunits), 40S or 60S subunits, and which are devoid of ribosomal material. Similar information can be obtained from Northern blots for 18S (40S subunit) and 23S (60S subunit) rRNAs.

14. An acetone wash helps to resuspend the pellet more easily by removing any residual TCA.
15. The polysome and monomer fractions (from 1 to 7) are very easy to resuspend in 20–30 μl of loading buffer, but it is better to resuspend the non polysome fraction (8, 9) in 100–200 μl of loading buffer.

References

1. Wullschleger, S., Loewith, R., and Hall, M.N. (2006) TOR signaling in growth and metabolism. *Cell*, **124,** 471–484.
2. Proud, C.G. (2007) Signalling to translation: how signal transduction pathways control the protein synthetic machinery. *Biochem. J*, **403,** 217–234.
3. Proud, C.G. (2009) mTORC1 signalling and mRNA translation. *Biochem. Soc. Trans*, **37,** 227–231.
4. Mayer, C. and Grummt, I. (2006) Ribosome biogenesis and cell growth: mTOR coordinates transcription by all three classes of nuclear RNA polymerases. *Oncogene*, **25,** 6384–6391.
5. Wang, X. and Proud, C.G. (2009) Nutrient control of TORC1, a cell-cycle regulator. *Trends Cell Biol.*, **19,** 260–267.
6. Feldman, M.E., Apsel, B., Uotila, A., Loewith, R., Knight, Z.A., Ruggero, D., and Shokat, K.M. (2009) Active-site inhibitors of mTOR target rapamycin-resistant outputs of mTORC1 and mTORC2. *PLoS. Biol.*, **7,** e38.
7. Jastrzebski, K., Hannan, K.M., Tchoubrieva, E.B., Hannan, R.D., and Pearson, R.B. (2007) Coordinate regulation of ribosome biogenesis and function by the ribosomal protein S6 kinase, a key mediator of mTOR function. *Growth Factors*, **25,** 209–226.
8. Gingras, A.-C., Raught, B., and Sonenberg, N. (2001) Regulation of translation initiation by FRAP/mTOR. *Genes Dev*, **15,** 807–826.
9. Shahbazian, D., Roux, P.P., Mieulet, V., Cohen, M.S., Raught, B., Taunton, J., Hershey, J.W., Blenis, J., Pende, M., and Sonenberg, N. (2006) The mTOR/PI3K and MAPK pathways converge on eIF4B to control its phosphorylation and activity. *EMBO J*, **25,** 2781–2791.
10. Meyuhas, O. and Hornstein, E. (2000) Translational control of TOP mRNAs. In Translational control of gene expression (Sonenberg, N., Hershey, J.W.B. and Mathews, M.B. Eds.), pp. 671–693. Cold Spring Harbor Laboratory Press, Cold Spring Harbor, NY.
11. Bradford, M.M. (1976) A rapid and sensitive method for the quantitation of microgram quantities of protein utilizing the principle of protein-dye binding. *Anal. Biochem*, 77, 248–254.
12. Wang, L. and Proud, C.G. (2002) Ras/Erk signaling is essential for activation of protein synthesis by Gq protein-coupled receptor agonists in adult cardiomyocytes. *Circ. Res*, **91,** 821–829.
13. Hayashi, A.A. and Proud, C.G. (2007) The rapid activation of protein synthesis by growth hormone requires signaling through mTOR. *Am. J. Physiol Endocrinol. Metab*, **292,** E1647–E1655.
14. Rappsilber, J., Ishihama, Y., and Mann, M. (2003) Stop and go extraction tips for matrix-assisted laser desorption/ionization, nanoelectrospray, and LC/MS sample pretreatment in proteomics. *Anal. Chem*, **75,** 663–670.
15. Sambrook, J., Fritsch, E.F. and Maniatis, T. (1989) Molecular cloning: a laboratory manual, Cold Spring Harbor Laboratory Press.
16. Huang, B.P., Wang, Y., Wang, X., Wang, Z., and Proud, C.G. (2009) Blocking eukaryotic initiation factor 4 F complex formation does not inhibit the mTORC1-dependent activation of protein synthesis in cardiomyocytes. *Am. J. Physiol Heart Circ. Physiol*, **296,** H505–H514.

Chapter 11

A Genome-wide RNAi Screen for Polypeptides that Alter rpS6 Phosphorylation

Angela Papageorgiou and Joseph Avruch

Abstract

Mammalian target of rapamycin (mTOR) is a giant protein kinase that controls cell proliferation, growth, and metabolism. mTOR is regulated by nutrient availability, by mitogens, and by stress, and operates through two independently regulated hetero-oligomeric complexes. We have attempted to identify the cellular components necessary to maintain the activity of mTOR complex 1 (mTORC1), the amino acid-dependent, rapamycin-inhibitable complex, using a whole genome approach involving RNAi-induced depletion of cellular polypeptides. We have used a pancreatic ductal adenocarcinoma (PDAC) cell line, Mia-PaCa for this screen; as with many pancreatic cancers, these cells exhibit constitutive activation of mTORC1. PDAC is the most common form of pancreatic cancer and the 5-year survival rate remains 3–5% despite current nonspecific and targeted therapies. Although rapamycin-related mTOR inhibitors have yet to demonstrate encouraging clinical responses, it is now evident that this class of compounds is capable of only partial mTORC1 inhibition. Identifying previously unappreciated proteins needed for maintenance of mTORC1 activity may provide new targets and lead to the development of beneficial therapies for pancreatic cancer.

Key words: rpS6, Phosphorylation, mTOR, Genome wide, RNAi, Screen, Immunofluorescence, Pancreatic cancer

1. Introduction

The phosphatidylinositol 3-kinase/protein kinase B/mammalian target of rapamycin-pathway (PI3K/AKT/mTOR-pathway) regulates cell proliferation, cell survival, angiogenesis, and resistance to antitumor treatments. In tumors, the PI3K/AKT/mTOR-pathway is activated through several different mechanisms, most commonly activating mutations in PI-3 kinase or loss of PTEN expression. mTOR functions through two independently regulated complexes, both of which are activated by PI-3 kinase (1). The mechanisms

Thomas Weichhart (ed.), *mTOR: Methods and Protocols*, Methods in Molecular Biology, vol. 821,
DOI 10.1007/978-1-61779-430-8_11, © Springer Science+Business Media, LLC 2012

underlying activation of mTOR complex 2 by PI-3 kinase, as well as the contribution of mTORC2 to the malignant phenotype are poorly understood. In contrast, based primarily on extensive work using rapamycin and its congeners, activated mTOR complex 1 is thought to support several aspects of the malignant phenotype; therefore, inhibition of mTORC1 is widely considered a promising antitumor treatment (2–7). The pathway from PI-3 kinase to TOR is well described, proceeding through Akt inhibition of the Tuberous Sclerosis complex, a GTPase activating protein for the ras-like GTPase Rheb, and Rheb activation of mTOR complex 1 (8). Independently, mTORC1 activity is also regulated by cellular amino acid levels such that depletion of intracellular amino acids, most notably leucine (9, 10), will inactivate mTORC1 despite the presence of a highly active PI-3 kinase pathway. The cellular elements that comprise this amino acid-dependent regulatory input are much less well known, save for the Rag GTPases, however, apart from their ability to bind raptor and thus modify mTORC1 subcellular localization, the regulation of the Rag GTPases is largely obscure (11). We undertook a genome-wide RNAi screen to uncover previously unknown components needed to sustain mTORC1 signaling, using pancreatic cancer cells. In addition to providing a more comprehensive understanding of this signaling pathway, the identification of polypeptides and pathways necessary for mTORC1 activity in such cells may identify "druggable" targets whose inhibition may complement and/or augment the efficacy of the direct mTOR and PI-3 kinase inhibitors currently available or in development (12).

We chose ribosomal S6 phosphorylation as the indicator of mTORC1 signaling activity. S6 is the single phosphopeptide of the mammalian 40S ribosomal subunit (13). Ribosomal S6 is phosphorylated *in vivo* on five serine residues (235, 236, 240, 244, and 247) located within the carboxylterminal 15 amino acids of the S6 polypeptide (14). Phosphorylation at each of these sites is catalyzed in a sequential, processive reaction (236 first, then 235, 240, 244, 247) by the p70S6 kinase/S6K1 (15). In support of the suitability of S6P to serve as a reporter of mTORC1 activity is the ability of rapamycin to inhibit S6 phosphorylation and S6K1 activity in essentially all mammalian cells examined. Thus, although S6 can also be phosphorylated by the Rsk family of kinases, which appear to sustain S6 phosphorylation in S6K1/2 deficient mice (16), Rsks are insensitive to rapamycin and appear to play no part in S6 phosphorylation in normal or malignant cells (17). S6K1 is a direct substrate of the mTOR complex 1, which phosphorylates several sites in an autoinihibitory region in the carboxylterminal tail, and most importantly, Thr389/412 in a hydrophobic motif just carboxylterminal to the S6K1 catalytic domain (18); the latter phosphorylation creates a binding site for the protein kinase PDK1, which then phosphorylates Thr229/252 in the S6K1 activation loop (19).

Together, the phosphorylations at Thr389/412 and Thr229/252 activate S6K1 in a synergistic manner (19). The critical importance of Thr389/412 phosphorylation in S6K1 activation makes mTORC1 essential for S6K1 activity. The phosphorylation of S6K1(Thr389/412) is completely inhibited by rapamycin (IC50 ~ 2 nM) in all cells (20).

Inasmuch as S6K1(Thr389/412-P) is a direct indicator of mTORC1 activity, we first examined the feasibility of using this modification as a readout for a high-throughput assay using antibody-based immunofluorescence (IF) detection. Although this phosphorylation is a reliable indicator of mTORC1 activity in a western blot format (20), conversion to high-throughput formats was not feasible. This is so because of the very low abundance of endogenous S6K1 allows the small background of nonspecific fluorescence associated with all S6K1(Thr389/412-P) antibodies in this mode of detection to swamp the specific signal generated from the low abundance S6K1(Thr389/412-P).

In contrast, ribosomes are highly abundant cellular constituents, and especially so in transformed cells. Although the physiological/functional significance of S6 phosphorylation remains obscure (21), ribosomal S6 protein is unquestionably the best characterized, most specific and abundant S6K1 substrate. Recent studies with rapamycin and the more potent mTOR catalytic site inhibitors have shown that a considerably higher mTORC1 activity is required to maintain S6K1(Thr389/412) phosphorylation than is needed to sustain 4E-BP(Thr37/46) phosphorylation (22); thus the loss of the S6K1 substrate, S6P provides a very sensitive indication of the inhibition of mTORC1. After screening a variety of monoclonal and polyclonal phospho-specific S6 antibodies, we selected a monoclonal antibody directed at S6(Ser235P/236P) and optimized the conditions necessary for detection of pS6 by IF that provide maximum sensitivity and minimal background signals in a high-throughput fashion (i.e., 384-well plate).

The following criteria are important in choosing a cell line: transfection efficiency, reproducibility across experiments, dynamic range (signal to background ratio), and z' score (see below). Seeking to favor identification of positive regulators of mTOR complex 1, we chose a cell line with a high level of mTORC1 activity, as reflected by the level of rpS6 phosphorylation (see below) under usual culture conditions, thereby obviating the need for a stimulation of the cells prior to analysis, and providing the largest possible dynamic range between the initial and inhibited levels of S6P.

The Mia-Paca 2 cells are considered a representative pancreatic cancer cell line based on their mutational profile (i.e., K-Ras mutant, P53 mutant, P16HD, and DPC4 wild-type) (23, 24). We chose them for this screen because the mTOR/PS6K/PS6 pathway is constitutively active (25), because the cells are reasonably transfectable, and because their morphology enabled reliable quantitation of the IF signals on a cell by cell basis.

Immunofluorescence-based HTP screens can be accomplished using either high content microscopes or plate readers with fluorescent capability. Plate readers, which sum fluorescence from an entire well or segment, may facilitate a higher throughput, but at the cost of cell-specific information, e.g., subcellular localization, cell cycle stage, etc., as well as an ability to focus detection on specific cell populations or subcellular compartments. After testing different plate readers (Envision, LICOR Aerius, M5) and reading methods (top and bottom read), we decided that high content microscopy, using the IXM apparatus, enabled analysis on a cell-by-cell basis and provided the best possible signal-to-background ratio. Only high content microscopy provides assurance that the IF signal in fact originates from the target, minimizing false positives.

The z'-score – used for assessment of assay conditions and for optimization of HTP screens – is given by the formula:

$$z' = 1 - \frac{(3\mathrm{SD}_{+\mathrm{control}} + 3\mathrm{SD}_{-\mathrm{control}})}{(\mathrm{AVG}_{+\mathrm{control}} - \mathrm{AVG}_{-\mathrm{control}})}.$$

The dynamic range, i.e., the difference among the positive and negative controls and the variance of the measurements are thus reflected by the z', providing a useful measure of overall assay performance.

The smart pool PLK1 oligo is used routinely to monitor transfection efficiency; knockdown of PLK1 gene expression induces a G2/M arrest and gives a cell-death phenotype. Five different commercially available transfection reagents and different ratios of the transfection reagent (amount of transfection reagent used) to the input RNAi (25, 50, 100 nM), as well as the duration of exposure to the transfection mix were evaluated. The reproducibility across experiments is monitored by the z'. Transfection efficiency proved to be strongly dependent on cell confluency/cell number at plating, with approximately 50% confluency proving optimal. At cell densities significantly above this level, transfection efficiency fell, perhaps due to competition for RNAi molecules, even at the practical upper limit of 100 nM. Moreover, imaging becomes problematic at very high cell numbers. At lower cell numbers, the abundance of cells appropriate for imaging may be insufficient even after optimizing transfection efficiency. As regards termination of the experiment, several timepoints (48, 72, and 96 h) posttransfection should be evaluated.

In selecting optimal conditions, the following were most important: z'-score (which incorporates the fold difference among positive and negative RNAi controls as well as the dynamic range of the signal), the ease and accuracy with which a cell line can be imaged (adherent spread-out cells are more easily imaged as compared to others), and whether a cell line behaved consistently in terms of its cell growth rate, the intensity of the primary readout, i.e., PS6 and the transfection efficiency.

2. Materials

2.1. Cell Culture

1. Dulbecco's Modified Eagle's Medium (Gibco/BRL, Bethesda, MD) supplemented with 10% fetal bovine serum (FBS), Hyclone (Hyclone Laboratories, Logan, Utah) (see Note 1), horse serum (Gibco/BRL), and penicillin/streptomycin.
2. Trypsin (0.05%) EDTA (1×) (0.53 mM EDTA in HBSS).
3. Countess® Cell Counting Chamber Slides (Invitrogen, Carlsbad, CA).
4. PCR Mycoplasma test: MycoSensor PCR Assay Kit (Stratagene, Santa Clara, CA).

2.2. Transfection

1. Whole si-genome SMARTpool RNAi library (Dharmacon, Lafayette, CO): Targets the whole genome and is consisted of RNAis against 21,176 genes. Each RNAis is a pool of four individual double-stranded RNAi oligonucleotides, mixed together.

   Controls:

   - Non-Targeting siGenome SMARTpool.
   - FRAP1 siGENOME SMARTpool.
   - PLK1 siGenome SMARTpool.
2. Transfection reagent (empirically determined, using the cell line of choice): Lipofectamine 2000 (Invitrogen, Carlsbad, CA).
3. Opti-MEM (GIBCO/BRL, Bethesda, MD).
4. Nuclease-free water (Qiagen, Valencia, CA).
5. 5× siRNA Buffer (Dharmacon, Lafayette, CO).
6. Two types of tubing cartridges: Eight-channel standard bore disposable tubing cartridge (presterilized) and small bore disposable tubing cartridge (Thermo Scientific, Hudson, NH).
7. Aspirator Tool – "The Wand": It is a 24-channel adaptor and is used for aspirating liquid from 384-well assay plates. Drummond Scientific stainless steel wand (VWR, West Chester, PA).
8. 30 μL 384-tip racks, nonsterile (Velocity11).
9. RNAase Zap Wipes (Ambion, Austin, TX); RNAase Zap (Ambion, Austin, TX).
10. 384-Well plates Assay Plates (Corning Incorporate, Corning, NY).
11. Bleach-Rite. BRSPRAY16 (Current Technologies Inc., Crawfordsville, IN).
12. Sharpie: Silver Metallic 39100 (website: http://www.sharpie.com/; available at Staples).
13. Instant Sealing Sterilization Pouches (7×13 in.). These are autoclave bags in which a matrix wellmate cartridge can be sterilized (Fisher Scientific, Pittsburg, PA).

2.3. Immunofluorescence

1. PBS (containing $CaCl_2$ and $MgCl_2$; GIBCO, BRL). Alternatively, prepare 10× PBS stock solution by the addition of 8 g NaCl, 0.2 g KCl, 1.15 g Na_2HPO_4 (sodium phosphate, dibasic), 0.2 g KH_2PO_4 (potassium phosphate, monobasic). Add water to 1 l. Adjust pH to 7.4.
2. Formaldehyde (37% solution).
3. Methanol.
4. Block solution (Containing 1% Goat serum and 5% Horse Serum).

   Goat Serum.
5. Primary Antibodies:

S6: Phospho-S6 Ribosomal protein (Ser 235/236) Rabbit monoclonal Antibody (Cell Signaling Technology). For longer term storage the antibody should be aliquoted and stored at −80°C. Avoid freeze–thaw of the aliquots. 4000 is a clone 2F9 4854 but produced recombinantly.

Phospho-S6 Ribosomal Protein (Ser235/236) (2F9) Rabbit mAb (Cell Signaling Technology).

Total S6S6 Ribosomal Protein (54D2) Mouse mAb (Cell Signaling Technology).

6. Secondary antibody: Alexa Fluor® 488 goat anti-rabbit IgG (H+L) highly cross-adsorbed; 2 mg/mL (Invitrogen, Carlsbad, CA).
7. Nuclear stain: DAPI (Sigma, St. Louis, MI). Make stock solution in water and then dilute 1:1,000.
8. Cover for plates: Thermowell Sealing Tape (Costar Corning Incorporate, Corning, NY).

3. Methods

3.1. Transfection

1. For all transfection and immunofluorescence steps: Use a Matrix-Wellmate (Thermofisher) to dispense all cells and for all subsequent liquid dispensing steps. The following steps are all performed in a TC hood (Class II Biohazard). Please refer to Notes 2–4 for comments on TC methods.
2. Thoroughly wipe the hood using 70% ethanol and kim-wipes. If needed preclean with either 10% bleach or Bleach-Rite. Anything that is placed in the hood should be cleaned with 70% ethanol. It is advisable that pipettes are also cleaned with RNAase zap and then rinsed with water.
3. Transfer of RNAis from the library stock plates to assay plates: The Velocity 11 Bravo automated liquid handling platform is used to transfer 4 μl of 1 μM RNA from the library plates to

the assay plate by use of the Velocity 11 384-well tips. The final RNAi concentration is 100 nM. This type of transfection is referred to as "dry," in that the RNA is added first to the plate and the transfection reagent in Opti-MEM (referred to as "reagent mix") is added thereafter. Alternatively, a "wet" transfection is one in which case the transfection reagent mixed with the Opti-MEM is added first to the assay plates followed by the RNAi oligonucleotides.

4. Label assay plates (Corning) using a silver pen (Shapiro).
5. The Wellmate small bore tubing cartridge, used to dispense small volumes of liquid (i.e., the transfection reagent mix and subsequently the cells), is used for all subsequent dispensing steps during the transfection. Before using the cartridge, rinse with nuclease-free water (3×), 70% ethanol (1×) followed by nuclease-free water (3×). Make sure that the cartridge is not clogged and that all liquid is removed by priming with air. Ensure sterility of the cartridge by autoclaving it. It can be autoclaved up to three times according to the manufacturer. Use a test plate to determine whether the alignment or the height of the Wellmate requires adjustment. Examine visually whether the liquid is dispensed onto the center of the well; this is necessary in order to ensure homogeneous addition of the cells and the transfection reagent mix. Adjust the height of the dispensing head by raising or lowering it to the required height. Adjust the x-stage position of the wellmate if necessary, following the user manual instructions.
6. Add Controls: The RNAi smart pool oligonucleotides are shipped in dry form. Resuspend the oligo in 1× RNAi Buffer to 100 μM and then to 1 μM. Prepare 1× buffer by diluting the 5× Buffer with nuclease-free water. Add controls to as many wells as possible (at least 16 wells for each control per plate).
7. Prewarm media, Opti-MEM and trypsin in a 37°C tissue culture – dedicated water bath before use. Prepare transfection reagent with medium: In a 50-ml sterile falcon tube mix, prepare a mix containing Lipofectamine 2000 with Opti-MEM at a ratio of 1:100 (e.g., for 20 ml of final volume use 200 μl of Lipofectamine 2000) (see Note 5). Incubate the transfection reagent mix at room temperature for 5 min. During this time, invert the tube containing the transfection reagent mix a couple of times and spin at 233 × *g* for 1 min. Repeat twice to ensure the transfection reagent mix is equally mixed (see Note 6). Set the dispensing volume of the wellmate to 10 μl and add the transfection reagent mix to each well. Be sure the tubing cartridge is devoid of bubbles and use the highest dispensing speed (S1) for dispensing the mix. After the transfection mix has been added subject the plates immediately to centrifugation at 233 × *g* for 1 min. It is preferable that the temperature of the centrifuge is set at room temperature.

8. Prepare cells: The cells should be passaged twice before the day of the transfection. The confluency of the cells at the time they are split is very critical as this will in turn affect the cell growth rate during the course of the experiment. At the time of harvest for transfection, cells should be approximately 80–90% confluent. A cell count is always used to determine cell number at the time of transfection.

9. Trypsinize cells. Remove media and wash off residual serum with trypsin–HBSS (see Note 7). Prepare media to be used for the transfection omitting the antibiotics. After addition of media containing 10% FBS to neutralize trypsin, spin down cells at 335 × *g* for 5 min in a 50-ml sterile tube. Remove the supernatant and resuspend the cell pellet in transfection media. Pipette up and down the cell pellet so that you have a homogeneous cell population. Count the cells by using a hematocytometer or an automated cell counter (as available). To ensure that the cell number is accurate, it is recommended that you assay four replicates. Prepare the final dilution of the cells: 95 cells/μl plated per well (×26 μl) or 2,500 cells/well. Please note that this number is dependent on the serum lot/catalog number since different serum lots can differentially affect the cell growth rate and therefore the cell number needs to be optimized accordingly. Use a 250-ml flask to make the final dilution from which cells are dispensed. Since the cells will settle quickly, it is critical to swirl the flask continuously while dispensing to ensure there will be a homogeneous distribution of the cells. Use a test plate first to visually inspect that the dilution is accurate. You should aim for a plating confluency of around 50–55%. A denser seeding density will lead to increased cell number which in turn will cause the cells to starve and decrease PS6 baseline levels (see Note 8). At 2 h after addition of cells to the transfection mix, spin down plates for 1 min at 233 × *g*. Use the aspirator wand to remove media. To prevent dislodging of the cells during all the aspiration steps, place tape around the edges of the wand. The wand can be autoclaved to ensure sterility. Before and after each use it is best to use distilled water so as to prevent clogging of the wand and deposition of particles. Decontaminate the wand with 70% ethanol. Next, add 70 μl of transfection media per well using the small-bore wellmate cartridge. Change the dispensing speed to S2 so as to prevent the cells from being dislodged. Spin down the plates for 1 min at 233 × *g*. Place in a 37°C tissue culture incubator for 72 h. During this time interval, the growth rate of the cells can be inspected visually.

10. Edge effects are a concern as they alter the outer wells; their occurrence is primarily dependent on the humidity and configuration of the incubator. Edge effects are usually evident

from the loss of volume from outer wells or by diminished cell viability in those wells. If the outer two columns and rows contain experimental samples the following approaches can be taken to minimize edge effects. (a) 48-h posttransfection add 10–15 μl of transfection media to the outer well and columns, (b) place moist paper towels in a container and place the assay plates on top of the paper towels.

3.2. Immunofluorescence

1. Harvest the cells 72 h after beginning the transfection (see Notes 9–11).
2. First wash the cells once with 1× PBS (Cat. No. 11490).
3. Fix the cells for 15 min at room temperature in 4% formaldehyde. Prepare a 4% formaldehyde solution fresh each time and dilute in PBS from the 37% formaldehyde stock. Note the plate order to which formaldehyde is added. When aspirating the fixation buffer keep the same order so that all plates are fixed for the same amount of time. Wash, Spin at 233 × *g*, and aspirate using the wand: this step is repeated five times.
4. Permeabilize the cells by addition of ice-cold methanol for 10 min. Keep the methanol at −80°C or −20°C for a short time prior to addition, to ensure that it is cold when added to the cells. Subsequent to permeabilization repeat the steps noted above: wash, spin, aspirate five times.
5. Prepare the blocking solution composed of 5% HS and 1% GS in PBS. Incubate with blocking solution for 1 h at room temperature.
6. Prepare the working antibody dilution at a final concentration of 2.185 μg/ml in block solution in a 50-ml falcon tube. Ensure that antibody is evenly mixed by inverting the tube several times and subsequent centrifugation at 233 × *g* for 1 min. Repeat this mix/spin step six times. Alternatively, you can prepare the working final antibody dilution and place it on a platform shaker for 30 min prior to adding the antibody.
7. Centrifuge plates (233 × *g* for 1 min). Antibody addition can be performed in one of the following two ways: (a) Aspirate the wash solution using the aspirator wand. Remove the residual volume using a fine tip (attached to a vacuum) from the edge of each well. This step needs to be performed rapidly (5 min or less for the whole 384-well plate) so that the cells do not dry out. If more time is needed, then adjust accordingly the number of wells that you process at a time. The antibody can be added either manually using a multiwell pipette or by using the small bore wellmate cartridge and the S2 dispension rate. Add 15 μl of the working antibody dilution per well. Spin down plates. Cover plates with seals (see Note 5). (b) Alternatively, the removal of the residual volume can be avoided; if a residual volume of 10–15 μl/well remains, use an antibody solution of

4.38 μg/ml (twice the above final concentration) and add 15 μl using the small bore Wellmate cartridge. Centrifuge plates and then cover with sealing tape. In practice, we found this method to give higher background and less well-defined staining (see Note 6). The primary antibody can be stored at 4°C for 1 month; for a longer storage times, aliquot and store the antibody at −80°C. Avoid freeze–thaw cycles of the antibody.

8. Incubate antibody-loaded plates overnight at 4°C for 16–20 h using a shaking platform at low speed.
9. Centrifuge plates and aspirate the primary antibody using the wand. Wash five times with PBS, spin down the plates, aspirate using the wand.
10. Before use, spin the secondary antibody in a microcentrifuge for 20 min at top speed and take care to use only the supernatant so as to eliminate addition of formed protein aggregates to your working solution; precipitated antibody will cause nonspecific staining. For this reason, the highly cross-absorbed secondary antibody is preferred over other types (see Note 12).
11. Prepare the secondary working antibody dilution (Alexa 488; 1:200 dilution of stock) in block solution at a final concentration of 0.01 mg/ml as described above for the final dilution of the primary antibody.
12. Addition of the secondary antibody: Add 15 μl of the working secondary antibody dilution per well using the small bore cartridge and the S2 dispension rate. Spin down the plates. Perform these steps involving the secondary antibody in the dark if possible and/or cover the plate with an aluminum seal cover. Cover plates with foil. Incubate for 2 h at room temperature in the dark.
13. Removal of secondary antibody: Spin down plates and aspirate the secondary antibody with the aspirator wand. Wash five times as above using the large bore cartridge and the S3 dispensing rate.
14. Addition of the nuclear dye: Dissolve DAPI in water (20 mg/ml). Aliquot and store at −20°C in the dark. For a DAPI working solution, add 20 μl of the DAPI stock per ml of D-PBS (1:1,000 dilution) and add 80 μl/well; incubate for 10 min at room temperature. Take note of the order of the plates to which DAPI was added. Remove DAPI from the plates using the wand in the same order. Wash three times with PBS as described above (wash, spin, aspirate). Seal the plates with sealing tape. The plates are now ready to be imaged (Fig. 1).

The washing steps can also be performed using a plate washer after determining settings that allow retention of both cell populations and the total cell number.

Secondary assays are performed with the goal of identifying where validated/confirmed hits map within the pathway or for functional analysis (e.g., cell death, cell proliferation). For example,

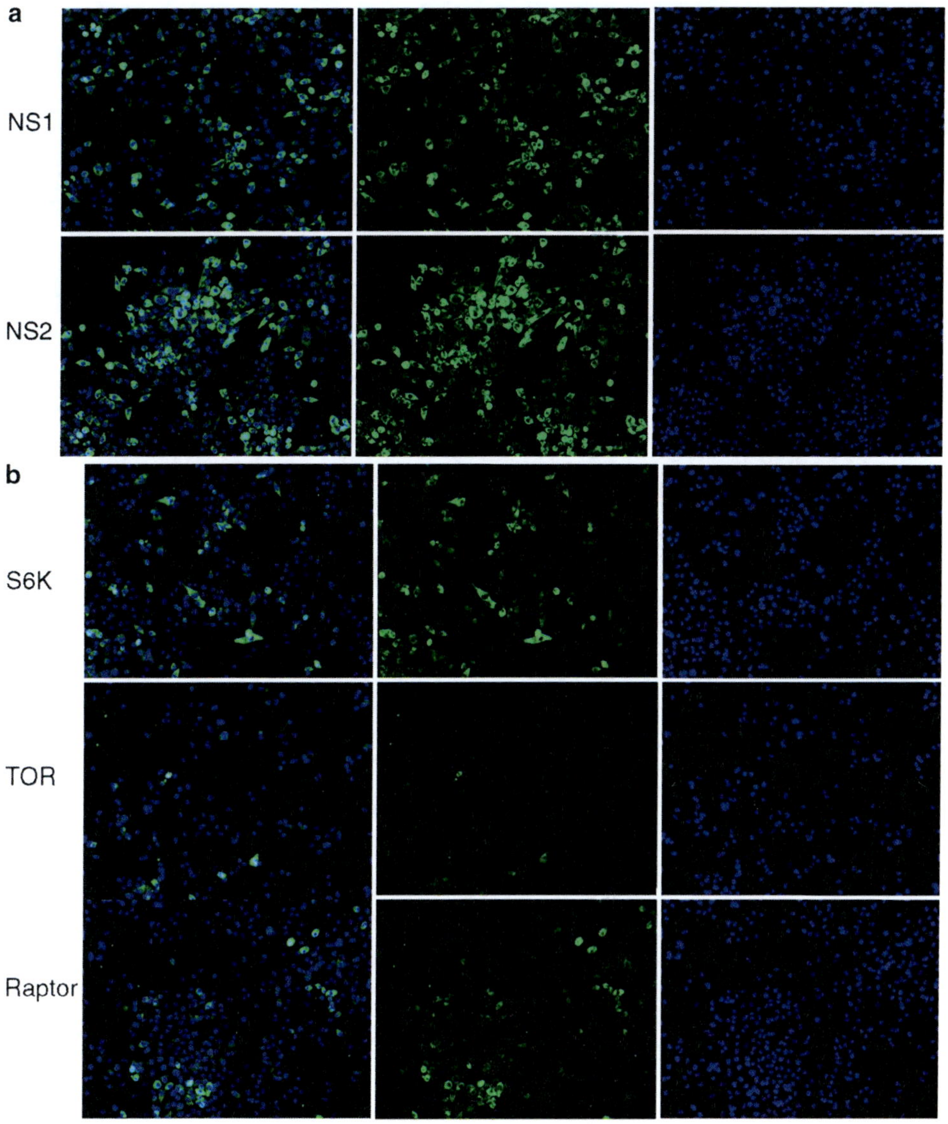

Fig. 1. Immunofluorescence analysis (IF) of rpS6Ser(235/236) phosphorylation. Mia-Paca 2 cells in 384-well plates were transfected with a control, nonspecific RNAis NS1 (*upper panel*) and (NS2) (*lower panel*) (**a**), S6K, TOR, and Raptor (**b**), TSC1, TSC2, and PTEN (**c**). The cells were then left unperturbed in 10% media for 72 h. Subsequently, they were fixed, permeabilized, and stained by using a rabbit monoclonal anti-PS6 primary antibody. Shown here are representative images; PS6 antibody (Ser 235/236) in green detected with secondary anti-rabbit Alexa 488 antibody. Nuclei are stained with DAPI. Samples were run on the High Content Imaging microscope ImageXpress®Micro and analyzed using the MetaXpress Imaging Software.

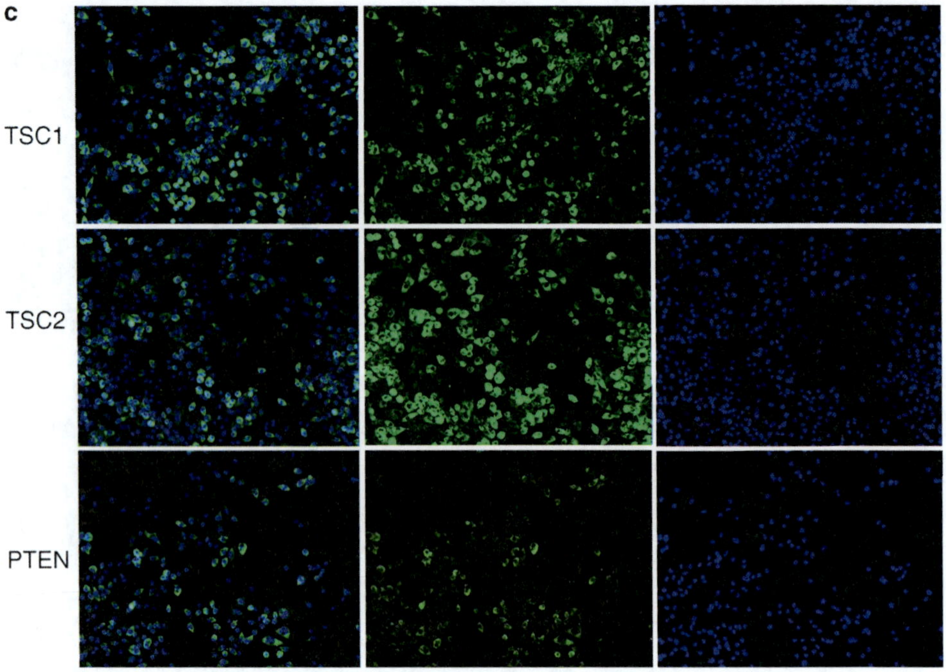

Fig. 1. (continued)

one can choose TSC1/2 null MEFs in order to delineate whether positive hits act downstream of the TSC1/2 complex. Using new cell lines, one needs to optimize the transfection conditions as described above. Additionally, other target readouts are needed to determine whether a hit acts through TORC1 or TORC2.

3.3. Image Acquisition

1. For image acquisition, run samples on the High Content Imaging microscope ImageXpress® Micro System (Molecular Devices). Configure and save the "Acquisition" settings that will be used for all subsequent experiments. Binning refers to the camera binning one binning is equal to one pixel; two binning combines a 2 by 2 set of pixels, three binning a 3 by 3 set of pixel, etc. Choose one binning which allows maximal resolution. Gain refers to the amplification applied to the camera signal, with "1" giving least gain and least background amplification; however at a binning of 1×, a gain of 2 is required to enable sufficient signal (Fig. 2a, screenshot).
2. Select the objective magnification at which the images will be acquired (10× Plan Fluor 0.3 NA objective).
3. Run the LAF (Laser AutoFocus) wizard by following the steps described so that the plate parameters are defined. The LAF Wizard defines the multiple settings required for each specific plate type and a particular objective so that the objective will

a

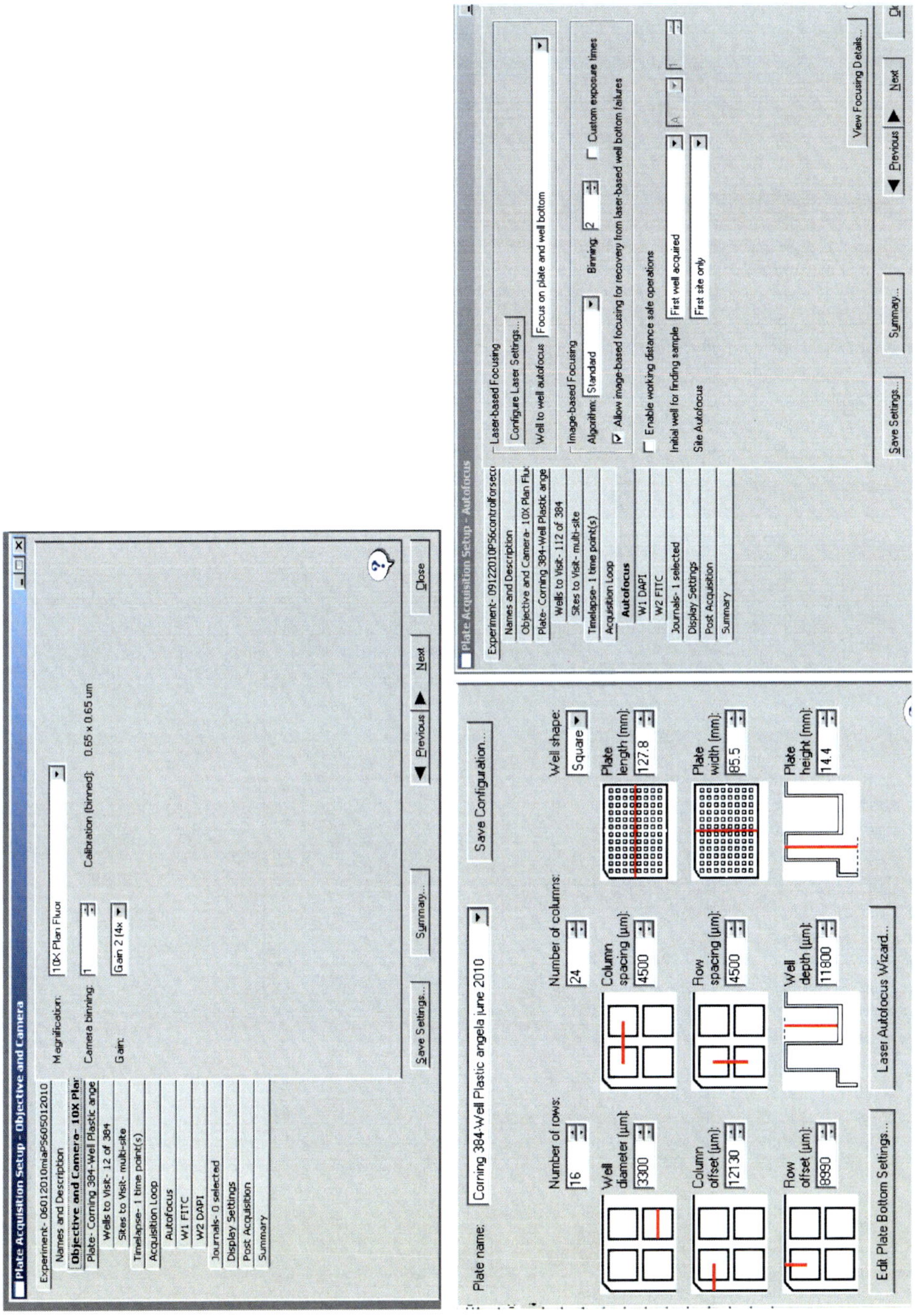

Fig. 2. Screenshots depicting the image acquisition settings as described in the main text. (**a**) Setup of general settings as described in the Image acquisition section. (**b**–**d**) Setup of specific settings: (**b**) Laser-based settings. (**c**) Image-based settings. (**d**) Use of a journal that allows image acquisition at different *z* heights.

b

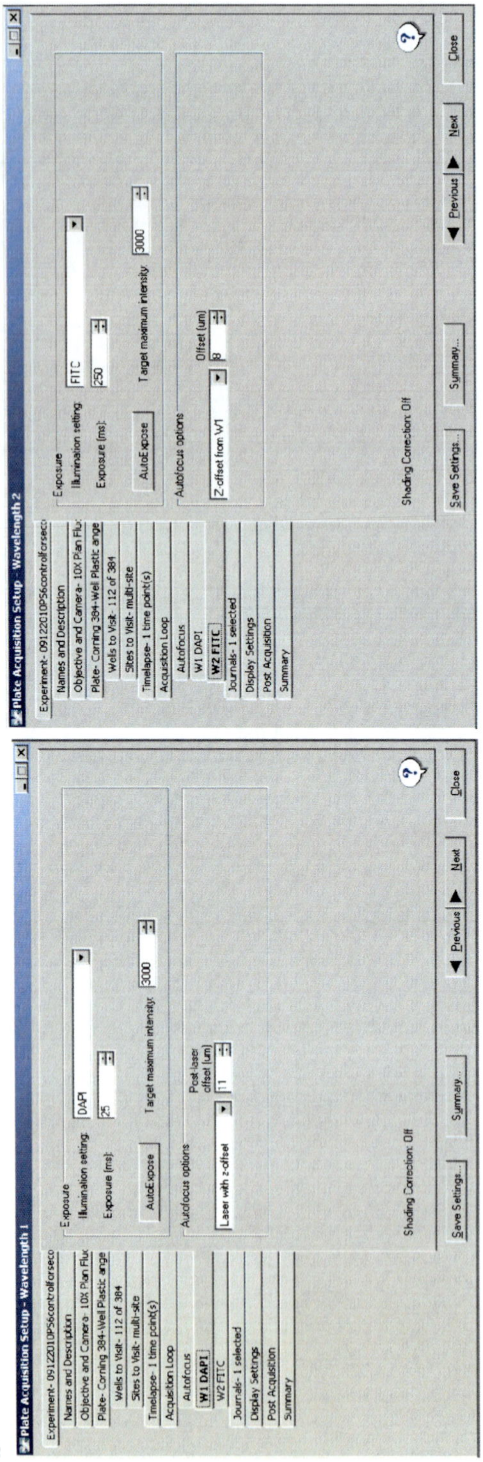

Fig. 2. (continued)

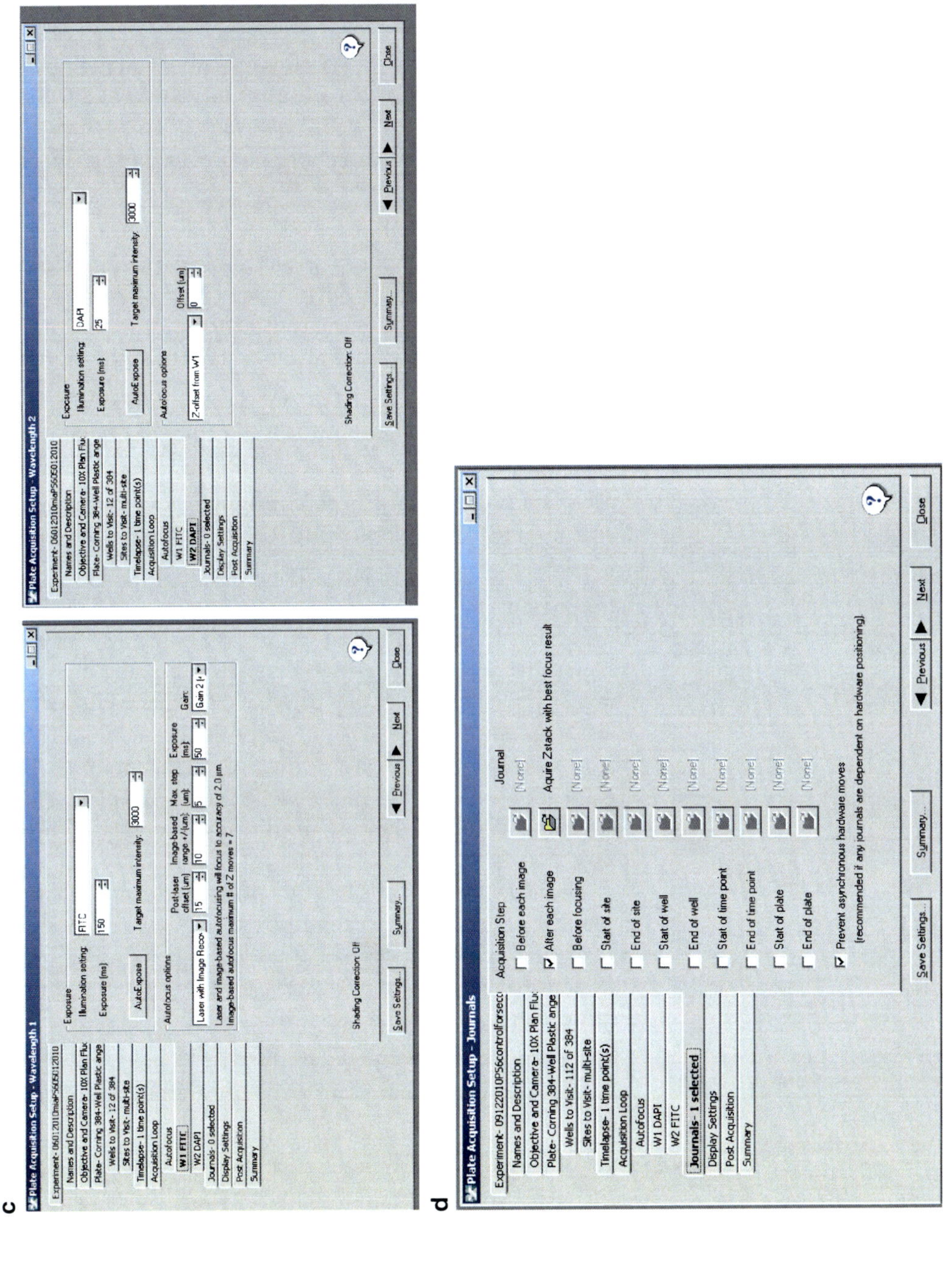

Fig. 2. (continued)

move to an appropriate position under the well and plate bottom, so that the cells will be in focus for image acquisition (Fig. 2a, lower screenshots). It is important that the plate parameters (plate bottom, bottom thickness, etc.) are accurately defined. If the plate parameters are not accurate, out-of-focus images result.

4. Choose the number of sites per well you wish to image. This is dependent on the total number of cells you would like to have imaged. We used four sites/well, however, up to 81 sites may be chosen; alternatively one can choose to image a fixed number of cells per well. A minimum of 1,200 cells/well are needed for this protocol.
5. Inasmuch as only one image plane is sought (i.e., not time-lapse images) "one" time point is selected; alternatively a *Z*-stack acquisition journal may be used to select the optimal image plane, in which case the time points reflect the number of stacks used.
6. Choose the number of wavelengths that you will image at (in this case, DAPI, FITC).
7. Once the plate settings have been configured, define the focus by choosing one of the following three options to acquire images.
   (a) Laser-based: Select a sample to be imaged and use the "Find Sample" option to adjust focus (Fig. 2b). For each wavelength, adjust the offsets from the *z* plane at which images are acquired. One can choose the well-bottom autofocus option (runs faster) or both the plate and well bottom option (takes longer but is more accurate), when variation of the *Z* height in the well exceeds the plate thickness; this is typically the case in thin-bottomed (<150 μm) plastic plates.

   The Mia-Paca 2 cells are not positioned on the single *z* plane because they are composed of two cell populations, adherent and globular. If laser-based focusing is not sufficient to enable one to observe a well-defined staining, choose either one of the following:

   (b) Laser plus Image-recovery: if the laser fails to find the image in some wells, an autofocus program will define an appropriate image (Fig. 2c).
   (c) Journal-based: this will allow a *z*-stack at different planes (Fig. 2d). You can use the "acquire *z* stack" and define the number of planes you need to acquire as well as the step size (e.g., if the number of planes is 10 and the step size 1 μm, this means that ten images 1 μm apart will be acquired). The software will retain the image with the best

focus as the final image. Alternate methods for image acquisition at different *z* planes also exist.

This last option is preferred when the cells grow at different planes. A *z*-stack allows a well-defined acquisition of images and is particularly useful when the cells' morphology is more globular than epithelial.

8. Once the settings have been saved you are ready to run the plates. First examine at a couple of different exposure times the control wells, those in which cells are transfected with RNAi against TOR and a nonspecific RNAi. The purpose is to obtain the exposure time that corresponds to the best z'. Always inspect images obtained from Control wells visually to access staining quality. After all plates have been run, use the cell scoring module for analysis (described in detail below). The primary readout for identifying hits is % positive cells.

3.4. Image Analysis

1. Images are analyzed and quantified by using the MetaXpress® imaging software. Use the cell scoring application module to detect, count and document measurements from the primary FITC readout. The specific module can be applied to the nucleus, cytoplasm, or both; this assay focused on cytoplasmic staining. The cell scoring module (a) identifies nuclei and segments them by using the nuclear DAPI marker, and (b) identifies positive cells according to a FITC (i.e., S6P) fluorescence intensity that exceeds a predetermined value, i.e., one that provides the highest signal-to-background/noise ratio, as well as the percent positive cells, i.e., the number of FITC positive cells/total number of DAPI positive cells (Fig. 3).
2. In configuring the analysis settings you choose the preview option to inspect and alter these as needed. To do so, go to "review plate data," select an image and open it. Define the minimum and max width for both the DAPI and FITC readouts which corresponds to the minimum and max width of the nuclei in the case of DAPI and of the cytoplasm in the case of FITC. You can manually count the number of nuclei from representative images ($n = 10$) by using the option "manually count objects." Then you can set different minimum and maximum widths and choose those that most accurately describe the average number of nuclei you obtained manually. For both readouts, define the intensity above background, i.e., the threshold above which cells will be considered positive. A threshold of FITC fluorescence to score a cell "positive" should be one that provides a signal-to-background ratio at least above three for cells that appear positive by visual inspection, although the dynamic range of images obtained from cells transfected with mTOR RNAi versus nonspecific RNAi may be tenfold or more.

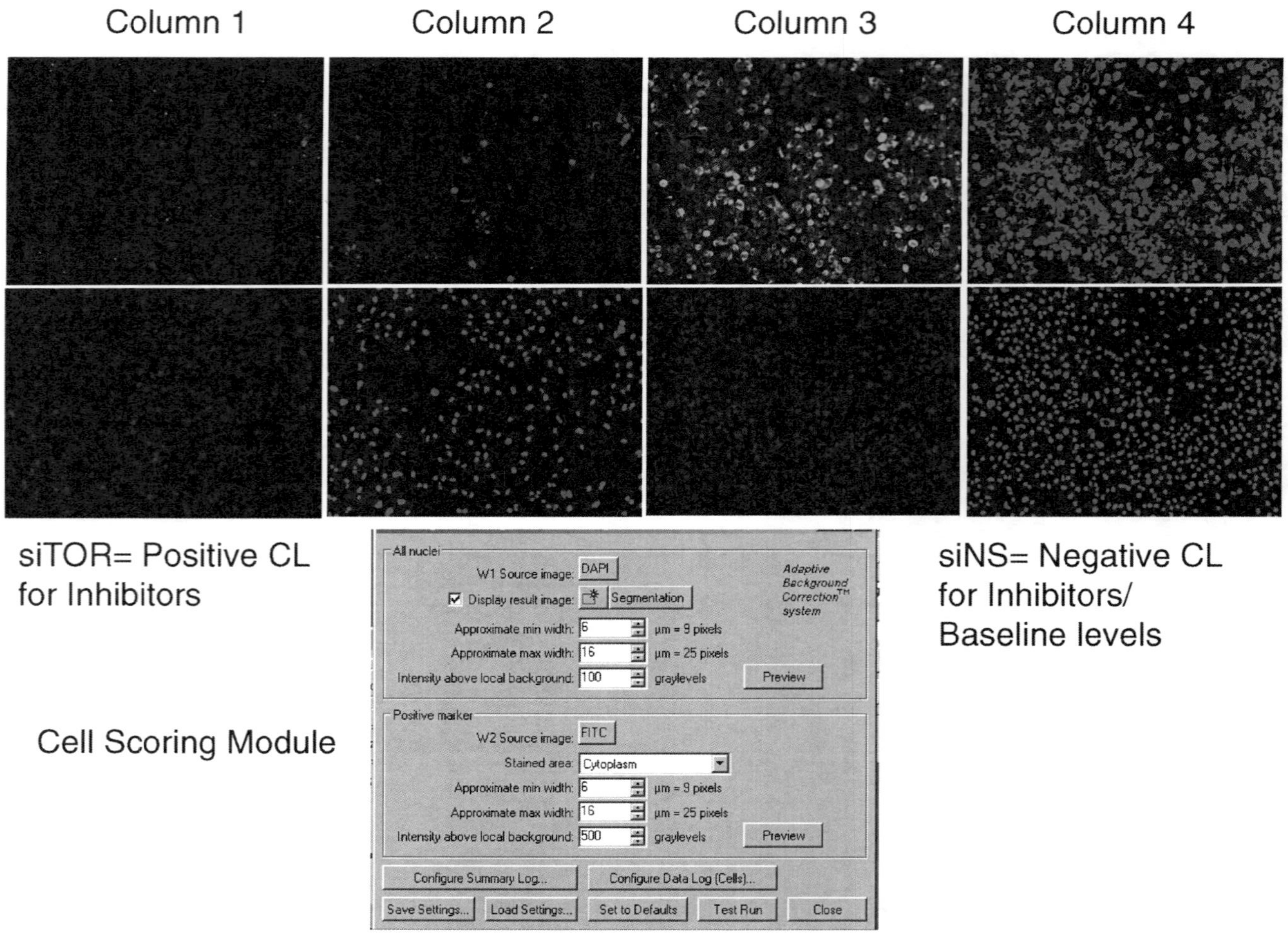

Fig. 3. A screenshot of the cell scoring module for analysis of the acquired images. Shown on the *left* (*column 1*) is the positive control (cells transfected with RNAi for mTOR) and on the *right* (*column 3*) is the negative control (cells transfected with a nonspecific targeting RNAi oligo). To the right of each set (*columns 2 and 4*) are the segmentation data; shown in *green* and *red* are the cells that are scored as positive when the threshold intensity is set at 100 and 500 for the DAPI and the primary FITC readout, respectively.

3. Select the specific measurements you wish to collect, here most essential are total cell number (as measured by DAPI), number of positive and/or percent positive cells as defined by the primary FITC readout. General procedures for collecting these values are described in the MetaXpress® 3.1 Analysis guide.

The following measurements may be used as secondary readouts:

Positive cells, total cell number/per site, total area of positive cells, mean area of positive cells.

Positive cells Integrated Intensity: By using the cell scoring module, you consider all positive cells for which the intensity is above a certain threshold. However, in the case of the Mia-Paca 2 cells (and possibly other cells) there are several cell populations of different FITC intensities, e.g., high, medium, and low. This parameter can be used to classify the primary hits according to the sum of the intensity. Alternatively, you can use the Multi-Wavelength Scoring Module to classify the cell populations and determine the number of cells that fall in each category.

Importantly, perform the screen in triplicate to confirm reproducibility of individual candidate genes, to reduce the number of false positives due to off-target effects, and to increase the robustness and confidence as to which are the potential positive hits. Select as "cherry picks" candidate genes that show a consistent response in at least two of the triplicates (i.e., either an increase or decrease of the basal P-S6 levels), and rescreen using the four deconvoluted siRNAs from each pool separately at 100 nM concentration.

3.5. Data Analysis

1. Exclude replicates in which the average total cell number was 2 SD below the average total cell number corresponding to the mTOR positive control; this eliminates proliferative failure for reasons other than inhibition of mTOR.
2. Obtain a normalized S6P value for each well: Divide the absolute number of cells scored as positive for S6P by the total cell number so as to obtain the percentage of positive cells corresponding to each transfected RNAi, i.e., "% positive cells."
3. Confirm visually the "% positive cells" values from the mTOR and the nonspecific targeting (NS1) RNAi controls. The correspondence between the visual images and the calculated "% positive cells" values reflect the data quality of each experiment.
4. To identify hits, use a level of "% positive cells" that is 2 or preferably 3 SD above or below that of the average value of "% positive cells" in all wells on the plate transfected with a scrambled RNAi, minimally 16 wells/plate.

For this calculation, convert "% positive cells" values to *z*-scores that correspond to each value/hit by using the Excel program. The *z* score is obtained by the formula:

$$z = \frac{X - \text{AVG}}{y * \text{SD}},$$

where X = "% positive cells" value corresponding to the specific well, AVG = average value obtained from all wells on the plate transfected with a scrambled RNAi, y = the number of SD chosen, i.e., 2 or 3, * means "times."

5. To obtain the primary hits, define the z score which you will use as a cut-off beyond which to identify hits. Most published screens use a z score of ±2 as a cut-off. This criterion is user-dependent and fairly arbitrary.
6. Exclude any wells with out-of-focus images, which are obvious by visual inspection (DAPI stain) and also evident when the average total cell number is discordant among the three plates.
7. For all of the wells for which the z was above the cut-off value (z = ±2), calculate the probability corresponding to each z by using the formula (in excel):

$$Q = (1 - x) * 100$$

where x = Absolute value (1 –[1 – NormalDistribution(z)]*2) (26) Subsequently, calculate the cumulative probability considering all three assay plates that corresponds to a specific hit by taking the average of all probabilities, since the events are independent. The strength of each RNAi, i.e., strong, medium, or weak is judged by the average Q of all replicates. You can define these ranges of by comparison with the values observed for known positive and negative mTOR regulators.

Sample calculation:

Example: % positive cells value for TOR = 4.5%

AVG % positive cells value for NS = 67.6%

SD corresponding to NS = 9.8%

% positive value for a specific hit in three replicates = 9.8, 7.6, 8.1

$$z = (67.6 - 9.8) / (3 * 9.15) = 2.10;$$
$$z = (67.6 - 7.6) / (3 * 9.15) = 2.18;$$
$$z = (67.6 - 8.1) / (3 * 9.15) = 2.16$$

$$\text{y1} = \text{ABS}\left(1 - (1 - \text{NORMSDIST}(\text{B19}))*2\right)$$
$$= 0.96; y20.97;\ \text{y3} = 0.97$$

$$1 - y = 1 - \text{ABS}\left(1 - (1 - \text{NORMSDIST}(\text{B19}))*2\right)$$
$$= 1 - 0.96 = 0.04$$

$$\text{Q1} = 100 * (1 - y) = 4\%;\ \text{Q2} = 3\%;\ \text{Q3} = 3\%$$

Average Q = 3.19

8. Rank RNAi SMART pools in order of average *Q*; similarly rank deconvoluted RNAi oligos in order of average *Q*. Decide on the average *Q* values that will define hits as strong, medium, and weak. In constructing your final list of primary hits, it is practical to exclude LOCs, hypothetical and retracted/retired genes.

3.6. Validation of Hits Obtained from the Primary Screen

1. Whereas in the primary screen, a pool of four RNAi oligos are used with a total concentration of 100 nM, validation uses the individual RNAis at 100 nM; otherwise methods are as in steps 3.1–3.7 above.

3.7. Data Summarization

1. Analysis of RNAi screening data, use the spotfire software program licensed through TIBCO Software Inc. http://www.tibco.com website; Spotfire® DecisionSite® 9.1.1 for Lead Discovery (TIBCO, Palo Alto, CA). It provides data summarization, data visualization, and clustering. This program is user friendly allowing graphing and data tabulation by drag-and-drop actions. Alternatively, software programs from Acuity Software, Inc. are also suitable.

   This process is best visualized while using the software, the following description will be most easily absorbed in that manner.

2. First create a worksheet in excel containing all the data you want. For example, it can contain the following parameters: % positive cells for replicates A, B, C, *z*-score for plate A/B/C, individual and average probability, total cell number, library plate number, well number. Import the data into Spotfire. A scatter plot is generated in the main graphing area. You can change the *X* and *Y* axis by dragging and dropping the different fields from the filter panel onto the *x*/*y* axis.

3. An easy and fast way to create charts, scatter plots, and bar graphs containing all data parameters is to use the Spotfire "HTS guide," which can automatically generate a variety of graphs. The guide creates visualizations to review your HTS data. To run the guide, first create a unique identifier which will represent the library plate number and the well number. To do so, select the "new column from expression" under Data and enter an expression in the window whose syntax will be accepted. To create a column based on plate and well ID you can enter, e.g., STOCK ID&WELL. This identifier is referred to as "calculated column." Select as a type of review to analyze, type "single run." Go to: Guides > Data Analysis > Review HTS data > pick single run > pick a results column. Select the parameter you want to have displayed on the results column. Among the choices are average % positive cells A/B/C, average % positive cells of all the triplicates, *z*-score for plates A/B/C, average *z* score created from the average % positive cells of all

triplicates, average *Q*. Select how the plate position will be identified in your data set. Identify row and column position (A1, A2). The row and column data can be shown in the same column. The results column can be any of the above fields (imported from Excel: *z* factor for plate A/B/C), % average for plate A/B/C. Select the sample ID column: Stock ID. Select as your plate ID the calculated column. Select same column for well positions. Choose well and click to create columns. This will lead to creation of six visualizations that will popup. You can select to view each one individually or all simultaneously. You can select the fields that you need to have included in the analysis: Examples include % positive cells, *z*, cumulative *Q*. The following visualizations will be displayed: pie plate sum by result: displays size by taking the sum of the average of % positive cells, results by standard deviation; two examples are shown in Fig. 4. The following are some of the program's applications in the context of the screen:

One can determine the degree of reproducibility between the plates by comparing the standard deviations from the mean for any particular normalized value among the triplicate plates (i.e., between replicate plates A vs. B, B vs. C, A vs. C). One can directly compare the % positive cells value and *z* scores between replicate plates A vs. B, B vs. C, A vs. C. One has the option of including data from all wells: experimental, positive and negative controls, empty wells or one can filter out the wells you wish to exclude (e.g., one can select only experimental wells and thus deselect all other types under the Filter Options Query Devices). One can select different colors to represent different library plates. Similarly, one can construct other graphs with all relevant fields. One can use the "details on demand" to examine a subset of data points in more detail. Similarly, one can select to visualize results from one or more library plates.

3.8. Bioinformatics/Computational Analysis

1. It is usually desirable to carry out a classification of hits into biological processes, molecular function, and categories by analyzing the validated hits using Ingenuity Pathway Analysis (IPA; Ingenuity Systems, Mountain View, CA; http://www.ingenuity.com) and Genego (St. Joseph, MI; http://www.genego.com). Both IPA and Genego are web-based software programs that use scientific information extracted from journal articles, textbooks, and other data sources to construct canonical pathways and functional categories (27, 28). These programs help define what is known about a hit or set of hits; the main questions that may be addressed through such computational analyses are the putative identities of the signaling, metabolic, and other functional categories represented in the hit list, what pathways are most prominently represented.
2. Genego: Create a list in excel containing the confirmed hits (i.e., those positive with two or more of the individual RNAis

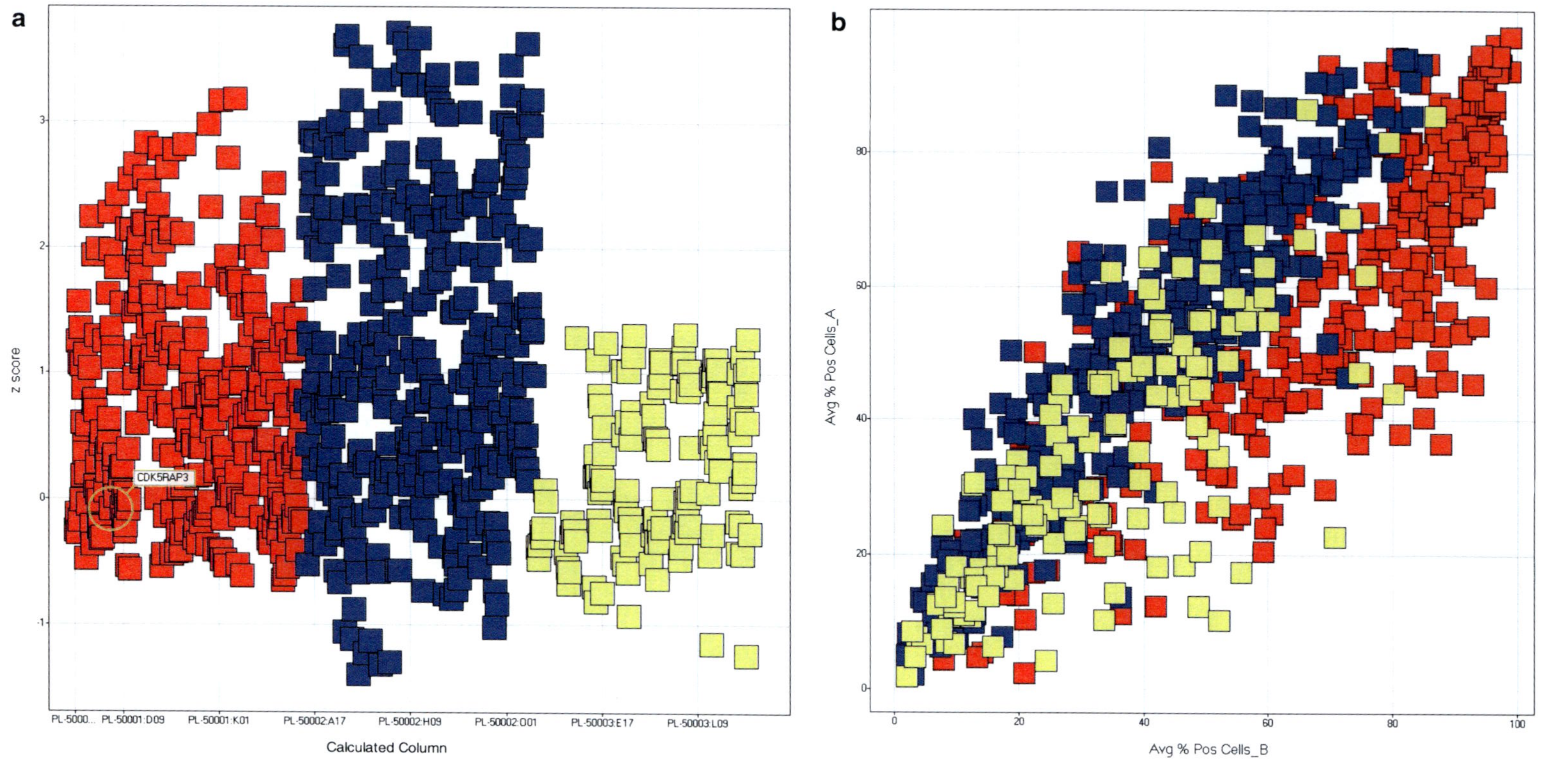

Fig. 4. Examples of results obtained by use of the spotfire program. (**a**) The *z*-score corresponding to each well for a total of three library plates containing different RNAi oligos. Each plate is depicted in a different color. (**b**) A representative scatter plot demonstrating the degree of reproducibility among two replicate plates containing identical RNAis; one can calculate the correlation coefficient, which ranged from ~0.67 to >0.9. Candidate genes that show consistent response in two of the three replicates (i.e., either an increase or decrease of the basal P-S6 levels) are selected as potential positives, and rescreened using each of the four individual siRNAs from each pool.

of your pool). Include the gene name, or the entrez gene ID or any of the other relevant gene identifiers recognized and listed as options, as well as the average probability that corresponds considering all replicates. Upload your experimental data using the EZ Start window so as to generate an experimental data set that is selectable (an "active" data set) for all subsequent analyses.

3. One can use the "Functional Enrichment by Ontology" function to perform ontology enrichment of the validated hit list for a selected ontology. The Functional Ontology Enrichment tool, which maps the hits to the built-in ontology, is comprised of GeneGo Maps (which represent biochemical or signaling pathways), network ontologies, and some gene ontologies (including processes, molecular functions, and localizations). The "Enrichment by Protein Function" tool can be used to determine the number of objects that belong to different protein classes (e.g., receptors, ligands, proteases, kinases, transcription factors, phosphatases, etc.). This tool uses the p-value to sort results; the p-value corresponds to the probability to have a given value of r, where r = number of network objects from the hit list for a given protein class (e.g., kinases). You can choose the default 5% false discovery rate (FDR) filter as the cut-off or a more stringent one. This program will generate a list of the maximum –log(p-value) for every term, a value that corresponds to the probability a specific gene set has occurred by chance based on its size (defined as the number of hits found present in the gene set). Stated differently, p-value is the probability of a match between the dataset and a specific ontology occurs by chance, given the size of the database. A low p-value indicates a larger number of hits belonging to a specific process/pathway and the higher the probability that there is a match between a group of hits and the ontology found.

4. Perform a "pathway analysis" to determine all the interactions associated with the validated hits and whether such interactions are part a known pathway.

5. Perform a "workflow data analysis report" with the dataset. Use the hit list to build networks of pathways and processes associated with the validated hits and sort pathways for importance by the G-score, which is based on the number of Canonical Pathways used to build the network. A network with a high G-score indicates that it contains many Canonical Pathways (i.e., set of genes that participate in a canonical biological process). Another option – referenced under the "data analysis workflows" – is to Perform an Enrichment Analysis which scores and ranks the most relevant cellular processes, disease targets, biomarkers, toxicity processes, and molecular functions for the hit list. Detailed specific information is available online at the Genego site.

6. Ingenuity Pathway Analysis (IPA) uses a knowledge database (the Ingenuity Pathways Knowledge Base) that contains information about genes, chemicals, cellular and disease processes, signaling, and metabolic pathways to identify the biological functions, canonical pathways, and networks that are most significant to the validated hit list. This curated database is comprised of protein interaction networks derived from published protein–protein interactions and/or associations, including direct and indirect protein interactions (such as those involved in physical binding, enzyme–substrate relationships, and transcriptional control). In IPA, networks are comprised of pathway network nodes (which are your hits/proteins) and edges (which are the biological relationship between the hits). First, create an Excel file containing the validated hit list that includes the gene name, Entrez Gene ID, accession number, and average probability corresponding to the z' from all three replicates.
7. Generate protein interaction networks, i.e., a dataset containing the validated hits (=H) by uploading into IPA the excel file containing the gene identification names. Compare the hits against a global molecular network derived from the information in the Ingenuity Knowledge Base (IKB). Generate networks of these hits algorithmically by including hits and other proteins from the IKB which enable the generation of a network based on connectivity. Identify canonical pathways and functions from the IPA library based on their significance to the dataset (hits). To generate the network(s) from the hit list and to define which networks interact with specific hits, upload the excel file containing the gene IDs into IPA, and ensure that each hit is uniquely and accurately recognized by the software. Protein interaction networks are then generated automatically and ranked by "score," which is the negative log of this probability estimate (p-value) and defined as score = $-\log_{10}(p\text{-value})$. To generate hypothetical protein networks you can include direct and/or indirect protein interactions, include all or specific species, tissues, and cell lines. The hits are compared against the IPA database. The significance of the association between the hit list and a canonical pathway is measured by the ratio of the number of hits that map to the pathway divided by the total number of proteins that exist in the canonical pathway and by the p-value that is obtained by comparing the number of hits to the total number of genes/proteins in all pathway annotations stored in the IPA database. A low p-value suggests that the association between the hits and these pathways is significant and not random. A $p\text{-value} \leq 0.05$ is considered statistically significant.
8. Perform a biomarker analysis to identify potential biomarker candidates.
9. You can also create/visualize new pathways which originate from the data set that correspond to the hit list, add novel

(newly identified) components to already known pathways and/or identify already existing canonical pathways not previously associated with a specific disease or pathway.

4. Notes

1. It is important to keep the same FBS lot the same through the screen. Different lots have differential effect on cell growth rate which in turn affects the PS6 readout. If there is need to change serum lots, then you should initially test the number of cells plated per well and adjust this number as a function of the optimal PS6.
2. Follow standard mammalian tissue culture guidelines (29).
3. Expand and freeze down as many vials as possible originating from the same lot and the same passage. It is advisable to use a fresh vial for each transfection so that all experiments are performed with cells corresponding to the same passage. This lowers the probability that mutations will accumulate while the cells are maintained in culture and prevent alterations in cell growth rate over time. In addition, this also facilitates performing the experiment at a defined schedule and allows each experiment to be independent of others. The cell growth rate of each experiment remains unaffected.
4. Test the cells for mycoplasma either by PCR or by DAPI staining; many mycoplasma PCR kits are commercially available. Mycoplasma can cause the cell responses to drift over time and is one of the causes of unreproducible results.
5. For Lipofectamine 2000: Keep Lipofectamine 2000 on ice and/or minimize the time that Lipofectamine is left at room temperature, inasmuch the exposure of Lipofectamine 2000 to nonoptimal temperatures can affect efficiency of the reagent.
6. We have observed empirically that further incubation of the transfection reagent mix decreases transfection efficiency.
7. Be careful to not allow trypsin to remain on cells longer than needed as this will decrease the baseline P-S6 signal.
8. Try to work as fast as possible. Mia-Paca 2 cells will start clumping after 50–60 min post-trypsinization.
9. Before harvest, inspect visually the phenotype of the cells in the control wells. Take of the confluency in wells transfected with a non-specific RNAi targeting oligo. If the confluency is less than approximately 85%, delay the harvest for a few more hours.

For all steps during the IF protocol other than the two antibody addition steps, use the white (large) bore Wellmate cartridge and the S3 dispension speed. Inasmuch as the Mia-Paca 2 cells are composed of an adherent epithelial cell population and a globular, more loosely attached population, all steps are preceded by a 2-min centrifugation step at 233 × *g*, which serves to preserve both cell populations. Similarly the S3 dispensing speed rate minimizes shear stress imposed on the cells, minimizing loss of the loosely attached cells during the course of the experiment. Liquid is removed using the aspirator wand. Note the strength of the vacuum and the residual volume throughout the course of the experiment, seeking to keep these uniform. If the vacuum strength is weak, increase the number of washes based on the residual volume. With a residual volume of 10–15 μl, five washes each after fixation and after permeabilization step are sufficient.

10. Using the small bore wellmate cartridge will lead to a loss of approximately 100–150 μl of the final working antibody solution.
11. The remaining primary antibody working dilution that remains in the cartridge can be recycled and used within a period of 15 days.
12. If you plan to use the secondary antibody within 1 month, the antibody can be stored at 4°C in the dark. For a longer storage, prepare single-use aliquots and store at −20°C in the dark. The manufacturer claims that the antibody is stable for up to 6 months at −20°C.

Acknowledgments

This work is supported by NIH awards CA73818 and DK17776 and the Massachusetts General Hospital "ECOR Fund for Medical Discovery."

References

1. Wullschleger S, Loewith R, Hall MN (2006) TOR signaling in growth and metabolism. Cell.124: 471–84.
2. Liu Q, Thoreen C, Wang J, Sabatini D, Gray NS (2009) mTOR Mediated Anti-Cancer Drug Discovery. Drug Discov Today Ther Strateg. 6, 47–55.
3. Dowling RJ, Topisirovic I, Fonseca BD, Sonenberg N (2010) Dissecting the role of mTOR: lessons from mTOR inhibitors. Biochim Biophys Acta. 1804, 433–9.
4. Dancey J (2010) mTOR signaling and drug development in cancer. Nat Rev Clin Oncol. 7, 209–19.

5. Lane HA, Breuleux M (2009) Optimal targeting of the mTORC1 kinase in human cancer. Curr Opin Cell Biol. 21, 219–29.
6. Dowling RJ, Pollak M, Sonenberg N (2009) Current status and challenges associated with targeting mTOR for cancer therapy. BioDrugs. 23, 77–91.
7. Baldo P, Cecco S, Giacomin E, Lazzarini R, Ros B, Marastoni S (2008) mTOR pathway and mTOR inhibitors as agents for cancer therapy. Curr Cancer Drug Targets. 8, 647–65.
8. Avruch J, Hara K, Lin Y, Liu M, Long X, Ortiz-Vega S, Yonezawa K (2006) Insulin and amino-acid regulation of mTOR signaling and kinase activity through the Rheb GTPase. Oncogene. 25:6361–72.
9. Hara K, Yonezawa K, Weng QP, Kozlowski MT, Belham C, Avruch J (1998) Amino acid sufficiency and mTOR regulate p70 S6 kinase and eIF-4E BP1 through a common effector mechanism. J Biol Chem. 273, 14484–94.
10. Avruch J, Long X, Ortiz-Vega S, Rapley J, Papageorgiou A, Dai N (2009) Amino acid regulation of TOR complex 1. Am J Physiol Endocrinol Metab. 296, E592–602.
11. Sancak Y, Sabatini DM (2009) Rag proteins regulate amino-acid-induced mTORC1 signalling. Biochem Soc Trans. 37, 289–90.
12. Iorns E, Lord CJ, Turner N, Ashworth A. (2007) Utilizing RNA interference to enhance cancer drug discovery. Nat Rev Drug Discov. 6, 556–68.
13. Gressner AM, Wool IG. (1974) The phosphorylation of liver ribosomal proteins in vivo. Evidence that only a single small subunit protein (S6) is phosphorylated. J Biol Chem. 249, 6917–25.
14. Bandi HR, Ferrari S, Krieg J, Meyer HE, Thomas G (1993) Identification of 40S ribosomal protein S6 phosphorylation sites in Swiss mouse 3T3 fibroblasts stimulated with serum. J Biol Chem. 268, 4530–3.
15. Ferrari S, Bandi HR, Hofsteenge J, Bussian BM, Thomas G (1991) Mitogen-activated 70KS6 kinase. Identification of in vitro 40S ribosomal S6 phosphorylation sites. J Biol Chem. 266, 22770–5.
16. Pende M, Um SH, Mieulet V, Sticker M, Goss VL, Mestan J, Mueller M, Fumagalli S, Kozma SC, Thomas G (2004) S6K1(–/–)/S6K2(–/–) mice exhibit perinatal lethality and rapamycin-sensitive 5'-terminal oligopyrimidine mRNA translation and reveal a mitogen-activated protein kinase-dependent S6 kinase pathway. Mol Cell Biol. 24, 3112–24.
17. Price DJ, Grove JR, Calvo V, Avruch J, Bierer BE (1992) Rapamycin-induced inhibition of the 70-kilodalton S6 protein kinase. Science 257, 973–7.
18. Isotani S, Hara K, Tokunaga C, Inoue H, Avruch J, Yonezawa K. (1999) Immunopurified mammalian target of rapamycin phosphorylates and activates p70 S6 kinase alpha in vitro. J Biol Chem. 274, 34493–8.
19. Alessi DR, Kozlowski MT, Weng QP, Morrice N, Avruch J. (1998) 3-Phosphoinositide-dependent protein kinase 1 (PDK1) phosphorylates and activates the p70 S6 kinase in vivo and in vitro. Curr Biol, 8, 69–81.
20. Weng QP, Kozlowski M, Belham C, Zhang A, Comb MJ, Avruch J. (1998) Regulation of the p70 S6 kinase by phosphorylation in vivo. Analysis using site-specific anti-phosphopeptide antibodies. J Biol Chem. 273, 16621–9.
21. Ruvinsky I, Katz M, Dreazen A, Gielchinsky Y, Saada A, Freedman N, Mishani E, Zimmerman G, Kasir J, Meyuhas O (2009) Mice deficient in ribosomal protein S6 phosphorylation suffer from muscle weakness that reflects a growth defect and energy deficit. PLoS One. 19; 4(5):e5618.
22. Thoreen CC, Kang SA, Chang JW, Liu Q, Zhang J, Gao Y, Reichling LJ, Sim T, Sabatini DM, Gray NS (2009) An ATP-competitive mammalian target of rapamycin inhibitor reveals rapamycin-resistant functions of mTORC1. J Biol Chem. 284, 8023–32.
23. Yunis AA, Arimura GK, Russin DJ (1977) Human pancreatic carcinoma (MIA PaCa-2) in continuous culture: sensitivity to asparaginase. Int J Cancer. 19, 128–35.
24. Sipos B, Möser S, Kalthoff H, Török V, Löhr M, Klöppel G (2003) A comprehensive characterization of pancreatic ductal carcinoma cell lines: towards the establishment of an in vitro research platform. Virchows Arch. 442, 444–52.
25. Grewe M, Gansauge F, Schmid RM, Adler G, Seufferlein T (1999) Regulation of cell growth and cyclin D1 expression by the constitutively active FRAP-p70s6K pathway in human pancreatic cancer cells. Cancer Res. 59, 3581–7.
26. http://www.measuringusability.com/pcalcz.php.
27. Ekins S, Nikolsky Y, Bugrim A, Kirillov E, Nikolskaya T (2007) Pathway mapping tools for analysis of high content data. Methods Mol Biol. 356, 19–50.
28. Ganter B, Giroux CN. (2008) Emerging applications of network and pathway analysis in drug discovery and development. Curr Opin Drug Discov Devel. 11, 86–94.
29. Phelan MC (2006) Techniques for Mammalian Cell Culture. Curr Protocols Hum. Genet. Appendix 3.

Chapter 12

Immunohistochemical Analysis of mTOR Activity in Tissues

Jinhee Kim, Nancy Otto, Claudio J. Conti, Irma B. Gimenz-Conti, and Cheryl L. Walker

Abstract

mTOR is a key regulator of cell growth and size, and its activity is often dysregulated in a wide variety of diseases. The mTOR signaling pathway is also a therapeutic target for many diseases, including cancer. Immunohistochemistry is a powerful method to assess mTOR activity in clinical/histological samples, however, care should be taken in choosing the targets for determining mTOR activity due to the complexity of its regulation. This chapter describes the most up-to-date methods for visualizing mTOR activity by immunohistochemistry using commercially available antibodies, including considerations for validating new antibodies for assessing mTOR signaling.

Key words: Immunohistochemistry, mTOR pathway, Paraffin-embedded section, Immunoperoxidase method, Antibody validation

1. Introduction

Immunohistochemistry (IHC) is a useful tool for studying activation of the mTOR pathway. It has been used extensively as a correlate to molecular/biochemical experiments, as well as to localize specific cells within a tissue or a tumor that have activated mTOR signaling (1–3). In addition, IHC can be used in settings where only archival samples are available (4). In the last few years, a wide variety of antibodies against mTOR/phospho-mTOR itself as well as antibodies recognizing both upstream activators and downstream targets of mTOR has become commercially available. These reagents have made it possible to use IHC for assessing mTOR activity and identifying mechanisms responsible for its activation in clinical samples (5, 6). In this chapter, we describe protocols to detect mTOR and its upstream activators and downstream

Thomas Weichhart (ed.), *mTOR: Methods and Protocols*, Methods in Molecular Biology, vol. 821,
DOI 10.1007/978-1-61779-430-8_12, © Springer Science+Business Media, LLC 2012

effectors, and provide examples of mTOR analysis using IHC in prostate cancer in which mTOR is activated as a consequence of loss of the tumor suppressor PTEN.

2. Materials

2.1. Antibodies Used for Measuring mTOR Signaling with IHC

Since mTOR can be regulated by multiple upstream components and activity of this pathway can be determined by multiple downstream readouts (Fig. 1, diagram), the target proteins for IHC should be selected cautiously and considered in the context of each experimental setting or disease mechanism.

Table 1 summarizes antibodies specific for mTOR signaling components with which we have experience, their manufacturers, and conditions for their use (see Notes 1 and 2).

2.2. Reagents

1. 10% Neutral Buffered Formalin (Thermo/Fisher).
2. 70% Ethanol (diluted from AAPER 200 Proof Ethanol).

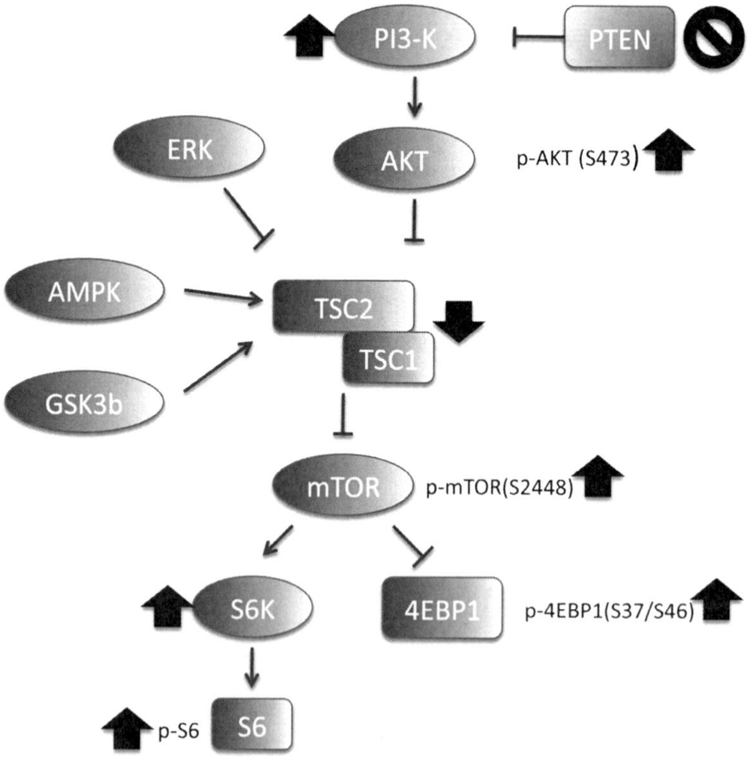

Fig. 1. mTOR signaling pathway and key proteins used for assessing mTOR activity with IHC The canonical pathway for PI3-K signaling to mTOR is shown. When activated, for example by loss of the PTEN tumor suppressor, AKT phosphorylation is increased (*denoted by solid arrow*) suppressing TSC1/2 activity and increasing activity and phosphorylation of mTOR, S6K, S6, and 4EBP1.

Table 1
Commonly used mTOR pathway antibodies and IHC conditions

Antibody	Manufacturer	Source/clone (see Note 1)	Species cross-reactivity[a]	Specied tested	Antibody dilution	Polymer detection
mTOR	Cell signaling #2983	Rabbit Mab (7C10)	H, M, R, Mk	H, M, R	1:100 ON at 4°C	Biocare #RMR622
p-mTOR (Ser2448)	Cell signaling #2976	Rabbit Mab (49F9)	H, M, R	H, M, R, Hm	1:50 ON at 4°C	Biocare #RMR622
p-p70 S6K (Ser371)	Cell signaling #9208	Rabbit Pab	H, M, R	M	1:25 ON at 4°C	Dako #K4011
Ribosomal S6	Cell signaling #2217	Rabbit Mab (5G10)	H, M, R, Mk	M, R	1:50 ON at 4°C	Dako #K4011
p-S6 (Ser235/236)	Cell signaling #2211	Rabbit Pab	H, M, R, Mk, Sc, C, X	M, R	1:50 1 h at RT	Dako #K4011
p-S6 (Ser240/244)	Cell signaling #2215	Rabbit Pab	H, M, R, Mk, Z, C, X	M, R	1:50 2 h at RT	Dako #K4011
p-4EBP1 (Thr37/46)	Cell signaling #2855	Rabbit Mab (236B4)	H, M, R, Mk, Dm	M	1:500 ON at 4°C	Dako #K4011
p-AMPK (Thr172)	Cell signaling #2535	Rabbit Mab (40H9)	H, M, R, Mk, Sc, Dm, Hm, C, B	M	1:50 ON at 4°C	Biocare #RMR622
Tuberin/TSC2 (see Note 2)	Epitomics #1613-1	Rabbit Mab (Y320)	H, M, R	R	1:50 2 h at RT	Dako #K4011
ERK	Cell signaling #9102	Rabbit Pab	H, M, R, Mk, Hm, B, Z	M, R	1:25 1 h at RT	Dako #K4011
p-ERK (Thr202/204)	Cell signaling #9101	Rabbit Pab	H, M, R, Mk, Mi, Pg, C, Hm, B, Dm, Z	M, R	1:100 1 h at RT	Dako #K4011
AKT	Cell signaling #4691	Rabbit Mab (C67E7)	H, M, R, Mk	M, R	1:100 ON at 4°C	Dako #K4011
p-AKT (Ser473)	Santa Cruz #sc7985R	Rabbit Pab	H, M, R, X	M, R	1:50 1 h at RT	Dako #K4011
GSK3-beta	Chemicon #AB8687	Rabbit Pab	H, M, R, B	M	1:200 1 h at RT	Dako #K4011
p-GSK3-beta (Ser9)	Santa Cruz #sc11757R	Rabbit Pab	H, M, R, X, Z	M	1:50 ON at 4°C	Dako #K4011

H Human, *M* Mouse, *R* Rat, *Mk* Monkey, *B* Bovine, *C* Chicken, *Hm* Hamster, *Z* Zebra fish, *Mi* Mink, *Pg* Pig, *Dm D. melanogaster*, *X* Xenopus, *Sc S. cerevisiae*
Abbreviations: *Mab* monoclonal antibody, *Pab* polyclonal antibody, *ON* overnight, *RT* room temperature
[a]Cross-reactivity data supplied by manufacturer

3. Reagent alcohol.
4. Xylene or xylene substitute (Thermo/Shandon).
5. Paraplast X-tra (McCormick Scientific).
6. RM2235 rotary microtome (Leica).
7. Superfrost/Plus microscope slides (Fisher Scientific #12-550-15).
8. 3% H_2O_2 (made from 30% stock, Sigma #H1009).
9. Background sniper (Biocare #BS966MM) or equivalent blocking reagent.
10. Dulbecco's phosphate-buffered saline (Sigma).
11. Bovine serum albumin.
12. Tween 20.
13. Super PAP pen (Ted Pella #22309).
14. DaVinci Green (Biocare #PD900) or equivalent antibody diluent.
15. 10 mM Citric acid buffer.
16. 3,3′ Diaminobenzidine (from Dako Kit #4011).
17. Copper II sulfate (Sigma #C1297) in saline.
18. Hematoxylin (Invitrogen).
19. Bluing Reagent (Thermo/Scientific).
20. Consul Mount (Thermo/Scientific).

3. Methods

Basically, there are two approaches for visualizing immunoreactivity by IHC: direct and indirect methods. The direct method uses labeled primary antibody, whereas indirect method uses unlabeled primary antibody with a labeled secondary antibody that is species-specific for the primary antibody used. Since we use several different antibodies for detection of mTOR signaling in tissues, the indirect method is a more convenient and versatile way of conducting immunohistochemistry, and is the approach we describe in this chapter.

3.1. Preparation of Tissues

1. Immediately after taking down animals, put the tissues into the cassette and fix with 10% NBF for 24–48 h (see Notes 3 and 4).
2. Drain the fixative and put the cassette into 70% grade ethanol (see Note 5).
3. Dehydrate in graded alcohol and embed in paraffin wax.
4. Make 4-μm thick sections on a rotary microtome and place onto charged microscope slides.

3.2. Staining by Immunoperoxidase Method (see Note 6)

1. Bake the slides at 60°C for 30 min before dewaxing to adhere the sample to the slide (see Note 6).
2. Deparaffinize sections in xylene or xylene substitute baths for two times, 5 min each.
3. Rehydrate for 5 min in 100% graded alcohol bath, twice. Then 5 min in 95% graded ethanol bath, and under running water for 1–2 min.
4. Block endogenous peroxidase activity by incubating with 3% H_2O_2 (diluted from 30% stock in distilled water) for 10 min followed by a 1-min wash in running water (see Note 7).
5. Retrieve antigen with 10 mM citrate buffer, pH 6.0 for 3 min on full power followed by 10 min at 50% power in a microwave oven (see Note 8).
6. After a 20 min cool down, wash the slides in running water for 2 min (see Note 9).
7. Working with one slide at a time to prevent the tissue from drying, wipe the slide around the tissue with a disposable lint-free wipe and circle the tissue with a Superfrost PAP Pen leaving a hydrophobic line (see Note 10). Without letting the tissue dry out, drop enough Buffer Wash (see Note 11) onto the slide to cover the tissue. Repeat until all the slides have a PAP-line and are in Buffer Wash (see Note 12).
8. Drain off the Buffer Wash and drop on enough Background Sniper blocking reagent to cover the tissue section (see Note 13). Incubate for 10 min.
9. Drain the slides, but do not wash, and add the primary antibody that has been diluted to the predetermined concentration in antibody diluent, such as DaVinci Green (see Note 14) and incubate as previously determined (see Table 1).
10. After the appropriate time in primary antibody, wash the slides in Buffer Wash for 5 min, changing once with fresh Buffer Wash.
11. Continue adding the detection reagents (see Table 1), incubating, and washing your slides in the same manner as you did for the primary antibody (see Note 15).
12. Allow to react with 3,3′ Diaminobenzidine (DAB) solution (see Note 16). Incubation times will vary, but usually 5 min is sufficient to develop your DAB chromogen (see Note 17). It might be necessary to check the reaction (brown color) under the microscope for initial setup.
13. Stop the reaction by washing the slides under running water for 2 min and then incubate the slides in copper sulfate–saline solution (see Note 18). Wash with water.
14. Counterstain the sections with hematoxylin, wash with running water, and then incubate with bluing solution (see Note 19).

15. After a final water wash, dehydrate with 95% and 100% ethanol for 3 min each, and clear slide in xylene before coverslipping with a permanent mounting media (see Note 20).

3.3. Validating a New mTOR Signaling Antibody

1. The antibodies shown in Table 1 have been assessed in our laboratory using standard techniques. It is critical for IHC that careful validation is carried out each time a new antibody is used, even if the same antibody has been used successfully in other laboratories. Also, it is common for a manufacturer to discontinue certain antibodies or replace them with new polyclonal and/or monoclonal antibodies. It is highly recommended that a stock be maintained of any batch/lot of good working antibodies sufficient to conduct multiple experiments, especially for those used regularly.
2. There are four major variables that need to be optimized to get reliable IHC results: (1) Fixation methods and time, (2) methods for antigen retrieval, (3) staining conditions for each antibody, and (4) species of tissue and the primary antibody used.
3. It is recommended that the conditions of fixation be within the standard protocols for IHC used in the clinical pathology laboratory. Care should be taken in choosing fixatives since some antibodies work only with certain fixation methods (i.e., formalin-fixed versus frozen sections) (7). Fixation time is usually more than 6 h and less than 72 h and the size or depth of sample might affect the proper fixation. If it is not possible to process the tissues in this time frame, the samples should be transferred and held in 70% ethanol.
4. In validating an antibody, it is essential to use positive and negative controls. The best negative controls in experimental material are knockout animals for the protein under study (e.g., ERKO mice for the estrogen receptor (ER) alpha or p53 knockout mice for p53 IHC). If these are not available, it is advisable to use tissues that have been previously shown to be negative for the targeted protein. Positive controls are also essential, particularly when negative results with the antibody/target of interest are obtained. Ideal positive controls are internal controls such as nonneoplastic breast tissue showing immunoreactivity for ER-specific antibodies when studying an ER negative breast cancer tumor.
5. In the case of the mTOR pathway, knockout mice or conditional knockout mice (animals deficient for a given gene in selected tissues) are available such as $TSC2^{+/-}$, $TSC1^{+/-}$ mice (8, 9), LKB1 (10), $PTEN^{+/-}$ (11), and TSC2-deficient Eker rats (12). If tissues with genetic inactivation of the gene of interest are not available, nonneoplastic and nonproliferative tissues can be considered to be negative controls for mTOR

and its downstream readouts such as phospho-S6 and phospho-4EBP1, since mTOR activity is generally low in these tissues. By contrast, upstream regulators of mTOR, such as AKT and AMPK, can be activated by normal proliferative signals or starvation, and is not uncommon to find activation of these enzymes in normal tissues. Therefore, caution should be taken when using normal tissues as control for mTOR signaling. A source of positive control for validation of tissues can be also transgenic animals overexpressing the target protein in a particular tissue.

6. Such mouse lines, either transgenic or knockout, can be found in the database of Jackson Laboratory Mice for cancer research (http://jaxmice.jax.org/research/cancer/index.html), and the NCI Mouse consortium for Cancer Research (http://emice.nci.nih.gov/).

3.4. Example and Interpretation

1. To illustrate mTOR IHC for this chapter, we have used the Pten mouse knockout model of prostate carcinogenesis. These mice are hemizygous (+/−) for the tumor suppressor *Pten* and develop prostate preneoplastic (prostate intraepithelial neoplasia, PIN) and neoplastic carcinoma lesions (11). As part of the carcinogenesis process the functional allele of *Pten* is lost, which results in activation of AKT and ultimately of mTOR and its downstream effectors (Fig. 1, solid arrows).
2. In Fig. 2, we show lesions in these mice stained with the protocols described in this chapter. We also show a conventional H&E staining to illustrate the morphological characteristics of the lesions. Antibodies against phospho-AKT (Ser473) produce a strong staining, both nuclear and cytoplasmic in these lesions (Fig. 2c). Interestingly, histologically normal tissues, particularly in areas adjacent to preneoplastic lesions, also show relatively strong staining. This is likely caused by paracrine factors secreted by the tumors. In the case of phospho-mTOR (S2448), staining is almost completely restricted to the neoplastic areas, where both nuclear and cytoplasmic staining is observed (Fig. 2b). For two downstream effectors, phospho-4EBP1 and phospho-S6, immunoreactivity is highly restricted to the PINs and not observed in adjacent normal cells. For both S6 and 4EBP1, the canonical pathway predicts a purely cytoplasmic localization. While this is observed for phospho-S6 (Fig. 2e), phospho-4EBP1 staining is predominantly cytoplasmic, but some nuclear immunoreactivity is also observed (Fig. 2d). It is not uncommon that immunohistochemistry detects variants to the canonical pathways, which are later confirmed biochemically; however, it is also possible that the anti-phospho-4EBP1 antibody is cross-reacting with another protein epitope in the nucleus rather than 4EBP1.

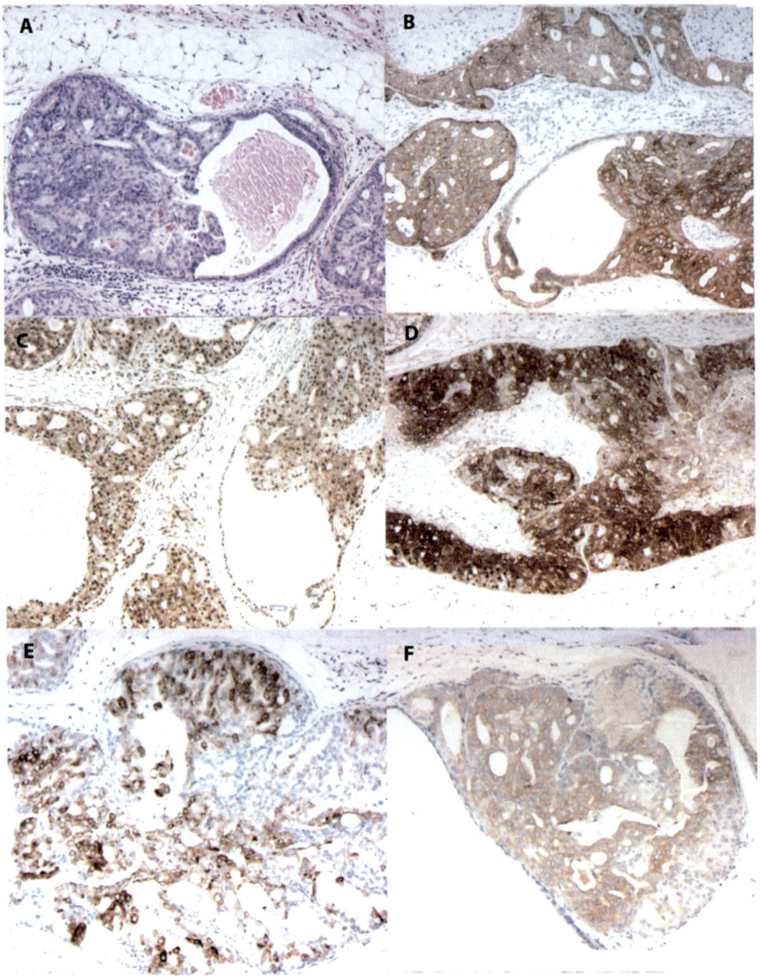

Fig. 2. Immunostaining of components of the mTor pathway. Mouse prostates from *Pten*$^{+/-}$ animals, presenting in situ prostate intraepithelial neoplasia (PIN) and microinvasive lesions were stained with H&E (**a**) or IHC using antibodies against p-mTOR (Ser2448) (**b**): p-AKT (Ser473) (**c**): p-4EBP1 (Thr37/46) (**d**): p-S6 (Ser235/236) (**e**) and p-AMPK (Thr172) (**f**). All staining was performed in formalin-fixed paraffin embedded sections following the procedures described in this chapter. All immunohistochemical staining was counterstained with hematoxylin and the pictures were taken with a ×10 objective.

4. Notes

1. Unlike clinical studies based on human tissue samples, our experience is primarily with rodent tissues, which makes it difficult to use mouse monoclonal antibodies due to species-reactivity. As a result, this chapter is based mostly on our experience with rabbit poly- or monoclonal primary antibodies. However, most antibodies from major manufacturers recognize human proteins, so using these antibodies on human tissues should not be a problem for most cases, though proper titration should be obtained for best results.

2. Some antibodies are not suitable for IHC even though the manufacturer's datasheet recommend it for this purpose. For example, commercially available TSC2 antibodies exhibit nonspecific cross-reactivity to epitopes on other endogenous proteins, as observed with false positive IHC in TSC2 null-tumor tissues. In our hands, we have confirmed by Western that an antibody from Epitomics is specific for TSC2 protein and is suitable for IHC use.
3. Controlled fixation is one of the most important steps in getting good, reliable, reproducible immunohistochemistry. Some antibodies only recognize their epitopes with a certain fixation method, so it is critical to choose the proper fixation method for your tissue and antibody combination. All antibodies referred in this chapter have been validated for use in formalin-fixed, paraffin-embedded tissues.
4. For mouse tissues, 24 h in fixative is usually sufficient, however, in case of rat tissues, change to fresh fixatives after 24 h of initial fixation, and wait another 24 h for completion. Tissues fixed in formalin for over 48 h can exhibit excessive protein cross-linking, making detection of some antigens difficult to impossible even after antigen retrieval.
5. These tissues can be stored in 70% ethanol until needed.
6. We use a two-step polymer based detection system utilizing horseradish peroxidase (HRP) whenever possible to eliminate endogenous biotin background. A standard three-step biotin based peroxidase system or alkaline phosphatase system should also work, but reagent titers will need to be optimized.
7. 30% hydrogen peroxide is very caustic and will burn skin. Hydrogen peroxide will quench endogenous peroxidases that are found in RBCs, mast cells, etc. This step can be left out if you are utilizing an alkaline phosphatase system.
8. This process is so called Heat Induced Epitope Retrieval (HIER). We use a 1,200-W microwave oven to perform our HIER.
9. Once slides have been boiled, care should be taken to keep tissue sections from detaching from the slides. Do not place the tissues directly under running water, but rather direct the stream of water to the end of the slides for all subsequent running water washes.
10. If possible, leave a 3–4 mm separation between your hydrophobic PAP-line and the tissue section to allow your reagents to spread evenly.
11. Our standard *Buffer Wash* is 1× Dulbecco's Phosphate Buffered Saline with 0.01% Tween 20 and 1 g/L Bovine Serum Albumin (BSA) added. This can also be used as an antibody diluent.
12. This is a good stop point if needed. Tissues can be held for several hours in buffer wash without loss of antigenicity.

13. Background Sniper is a serum-free, casein-based blocking reagent. Other blocking agents, such as 1% BSA or normal serum from the same species of the antibody, can be used for certain antibodies. Extended time in Background Sniper can cause a reduction in antibody binding.
14. DaVinci Green is a phosphate-based antibody diluent. It contains a preservative so most antibodies can be stored diluted for up to 6 months. Antibodies diluted in buffer wash should be used within 24 h unless a preservative such as NaN_3 is added.
15. For the mTOR pathway antibodies, we use a two-step polymer based detection system with an incubation time of 30 min at room temperature. The Dako EnVision™ Plus system gives very clean results. The Biocare polymer system can give a higher background, but we use it when a more sensitive detection system is needed. We use the Rabbit-on-Rodent polymer; however, other Biocare polymer reagents are available for human tissue.
16. DAB is a chromogen used to visualize a dark brown end product of peroxidase-based immunohistochemistry. For the mTOR pathway antibodies, we use the DAB that comes with the Dako EnVision™Plus polymer kit (Dako #K4011). We also use the Sigma Fast tablet DAB (Sigma, St. Louis, MO #D4239) regularly, but not for the mTOR pathway antibodies. The Sigma DAB is not as strong, but it produces a crisp clean signal and works well for robust antibodies.
17. If endogenous melanin or other brown pigment is a problem, we have achieved good results using Vector VIP Purple (Vector Labs, Burlingame, CA # SK-4600) instead of DAB. Alternatively you could use an alkaline-phospatase system instead of a peroxidase based system and use Vector Red (Vector, SK-5100) as your permanent chromogen. Important note: Switch to Tris Buffer instead of PBS if using an alkaline-phospatase based system.
18. Enhancing DAB in a copper sulfate solution is not necessary but it will darken the DAB end product from reddish-brown to brownish-black. Our copper sulfate solution is made by mixing 5 g of copper sulfate with 9 g of sodium chloride in 1 L of distilled water. This solution is shelf stable. If using a chromogen other than DAB, this step can be omitted.
19. Bluing the hematoxylin can be accomplished with a commercially available bluing reagent, or a weak basic solution of dilute ammonia or lithium for 1–2 min. If using Vector VIP or Vector Red, do not use a bluing solution containing alcohol.
20. Real xylene will effectively remove the PAP-line ensuring clear coverslipping. If using a xylene substitute, make sure that your mounting medium is miscible with it. Some xylene substitutes react with toluene-based mounting media forming a white precipitate.

References

1. Evren, S., Dermen, A., Lockwood, G., Fleshner, N., and Sweet, J. 2010 Immunohistochemical examination of the mTORC1 pathway in high grade prostatic intraepithelial neoplasia (HGPIN) and prostatic adenocarcinomas (PCa): a tissue microarray study (TMA), *Prostate 70*, 1429–1436.
2. Clark, C., Shah, S., Herman-Ferdinandez, L., Ekshyyan, O., Abreo, F., Rong, X., McLarty, J., Lurie, A., Milligan, E. J., and Nathan, C. O. 2010 Teasing out the best molecular marker in the AKT/mTOR pathway in head and neck squamous cell cancer patients, *Laryngoscope 120*, 1159–1165.
3. Hirashima, K., Baba, Y., Watanabe, M., Karashima, R., Sato, N., Imamura, Y., Hiyoshi, Y., Nagai, Y., Hayashi, N., Iyama, K., and Baba, H. 2010 Phosphorylated mTOR expression is associated with poor prognosis for patients with esophageal squamous cell carcinoma, *Ann Surg Oncol 17*, 2486–2493.
4. Andersen, J. N., Sathyanarayanan, S., Di Bacco, A., Chi, A., Zhang, T., Chen, A. H., Dolinski, B., Kraus, M., Roberts, B., Arthur, W., Klinghoffer, R. A., Gargano, D., Li, L., Feldman, I., Lynch, B., Rush, J., Hendrickson, R. C., Blume-Jensen, P., and Paweletz, C. P. 2010 Pathway-based identification of biomarkers for targeted therapeutics: personalized oncology with PI3K pathway inhibitors, *Sci Transl Med 2*, 43 ra55.
5. Kim, J., Jonasch, E., Alexander, A., Short, J. D., Cai, S., Wen, S., Tsavachidou, D., Tamboli, P., Czerniak, B. A., Do, K. A., Wu, K. J., Marlow, L. A., Wood, C. G., Copland, J. A., and Walker, C. L. (2009) Cytoplasmic sequestration of p27 via AKT phosphorylation in renal cell carcinoma, *Clin Cancer Res 15*, 81–90.
6. Baba, Y., Nosho, K., Shima, K., Meyerhardt, J. A., Chan, A. T., Engelman, J. A., Cantley, L. C., Loda, M., Giovannucci, E., Fuchs, C. S., and Ogino, S. 2010 Prognostic significance of AMP-activated protein kinase expression and modifying effect of MAPK3/1 in colorectal cancer, *Br J Cancer 103*, 1025–1033.
7. Werner, M., Chott, A., Fabiano, A., and Battifora, H. (2000) Effect of formalin tissue fixation and processing on immunohistochemistry, *Am J Surg Pathol 24*, 1016–1019.
8. Onda, H., Lueck, A., Marks, P. W., Warren, H. B., and Kwiatkowski, D. J. (1999) Tsc2(+/−) mice develop tumors in multiple sites that express gelsolin and are influenced by genetic background, *J Clin Invest 104*, 687–695.
9. Kwiatkowski, D. J., Zhang, H., Bandura, J. L., Heiberger, K. M., Glogauer, M., el-Hashemite, N., and Onda, H. (2002) A mouse model of TSC1 reveals sex-dependent lethality from liver hemangiomas, and up-regulation of p70S6 kinase activity in Tsc1 null cells, *Hum Mol Genet 11*, 525–534.
10. Shackelford, D. B., Vasquez, D. S., Corbeil, J., Wu, S., Leblanc, M., Wu, C. L., Vera, D. R., and Shaw, R. J. (2009) mTOR and HIF-1alpha-mediated tumor metabolism in an LKB1 mouse model of Peutz-Jeghers syndrome, *Proc Natl Acad Sci USA 106*, 11137–11142.
11. Blando, J., Portis, M., Benavides, F., Alexander, A., Mills, G., Dave, B., Conti, C. J., Kim, J., and Walker, C. L. (2009) PTEN deficiency is fully penetrant for prostate adenocarcinoma in C57BL/6 mice via mTOR-dependent growth, *Am J Pathol 174*, 1869–1879.
12. Yeung, R. S., Xiao, G. H., Everitt, J. I., Jin, F., and Walker, C. L. (1995) Allelic loss at the tuberous sclerosis 2 locus in spontaneous tumors in the Eker rat, *Mol Carcinog 14*, 28–36.

Chapter 13

Assessing Cell Size and Cell Cycle Regulation in Cells with Altered TOR Activity

Megan Cully and Julian Downward

Abstract

Target of rapamycin (TOR) regulates the growth of cells and organisms. Numerous growth-promoting and growth-arresting pathways converge on TOR; TOR acts as an important hub, balancing the pro- and anti-growth signals within a cell. Since it regulates growth at the cellular level, cell size can be used as an indirect readout of TOR activity. Here, we describe methods used to analyze cell size in cell culture and in the *Drosophila* wing.

Key words: mTOR, Cell size, FACS analysis, Coulter counter, Cell volume, Drosophila, siRNA

1. Introduction

One of the hallmarks of changes in TOR activity is a change in cell size (1–4). Increased TOR activity increases cell size while decreased TOR activity decreases cell size. The mechanism through which this occurs remains unclear, but growth is associated with the increased activity of components of the translational machinery (reviewed in ref. 5). Cell size can provide a fast, high-throughput, and accurate readout of TOR activity (6).

As cells progress through the cell cycle, they also double in volume. Altering the overall cell cycle composition within a population of cells, therefore, skews both mean and median cell volume measurements independently of TOR activity. Thus, it is crucial to measure cell size within each phase of the cell cycle in order for size to accurately reflect mTOR activity. There are a number of different methods that can be used to analyze cell size. As outlined below, a few key factors should be considered when determining which method is most appropriate for a given application.

Thomas Weichhart (ed.), *mTOR: Methods and Protocols*, Methods in Molecular Biology, vol. 821,
DOI 10.1007/978-1-61779-430-8_13, © Springer Science+Business Media, LLC 2012

The ideal sample for cell size analysis would contain uniform cells of a single type in a single phase of the cell cycle. In practice, this is rarely the case, with the notable exception of the adult *Drosophila* wing (7). Cells grown in culture can be arrested in a given cell cycle phase, but often this also affects cell size and TOR activity. Cultured cells are best analyzed by side-by-side Coulter counter and flow cytometry analysis for reasons discussed in more detail below.

As cells grow, they increase in both volume and in cross-sectional area. Light microscopy-based techniques, including the Beckman Coulter Vi-CELL or measurements of cells incorporating fluorescent dyes, measure the cross-sectional area of a cell while Coulter counters measure volume. If we approximate a cell to be a sphere, then the volume will increase according to the equation,

$$\text{Volume} = 4 / 3\pi r^3,$$

while the cross-sectional area of that sphere will increase according to the equation,

$$\text{Area} = \pi r^2,$$

where r is the radius of the sphere. This means that if one cell has twice the volume of another, it will only have approximately 1.6 times the cross-sectional area. This becomes a potential problem when the difference in size is small; cells with a 20% increase in cell volume have only a 13% increase in cross-sectional area. Light microscopy-based techniques are most useful when cells cannot be placed in suspension for other types of analysis. These techniques can also be labour intensive, and are therefore not useful for high-throughput applications.

A second factor to consider is how many cells can be analyzed per sample. Within any population of cells, there is a distribution of cell sizes. In most cell types, this distribution is approximately Gaussian, where the mean is equal to the median and the curve is symmetrical. A meaningful data set must include enough data points to allow for statistical analysis of the distribution of cell sizes within a population. Ideally, 10,000 cells should be analyzed in order to be able to examine the population, although in some cases as few as 1,000 cells can still give good data. Light microscopy-based techniques are limited by the number of cells per visual field, and in practice it is difficult to measure more than a couple of hundred cells per sample. This gives a high standard deviation for the size measurements, especially if cells are not pre-selected based on cell cycle profile. In tissues with quiescent cells (and therefore a narrower distribution of cell sizes), such as the adult *Drosophila* wing, light microscopy is still useful to compare cell size.

There are two robust methods for measuring cell volume quickly and efficiently, and which can be used to measure 1,000–10,000 wcells per sample. Coulter counters measure the change in an electrical field that occurs as a cell passes between two electrodes (see the Beckman Coulter Web site for more details http://www.beckmancoulter.com/coultercounter/homepage_tech_coulter_principle.jsp?id=frombec&source=301redirect). This electrical impedance is proportional to the volume of the cell. Coulter counters can be used to analyze live cells in suspension, so the sample preparation time required is minimal. The main disadvantage of the Coulter counter is that it cannot be used to distinguish between cells in different phases of the cell cycle.

Flow cytometry can be used to compare the relative sizes of two populations. The excitatory laser is scattered in two directions, forward and to the side. The forward scatter (fsc) of the incident laser correlates with cell size while side scatter (ssc) correlates with cell granularity. It should be noted that fsc is not an actual measurement of cell size and that other factors, such as particle composition and membrane integrity, can affect fsc measurements. Furthermore, the fixation process itself can change cell size. However, when comparing similar samples processed simultaneously, such as different drug or RNAi treatments of the same cell type, fsc measurements can be useful in determining relative cell size. The main advantage of flow cytometry is that cells can be stained with DNA-intercalating dyes, such as propidium iodide, so cells in the various stages of the cell cycle can be examined independently. Flow cytometry machines with UV lasers are particularly useful, since the DNA of live cells can be stained with Hoechst dyes and visualized using the UV laser, thus eliminating the need for ethanol fixation.

2. Materials

2.1. Egg Lay and Drosophila Culture for Wing Measurements

1. Low nutrient media (8): Boil 500 ml of water in a beaker. Add 17 g yeast, 83 g dried corn, 10 g agar, 60 g sugar, and 4.6 g Moldex (solubilized in a small amount of ethanol). Fill the beaker to 1 L. Microwave for a total of 10 min, ensuring that the beaker does not boil over. Remove the beaker every 2–3 min and mix well. Pour 5–7 media into each vial, and allow it to cool.
2. Apple juice plates: Autoclave 25–30 g of agar with 700 ml of water for 40 min. In a second container, add 20 ml of 95% ethanol to 0.5 g of p-hydroxymethylbenzoate. Add this solution to 300 ml of apple juice concentrate. Add the autoclaved agar and pour into Petri dishes.
3. Resuspended yeast: Dried yeast is added to water and mixed into a paste. This mixture can be stored, covered, at 4°C for up to a week.
4. Fine-tipped paintbrush.

2.2. Analysis of Adult Drosophila Wings Using Light Microscopy

1. 95% ethanol.
2. Microscope slides.
3. Microscope coverslips (22 × 50 mm).
4. Euparal mounting media (Agar Scientific).

2.3. Analysis of Drosophila Wing Disc Cells Using the LSR

1. Phosphate-buffered saline (PBS): 8 g/L NaCl, 0.25 g/L KCl, 1.43 g/L Na_2HPO_4, 0.25 g/L KH_2PO_4.
2. Cell suspension buffer: In PBS, dissolve 4.5 mg/ml of trypsin (Invitrogen 7001–75) (see Note 1), 0.5 μg/ml Hoechst 33342 (Sigma), and 1 mM EDTA. This solution must be made fresh every time. The Hoechst can be stored as a 250 μg/ml stock solution (in water) at 4°C in the dark. The EDTA can be stored as a 500 mM stock solution at room temperature in water with a pH >8.0.
3. Glass Pasteur pipette with a rubber suction device.
4. FACS tubes.
5. Aluminium foil.

2.4. Drosophila RNAi Synthesis

1. Solution A: 100 mM Tris, pH 7.4, 100 mM EDTA, 1% SDS.
2. Pestle for 1.5-ml tubes.
3. 8 M potassium acetate solution: Made in water and stored at room temperature.
4. Isopropanol.
5. 70% ethanol.
6. Gene-specific primers (see Note 2) at a concentration of 5 μM. Both primers should start with the T7 binding site (TAATACGACTCACTATAGG).
7. 100 mM dNTPs.
8. HotStarTaq polymerase (Qiagen).
9. T7 Megascript kit (Ambion).
10. RNAse-free water.
11. 3 M sodium acetate made in RNAse-free water.
12. 95% ethanol.

2.5. RNAi Treatment of Drosophila S2 Cells

1. Schneider's Drosophila medium (Invitrogen).
2. Schneider's Drosophila medium (Invitrogen), supplemented with 10% fetal bovine serum (FBS, Sigma F7524).

2.6. Cell Culture

1. Schneider's Drosophila medium (Invitrogen), supplemented with 10% FBS for S2 cells.
2. Dulbecco's Modified Eagle's Medium (DMEM) (Gibco/BRL, Bethesda, MD) supplemented with 10% FBS for 293 and 293-ER cells.

3. 4-hydroxytamoxifen (4-OHT, Calbiochem) is dissolved in ethanol at a concentration of 0.5 mM and stored in aliquots at −20°C.
4. SB202190 (Calbiochem) is dissolved in DMSO at a concentration of 20 mM and stored in the dark at −20°C.
5. Rapamycin (Calbiochem) is dissolved in ethanol at a concentration of 1 mM and stored in aliquots at −20°C.
6. Trypsin/EDTA: 2.5 g/L trypsin (Difco), 8 g/L NaCl, 1.15 g/L Na_2HPO_4, 0.3 g/L KH_2PO_4, 1 g/L EDTA.

2.7. Coulter Counter Analysis of Cultured Cells

1. Trypsin/EDTA.
2. DMEM supplemented with 10% FBS.
3. Isoton buffer (Beckman Coulter).
4. Tubes for Z2 Coulter counter.

2.8. Flow Cytometry Analysis of Cultured Cells

1. Ice-cold 70% ethanol.
2. PBS.
3. Ribonuclease A: 100 μg/ml, stored at 4°C.
4. Propidium iodide (50 μg/ml, stored at room temperature in the dark). Propidium iodide is a DNA-intercalating agent and therefore a potent carcinogen. It should be handled with care, and gloves should be worn at all times.

3. Methods

3.1. Egg Lay and Drosophila Culture for Wing Measurements

1. The night before the egg lay, at least 20 male and 20 female adult flies of the appropriate genotype are placed in vials together and incubated at 25°C.
2. The following morning, the flies are transferred to containers with apple juice plates on the bottom. A small amount of resuspended yeast is placed in the centre of the plate. Flies are allowed to lay eggs for 4–6 h, at which point the Petri dishes are changed.
3. Approximately 24 h after each egg lay, the eggs will have hatched and the larvae will be wandering around the plate. Most of them can be found in the yeast in the centre. Using a fine-tipped paintbrush, collect the larvae under a microscope and transfer them to the vials containing low-nutrient food. Transfer 50 larvae to each tube and incubate at 25°C.

3.2. Analysis of Adult Drosophila Wings Using Light Microscopy

1. Collect 3-day-old adults from the vials containing low nutrient food. Separate males and females. Flies from different vials may be pooled, as long as each vial had 50 larvae in it. At least ten males and ten females of each genotype should be used.

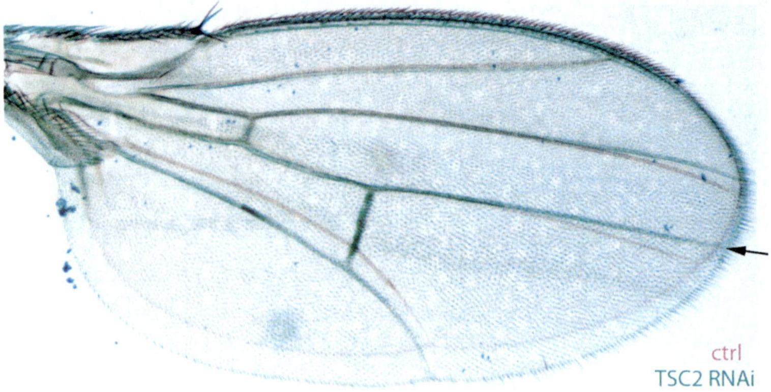

Fig. 1. Analysis of wing area. *TSC2 RNAi* was overexpressed in the dorsal compartment of the wing using the *UAS-Gal4* driver system under the control of the *engrailed* promoter. This promoter drives expression from the 4th inguinal vein (*indicated by the arrow*) to the dorsal edge of the wing. The wing containing both the *engrailed-Gal4* driver transgene and the *UAS-TSC2 RNAi* transgene is shown in *green*. The wing containing only the *engrailed-Gal4* transgene is shown in *red*.

2. Using fine-tipped forceps, remove one wing from each fly and place the wing in a glass dish containing ethanol.
3. Once you have collected ten wings per genotype, remove them from the ethanol one by one and lay them in rows on a microscope slide. Leave them to dry (for about 10 min).
4. Add 20 μl of euparal, and gently place a coverslip over them (see Note 3).
5. Place on a 60°C hotplate overnight. The slides can now be stored for months. Eventually, euparal will dry out; to store them indefinitely, seal them by putting a layer of nail polish along the edge of the coverslip.
6. Take pictures of each wing using the same magnification each time.
7. The wings can be analyzed in Photoshop or a similar program (Fig. 1). In Photoshop, the "magnetic lasso" tool is particularly useful to measure total wing area. Once the wing has been lassoed, the number of pixels within that lasso can be found in the histograms. Since each cell in the wing blade secretes a single hair or trichome, the density of these hairs reflects cell size in the wing. These hairs simply need to be counted per unit area. Choosing an area with approximately 200 hairs tends to give reliable data.

3.3. Analysis of Drosophila Wing Disc Cells Using the LSR

1. Perform an egg lay on apple juice plates and collect L1 larvae as in Subheading 3.1. If clones are to be induced using the heat-shock promoter, the larvae should be heat shocked 48 h after the egg lay by placing them in a 42°C water bath for 1 h.

2. Ninety-six to one hundred and twenty hours after the egg lay, the larvae should be wandering up the side of the vial, preparing to pupate. Collect these larvae. Dissect out the wing discs.
3. Remove all pieces of trachea and other tissue. Move the discs through 2–3 drops of PBS to clean them up. About 20 wing discs are needed per sample.
4. Put 0.5 ml of cell suspension buffer in each FACS tube.
5. Coat the inside of a glass Pasteur pipette with cell suspension buffer by pipetting up and down.
6. Use this coated pipette to transfer the wing discs from the PBS drop into the FACS tube.
7. Wrap the tubes in foil, and incubate them on an agitation table for 2–3 h (see Note 4).
8. Analyze the cells on an LSR or similar flow cytometry machine (see Note 5). Live cells should then be gated based on cell cycle (see Note 6). Forward scatter is used as an indicator of cell size.

3.4. Drosophila RNAi Synthesis

1. Collect ten wild-type flies in Eppendorf tube.
2. Add 400 μl of solution A.
3. Homogenize with a pestle for about 1 min.
4. Incubate at 70°C for 30 min with shaking.
5. Add 56 μl of potassium acetate. Incubate on ice for 30 min.
6. Spin for 15 min at 4°C at 140,000 × *g*.
7. Transfer the supernatant to a new Eppendorf. Add 240 μl of isopropanol. Shake well and incubate for 5 min at room temperature (see Note 7).
8. Spin for 10 min at 14,000 × *g* at room temperature.
9. Aspirate the supernatant and wash the pellet once with 500 μl of 70% ethanol.
10. Dry the pellet at room temperature until the pellet starts to appear transparent.
11. Resuspend the pellet in 200 μl of water. Measure the DNA content and dilute to 1 μg/μl. This is now a genomic DNA prep, and can be used for a number of applications including the synthesis of RNAi. It should be stored at -20°C for up to 6 months.
12. Perform PCR amplification from the genomic DNA. Add 0.5 μl of the genomic DNA, 1 μl of dNTPs, 5 μl of 10× buffer (supplied with the polymerase), 0.5 μl HotStarTaq polymerase, 1 μl of each primer, and 35 μl of distilled water per reaction. In the PCR machine, first release the enzyme by incubating at 95°C for 15 min; then, perform 35 cycles of 1 min at 95°C, 1 min at 50–60°C, and 1 min at 72°C; finally, perform a single 10-min incubation at 72°C to complete the elongation. Run

out the PCR products on a gel to be sure that you have good amplification.

13. Make dsRNA from the PCR product using the T7 Megascript kit. Add 16 μl of the PCR product, 4 μl of each nucleotide, 4 μl of the 10× buffer, and 4 μl of the T7 enzyme (everything, but the PCR product, is included in the kit). Incubate at 37°C for 12–36 h.
14. Precipitate the dsRNA by adding 40 μl RNAse-free water, 10 μl of 3 M sodium acetate, and 250 μl of ethanol. Incubate at –20°C for 15 min.
15. Spin at 14,000 × *g* for 15 min.
16. Remove supernatant and air dry in a sterile tissue culture hood until the pellet is almost transparent (approximately 30 min).
17. Resuspend in 100 μl of RNAse-free water (see Note 8).
18. Anneal the RNA by putting the tubes in a large beaker of water at 68°C (or a heating block) and allowing the tubes to cool slowly to room temperature. Between 68 and 50°C, the rate of change should be no greater than 1°C/min.
19. Check the concentration and dilute to 1 μg/ml (see Note 9).

3.5. RNAi Treatment of Drosophila S2 Cells

1. Collect cells by tapping the bottom of the flask.
2. Centrifuge at low speed (e.g. 1,000–1,500 × *g* for 5 min) to collect the cells.
3. Remove the supernatant and resuspend the cells at a concentration of 2×10^6 cells/ml in serum-free Schneider's medium.
4. Add 1 ml of cells to each well of a 6-well plate.
5. Add 25 μg of dsRNA, and incubate for 1 h (see Note 10).
6. Add 2 ml of Schneider's medium containing 10% FBS.
7. Incubate for 3–5 days.

3.6. Mammalian Cell Culture and Preparation of Samples for Size Analysis

1. 293 and 293-ER cells are maintained in DMEM with 10% FBS. The day before the start of the experiment, split the cells to approximately 30% confluency.
2. Treat cells with inhibitors. SB202190 is used at a final concentration of 20 μM, 4-hydroxytamoxifen is used at 100 nM, and rapamycin is used at 100 nM. Incubate for 24 h.
3. Harvest the cells by trypsinization. Inactivate the trypsin by adding 5 volumes of DMEM with 10% FBS, and then spin the cells at low speed (e.g. 1,000–1,500 × *g* for 5 min). Remove the supernatant and proceed immediately to Subheadings 3.7 or 3.8.
4. Following a 3–5-day treatment of the S2 cells with dsRNA, harvest the cells by pipetting the media up and down and washing the cells off the bottom of the dish. Leave them suspended in this media and transfer them to 15-ml tubes.

3.7. Coulter Counter Analysis of Cultured Cells

1. Resuspend the pelleted cells (for mammalian cell culture) in DMEM with 10% FBS. S2 cells are already in suspension in media following Subheading 3.6.
2. Add 1 ml of the cell suspension and 9 ml of Isoton buffer to a Z2 Coulter counter analysis tube.
3. Set up the parameters according to the cell type and size. Analyze the sample.

3.8. Flow Cytometry Analysis of Cultured Cells

1. Resuspend the cell pellet in PBS, and centrifuge again at low speed. Remove the supernatant.
2. Fix the cells in cold 70% ethanol. Add the ethanol dropwise to the cell pellet while vortexing. This ensures fixation of all cells and minimizes clumping.
3. Incubate the cells at 4°C for as little as 30 min and as long as 24 h (see Note 11).
4. Centrifuge the cells at low speed and remove the supernatant.
5. Wash the cells in PBS. Flick the bottom of tube to resuspend the pellet in the small amount of residual supernatant. Add 3 ml of PBS, spin at low speed, and remove the supernatant.
6. Repeat step 5, transferring the sample to a FACS analysis tube using a 5-ml pipette before spinning.
7. Resuspend the pellet by gently flicking the bottom of the tube.
8. Treat cells with ribonuclease. Add 50 μl of RNAse to the pellet. Flick the bottom of the tube to mix.
9. Add 200 μl of propidium iodide. Flick the bottom of the tube to mix.
10. Analyze by flow cytometry (see Note 6) (Fig. 2).

4. Notes

1. The concentration and purity of the trypsin are critical. Before starting this protocol, we recommend trying out a couple of different brands and/or concentrations of trypsin with some wild-type discs to be sure that a suspension of live cells can be obtained.
2. In designing the primers, there are a few key points in addition to the standard primer design guidelines. For dsRNA synthesis, PCR products of approximately 500–1,000 bp should be generated. Since genomic DNA is used as a template, design the primers based on the genomic sequence, not based on the cDNA. It is all right if the PCR product contains an intron, but it should be mainly exonic sequence. Remember that the

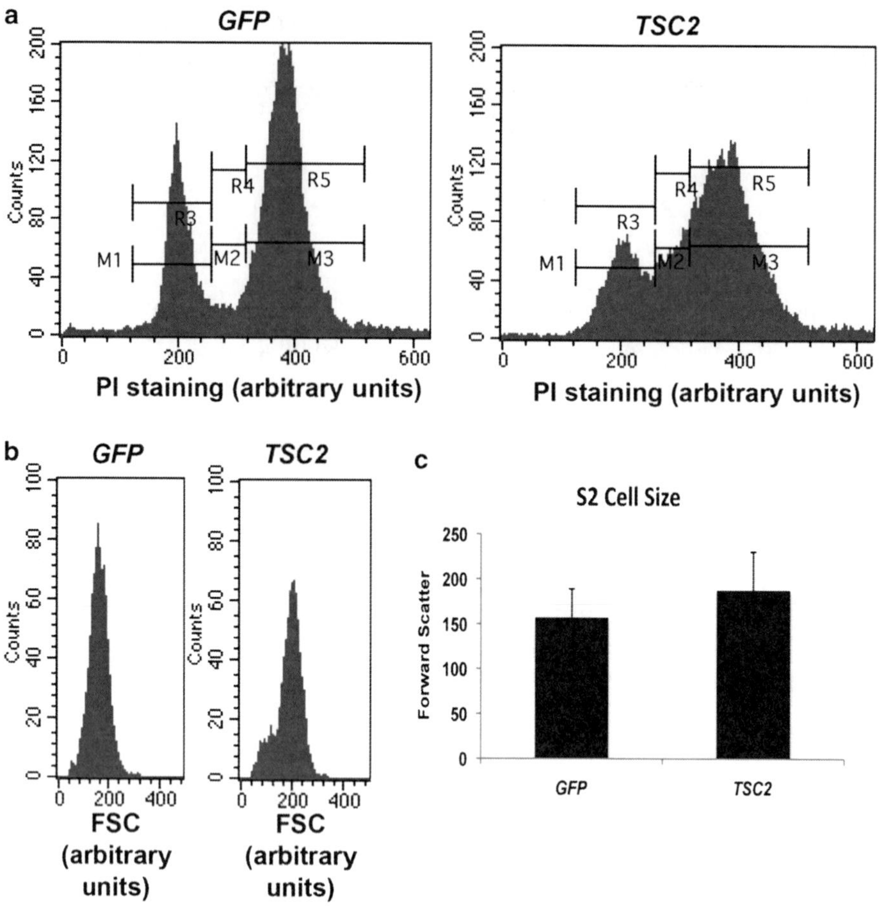

Fig. 2. Analysis of cell size using flow cytometry. S2 cells were treated with RNAi targeting the indicated gene for 5 days. (**a**) G1 cells were selected using propidium iodide (PI) staining. M1 and R3 indicate the G1 population. (**b**) Forward scatter (FSC) measurements for the G1 populations gated in (**a**). (**c**) Mean FSC for the populations in (**b**). *Error bars* indicate standard deviation.

dsRNA is cut into 22–23mers once inside the cell. We often use Primer 3 (http://frodo.wi.mit.edu/primer3/input.htm) (9) or E-RNAi (http://www.dkfz.de/signaling/e-rnai3/idseq.php) to design primers.

3. It is important not to introduce air bubbles. Place euparal on one side of the wings (make sure that there are no air bubble in the euparal drop). Place the edge of the coverslip beside the drop on the side opposite to the wings. Slowly bring the coverslip down onto the wings, allowing euparal to spread.
4. It is important that these tubes do not get inverted or overshaken while they are incubating, as this damages the cells. Prop the tops of the tubes up so that the tubes are sitting on the shaker at an angle.

5. The cells must be analyzed on a flow cytometer with a UV laser in order to see the Hoechst. Please note that these cells are very small. If there is a lot of debris in the sample, 7-aminoactinomycin D (7-AAD) exclusion can be used to differentiate live cells from dead cells as it does not interfere with GFP fluorescence that may be used for clonal analysis. Cell size is best compared within a sample. As such, clonal analysis is the most appropriate way to compare the size of cells with two different genotypes.
6. G1 cells tend to have the narrowest distribution of cell size, so we normally use them for cell size analysis.
7. Once isopropanol has been added, these samples can be stored at –20°C for up to 1 week.
8. Resuspending the RNA can be difficult if the RNA is over-dried. Incubating at 68°C with agitation works and does not damage the RNA.
9. To check the quality of the dsRNA, it can be run on a 2% agarose gel. It should be a single band with no smears.
10. If there is a limited supply of the RNA, Fugene can be used to increase the efficiency of RNA uptake. For our screen, we used 0.3 µl of Fugene and 0.1 µg of dsRNA per well of a 96-well plate (6).
11. Once fixed, the cells are fragile. Avoid pipetting the cells with micropipettes.

References

1. Stocker, H., Radimerski, T., Schindelholz, B., Wittwer, F., Belawat, P., Daram, P., Breuer, S., Thomas, G., and Hafen, E. (2003) Rheb is an essential regulator of S6K in controlling cell growth in Drosophila, *Nat Cell Biol 5*, 559–565.
2. Montagne, J., Stewart, M. J., Stocker, H., Hafen, E., Kozma, S. C., and Thomas, G. (1999) Drosophila S6 kinase: a regulator of cell size, *Science 285*, 2126–2129.
3. Potter, C. J., Huang, H., and Xu, T. (2001) Drosophila Tsc1 functions with Tsc2 to antagonize insulin signaling in regulating cell growth, cell proliferation, and organ size, *Cell 105*, 357–368.
4. Tapon, N., Ito, N., Dickson, B. J., Treisman, J. E., and Hariharan, I. K. (2001) The Drosophila tuberous sclerosis complex gene homologs restrict cell growth and cell proliferation, *Cell 105*, 345–355.
5. Guertin, D. A., and Sabatini, D. M. (2007) Defining the role of mTOR in cancer, *Cancer Cell 12*, 9–22.
6. Cully, M., Genevet, A., Warne, P., Treins, C., Liu, T., Bastien, J., Baum, B., Tapon, N., Leevers, S. J., and Downward, J. (2010) A role for p38 stress-activated protein kinase in regulation of cell growth via TORC1, *Mol Cell Biol 30*, 481–495.
7. Weinkove, D., Neufeld, T. P., Twardzik, T., Waterfield, M. D., and Leevers, S. J. (1999) Regulation of imaginal disc cell size, cell number and organ size by Drosophila class I(A) phosphoinositide 3-kinase and its adaptor, *Curr Biol 9*, 1019–1029.
8. Arquier, N., Bourouis, M., Colombani, J., and Leopold, P. (2005) Drosophila Lk6 kinase controls phosphorylation of eukaryotic translation initiation factor 4E and promotes normal growth and development, *Curr Biol 15*, 19–23.
9. Rozen, S., and Skaletsky, H. (2000) Primer3 on the WWW for general users and for biologist programmers, *Methods Mol Biol 132*, 365–386.

Chapter 14

Quantitative Visualization of Autophagy Induction by mTOR Inhibitors

Beat Nyfeler, Philip Bergman, Christopher J. Wilson, and Leon O. Murphy

Abstract

Autophagy is a catabolic pathway that degrades bulk cytosol in lysosomal compartments enabling amino acids and fatty acids to be recycled. One of the key regulators of autophagy is the mammalian target of rapamycin (mTOR), a conserved serine/threonine kinase which suppresses the initiation of the autophagic process when nutrients, growth factors, and energy are available. Inhibition of mTOR, e.g., by small molecules such as rapamycin, results in activation of autophagy. To quantify autophagy induction by mTOR inhibitors, we use an mCherry-GFP-LC3 reporter which is amenable to retroviral delivery into mammalian cells, stable expression, and analysis by fluorescence microscopy. Here, we describe our imaging protocol and image recognition algorithm to visualize and measure changes in the autophagic pathway.

Key words: Autophagy, mTOR, Rapamycin, Ku-0063794, LC3, High-content imaging, Image recognition algorithm

1. Introduction

The mammalian target of rapamycin (mTOR) regulates anabolic and catabolic pathways in response to upstream signals, such as nutrients, energy levels, growth factors, or cytotoxic stress (1, 2). When growth conditions are favorable, mTOR phosphorylates ribosomal protein S6 kinase 1 (S6K1) and eukaryotic translation initiation factor 4E-binding protein 1 (4E-BP1), thereby promoting protein synthesis and cell growth. At the same time, mTOR suppresses protein turnover by inhibiting autophagy, a cellular degradation pathway which is characterized by the sequestration of bulk cytoplasmic material into the so-called autophagosomes which then fuse with lysosomes to form autolysosomes (Fig. 1a). In the acidic luminal environment of the autolysosomes, various proteases degrade and recycle the sequestered macromolecules. In contrast to proteasomal

Thomas Weichhart (ed.), mTOR: Methods and Protocols, Methods in Molecular Biology, vol. 821,
DOI 10.1007/978-1-61779-430-8_14, © Springer Science+Business Media, LLC 2012

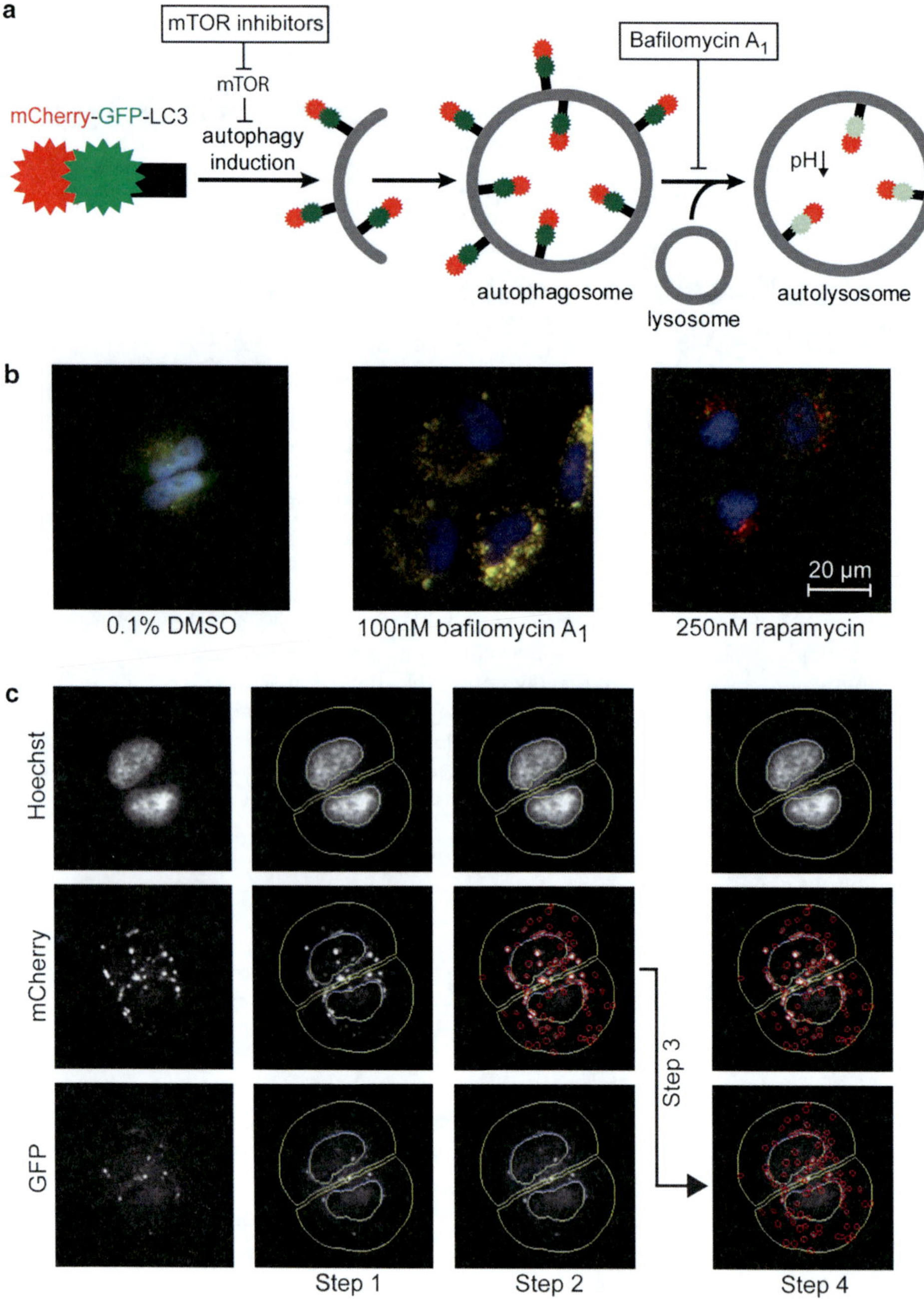

Fig. 1. mCherry-GFP-LC3 assay. (**a**) Schematic representation of the mCherry-GFP-LC3 assay. During the autophagic process, mCherry-GFP-LC3 is recruited to a double-membrane structure that elongates and forms autophagosomes which can be detected as both GFP- and mCherry-positive puncta. The fusion of autophagosomes and lysosomes generates autolysosomes which are predominantly positive for mCherry fluorescence since the GFP signal is pH sensitive and lost in the acidic luminal environment of the autolysosomes. (**b**) H4 mCherry-GFP-LC3 cells were grown in 4-well culture slides, treated for 18 h with 0.1% DMSO, 250 nM rapamycin, or 100 nM bafilomycin A_1, fixed, and images were acquired by fluorescence microscopy. Representative images are shown and GFP (*green*), mCherry (*red*), and DAPI (*blue*) channels were merged. (**c**) Illustration of the image recognition algorithm as described in Subheading 3.4. The Hoechst, mCherry, and GFP images are acquired by automated high-content microscopy and represent rapamycin-treated H4 mCherry-GFP-LC3 cells. Step 1: Nuclei are detected in the Hoechst image and cells defined using a 10-μm collar. Step 2: Puncta (organelles) are identified in the mCherry image. Step 3: Mask of mCherry-positive puncta is transferred onto GFP image. Step 4: GFP fluorescence intensity is measured inside the puncta mask (reference intensity).

degradation, autophagy has the capacity to degrade whole organelles as well as aggregated proteins. The autophagic process is regulated and facilitated by a heterogeneous class of autophagy-related (ATG) proteins which were originally identified in yeast and have been comprehensively reviewed (3–6). mTOR suppresses the autophagic process at the initiation step presumably through direct phosphorylation of the Ulk1–mAtg13–FIP200 complex (7).

Low levels of basal autophagy constantly regulate protein and organelle turnover, but can be further activated to provide cells with nutrients and energy during periods of prolonged starvation and stress. Furthermore, pharmacological activation of autophagy is considered a therapeutic strategy to treat various proteinopathies by increasing the clearance of aggregation-prone protein species and thereby reducing cytotoxicity (8). Releasing autophagy from mTOR suppression is one way to activate the pathway (9) and mTOR inhibitors, such as rapamycin, have indeed been shown to induce autophagy and reduce the amount of aggregated mutant huntingtin protein in cell culture and mouse models of Huntington's disease (10).

What methods are available to study and monitor the activation of autophagy? Most current techniques rely on LC3, the mammalian homologue of yeast Atg8. Upon autophagy induction, cytosolic LC3 is conjugated to phosphatidylethanolamine in the autophagosomal membrane and this can be detected by western blotting based on a shift in molecular weight (11). In addition, imaging of GFP-tagged LC3 is often used to visualize the redistribution of cytosolic LC3 to autophagosomes which are apparent as puncta by fluorescence microscopy (12). These methods reveal a snapshot of the number of autophagosomes but are not a reliable readout for autophagic flux. Rapamycin and bafilomycin A_1, for example, both increase the amount of lipidated LC3 and the number of GFP-LC3 puncta but have two opposing effects on autophagic flux. While rapamycin increases autophagic flux, bafilomycin A_1 stalls autophagy by inhibiting the vacuolar ATPase (13). In order to measure true activation of autophagy, several methods have been refined and two excellent recent guidelines summarize how to best measure autophagic flux (14, 15). One of the improvements consists of using tandem fluorescent-tagged LC3, such as mRFP-GFP-LC3 (16) or mCherry-GFP-LC3 (17), and takes advantage of the differential pH sensitivities of red and green fluorescent proteins. The fluorescent signals of mRFP and mCherry are relatively pH resistant while GFP fluorescence is pH sensitive and significantly reduced in acidic environments. As a consequence, mRFP-GFP-LC3 and mCherry-GFP-LC3 reporters lose their GFP signal upon reaching the acidic environment of autolysosomes (Fig. 1a). Increased autophagic flux results in more reporter molecules being delivered to autolysosomes and this can be detected by more puncta which appear red in color due to diminished GFP fluorescence in the acidified lysosomal compartment. Prelysosomal stalling of autophagy, on the other hand, results in the accumulation of the reporter in a compartment which appears yellow colored due to equal red and green fluorescence.

In this chapter, we describe the mCherry-GFP-LC3 assay (18) and focus on how imaging-based analyses can be applied to measure autophagic flux. mCherry-GFP-LC3 is expressed from a retroviral vector which allows delivery and stable integration into a wide variety of mammalian cells. Stable cell lines that express mCherry-GFP-LC3 can be genetically and chemically modulated and autophagy analyzed by fluorescence microscopy. By combining high-content microscopy with an image recognition algorithm, changes in autophagic flux can be readily measured and quantified.

2. Materials

2.1. Retroviral Packaging

1. pL(mCherry-GFP-LC3] (18) is based on the pLEGFP-C1 vector (Clontech, Mountain View, CA). The coding sequence of human LC3A derives from pCMV6-XL5[MAP1LC3A] (NM_032514.2, Origene, Rockville, MD) and is inserted downstream of GFP. The coding sequence of mCherry (Clontech) was inserted upstream of GFP. The amino acid sequence of the mCherry-GFP-LC3 reporter is depicted in Fig. 2.
2. GP2-293 cells stably express the viral *gag* and *pol* genes.
3. Dulbecco's modified Eagle's medium (DMEM) containing 4.5 g/L D-glucose, 4 mM L-glutamine, and 110 mg/mL sodium pyruvate.
4. Fetal bovine serum (FBS).
5. Phosphate-buffered saline (PBS).
6. 0.05% Trypsin containing EDTA.
7. Opti-MEM.
8. Lipofectamine2000 (Invitrogen, Carlsbad, CA).
9. pVPackVSV-G (Stratagene, La Jolla, CA).
10. Poly-D-Lysine coated 10 cm dishes (BD Biosciences, San Jose, CA).

2.2. Retroviral Transduction and Selection of Stable Cell Line

1. See Subheading 2.1 for cell culture reagents.
2. H4 neuroglioma cells (ATCC, Manassas, VA).
3. Polybrene (American Bioanalytical, Natick, MA) is stored at –20°C as 10 mg/mL stock and used at 8 μg/mL.
4. G418 (Invitrogen) is dissolved at 500 μg/mL in growth medium and used to select H4 cells. Cells which have stably integrated the pL[mCherry-GFP-LC3] construct express the neomycin resistance gene and are resistant to G418.

MVSKGEEDNMAIIKEFMRFKVHMEGSVNGHEFEIEGEGEGRPYEGTQTAK
LKVTKGGPLPFAWDILSPQFMYGSKAYVKHPADIPDYLKLSFPEGFKWER
VMNFEDGGVVTVTQDSSLQDGEFIYKVKLRGTNFPSDGPVMQKKTMGWEA
SSERMYPEDGALKGEIKQRLKLKDGGHYDAEVKTTYKAKKPVQLPGAYNV
NIKLDITSHNEDYTIVEQYERAEGRHSTGGMDELYKPVAT**MVSKGEELFT**
GVVPILVELDGDVNGHKFSVSGEGEGDATYGKLTLKFICTTGKLPVPWPT
LVTTLTYGVQCFSRYPDHMKQHDFFKSAMPEGYVQERTIFFKDDGNYKTR
AEVKFEGDTLVNRIELKGIDFKEDGNILGHKLEYNYNSHNVYIMADKQKN
GIKVNFKIRHNIEDGSVQLADHYQQNTPIGDGPVLLPDNHYLSTQSALSK
DPNEKRDHMVLLEFVTAAGITLGMDELYKSGLRSRAQASNSAVDMPSDRP
FKQRRSFADRCKEVQQIRDQHPSKIPVIIERYKGEKQLPVLDKTKFLVPD
HVNMSELVKIIRRRLQLNPTQAFFLLVNQHSMVSVSTPIADIYEQEKDED
GFLYMVYASQETFGF

Fig. 2. mCherry-GFP-LC3 reporter. Amino acid sequence of the mCherry-GFP-LC3 construct. mCherry sequence is *underlined*, GFP sequence is in *bold*, and LC3A sequence is *boxed*.

2.3. Compound Treatment and Fluorescence Microscopy

1. Bafilomycin A_1 (Tocris, Ellisville, MO) is an inhibitor of vacuolar ATPases (13). Bafilomycin A_1 is dissolved at 100 μM in DMSO (Sigma, St. Louis, MO) and aliquots are stored at −20°C. We use bafilomycin A_1 at saturating concentrations of 50–100 nM.
2. Rapamycin (Calbiochem, Gibbstown, NJ) is an allosteric inhibitor of mTOR complex 1. Rapamycin is dissolved at 10 mM in DMSO and aliquots are stored at −20°C. We use rapamycin at saturating concentrations of 100–250 nM.
3. Ku-0063794 (Chemdea, Ridgewood, NJ) is an ATP-competitive catalytic inhibitor of mTOR [19). Ku-0063794 is dissolved at 10 mM in DMSO and aliquots are stored at −20°C. We use Ku-0063794 at saturating concentrations of 3–5 μM.
4. Poly-D-Lysine coated 4-well culture slides (BD Biosciences) and 22 × 50 mm glass cover slides (Corning, Corning, NY).
5. Mirsky's fixative (National Diagnostics, Atlanta, GA) is provided as a two-bottle system containing 10× concentrate and 10× buffer. Prior to use, one part concentrate is mixed with one part buffer resulting in a 5× fixation solution.
6. ProLong Gold Antifade reagent containing DAPI (Invitrogen) is used to embed cells.

2.4. High-Content Analysis and Image Recognition

1. See Subheading 2.3 for DMSO, bafilomycin A_1, rapamycin, Ku-0063794, and Mirsky's fixative.
2. Black-colored, clear-bottom 384-well plates (Corning).
3. Hoechst33342 (Invitrogen) is used to stain cellular nuclei. Hoechst33342 is added to Mirsky's fixative prior to cell fixation and used at a final concentration of 5 μg/mL.
4. 384-well plate washer (BioTEK, Winooski, VT).
5. Tris-buffered saline (TBS, Boston BioProducts, Ashland, MA) supplemented with 0.02% sodium azide (Sigma).
6. Adhesive PCR foil (Thermo Scientific, Waltham, MA) is used to seal 384-well plates prior to imaging.
7. InCell Analyzer 1000 instrument and InCell Investigator software (GE Healthcare, Piscataway, NJ).
8. Quadruple band pass mirror (Semrock, Rochester, NY; part: FF410/504/582/669-Di01-25×36).

3. Methods

3.1. Production of mCherry-GFP-LC3 Retroviral Particles

1. GP2-293 cells are maintained at 37°C and 5% CO_2 in T75 flasks in growth medium (DMEM supplemented with 10% FBS). Cells are always split prior to reaching confluence by washing once with PBS, trypsinizing in 0.05% trypsin, resuspending in growth medium, and transferring to new flasks.
2. Day 0: Cell plating. Subconfluent GP2-293 cells are harvested by trypsinization, resuspended in growth medium, and counted. 4×10^6 cells are seeded into a poly-D-Lysine-coated 10-cm dish in 10 mL growth medium.
3. Day 1: Transfection. 72 μL Lipofectamine2000 are added to 750 μL Opti-MEM in a 1.5-mL microcentrifuge tube and incubated at room temperature for 5 min. Lipofectamine2000-containing Opti-MEM is transferred to a new 1.5-mL microcentrifuge tube containing 750 μL Opti-MEM, 6 μg pVPackVSV-G, and 6 μg pL[mCherry-GFP-LC3]. Mixture is pipetted up and down, incubated at room temperature for 15 min, and added dropwise to GP2-293 cells. Cells are placed at 37°C and 5% CO_2.
4. Day 2: Media is replaced with 10 mL fresh growth medium (see Notes 1 and 2).
5. Day 4: Virus harvesting. The media supernatant (containing the retroviral particles) is transferred into a 15-mL conical tube and centrifuged at 500 × *g* for 5 min to remove GP2-293 cells and debris (see Note 3). 1.5-mL aliquots of the retrovirus are then transferred to cryovials and stored at –70°C (see Note 4).

3.2. Generation of H4 mCherry-GFP-LC3 Stable Cell Line

1. H4 cells are maintained at 37°C and 5% CO_2 in T75 flasks in growth medium (DMEM supplemented with 10% FBS). Cells are split prior to reaching confluence by washing once with PBS, trypsinizing in 0.05% trypsin, resuspending in growth medium, and transferring to new flasks.
2. Day 0: Cell plating. H4 cells are harvested by trypsinization, resuspended in growth medium, and counted. 5×10^4 cells are seeded into the wells of a 6-well plate in 2 mL media per well.
3. Day 1: Infection. A 1.5-mL retrovirus aliquot is quickly thawed in a 37°C water bath. 0, 0.1, and 1 mL of retrovirus are transferred into 15-mL conical tubes (see Notes 1 and 5). Prewarmed growth medium is added to a final volume of 2 mL per tube. 1.6 μL of 10 mg/mL polybrene is then added per tube which results in a final polybrene concentration of 8 μg/mL. The samples are gently mixed and incubated for 5 min at room temperature. In the meantime, the 6-well plate of H4 cells is removed from the incubator and the growth medium aspirated. The retrovirus + polybrene mixtures are then added to individual wells of the 6-well plate (see Note 6).
4. Day 2: Start of selection. Medium is replaced with 2 mL per well growth medium containing 500 μg/mL G418.
5. G418-containing selection medium is replaced every 2 days. Cell killing is monitored daily using a phase-contrast microscope. Selection is complete when all nontransduced cells (0 mL retrovirus) are killed. This takes around 8 days for H4 cells.
6. For the infected cells, one well is chosen for expansion and future experiments. We typically choose the well in which 50–80% of the cells have survived G418 selection (see Note 5). Cells are analyzed under an inverted fluorescence microscope for mCherry and GFP fluorescence to confirm expression of the mCherry-GFP-LC3 reporter. Cells are grown to confluence and are sequentially expanded into a T150 flask and frozen down in aliquots.

3.3. Visualization of Autophagy by Fluorescence Microscopy

1. H4 mCherry-GFP-LC3 cells are cultured and analyzed as bulk population (see Note 7). 500 μg/mL G418 is maintained during the culture of H4 mCherry-GFP-LC3 cells but omitted when cells are split for an experiment.
2. Day 0: Cell plating. Subconfluent H4 mCherry-GFP-LC3 cells are harvested by trypsinization, resuspended in growth medium, and counted. Fifty thousand cells are added into the wells of poly-D-Lysine-coated 4-well culture slides in 1 mL medium per well.
3. Day 1: Compound treatment. H4 mCherry-GFP-LC3 cells are treated with 0.1% DMSO, 100 nM bafilomycin A_1, 250 nM rapamycin, and 3 μM Ku-0063794. Compound aliquots are

thawed at room temperature, diluted in DMSO to 1,000× of their final concentration, and then diluted 1:1,000 in growth medium. This results in a final concentration of 0.1% DMSO for all treatments. The 4-well culture slide of H4 mCherry-GFP-LC3 cells is removed from the incubator, the media is aspirated, and 1 mL/well compound-containing media is added. Compound treatment is performed for 16–18 h (see Note 8).

4. Day 2: Cell fixation and embedding. Cells are fixed for 1 h at room temperature by directly adding 250 µL 5× concentrated Mirsky's fixative to the media-containing wells of the 4-well culture slide (see Note 9). Mirsky's fixative is then aspirated and cells are washed four times with 1 mL/well PBS. After the last washing step, the plastic chamber is removed and excess PBS drained off. Several drops of ProLong Gold Antifade reagent containing DAPI are added onto the culture slide which is then capped with a glass cover slide and allowed to dry.
5. The slide is analyzed by fluorescence microscopy. We use a Zeiss Axiovert 200 M microscope with 40× magnification, and DAPI, FITC, and Cy3 filter sets to visualize DAPI, GFP, and mCherry, respectively (see Note 10). Images are acquired using the Zeiss AxioVision LE software and are autocontrasted and overlaid using Adobe Photoshop. Examples of representative images are shown in Fig. 1b.

3.4. Quantification of Autophagy Using High-Content Imaging and Analysis

1. Day 0: Cell plating. Subconfluent H4 mCherry-GFP-LC3 cells are harvested by trypsinization, resuspended in growth medium, and counted. A cell suspension of 66,000 cells/mL is prepared and 30 µL are added into the wells of a 384-well plate using an electronic multichannel pipette or bulk dispenser. This results in 2,000 cells/well being plated. The 384-well plate is briefly spun down (see Note 11) and placed at 37°C and 5% CO_2.
2. Day 1: Compound treatment. H4 mCherry-GFP-LC3 cells are treated in 384-well plates by adding 10 µL/well growth medium containing 4× concentrated compounds. This results in a total volume of 40 µL per well and a 1× final compound concentration. For this purpose, compound aliquots are thawed at room temperature, diluted in DMSO to 1,000× of their final concentration, and diluted 1:250 (4× final) in growth medium. 10 µL is then added to individual wells of the 384-well plate using an electronic multichannel pipette, resulting in a final DMSO concentration of 0.1% (see Note 12). Compound treatments are performed at least in triplicates. The 384-well plate is briefly spun down (see Note 11) and placed at 37°C and 5% CO_2. Compound treatment is performed for 16–18 h (see Note 8).
3. Day 2: Cell fixation. Cells are fixed by adding 10 µL/well 5× concentrated Mirsky's fixative supplemented with 25 µg/mL

Hoechst33342. This results in a total volume of 50 μL per well and a concentration of 1× Mirsky's fixative and 5 μg/mL Hoechst33342. The 384-well plate is briefly spun down (see Note 11) and incubated for 1 h at room temperature. Cells are then washed using a 384-well plate washer using a protocol which aspirates the volume down to 30 μL/well before dispensing 70 μL/well TBS supplemented with 0.02% sodium azide. Aspiration and dispensing steps are repeated seven times and a final volume of 100 μL/well is left. The plate is sealed using an adhesive PCR foil (see Note 13).

4. Imaging: The bottom of plate is cleaned with 70% ethanol and then imaged using the InCell 1000 automated epifluorescence microscope. 20× magnification is used and 4 different areas (fields) are imaged per well; this typically captures a total of around 400 cells per well. Hoechst33342 images are acquired using an excitation of 360 nm (D360_40x filter), an emission of 460 nM (HQ460_40M filter), and an exposure time of 150 ms. GFP images are acquired using an excitation of 475 nm (S475_20× filter), an emission of 535 nM (HQ535_50M filter), and an exposure time of 1 s. mCherry images are acquired using an excitation of 535 nm (HQ535_50× filter), an emission of 620 nM (HQ620_60M filter), and an exposure time of 1 s. A quadruple band-pass mirror is used for all images.
5. Image analysis: The InCell Investigator software is used to analyze the images using the Multi Target Analysis algorithm as illustrated in Fig. 1c. First, nuclei are detected in the Hoechst33342 image using top-hat segmentation and a minimal nuclear area of 50 μm^2. Cells are defined using a collar of 10 μm around the nuclei. Second, puncta (organelles) are identified in the mCherry image inside the cells using multitop-hat segmentation. Third, the mask of the mCherry puncta is transferred onto the GFP image. Fourth, the GFP fluorescence intensity inside the mCherry puncta mask is measured (reference intensity).
6. The "organelles" parameter (described in Subheading 3.4, step 5) reflects mCherry-positive puncta of the mCherry-GFP-LC3 reporter and is used to calculate "LC3 puncta/cell" (Fig. 3). For this purpose, the number of organelles is calculated per cell and averaged over all the cells in a given well. The "LC3 puncta/cell" bars in Fig. 3 are the average over four similarly treated wells with the error representing the standard deviation across those wells.
7. The "reference intensity" parameter (described in Subheading 3.4, step 5) reflects the GFP fluorescence intensity in the mCherry-positive puncta mask and is used to calculate "Autophagic flux" (Fig. 3). For this purpose, the reference intensity values are averaged over all the cells in a given well, and changes between

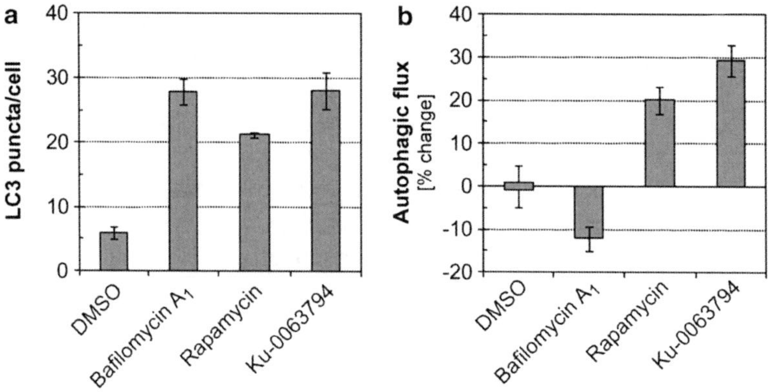

Fig. 3. Quantification of autophagy. H4 mCherry-GFP-LC3 cells were grown in 384-well plates and treated for 18 h with 0.1% DMSO, 100 nM bafilomycin A_1, 250 nM rapamycin, or 3 μM Ku-0063794. Cells were fixed, stained with Hoechst33342, imaged by high-content microscopy, and analyzed as described in Subheading 3.4. (**a**) The number of cellular LC3 puncta was averaged from 4 wells and is depicted as mean ± SD. (**b**) Autophagic flux was averaged from 4 wells and is depicted as mean ± SD.

DMSO- and compound-treated cells are calculated in % and depicted as "autophagic flux." Negative changes for the reference intensity reflect enhanced autophagic flux (GFP intensity in mCherry-positive puncta mask is lower than in control condition due to delivery of the reporter to autolysosomes). Positive changes in the reference intensity reflect decreased autophagic flux (GFP intensity in mCherry-positive puncta mask is higher than in control condition due to accumulation in a preautolysosomal compartment).

8. Autophagy activation, e.g., upon treatment with RAD001 or Ku-0063794, is characterized by an increase in both LC3 puncta/cell and autophagic flux. Autophagy stalling, e.g., upon treatment with bafilomycin A_1, is characterized by an increase in LC3 puncta/cell but an unchanged or decreased autophagic flux (Fig. 3).

4 Notes

1. Always use caution when handling recombinant retrovirus. NIH guidelines require that retroviral production and infection are performed in a Biosafety Level 2 facility. In addition, follow regional and institutional guidelines.
2. An inverted fluorescence microscope can be used to evaluate transfection efficiency of GP2-293 cells. Transfected cells are positive for mCherry and GFP fluorescence.
3. An alternative way of removing GP2-293 cells and debris is by filtration through a 0.45-μM filter.

4. Avoid multiple freeze/thaw cycles since this decreases viral titers.
5. To ensure infection of a representative number of cells without overinfection, two different retrovirus concentrations are used. We aim at using a concentration for which 50–80% of the cells survive G418 selection.
6. Prolonged exposure to polybrene can be toxic to some cell lines. We have not observed cytotoxicity in H4 cells for treatments with 8 μg/mL polybrene up to 24 h. If cytotoxicity is observed, it is suggested to decrease polybrene concentration (2–4 μg/mL) and exposure time (4–6 h).
7. The bulk population of H4 cells stably expressing mCherry-GFP-LC3 shows some heterogeneity in basal and compound-modulated autophagy. If a homogenous cell population and autophagy response is required, single cells can be sorted into 96-well plates using FACS, expanded, and tested for their response to autophagy-modulating compounds.
8. Autophagy modulation and redistribution of mCherry-GFP-LC3 can be already observed after a compound treatment time of 3–4 h. However, we see more robust effects with 16–18-h treatment times.
9. Paraformaldehyde works equally well to fix H4 cells and visualize mCherry-GFP-LC3, but unlike Mirsky's fixative requires special handling and safety considerations.
10. mCherry fluorescence bleaches more quickly than GFP and is imaged first.
11. The centrifugation step is essential for obtaining consistent results and ensures that all fluid (cell suspension, compound-containing media, Mirsky's fixative) reaches the plate bottom and is not lost at the walls of the plate.
12. Effects of DMSO concentration within the assay should be empirically tested and typically should not be greater than 0.5% for a cell-based assay.
13. Plates can be stored at 4°C for 2–3 days before imaging without affecting data quality.

Acknowledgments

We would like to thank Kristina Fetalvero, Fred Harbinski, and Emilia Temple for comments and suggestions.

References

1. Laplante M, Sabatini DM (2009) mTOR signaling at a glance. *J. Cell Sci. 122*, 3589–3594.
2. Wullschleger S, Loewith R, Hall MN (2006) TOR signaling in growth and metabolism. *Cell 124*, 471–484.
3. He C, Klionsky DJ (2009) Regulation mechanisms and signaling pathways of autophagy. *Annu. Rev. Genet. 43*, 67–93.
4. Klionsky DJ, Codogno P, Cuervo AM et al (2010) A comprehensive glossary of autophagy-related molecules and processes. *Autophagy. 6.*
5. Xie Z, Klionsky DJ (2007) Autophagosome formation: core machinery and adaptations. *Nat. Cell Biol. 9*, 1102–1109.
6. Yang Z, Klionsky DJ (2009) An overview of the molecular mechanism of autophagy. *Curr. Top. Microbiol. Immunol. 335*, 1–32.
7. Jung CH, Ro SH, Cao J et al (2010) mTOR regulation of autophagy. *FEBS Lett. 584*, 1287–1295.
8. Rubinsztein DC, Gestwicki JE, Murphy LO et al (2007) Potential therapeutic applications of autophagy. *Nat. Rev. Drug Discov. 6*, 304–312.
9. Noda T, Ohsumi Y (1998) Tor, a phosphatidylinositol kinase homologue, controls autophagy in yeast. *J. Biol. Chem. 273*, 3963–3966.
10. Ravikumar B, Vacher C, Berger Z et al (2004) Inhibition of mTOR induces autophagy and reduces toxicity of polyglutamine expansions in fly and mouse models of Huntington disease. *Nat. Genet. 36*, 585–595.
11. Mizushima N, Yoshimori T (2007) How to interpret LC3 immunoblotting. *Autophagy. 3*, 542–545.
12. Kabeya Y, Mizushima N, Ueno T et al (2000) LC3, a mammalian homologue of yeast Apg8p, is localized in autophagosome membranes after processing. *EMBO J. 19*, 5720–5728.
13. Bowman EJ, Siebers A, Altendorf K (1988) Bafilomycins: a class of inhibitors of membrane ATPases from microorganisms, animal cells, and plant cells. *Proc. Natl. Acad. Sci. USA 85*, 7972–7976.
14. Klionsky DJ, Abeliovich H, Agostinis P et al (2008) Guidelines for the use and interpretation of assays for monitoring autophagy in higher eukaryotes. *Autophagy. 4*, 151–175.
15. Mizushima N, Yoshimori T, Levine B (2010) Methods in mammalian autophagy research. *Cell 140*, 313–326.
16. Kimura S, Noda T, Yoshimori T (2007) Dissection of the autophagosome maturation process by a novel reporter protein, tandem fluorescent-tagged LC3. *Autophagy. 3*, 452–460.
17. Pankiv S, Clausen TH, Lamark T et al (2007) p62/SQSTM1 binds directly to Atg8/LC3 to facilitate degradation of ubiquitinated protein aggregates by autophagy. *J. Biol. Chem. 282*, 24131–24145.
18. Nicklin P, Bergman P, Zhang B et al (2009) Bidirectional transport of amino acids regulates mTOR and autophagy. *Cell 136*, 521–534.
19. Garcia-Martinez JM, Moran J, Clarke RG et al (2009) Ku-0063794 is a specific inhibitor of the mammalian target of rapamycin (mTOR). *Biochem. J. 421*, 29–42.

Chapter 15

The *In Vivo* Evaluation of Active-Site TOR Inhibitors in Models of BCR-ABL+ Leukemia

Matthew R. Janes and David A. Fruman

Abstract

Preclinical evaluation of candidate anticancer compounds requires appropriate animal models. Most commonly, solid tumor xenograft systems are employed in which immunocompromised mice are implanted with human cancer cell lines. Genetically engineered mouse models of solid tumors are also frequently employed. Both of these approaches can also be applied to studies of hematological malignancies. In this chapter, we describe three types of mouse models of leukemia driven by the human *BCR-ABL* oncogene. We also discuss the application of these models to preclinical testing of active-site TOR inhibitors, a novel class of compounds that selectively target the ATP-binding pocket of the target of rapamycin (TOR) kinase.

Key words: Mammalian target of rapamycin, TORC1, TORC2, Acute lymphoblastic leukemia, BCR-ABL, PP242

1. Introduction

The target of rapamycin (TOR) is a serine/threonine kinase that is an important drug target in various therapeutic areas including transplantation, autoimmune disease, and oncology (1–6). TOR (often referred to as "mTOR") is encoded by a single gene in mammals (*mTOR* in humans) but functions in two distinct cellular complexes, TORC1 and TORC2. These complexes are regulated differently and have a mostly distinct spectrum of substrates. The classical TOR inhibitor, rapamycin, binds to FKBP12 and the complex of rapamycin/FKBP12 inhibits TOR through an allosteric mechanism. It is now appreciated that TORC1 has both rapamycin-sensitive and rapamycin-insensitive substrates and that TORC2 is unaffected by acute cellular exposure to rapamycin (7, 8).

Thomas Weichhart (ed.), *mTOR: Methods and Protocols*, Methods in Molecular Biology, vol. 821,
DOI 10.1007/978-1-61779-430-8_15,

Partial TORC1 inhibition by rapamycin also blocks a negative feedback loop, leading to elevated upstream signaling in many cancer cell contexts. A major recent advance has been the synthesis and development of active-site TOR inhibitors (asTORi) that block TOR activity through an ATP-competitive mechanism (9, 10). This mechanistic difference enables asTORi to fully block all outputs of TORC1 and TORC2. asTORi have improved anticancer activity compared to rapamycin in a variety of solid tumor models (11–13), and at least three candidate compounds have entered clinical trials. Our laboratory has shown that asTORi might be particular effective in leukemia, in that the compounds strongly induce apoptosis and synergize with tyrosine kinase inhibitors that are currently in clinical use (14). Remarkably, asTORi appear to be less immunosuppressive than rapamycin (14). asTORi now represent an attractive therapeutic option in oncology that deserve careful comparison with other compounds that target prosurvival signals mediated by phosphoinositide 3-kinase (PI3K)/AKT/TOR pathway (15).

To model the constitutive activation of the PI3K/AKT/TOR pathway in a system that closely resembles a human disease, we describe mouse models of leukemia driven by the *BCR-ABL* fusion oncogene. BCR-ABL is encoded by the Philadelphia chromosome (Ph), an oncogenic lesion caused by the fusion of two chromosomes, abbreviated as t(9;22)(q34;q11.2). The chimeric fusion of BCR-ABL activates a multitude of progrowth, proliferation and survival signals including PI3K/AKT/TOR and several other parallel pathways (i.e., RAS/MAPK and JAK/STAT). The *BCR-ABL* oncogene is a defining hallmark of the myeloproliferative disorder, chronic myelogenous leukemia (CML), and is also present in a fraction of acute lymphoid leukemias of the B cell lineage, known as Philadelphia-positive pre-B cell acute lymphoblastic leukemia (Ph+ B-ALL) (16). In this chapter we will focus exclusively on three models of Ph+ B-ALL:

1. Mouse p190 BCR-ABL$^+$ progenitor B cell leukemia.
2. Human Ph+ (p190 form) B-ALL SUP-B15 bioluminescent xenograft.
3. Primary human Ph+ acute leukemia xenograft model.

Dependence upon the PI3K/AKT/TOR pathway for survival and maintenance of the transformed phenotype is very strong in all three models we present. Thus, the models can be used to evaluate and compare the efficacy of different drug classes targeted to distinct nodes in the pathway. We highlight the use of PP242 and rapamycin, which represent the active-site and allosteric TOR inhibitor classes, respectively.

2. Materials

2.1. Cell Culture and Virus Preparation

1. p190 BCR-ABL recombinant viral particles (preprepared; see below).
2. RPMI1640 (Gibco) supplemented with 20% fetal bovine serum (FBS, Hyclone). *Abbreviated RP20 media.*
3. Polybrene (hexadimethrine bromide, resuspended in distilled water; Sigma).
4. Recombinant mouse IL-7 (rIL-7; Invitrogen).
5. Human Ph+ ALL cell line SUP-B15 (ATCC# CRL-1929).
6. Iscove's Modified Dulbecco's Medium (IMDM; Gibco).
7. Firefly luciferase recombinant viral particles (preprepared; see below).
8. Cryopreserved bone marrow or peripheral blood from leukemia patients.

2.2. Mice

1. Production of p190 BCR-ABL transformed bone marrow cultures (p190 cells): 3–5 week-old BALB/cJ mice (Jackson Laboratory).
2. Recipient mice for syngeneic transplantation of p190 cells: 8 week-old BALB/cJ mice (Jackson Laboratory).
3. Recipient NSG mice for human leukemia xenografts: 6–8 week-old NOD.Cg-$Prkdc^{scid}$ $Il2rg^{tm1Wjl}$/SzJ (Jackson Laboratory).
4. ACK lysis buffer (0.15 M NH_4Cl, 10 mM $KHCO_3$, 0.1 mM Na_2EDTA in distilled water, adjust to pH 7.2–7.4, sterile filter and store at RT).
5. Ketamine (a controlled substance), at 100 mg/kg (i.p.), Xylazine, at 10 mg/kg (i.p.) in sterile saline.
6. PBS with 5 mM Na_2EDTA.
7. Sterile PBS with 1% FBS.
8. Heparinized microhematocrit capillary tubes (Fisher).
9. (Optional) Gamma irradiator.

2.3. Inhibitors and In Vivo Drug Preparation

1. 2-(4-Amino-1-isopropyl-1 H-pyrazolo[3,4-d]pyrimidin-3-yl)-1 H indol-5-ol (PP242; any commercial source such as Sigma or Chemdea).
2. Imatinib and Dasatinib (ABL tyrosine kinase inhibitors), obtained from LC Labs.
3. Rapamycin (LC Labs).

4. PEG400 (Carbowax polyethylene glycol; Fisher-Scientific).
5. Carboxymethylcellulose sodium salt (medium viscosity, CMC; Sigma-Aldrich).

2.4. In Vivo Bioluminescent Imaging of Disease Burden

1. D-Luciferin firefly, potassium salt (any commercial source).
2. Optical imaging cabinet for rodents (recommend IVIS Lumina or Spectrum, Caliper Life Sciences).

3. Methods

3.1. Mouse Bone Marrow Progenitor Cell Transduction and Establishment of p190 BCR-ABL Transformed Cells

These instructions describe a primary mouse leukemia model that uses retroviral transduction of mouse bone marrow with the p190 form of BCR-ABL to produce a transformed oligoclonal population of progenitor/precursor B cells (also termed p190 cells). p190 cells are a highly tractable *in vitro* system to model Ph+ disease; for example, bone marrow from gene-targeted mice can be used as the starting material. Furthermore, p190 cells efficiently transplant into recipient mice to initiate an aggressive pro-B lymphoid leukemia that closely resembles human Ph+ B-ALL. The model we employ has been adapted from a previously established system described by Van Etten's group and was modified to promote the emergence of a lymphoid leukemia (no 5-FU priming) with elimination of lethal radiation of recipient mice (17). Using these instructions, p190 cells will engraft and emerge with a fatal leukemia in both immunocompromised mice (NOD/SCID), or immunocompetent syngeneic recipients without any preparative radiation. Note that the Van Etten paper also reports a method that is now commonly used to model myeloproliferative disease resembling human CML. In this case, p210 form of BCR-ABL is delivered by retroviruses to bone marrow of 5-FU-treated mice (17).

1. To collect bone marrow for p190 transformation, euthanize a 3-weeks-old BALB/cJ mouse by CO_2 asphyxiation and cervical dislocation. Transformation efficiency decreases with older mice, especially after 6 week of age. Soak mouse with 70% EtOH. Remove skin on legs, cut tibia and femur bones, and dissect off muscle fascia.
2. Using 2 ml of ice-cold PBS, flush the long bones of the tibia and femur using a 22 G needle and syringe by careful ejection of the marrow plug by insertion of the needle into the apical or distal end of the long bone.
3. Centrifuge cell suspension 5 min at $280\times g$, 4°C. Remove supernatant and resuspend BM cells in 0.5 ml of ACK lysis buffer for 0.5–1 min at RT to lyse red blood cells (RBCs).

Typically, there are very few mature RBCs in the young mouse BM, so only a short incubation time is necessary.

4. After ACK lysis, wash BM cells with 1 ml of RP20 media and centrifuge the cell suspension for 5 min at 280 × *g*, 4°C. Remove supernatant and resuspend 0.5–10^7 BM cells in 1 ml of RP20 media in a one well of 6-well plate.
5. Prepare replication-defective viral particles from retroviral vectors carrying p190 BCR-ABL using standard methods. It is optimal to use vectors such as MSCV-IRES-GFP (MIG) in which the retroviral long-terminal repeat (in this case, MSCV-LTR) drives expression of both the BCR-ABL and a marker gene (eGFP) linked by an internal ribosome entry site (IRES). It is not necessary for the marker gene to encode a drug resistance gene, since only BCR-ABL-expressing cells survive long-term in the cultures. The marker gene should encode a fluorescent protein or a surface marker (e.g., human CD4) that can be stained and identified by FACS and fluorescence microscopy.
6. Using previously prepared and cryopreserved p190 BCR-ABL viral particles, thaw and add 2 ml of BCR-ABL retroviral supernatant to the BM suspension. Add 5 μg/ml polybrene, add 10 ng/ml rIL-7, and cosediment the cell/viral suspension by centrifugation for 45 min at 450 × *g* at 37°C. Following cosedimentation of cells and viral suspension, place plate in incubator overnight at 37°C, 5% CO_2.
7. The next morning collect and wash the cell suspension to remove viral supernatant. Centrifuge cell suspension 5 min at 280 × *g*. Aspirate supernatant and resuspend BM cells in 3 ml of RP20 media supplemented with 10 ng/ml rIL-7.
8. Monitor cell culture daily. After 2–3 days split the culture 1:2 into another well in the 6-well plate. Monitor cell cultures every 2–3 days by flow cytometry until a pure 100% marker-positive (eGFP or human CD4), cytokine- and stromal-independent culture is established as previously described (18). See Subheading 4 below for further details.

3.2. In Vivo *Syngeneic Transplant Experiments with Murine p190 BCR-ABL Transformed Cells*

1. Mouse p190 cells are used to initiate leukemia in either sublethally-irradiated (conditioned with 450 radians) or nonirradiated (immunocompetent) syngeneic (Balbc/J) recipients.
2. Prior to injection wash p190 cells with sterile PBS containing 1% FBS.
3. Prepare final cell suspension in sterile PBS at 10^7 cells/ml. Adjust cell concentration accordingly (*see details below*).
4. Anesthetize mouse (or multiple recipients) using ketamine/xylazine.

5. Position the mouse on its side and while restraining it, draw back the skin on either side of the eye with enough pressure so the eye slightly protrudes. Insert needle through the conjunctival membrane of the eye positioning it into the retrobulbar sinus. See Subheading 4 for additional details.
6. Inject 100 μl or 10^6 cells of cell suspension and remove needle gently.
7. Monitor the mouse during the recovery process and daily thereafter.
8. Determine leukemic engraftment in anesthetized animals by blood draws and analyze by flow cytometry following 5–7 days. Injecting 10^6 cells, clinical symptoms should appear within 7–9 days (conditioned recipients) or 12–18 days (nonconditioned recipients). Leukemic cells can be detected in the blood within 5–7 days. Typically within 7 days mice will have 10–30% bone marrow involvement, see Fig.1. Splenomegaly and lymphoadenopathy quickly develop, followed by weight loss, piloerection, and labored breathing. These symptoms can be followed by sudden death, so it is important to monitor mice daily.
9. To confirm leukemia and measure leukemic cell percentages and properties, anesthetize mouse as in step 4.
10. Collect 50 μl of blood from the retroorbital venous sinus plexus with a heparinized microhematocrit capillary tube.
11. Dispense blood into 300 μl of PBS with 5 mM EDTA. Immediately shake.
12. Samples can be stored at RT (for 24 h) until analyzed by flow cytometry or automated hematology analyzer.
13. Euthanize mice when displaying signs of extreme hunched posture with piloerection, labored breathing, or >20% body weight loss.
14. All euthanized animals bearing clinical hallmarks listed above should be necropsied to confirm leukemic engraftment (i.e., splenomegaly, hypercellular bone marrow).

3.3. Human Ph+ (p190 Form) B-ALL SUP-B15 Bioluminescent Xenograft Model

This model uses a human acute lymphoblastic leukemia cell line, SUP-B15, derived from an 8-year-old child with Ph+ B-ALL (ATCC# CRL-1929). SUP-B15 cells propagate well *in vitro* and efficiently engraft into adult NOD/SCID-$\gamma c^{-/-}$ (NSG) recipient mice to produce a fatal B-lymphoid leukemia that readily presents in both primary and secondary lymphoid organs including the central nervous system, which is a common feature of human Ph+ B-ALL. Retroviral transduction of this cell line with the firefly luciferase gene provides an excellent reporter for repeated bioluminescent imaging.

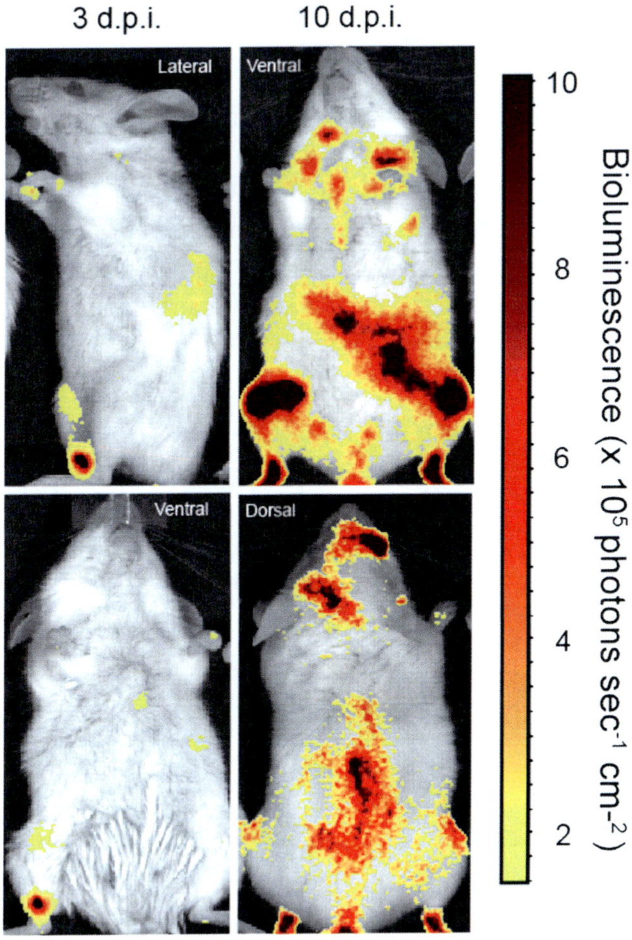

Fig. 1. Temporal and spatial engraftment of syngeneic p190 BCR-ABL leukemia model. p190 cells were infected with a retrovirus containing luciferase to track *in vivo* engraftment, and sites of leukemic involvement. Bioluminescent imaging was used as described in Subheading 3.3. Note the initial engraftment in spleen, bone marrow, and thymus, with disease progression to local lymph nodes, liver, and central nervous system. The right orbital plexus was used for the injection site, where typically a local ocular lymphoma develops behind the eye. Shown is a representative mouse at 10 days postinjection (d.p.i.) with systemically advanced disease.

1. It is possible to generate luciferase-expressing SUP-B15 cells by infection with lentiviruses or with amphotropic retroviruses. In our published bioluminescent imaging studies we have used SUP-B15 cells infected with an eGFP–ffLuc lentiviral (pHIV7 backbone) plasmid (a kind gift from Michael Jensen, City of Hope, Duarte, CA), termed SUP-B15^{ffLuc} cells.
2. SUP-B15^{ffLuc} cells are cultured in RP20. Typically, 1:5 splits every 3–4 days are suitable for propagation.
3. Inject 6–8 week-old female unconditioned NSG mice (i.v., retroorbital) with 2.5×10^6 SUP-B15^{ffLuc} cells.

4. Assess total body leukemia burden by bioluminescent imaging (BLI) 2–3 weeks post transplant. *Typically at 14–21 days disease is detectable in the bone marrow, CNS, spleen, thymus, and peripheral lymph nodes. Advanced disease ensues within 4–6 weeks following transplantation.*
5. Imaging is performed by injection of D-Luciferin, potassium salt at 150 mg/kg (i.p.) in awake animals.
6. Following 7 min of distribution, anesthetize mice with ketamine/xylazine solution. Within 2–3 min mice should be anesthetized.
7. Place animals on the imaging platform (IVIS Lumina, Caliper Life Sciences) in either prone or supine position. Image in groups of three animals or more depending on imaging apparatus. Following 10–12 min after D-Luciferin injection, image mice using a 15 s exposure. After image, mice can be flipped to obtain measurements in both angles. see Subheading 4 for further details.
8. Determine regions of bioluminescence (BLI) around each corresponding animal and the total photon flux/sec using Living Image software (Caliper Life Sciences).
9. Following assessment of total body BLI, randomize mice into treatment groups.
10. Initiate treatment design.
11. Perform sequential total body BLI again every 3–7 days by repeating steps 5–8, or as shown in Fig. 2.
12. Euthanize mice when displaying signs of extreme hunched posture with piloerection, labored breathing, or >20% body weight loss.

3.4. Primary Human Ph+ Acute Leukemia Xenograft Model

This model uses primary human patient samples obtained either from cryopreserved bone marrow aspirates or from the peripheral blood (e.g., leukophoresis or whole blood) of patients collected during the course of routine clinical care or as part of an IRB-approved specimen banking protocol. Using adult NOD/SCID-$\gamma c^{-/-}$ (NSG) mice as recipients, injection of purified leukemia cells produces a disease that is readily detectable within 30–90 days (average of 40 days) following transplantation. NSG mice that succumb to disease exhibit the same morphologic and phenotypic features of the original patient's disease with systemic involvement of both primary and secondary lymphoid organs as displayed in Fig. 3. The severity, aggressiveness, and latency of the disease varies per patient specimen, but is highly reproducible across multiple recipients capable of achieving a fatal leukemic burden. Assuming good viability of the cells post-thaw, on average ~80% of Ph+ B-ALL patient samples engraft using the methods we describe. Serial transplantation of the disease can also be used to propagate the patient

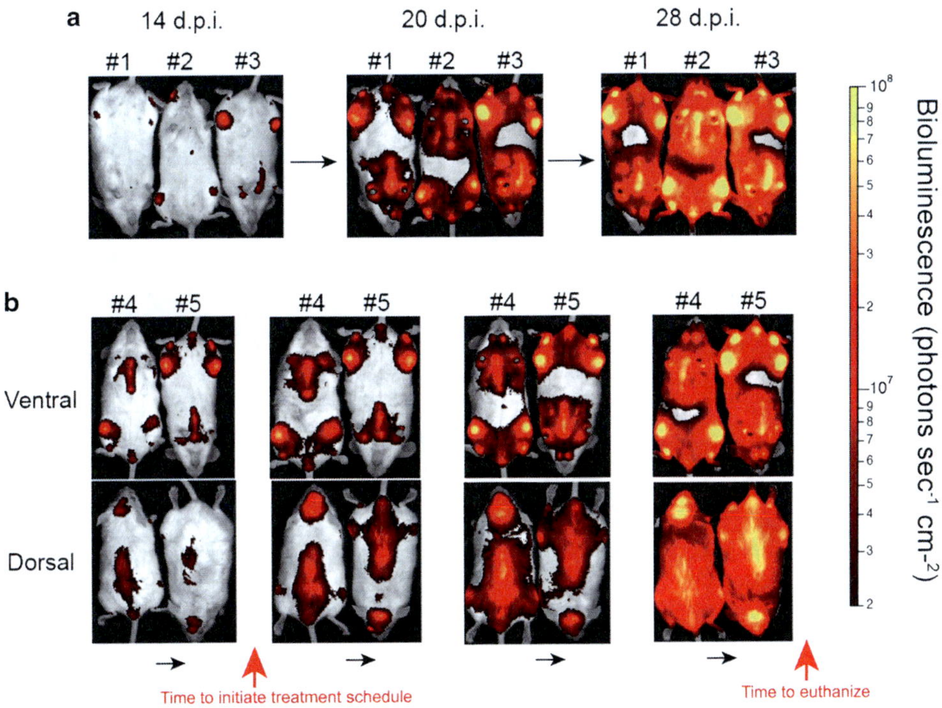

Fig. 2. *In vivo* dynamics of SUP-B15 xenograft model. (**a**) Temporal progression of disease in three representative NSG recipients in supine position (*ventral view*). (**b**) An example of experimental design to incorporate a treatment schedule in two representative NSG recipients. Following confirmation of leukemic burden, assign treatment groups and initiate treatment (*first red arrow*), and monitor leukemic burden through successive bioluminescent imaging. Euthanize mice that display signs of advanced disease.

sample *in vivo*. Thus, it is sometimes convenient to inject a small cohort of mice to maintain and expand the samples and then perform secondary transplants on those specimens that engraft. This model can also be adapted for other acute leukemias harboring other chromosomal translocations.

1. Thaw mononuclear cells isolated from human bone marrow or peripheral blood samples in a 37°C water bath for 2–3 min or until thawed.
2. Wash samples with IMDM containing 5% FCS.
3. Centrifuge cell suspension 5 min at 280×*g*, 4°C. Remove supernatant and resuspend cells in IMDM containing 5% FCS.
4. Perform manual cell counts. Samples with <80% cell viability should be centrifuged over Ficoll to remove dead cell debris (Optional).
5. (Optional for Peripheral blood samples) centrifuge cell suspension over a Ficoll-paque plus (GE Healthcare) layer using standard procedures.

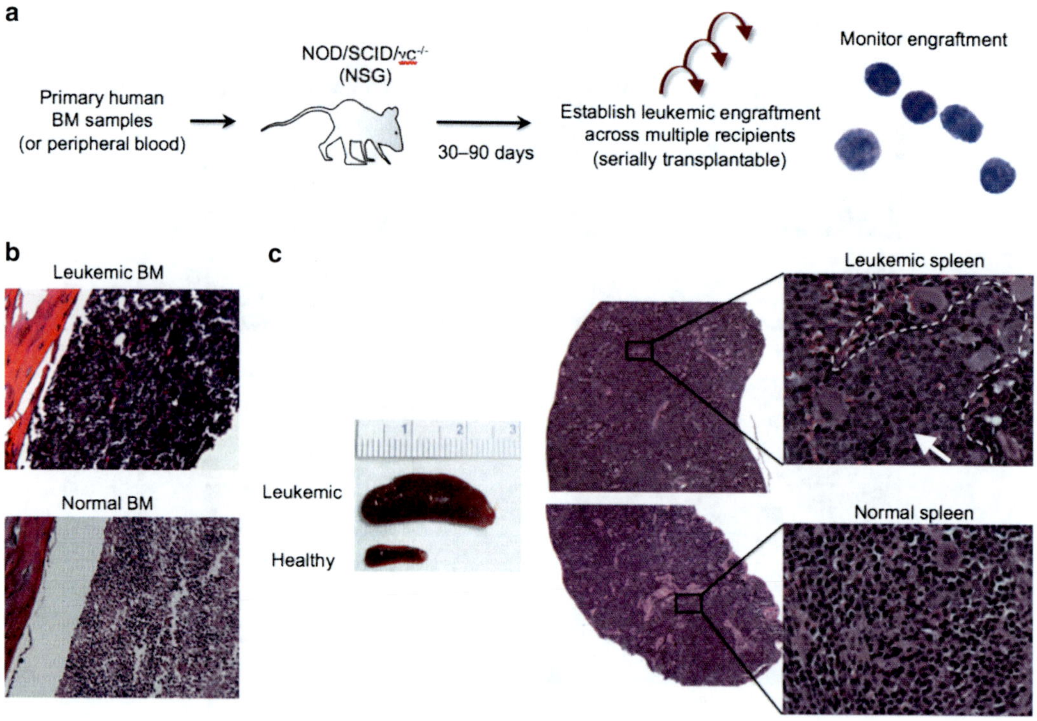

Fig. 3. Development of primary human ALL xenograft model. (**a**) Depicts the basic experimental setup for engraftment and monitoring of disease while maintaining the patient specimen by serial transplantation. (**b**) Representative images of bone marrow (BM) involvement. Note the hypercellularity of leukemic occupancy in the femur. (**c**) Representative gross pathology of the spleen depicting splenomegaly. *White arrow* depicts cytologic features of leukemic cells displaying large nuclei containing open chromatin.

6. Analyze samples by flow cytometry to determine the percentage of leukemia in sample. CD34+ CD19+ cells generally demark a consistent leukemic population in Ph+ B-ALL disease. See Subheading 4 for additional details.
7. (OPTIONAL) Removal of CD3+ T cells by positive selection using standard magnetic methods. The CD3 antigen is expressed on T cells and NKT cells.
8. Centrifuge cell suspension 5 min at 280×*g*, 4°C. Remove supernatant and resuspend cells in PBS. Keep samples on ice until ready for injection into NSG recipients. *Inject recipients within 1 h of thaw.*

3.4.1. Animal Preparation

1. NSG mice (or RAG-deficient, IL2R$\gamma^{-/-}$) are the best recipients for primary human samples. Using nonirradiated NSG mice, inject leukemic samples (an equivalent amount of $0.3–1\times10^6$ cells per recipient) resuspended in PBS retroorbitally as described in Subheading 3.2 step 4.
2. Following 30 days, check for leukemic engraftment in anesthetized animals by retroorbital bleeds or in awake animals by

puncture of the submandibular facial vein (see Subheading 4 for further details on facial vein access).

3. Analyze by flow cytometry as mentioned in step 5.
4. Euthanize mice upon signs of extreme hunched posture with piloerection, labored breathing, or weight loss.

3.5. (Optional) In Vivo Drug Preparations

These instructions describe the formulation and dosing regimens of PP242 and rapamycin. PK/PD and efficacy are further described in Janes et al. (14).

1. Calculate the quantity of PP242 to formulate for the entire interim of 1 week. (Note upon formulation of PP242, it will be stable for 1 week).
2. Dissolve PP242 in PEG400 (Carbowax polyethylene glycol; Fisher-Scientific).
3. Bath-sonicate the material (up to 5 min) to aid it going into suspension.
4. Aliquot suspension into 5 ml conical tubes and protect from light, stored at RT.
5. Dose PP242 by oral gavage at 10 ml/kg of body weight with sterile disposable 20 G-1.5" feeding needles at efficacious doses.
6. Calculate the quantity of rapamycin to formulate for the entire interim of 1 week. (Note upon formulation of rapamycin, it will be stable for 1 week).
7. Dissolve rapamycin in 0.5–2% CMC diluted in PBS.
8. Extensive bath-sonication will be required to bring rapamycin into solution.
9. Aliquot suspension into 5 ml conical tubes and protect from light, stored at RT.
10. Dose rapamycin by intraperitoneal injection at 10 ml/kg of body weight with a sterile insulin syringe at efficacious doses.

4. Notes

1. Approximately 0.5–10^7 bone marrow (BM) cells are usually harvested from one mouse.
2. Following the initial plating of bone marrow cell suspension and p190 BCR-ABL retroviral infection, within 24–48 h an adherent stromal layer will develop and support the precursor lymphoblasts during the transformation process. Typically within 1–2 weeks and two to three replatings (passages), a pure polyclonal suspension culture will be established. The emerging

leukemic (p190+) cells will outcompete the growth of the adherent stromal layer. Eventually through a couple of cell culture passages, the stromal layer will be diluted out – achieving a pure lymphoid leukemic suspension culture.

3. Supplementing rIL-7 during the initial stages of p190 transformation is not essential for the generation of p190 cells, but it accelerates the leukemic transformation.

4. We recommend to passage p190 cells using RP20 media every 2–3 days. An initial plating density of 0.5–10^6 cells per ml is suitable for continuous log phase growth and a cell density of $3–4 \times 10^6$ per ml is maximally achievable without significant effects on viability. p190 cells typically double every 18–24 h under optimal conditions in log phase growth. RPMI supplemented with 10% FBS can replace the 20% supplemented media after a stromal-independent culture has established (after ~2–3 weeks).

5. It is recommended to use freshly prepared p190 cultures (<4 weeks old) to initiate leukemia in recipients. Using sublethally irradiated syngeneic recipients allows for both quicker and more aggressive disease compared to nonirradiated. Note that C57/Bl6 mice can also be substituted for generation of p190 cells and recipient mice accordingly.

6. For further information regarding retroorbital injection, and IACUC/LARC standard operating procedures see the following Web site for details: http://www.iacuc.ucsf.edu/policies/awspretroorbitalinjection.asp.

7. Less than 10^6 can be injected for a more latent disease. Injecting 100–500 cells is sufficient to initiate disease. Although not recommended, injecting more than 10^6 cells will initiate an aggressive leukemia within 4–5 days.

8. To ensure total body widespread engraftment of SUP-B15 cells, NSG mice must be used over standard NOD-SCID mice. Even using sublethally irradiated (250 radians) NOD-SCID mice, only localized cervical lymphomas with minor periocular involvement develop (data not shown).

9. Exposure time for measuring total body bioluminescence in the SUP-B15^{ffLuc} model may vary depending on imaging platform, calibration, and/or luciferase expression. We recommend performing a luciferin substrate kinetic study with the SUP-B15^{ffLuc} cell line to determine peak signal time after luciferin administration. For further details on performing a kinetic study, refer to the following Web site for details: http://www.caliperls.com/assets/017/7559.pdf.

10. For further information regarding taking facial vein and submandibular blood samples from mice, refer to the following

Web site: http://www.medipoint.com/html/for_use_on_mice.html. Note that goldenrod animal lancets will be required.

11. Staining for the presence of human T lymphocytes within primary leukemic samples should be performed. Staining with antibodies to CD3, CD4, and CD8 using standard flow cytometry procedures is recommended. Using the above instructions with samples containing less than 5% of human T cells will be sufficient to not produce graft versus host disease in the recipient mice.
12. Surface phenotypic markers that demark human leukemias can be variable. We recommend using the side scatter (or orthogonal) parameter by CD45 staining to reveal lymphoblastic populations. Ph+ B-ALL blasts typically express CD34, CD10, CD19, with occasional expression of CD38 and CD20 (19–21). Surface immunoglobulin receptors should be absent. Following injection of leukemic sample, monitor leukemic engraftment by staining for similar immunophenotypic (antihuman) markers by flow cytometry. We recommend staining mouse blood draws or mouse bone marrow samples at necropsy with anti-mouse CD45 together with anti-mouse Ter119 to gate out all mouse cells to reveal any human cell presence. To take precaution, we also recommend to use a dead-cell indicator (i.e., 7-AAD or propidium iodide) and acquire an appropriate amount of cells to discern any minimal residual disease where appropriate at the time of necropsy.
13. Note that if mice display ill posture, and loss of weight within 2 weeks of injection, draw their blood and analyze by flow cytometry to confirm leukemic engraftment. The animals lacking engraftment may either (a) be sick with infection or parasite or (b) exhibit graft versus host disease. In either case, the animals should be euthanized.

Acknowledgments

The authors would like to thank Christian Rommel, Yi Liu, Pingda Ren, and Troy Wilson for the chemical synthesis and technical support with PP242 before it was commercially available. We also thank Andrew Miller for technical advice with NSG mice and Collin Vu for experimental assistance with primary human xenografts. We thank Marina Konopleva and Michael Lilly for access to primary human leukemia samples. Studies of TOR inhibitors in our laboratory have been supported by Intellikine, Inc., and by a Discovery Grant from the University of California Industry-University Cooperative Research Program.

References

1. Sparks, C. A., and Guertin, D. A. (2010) Targeting mTOR: prospects for mTOR complex 2 inhibitors in cancer therapy, *Oncogene 29*, 3733–3744.
2. Janes, M. R., and Fruman, D. A. (2010) Targeting TOR dependence in cancer, *OncoTarget 1*, 69–76.
3. Bhagwat, S. V., and Crew, A. P. (2010) Novel inhibitors of mTORC1 and mTORC2, *Curr Opin Investig Drugs 11*, 638–645.
4. Guertin, D. A., and Sabatini, D. M. (2009) The pharmacology of mTOR inhibition, *Sci Signal 2*, pe24.
5. Thomson, A. W., Turnquist, H. R., and Raimondi, G. (2009) Immunoregulatory functions of mTOR inhibition, *Nat Rev Immunol 9*, 324–337.
6. Weichhart, T., and Saemann, M. D. (2009) The multiple facets of mTOR in immunity, *Trends Immunol 30*, 218–226.
7. Sarbassov, D. D., Ali, S. M., Kim, D. H., Guertin, D. A., Latek, R. R., Erdjument-Bromage, H., Tempst, P., and Sabatini, D. M. (2004) Rictor, a novel binding partner of mTOR, defines a rapamycin-insensitive and raptor-independent pathway that regulates the cytoskeleton, *Curr Biol 14*, 1296–1302.
8. Thoreen, C. C., and Sabatini, D. M. (2009) Rapamycin inhibits mTORC1, but not completely, *Autophagy 5*, 725–726.
9. Feldman, M. E., Apsel, B., Uotila, A., Loewith, R., Knight, Z. A., Ruggero, D., and Shokat, K. M. (2009) Active-site inhibitors of mTOR target rapamycin-resistant outputs of mTORC1 and mTORC2, *PLoS Biol 7*, e38.
10. Thoreen, C. C., Kang, S. A., Chang, J. W., Liu, Q., Zhang, J., Gao, Y., Reichling, L. J., Sim, T., Sabatini, D. M., and Gray, N. S. (2009) An ATP-competitive mammalian target of rapamycin inhibitor reveals rapamycin-resistant functions of mTORC1, *J Biol Chem 284*, 8023–8032.
11. Yu, K., Shi, C., Toral-Barza, L., Lucas, J., Shor, B., Kim, J. E., Zhang, W. G., Mahoney, R., Gaydos, C., Tardio, L., Kim, S. K., Conant, R., Curran, K., Kaplan, J., Verheijen, J., Ayral-Kaloustian, S., Mansour, T. S., Abraham, R. T., Zask, A., and Gibbons, J. J. (2010) Beyond rapalog therapy: preclinical pharmacology and antitumor activity of WYE-125132, an ATP-competitive and specific inhibitor of mTORC1 and mTORC2, *Cancer Res 70*, 621–631.
12. Yu, K., Toral-Barza, L., Shi, C., Zhang, W. G., Lucas, J., Shor, B., Kim, J., Verheijen, J., Curran, K., Malwitz, D. J., Cole, D. C., Ellingboe, J., Ayral-Kaloustian, S., Mansour, T. S., Gibbons, J. J., Abraham, R. T., Nowak, P., and Zask, A. (2009) Biochemical, cellular, and in vivo activity of novel ATP-competitive and selective inhibitors of the mammalian target of rapamycin, *Cancer Res 69*, 6232–6240.
13. Chresta, C. M., Davies, B. R., Hickson, I., Harding, T., Cosulich, S., Critchlow, S. E., Vincent, J. P., Ellston, R., Jones, D., Sini, P., James, D., Howard, Z., Dudley, P., Hughes, G., Smith, L., Maguire, S., Hummersone, M., Malagu, K., Menear, K., Jenkins, R., Jacobsen, M., Smith, G. C., Guichard, S., and Pass, M. (2010) AZD8055 is a potent, selective, and orally bioavailable ATP-competitive mammalian target of rapamycin kinase inhibitor with in vitro and in vivo antitumor activity, *Cancer Res 70*, 288–298.
14. Janes, M. R., Limon, J. J., So, L., Chen, J., Lim, R. J., Chavez, M. A., Vu, C., Lilly, M. B., Mallya, S., Ong, S. T., Konopleva, M., Martin, M. B., Ren, P., Liu, Y., Rommel, C., and Fruman, D. A. (2010) Effective and selective targeting of leukemia cells using a TORC1/2 kinase inhibitor, *Nat Med 16*, 205–213.
15. Vu, C., and Fruman, D. A. (2010) Target of rapamycin signaling in leukemia and lymphoma, *Clin Cancer Res 16*, 5374–5380.
16. Sawyers, C. L. (1999) Chronic myeloid leukemia, *N Engl J Med 340*, 1330–1340.
17. Li, S., Ilaria, R. L., Jr., Million, R. P., Daley, G. Q., and Van Etten, R. A. (1999) The P190, P210, and P230 forms of the BCR/ABL oncogene induce a similar chronic myeloid leukemia-like syndrome in mice but have different lymphoid leukemogenic activity, *The J Exp Med 189*, 1399–1412.
18. Kharas, M. G., Janes, M. R., Scarfone, V. M., Lilly, M. B., Knight, Z. A., Shokat, K. M., and Fruman, D. A. (2008) Ablation of PI3K blocks BCR-ABL leukemogenesis in mice, and a dual PI3K/mTOR inhibitor prevents expansion of human BCR-ABL+ leukemia cells, *J Clin Invest 118*, 3038–3050.
19. Cox, C. V., Evely, R. S., Oakhill, A., Pamphilon, D. H., Goulden, N. J., and Blair, A. (2004) Characterization of acute lymphoblastic leukemia progenitor cells, *Blood 104*, 2919–2925.

20. Castor, A., Nilsson, L., Astrand-Grundstrom, I., Buitenhuis, M., Ramirez, C., Anderson, K., Strombeck, B., Garwicz, S., Bekassy, A. N., Schmiegelow, K., Lausen, B., Hokland, P., Lehmann, S., Juliusson, G., Johansson, B., and Jacobsen, S. E. (2005) Distinct patterns of hematopoietic stem cell involvement in acute lymphoblastic leukemia, *Nat Med 11*, 630–637.

21. le Viseur, C., Hotfilder, M., Bomken, S., Wilson, K., Rottgers, S., Schrauder, A., Rosemann, A., Irving, J., Stam, R. W., Shultz, L. D., Harbott, J., Jurgens, H., Schrappe, M., Pieters, R., and Vormoor, J. (2008) In childhood acute lymphoblastic leukemia, blasts at different stages of immunophenotypic maturation have stem cell properties, *Cancer Cell 14*, 47–58.

Chapter 16

Inducible *raptor* and *rictor* Knockout Mouse Embryonic Fibroblasts

Nadine Cybulski, Vittoria Zinzalla, and Michael N. Hall

Abstract

The mammalian Target of Rapamycin (mTOR) kinase functions within two structurally and functionally distinct multiprotein complexes termed mTOR complex 1 (mTORC1) and mTORC2. The immunosuppressant and anticancer drug rapamycin is commonly used in basic research as a tool to study mTOR signaling. However, rapamycin inhibits only, and only incompletely, mTORC1, and no mTORC2-specific inhibitor is available. Hence, a full understanding of mTOR signaling *in vivo*, including the function of both complexes, requires genetic inhibition in addition to pharmacological inhibition. Taking advantage of the Cre/LoxP system, we generated inducible knockout mouse embryonic fibroblasts (MEFs) deficient for either the mTORC1-specific component raptor (iRapKO) or the mTORC2-specific component rictor (iRicKO). Inducibility of the knockout was important because mTOR complex components are essential. Induction of either *raptor* or *rictor* knockout eliminated raptor or rictor expression, respectively, and impaired the corresponding mTOR signaling branch. The described knockout MEFs are a valuable tool to study the full function of the two mTOR complexes individually.

Key words: mTOR, Raptor, Rictor, Signaling, Mouse embryonic fibroblasts, Knockout, iRapKO, iRicKO

1. Introduction

The Target of Rapamycin (TOR) is an evolutionary conserved serine/threonine kinase that plays a central role in the control of cell growth and metabolism (1). TOR functions in two distinct multiprotein complexes termed TOR complex 1 (TORC1) and TORC2 (2). The core components of mammalian TORC1 (mTORC1) are mTOR, mLST8, and raptor (2–4). mTORC1 is activated by nutrients (amino acids), growth factors (e.g., insulin and IGF) and cellular energy status to control several cellular processes including transcription, protein synthesis, ribosome biogenesis, lipid synthesis,

Thomas Weichhart (ed.), *mTOR: Methods and Protocols*, Methods in Molecular Biology, vol. 821,
DOI 10.1007/978-1-61779-430-8_16, © Springer Science+Business Media, LLC 2012

nutrient transport, and autophagy (5, 6). The best-characterized substrate of mTORC1 is ribosomal protein S6 kinase (S6K), which mTORC1 phosphorylates at Thr389 within a hydrophobic motif to activate S6K and protein synthesis. mTORC1 also phosphorylates four sites in the translational inhibitor 4E-BP. The immunosuppressant and anticancer drug rapamycin inhibits mTORC1 kinase activity, but it has been suggested that this inhibition is incomplete as rapamycin fails to inhibit mTORC1 toward all its target sites in 4E-BP (7–9). mTORC2 contains mTOR, mLST8, rictor, and mSin1 and is resistant to rapamycin (10–15). mTORC2 controls cell survival and actin cytoskeleton organization in response to growth factors. The best-characterized substrates of mTORC2 are AGC kinase family members including Akt/PKB, PKCα and SGK1. mTORC2 phosphorylates Akt at Ser473 (16–18).

To date, the study of mTOR function in cultured cells has relied on rapamycin treatment or RNAi-mediated knockdown to inhibit mTOR signaling. However, neither approach is ideal because neither provides complete inhibition. Knockdown is never complete and residual protein levels could be sufficient to provide at least partial function. As discussed above, rapamycin appears to inhibit only mTORC1 and only incompletely. Consequently, experiments that have relied on rapamycin or knockdown to query mTORC1 or mTORC2 function might have missed important insights.

In contrast to rapamycin treatment and RNAi-mediated knockdown, genetic knockout completely eliminates a specific protein and its associated function. Thus, knockout MEFs lacking an mTORC1- or mTORC2-specific component are particularly valuable to study mTOR signaling in cultured cells. However, the isolation of such MEFs is problematic. The mTORC1 and mTORC2 components mTOR, mLST8, raptor, rictor, and mSin1 are all essential. In mice, knockout of *mLST8*, *rictor*, and *mSin1* results in developmentally delayed embryos that die by embryonic day 10.5–11.5. Mice with a deletion of *mTOR* or *raptor* die even earlier, by embryonic day 5.5 (11, 15, 19–21). This complicates the isolation of viable knockout MEFs. Furthermore, MEFs are isolated as early as possible, which, according to standard practice, is embryonic day 10.5–13.5. This precludes obtaining the desired mTOR-related knockout MEFs even if they could be viably maintained. To circumvent these problems, we resorted to isolating inducible knockout MEFs.

To study the roles of mTORC1 and mTORC2 in specific tissues, we generated tissue-specific *raptor* or *rictor* knockout mice using the Cre/LoxP system (22–25). In addition to studying the physiological roles of mTORC1 and mTORC2, we used the "floxed" *raptor* and *rictor* mice to generate inducible knockout MEFs. A so-called floxed allele is a functional version of a gene flanked by loxP sites that are target sites for Cre recombinase. Expression of Cre recombinase results in targeted deletion of the floxed gene. To generate inducible knockout MEFs, primary MEFs

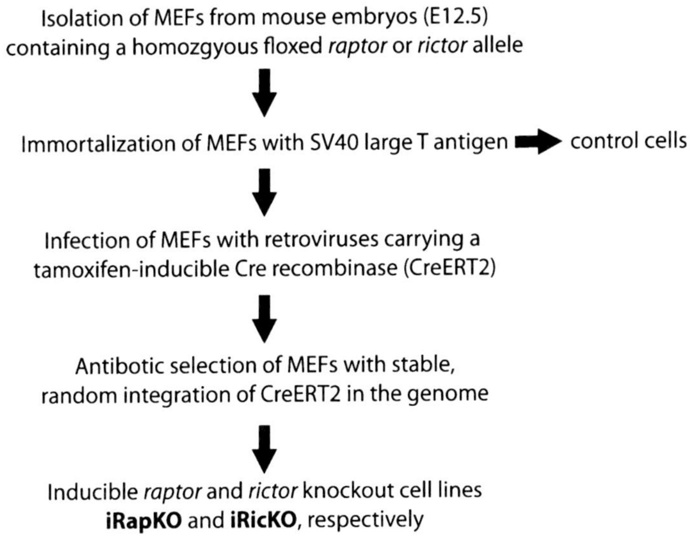
Isolation of MEFs from mouse embryos (E12.5)
containing a homozgyous floxed *raptor* or *rictor* allele
Immortalization of MEFs with SV40 large T antigen
control cells
Infection of MEFs with retroviruses carrying a
tamoxifen-inducible Cre recombinase (CreERT2)
Antibotic selection of MEFs with stable,
random integration of CreERT2 in the genome
Inducible *raptor* and *rictor* knockout cell lines
iRapKO and **iRicKO**, respectively

Fig. 1. Workflow illustrating the generation of the inducible *raptor* and *rictor* knockout MEF cell lines, which were named iRapKO and iRicKO, respectively.

were isolated from mice containing a homozygous floxed *raptor* or *rictor* allele. After immortalization with simian virus 40 (SV40) large T antigen, MEFs were infected with a retrovirus carrying a tamoxifen-inducible Cre recombinase (CreERT2), and MEFs containing a stable integration of the retrovirus were selected (Fig. 1). CreERT2 is a hybrid protein containing mutated estrogen receptor fused to Cre recombinase. Expressed CreERT2 is activated within the cell upon addition of the synthetic estrogen receptor ligand 4-hydroxytamoxifen (4-OHT) (26, 27). Thus, deletion of a floxed gene can be induced by addition of 4-OHT to cells. The inducible *raptor* and *rictor* MEF lines were named iRapKO and iRicKO, respectively.

Here, we describe in detail the generation and validation of the iRapKO and iRicKO cell lines. We show induced loss of mTORC1 and mTORC2 signaling in iRapKO and iRicKO cells, respectively. These cell lines are a powerful tool to study the functions of mTORC1 and mTORC2 (28).

2. Materials

2.1. Isolation of MEFs

1. Pregnant female mice at day 13 postcoitum (p.c.). Mice are homozygous for floxed *raptor* or *rictor* (23–25).
2. Sterile scissors and forceps for dissection.
3. 70% (v/v) ethanol.

4. Phosphate buffered saline (PBS) (Sigma), sterile.
5. MEF medium: Dulbecco's Modified Eagle's Medium (DMEM) (Sigma), supplemented with 10% fetal bovine serum (FBS, HyClone), penicillin/streptomycin (Gibco), and L-glutamine (Gibco).
6. 100 mm culture dishes.
7. 1 ml syringes and needles of the following size: 25 G × 5/8 in.; 0.5 mm × 16 mm.
8. 0.25% Trypsin–EDTA solution (Gibco).

2.2. Immortalization of MEFs

1. MEF medium (see Subheading 2.1.)
2. Six-well culture plates.
3. SV40 Large T-antigen expression vector.
4. jetPEI™ transfection reagent (Polyplus transfection).
5. 150 mM NaCl (autoclaved, tissue culture quality).

2.3. Retrovirus Production

1. Phoenix Cell line: ecotropic Phoenix retroviral producer line.
2. MEF medium (see Subheading 2.1.5).
3. CreERT2 cloned into pMSCVpuro expression vector (called CreERT2-pMSCVpuro vector).
4. Lipofectamine 2000 transfection reagent (Invitrogen, see Note 7).
5. Opti-MEM Reduced Serum Medium (Gibco).
6. 10-ml syringe, 0.45-μm filter.

2.4. Retroviral Infection of MEFs and Selection

1. Polybrene (also known as Hexadimethrine bromide) stock solution (5 mg/ml): dissolve 10 mg of polybrene (Sigma) in 2 ml 0.9% NaCl and store at 2–8°C. The solution is stable for 1 year. The final concentration of polybrene is 5 μg/ml.
2. Puromycin stock solution (3 mg/ml): dissolve 10 mg of puromycin (Sigma) in 3.33 ml of deionized water. Sterilize by filtration. Aliquot and store at −20°C. The final concentration of puromycin in the MEF medium is 3 μg/ml.

2.5. Induction of Knockout

1. 4-Hydroxytamoxifen (4-OHT) stock solution (2 mM): dissolve 5 mg of 4-OHT (Sigma) in 6.45 ml of 95% ethanol (a warm water bath may be needed to dissolve). Aliquot and store at −20°C. Keep protected from light. The final concentration of 4-OHT in the MEF medium is 1 μM.

2.6. Protein Extraction

1. Phosphate buffered saline (PBS) (Sigma).
2. TNE lysis buffer: 50 mM Tris–HCl, pH 8.0, 150 mM NaCl, 0.5 mM EDTA, 1% Triton X-100. Store at 2–8°C.

3. Protease Inhibitors stock solution: Dissolve one protease inhibitor cocktails tablet (Roche) in 2 ml of deionized water. Store at –20°C.
4. TNE lysis buffer with protease inhibitors: Add 1 volume of Protease Inhibitor stock solution to 25 volumes of TNE lysis buffer. Prepare freshly when needed and keep on ice.
5. Cell scraper.

3. Methods

3.1. Isolation of MEFs

1. Sacrifice pregnant female mouse at day 13 p.c. by cervical dislocation. Remove the uterus containing the embryos with sterile scissors and forceps. Briefly rinse the isolated uterus in 70% ethanol and place in a 100 mm dish containing PBS.
2. Dissect out the individual embryos and remove any extra-embryonic tissues. Using forceps decapitate the embryos and remove the liver (dark red organ). Rinse in PBS to remove any blood. The following steps should be performed in a sterile tissue culture hood. Each embryo should be processed individually (see Note 1).
3. Transfer the embryo to a 1-ml syringe filled with MEF medium. Put the needle on to the syringe and gently push with the plunger the contents of the syringe into a new 100-mm culture dish. Remove the needle. To disperse the embryo, draw up the MEF medium containing the embryo into the syringe, put back the needle on to the syringe and push the contents into the culture dish. Repeat these steps two to three times until the clumps are dissociated and a homogenous suspension is made. Add MEF medium for a total volume of 10 ml. Place the dish into a humidified 37°C tissue culture incubator. The MEFs are now at passage 0.
4. On the following day, change the medium to remove any cell debris and nonfibroblast cells that do not adhere to the culture dish. Medium may be changed frequently in the beginning to help remove the cell debris.
5. Cells must be monitored to determine when splitting is required. The cells should be split when they are ~80–100% confluent (usually within 2 days after the isolation).
6. For splitting cells, remove the medium, rinse with PBS, and then trypsinize (add 2 ml of 0.25% Trypsin–EDTA solution per 100 mm dish). Incubate for approximately 5 min in the 37°C incubator until cells start to detach from the plate. Inactivate trypsin by adding 3 ml of MEF medium. Pipette up

and down several times to fully detach the cells from the plate and to dissociate cells from each other. Centrifuge the cells for 5 min at 800 × *g*. Aspirate supernatant, resuspend cells in fresh MEF medium and replate in 10 mm dishes at a 1:3 and 1:5 dilution. The total volume should be 10 ml per dish. The MEFs are now at passage 1 after isolation.

3.2. Immortalization of MEFs

1. Immortalization is performed with MEFs at passage 2–4 after isolation (see Note 2). One day prior to transfection split primary MEFs into 6 well plates so that the cells will be 50–70% confluent at the time of transfection.
2. Transfect one well with 2 μg of SV40 Large T-antigen expression vector using jetPEI™ transfection reagent and following the manufacturer's protocol (see Notes 3 and 5):
   - Dilute 2 μg of SV40 Large T-antigen expression vector in 150 mM NaCl to a final volume of 100 μl (DNA solution).
   - Dilute 4 μl of jetPEI™ transfection reagent in 150 mM NaCl to a final volume of 100 μl (jetPEI™ solution).
   - Add the 100 μl jetPEI™ solution to the 100 μl DNA solution and mix gently.
   - After 15–30 min of incubation at room temperature add the combined jetPEI™/DNA mix (200 μl) dropwise to the MEFs. Mix gently by rocking the plate back and forth. Incubate cells in the 37°C incubator.
   - Change the medium 6 h after transfection.
3. When cells are confluent (~2 days after transfection) split all the cells into a 100-mm dish. Now, the cells are at passage 1 after immortalization.
4. The cells should be split when they are ~80% confluent. The goal is to have cells that have been split 1:10 at least five times (passage 6 after immortalization), which corresponds to a 1:100,000-fold splitting of the MEFs at passage 1 after immortalization (see Note 4). By passage 6 after immortalization, the cells should reach 80–100% confluency within 3–4 days after a 1:10 split.

3.3. Retrovirus Production

1. One day prior to transfection, plate ~5 million Phoenix cells in 10 ml of MEF medium in to a 100-mm dish (The cells should be ~70% confluent at the time of transfection).
2. Transfect one 100 mm dish with 24 μg of CreERT2-pMSCVpuro vector using lipofectamine 2000 transfection reagent and following the manufacturer's protocol (see Note 5):
   - Dilute 24 μg of CreERT2-pMSCVpuro vector in Opti-MEM to a final volume of 1.5 ml.
   - Dilute 60 μl of lipofectamine 2000 in Opti-MEM to a final volume of 1.5 ml. Incubate for 5 min at room temperature.

- Combine the diluted DNA with the diluted lipofectamine 2000. Mix gently and incubate for 20 min at room temperature.
- Add the 3 ml of DNA–lipofectamine solution dropwise to the cells. Mix gently by rocking the plate back and forth.
- Incubate cells in the 37°C incubator.
- Change the medium 6 h after the transfection.

3. Optional: 1 day after transfection, transfer the plates to a 32°C incubator (virus particles are more stable at lower temperature).
4. Two days after transfection, collect the retroviral supernatant from the transfected Phoenix cells. Filter the retroviral supernatant through a 0.45 μm filter to remove any cell debris (see Note 6).

3.4. Retroviral Infection of MEFs and Selection

1. One day prior to infection, plate ~1–2 million immortalized MEFs at passage 6 or higher after immortalization in 10 ml of MEF medium in a 100 mm dish. The cells should be ~70% confluent at the time of infection.
2. On the day of infection, remove the medium from the MEFs. Add 5 ml of the retroviral supernatant and 5 μl of polybrene (the final concentration of polybrene is 5 μg/ml) (see Note 7).
3. Optional: centrifuge the dish containing the retroviral supernatant and the MEFs at 1,000 × *g* for 1 h at room temperature.
4. Incubate the infected MEFs in a 37°C incubator for 6 h before adding 5 ml of MEF medium to have a final volume of 10 ml.
5. The next day, remove the supernatant and add fresh MEF medium. Further incubate the infected MEFs in a 37°C incubator for an additional day.
6. Three days after infection, start puromycin treatment to select for MEFs that have been infected. Add 10 μl of puromycin stock solution to the infected MEFs (the final concentration in the MEF medium is 3 μg/ml). Add puromycin also to noninfected MEFs as a control. Noninfected control MEFs must all be dead within 2 or 3 days after puromycin selection, while infected MEFs should survive under puromycin selection (~90% efficiency of infection is expected; see Note 8). Infected MEFs have now stably integrated the CreERT2.
7. After puromycin selection, stocks of the infected MEFs should be prepared and stored in liquid nitrogen.
8. While culturing the infected MEFs, they should be regularly treated with puromycin to ensure stable integration of CreERT2.

3.5. Induction of Knockout

Time course analysis after 4-OHT treatment of the iRapKO and iRicKO MEFs revealed that raptor and rictor proteins were no longer detectable after 3 days of 4-OHT treatment (Fig. 2). Therefore,

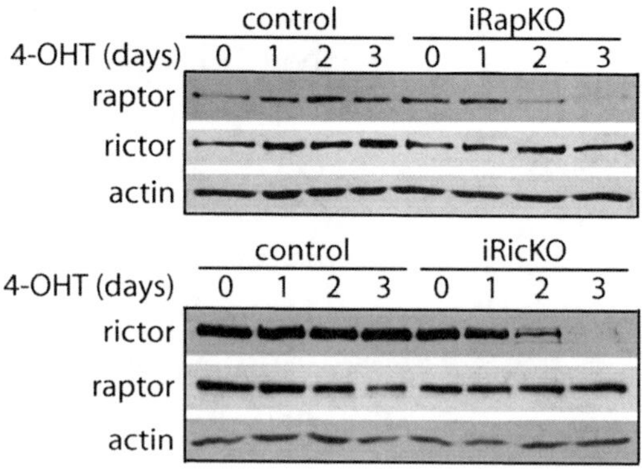

Fig. 2. Time course analysis of 4-OHT-induced loss of raptor and rictor protein expression in iRapKO (top) and iRicKO (bottom) MEFs, respectively. Controls correspond to lysates from cells that contain either *raptor* or *rictor* floxed alleles, but do not contain CreERT2. The cells were treated with 4-OHT for 0, 1, 2, or 3 days, respectively, and subsequently lysed with TNE lysis buffer containing protease inhibitors. Equal amounts of denatured protein lysate (normally 20–30 μg) were analyzed by SDS-PAGE and proteins investigated by Western blot analysis. The following antibodies were used: raptor, rictor (both Cell Signaling), and actin (Chemicon).

it is recommended to pretreat the MEFs with 4-OHT for at least 3 days to induce the knockout. Untreated iRapKO or iRicKO MEFs (no 4-OHT treatment) should be used as a control.

Since 4-OHT can also bind to the endogenous estrogen receptor, it is important to perform a control experiment to exclude any 4-OHT-induced effect that is not related to the induction of the knockout of *raptor* or *rictor*. Immortalized MEFs that were not infected with the CreERT2 expressing retroviruses should also be used as a control.

Induce the knockout 3 days before the planned experiment. Split cells to have the desired confluency after 3 days of 4-OHT treatment. Add the appropriate amount of 4-OHT (the final concentration of 4-OHT in the MEF medium should be 1 mM) to the cells directly after splitting. The medium does not need to be changed afterward and no additional 4-OHT needs to be added during the next 3 days (see Notes 9 and 10).

3.6. Protein Extraction

1. Remove the medium from the MEFs and wash with cold PBS.
2. Add ice-cold TNE lysis buffer with protease inhibitors to the cells (add 200–400 μl of lysis buffer to one 100 mm dish). Harvest the cells using a cell scraper and transfer the cell lysate to a 1.5-ml Eppendorf tube. Always keep the cell lysate on ice.
3. Rotate tubes gently for 10–20 min at 4°C.
4. Remove cell debris by centrifugation at 1,000 × *g* at 4°C for 5 min.
5. Transfer the supernatant to a new tube. If protein lysate is not used immediately, store at –80°C.

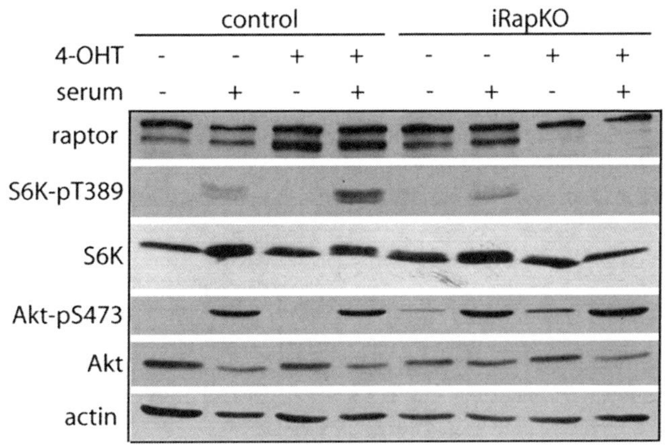

Fig. 3. Serum stimulated mTORC1 signaling is impaired in iRapKO MEFs after induction of *raptor* knockout. 4-OHT treated cells were cultured in the presence of 4-OHT for 3 days. On day 3 cells were serum-starved for 6 h and subsequently treated with 10% serum, or not, for 30 min. The cells were lysed with TNE lysis buffer containing protease inhibitors. Equal amounts of denatured protein lysate (normally 20–30 μg) were analyzed by SDS-PAGE and proteins investigated by Western blot analysis. The following antibodies were used: raptor, S6K-pT389, S6K, Akt-pS473 (all Cell Signaling), Akt (Santa Cruz), and actin (Chemicon). 4-OHT treated iRapKO cells show a strong reduction in S6K Thr389 phosphorylation confirming the specific loss of mTORC1 activity in iRapKO cells. An unspecific band (*upper band*) was detected with the anti-raptor antibody.

6. For western-blot analysis, 20–30 μg of denatured protein lysate is analyzed by SDS-PAGE followed by immunoblotting with the appropriate antibody (Figs. 2–4).

4. Notes

1. For isolation of MEFs, each embryo is initially cultured separately. At later stages, MEFs deriving from different embryos, but with the same genetic background, can be pooled.
2. After initial passages, the growth rate of primary MEFs can be strongly reduced. Replace medium every other day if not passaged.
3. To be sure that transfection with the SV40 Large T-antigen expression vector has been efficient, a control transfection can be performed in parallel with the same procedure using a plasmid expressing GFP. If cells are transfected they should express GFP the following day, which can be visualized by fluorescent microscopy.
4. If only a low number of cells were initially transfected with SV0 large T antigen, only few surviving cells may be obtained from the first 1:10 splits. Therefore, it is recommended to split the MEFs at a low density (1:10) and high density (1:4) at each

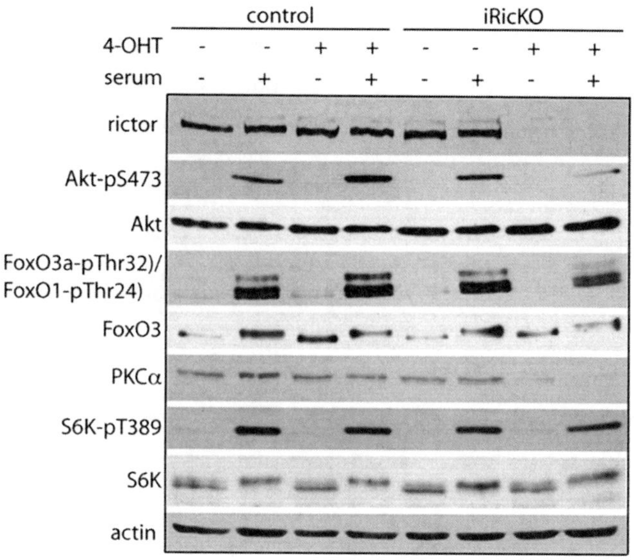

Fig. 4. Serum stimulated mTORC2 signaling is impaired in iRicKO MEFs after induction of *rictor* knockout. 4-OHT treated cells were cultured in the presence of 4-OHT for 3 days. On day 3 cells were serum-starved for 6 h and subsequently treated with 10% serum, or not, for 30 min. The cells were lysed with TNE lysis buffer containing protease inhibitors. Equal amounts of protein lysate (normally 20–30 μg) were analyzed by SDS-PAGE and proteins investigated by Western blot analysis. The following antibodies were used: rictor, S6K-pT389, S6K, Akt-pS473, FoxO3a-pT32/FoxO1-pT24, FoxO3, PKCα (all Cell Signaling), Akt (Santa Cruz), and actin (Chemicon). Loss of *rictor* results in a strong reduction of Akt Ser473 phosphorylation that also leads to a decrease in phosphorylation of FoxO1/3a, two Akt substrates, whose phosphorylation is affected when Akt is not phosphorylated at the hydrophobic motif (20). In agreement with previous findings, loss of mTORC2 activity also leads to a decrease in PKCα protein levels (11, 20, 29). As expected, in 4-OHT-treated iRicKO cells mTORC1 activity is unaffected upon insulin stimulation as observed by unchanged S6K phosphorylation.

passage. During the immortalization process the 1:10 split is a strong selection against nontransformed cells which can usually survive many splits at higher density (1:4). If the MEFs in the low density (1:10) plate are growing, then the cells of the high density (1:4) plate can be frozen as a backup or discarded. Often the MEFs in the high density (1:4) plate must be split before you can be sure that the MEFs in the low density (1:10) plate are growing (the MEFs in the low density (1:10) plates of the first two passages after immortalization may require 3–10 days to reach 80% confluency). Even if you keep high density (1:4) plates as a backup, you will only select for the nontransformed cells by multiple rounds of 1:10 splitting.

5. Any other lipid based transfection reagent you are familiar with can be used.

6. Supernatant can be stored at −80°C for later infection, however, the virus titer decreases by half for each freeze–thaw cycle.

Aliquots should be made to avoid freeze–thaw cycles. The retroviral supernatants collected from different dishes transfected with the same retroviral vector can be pooled before aliquoting.

7. Polybrene strongly increases the retroviral infection efficiency. If infection efficiency is low, MEFs can also be pretreated with polybrene for 1–4 h prior to the viral infection.
8. It is possible that the initial puromycin concentration (3 μg/ml) is too high, causing also death of infected cells. Therefore, the concentration in the medium should be adjusted from MEFs to MEFs (usually between 1.5 and 3 μg/ml). The minimal concentration to use can be determined by treating noninfected cells with several concentrations of puromycin. The concentration, which causes death of all noninfected cells within 2–3 days, should be taken.
9. In the case of the iRapKO and iRicKO MEFs, the cells can be trypsinized and split up to 2 days after 4-OHT treatment. At later time points, cells will not adhere anymore.
10. Keep in mind that MEFs will have reduced growth rates after 4-OHT treatment due to loss of mTORC1 or mTORC2 function, and will consequently grow slower than the untreated control cells. Therefore, it is necessary to split the MEFs for 4-OHT treatment at higher density than the untreated MEFs. Splitting conditions need to be adapted to the individual experimental setup.

Acknowledgments

We thank Dr. David Ron (New York University School of Medicine, New York, USA) and Dr. Patrick Matthias (Friedrich-Miescher-Institute Basel, Switzerland) for providing plasmids. We acknowledge support from the Swiss National Science Foundation, the Swiss Cancer League, the Louis Jeantet Foundation, and the Canton of Basel.

References

1. Wullschleger S, Loewith R, Hall MN (2006) TOR signaling in growth and metabolism. Cell 124:471–484
2. Loewith R, Jacinto E, Wullschleger S *et al* (2002) Two TOR complexes, only one of which is rapamycin sensitive, have distinct roles in cell growth control. Mol Cell 10:457–468
3. Kim D H, Sarbassov D D, Ali SM *et al* (2002) mTOR interacts with raptor to form a nutrient-sensitive complex that signals to the cell growth machinery. Cell 110:163–175
4. Hara K, Maruki Y, Long X *et al* (2002) Raptor, a binding partner of target of rapamycin (TOR), mediates TOR action. Cell 110: 177–189
5. Soulard A, Hall M N (2007) SnapShot: mTOR signaling. Cell 129:434
6. Laplante M, Sabatini D M (2009) mTOR signaling at a glance. J Cell Sci 122:3589–3594
7. Choo A Y, Yoon S O, Kim S G *et al* (2008) Rapamycin differentially inhibits S6Ks and 4E-BP1 to mediate cell-type-specific repression

of mRNA translation. Proc Natl Acad Sci USA 105:17414–17419

8. Feldman M E, Apsel B, Uotila A *et al* (2009) Active-site inhibitors of mTOR target rapamycin-resistant outputs of mTORC1 and mTORC2. PLoS Biol 7:e38
9. Thoreen C C, Kang S A, Chang J W *et al* (2009) An ATP-competitive mammalian target of rapamycin inhibitor reveals rapamycin-resistant functions of mTORC1. J Biol Chem 284:8023–8032
10. Cybulski N, Hall M N (2009) TOR complex 2: a signaling pathway of its own. Trends Biochem Sci 34:620–627
11. Jacinto E, Facchinetti V, Liu D *et al* (2006) SIN1/MIP1 maintains rictor-mTOR complex integrity and regulates Akt phosphorylation and substrate specificity. Cell 127:125–137
12. Jacinto E, Loewith R, Schmidt A *et al* (2004) Mammalian TOR complex 2 controls the actin cytoskeleton and is rapamycin insensitive. Nat Cell Biol 6:1122–1128
13. Sarbassov D D, Ali S M, Kim D H *et al* (2004) Rictor, a novel binding partner of mTOR, defines a rapamycin-insensitive and raptor-independent pathway that regulates the cytoskeleton. Curr Biol 14:1296–1302
14. Frias M A, Thoreen C C, Jaffe J D *et al* (2006) mSin1 is necessary for Akt/PKB phosphorylation, and its isoforms define three distinct mTORC2s. Curr Biol 16:1865–1870
15. Yang Q, Inoki K, Ikenoue T *et al* (2006) Identification of Sin1 as an essential TORC2 component required for complex formation and kinase activity. Genes Dev 20:2820–2832
16. Garcia-Martinez J M, Alessi D R (2008) mTOR complex 2 (mTORC2) controls hydrophobic motif phosphorylation and activation of serum- and glucocorticoid-induced protein kinase 1 (SGK1). Biochem J 416:375–385
17. Hresko R C, Mueckler M (2005) mTOR. RICTOR is the Ser473 kinase for Akt/protein kinase B in 3T3-L1 adipocytes. J Biol Chem 280:40406–40416
18. Sarbassov D D, Guertin D A, Ali SM *et al* (2005) Phosphorylation and regulation of Akt/PKB by the rictor-mTOR complex. Science 307:1098–1101
19. Gangloff Y G, Mueller M, Dann S G *et al* (2004) Disruption of the mouse mTOR gene leads to early postimplantation lethality and prohibits embryonic stem cell development. Mol Cell Biol 24:9508–9516
20. Guertin D A, Stevens D M, Thoreen C C *et al* (2006) Ablation in mice of the mTORC components raptor, rictor, or mLST8 reveals that mTORC2 is required for signaling to Akt-FOXO and PKCalpha, but not S6K1. Dev Cell 11:859–871
21. Shiota C, Woo J T, Lindner J *et al* (2006) Multiallelic disruption of the rictor gene in mice reveals that mTOR complex 2 is essential for fetal growth and viability. Dev Cell 11:583–589
22. Nagy A (2000) Cre recombinase: the universal reagent for genome tailoring. Genesis 26:99–109
23. Bentzinger C F, Romanino K, Cloetta D *et al* (2008) Skeletal muscle-specific ablation of raptor, but not of rictor, causes metabolic changes and results in muscle dystrophy. Cell Metab 8:411–424
24. Cybulski N, Polak P, Auwerx J *et al* (2009) mTOR complex 2 in adipose tissue negatively controls whole-body growth. Proc Natl Acad Sci USA 106:9902–9907
25. Polak P, Cybulski N, Feige J N *et al* (2008) Adipose-specific knockout of raptor results in lean mice with enhanced mitochondrial respiration. Cell Metab 8:399–410
26. Feil R, Brocard J, Mascrez B *et al* (1996) Ligand-activated site-specific recombination in mice. Proc Natl Acad Sci USA 93:10887–10890
27. Leone D P, Genoud S, Atanasoski S *et al* (2003) Tamoxifen-inducible glia-specific Cre mice for somatic mutagenesis in oligodendrocytes and Schwann cells. Mol Cell Neurosci 22:430–440
28. Patursky-Polischuk I, Stolovich-Rain M, Hausner-Hanochi M *et al* (2009) The TSC-mTOR pathway mediates translational activation of TOP mRNAs by insulin largely in a raptor- or rictor-independent manner. Mol Cell Biol 29:640–649
29. Facchinetti V, Ouyang W, Wei H *et al* (2008) The mammalian target of rapamycin complex 2 controls folding and stability of Akt and protein kinase C. EMBO J 27:1932–1943

Chapter 17

Expanding Human T Regulatory Cells with the mTOR-Inhibitor Rapamycin

Manuela Battaglia, Angela Stabilini, and Eleonora Tresoldi

Abstract

$CD4^+CD25^+FOXP3^+$ T regulatory (Treg) cells are pivotal for the induction and maintenance of peripheral tolerance in both mice and humans. The possibility to use Treg cells for the treatment of T-cell-mediated diseases has recently gained increasing momentum. However, given the limited amount of circulating $FOXP3^+$ Treg cells, efficient methods for their ex vivo expansion are highly desirable. Rapamycin allows for *in vitro* expansion of murine and human $FOXP3^+$ Treg cells, which maintain their regulatory phenotype and suppressive capacity. Here, we describe in detail the powerful methods for enriching human $FOXP3^+$ Treg cells starting from unfractionated $CD4^+$ T cells or for expanding $CD25^+$-enriched Treg cells in the presence of rapamycin.

Key words: Immunological tolerance, Regulatory T cells, Rapamycin

1. Introduction

The immune system is a rather complicated network of many different players who interact with each other and cooperate to protect against diseases. The immune system is fine-tuned to distinguish antigens that belong to the body from those that do not, allowing it to swiftly deploy a potent array of defense mechanisms whenever evidence of a foreign invasion is found. Among the many immune-system players, there are the T regulatory (Treg) lymphocytes, which act to suppress immune activation and thereby maintain immune homeostasis and tolerance to self- and nonself nonharmful antigens. The "Treg-cell team" comprises many players (1) including the $CD4^+$ naturally occurring Treg cells, which represent about 5–10% of total $CD4^+$ T lymphocytes in the periphery both in mice

Thomas Weichhart (ed.), *mTOR: Methods and Protocols*, Methods in Molecular Biology, vol. 821,
DOI 10.1007/978-1-61779-430-8_17, © Springer Science+Business Media, LLC 2012

and men. This cell subset is defined by the high expression of the IL-2Rα-chain (CD25), the transcription factor forkhead box P3 both in mice (Foxp3) and humans (FOXP3) the inability to produce interleukin-2 (IL-2) and to proliferate *in vitro* (2). IL-2 is crucial for the generation, expansion, survival, and effector function of the FOXP3^{+} Treg cells (3).

The possibility to use Treg cells for the treatment of T-cell-mediated diseases has recently gained increasing momentum (4). Given the limited amount of circulating FOXP3^{+} Treg cells, efficient and safe procedures for the ex vivo expansion of this cell subset is of crucial importance. However, these cell-culture conditions are also highly advantageous for the expansion of effector T-cell populations, which can contaminate the purified Treg-cell population. In humans, even the CD4^{+}CD25^{++} T-cell subset contains ~50% of *in vivo* recently activated T cells, which ultimately outgrow Treg cells after prolonged culturing *in vitro*. Thus, the risk of coexpanding potentially harmful cells is the greatest concern regarding the clinical use of ex vivo expanded Treg cells. We have found that the mTOR inhibitor rapamycin significantly reduces the undesired expansion of effector T cells, as rapamycin selectively allows the proliferation of Treg cells but inhibits the proliferation of effector T cells (5, 6). This selective activity of rapamycin seems to be connected to two effects. First, FOXP3^{+} Treg cells do not downregulate phosphatase and tensin homologue (PTEN) expression after TCR engagement, which impedes the activation of the rapamycin-susceptible PI3K–AKT–mTOR pathway. Second, FOXP3 drives the expression of PIM2, reinforced by IL-2- and TCR-mediated activation of signal transducer and activator of transcription 5 (STAT5); PIM2 compensates for AKT inactivity and promotes cell cycle progression (7).

2. Materials

1. D-PBS 1× w/o Calcium and Magnesium.
2. Lymphoprep density 1.077 g/ml and osmolality 290 mOsm (Axis-shield).
3. AC SOLUTION: Ammonium chloride solution (Stem Cell).
4. Store Buffer: D-PBS 1× + 2% pooled AB Human Serum.
5. ACD SOLUTION: Store Buffer + 1% Natrium Citrate 4%.
6. CD4^{+} T cell isolation kit II (Miltenyi Biotec).
7. Anti human CD8^{+} and CD25^{+} MicroBeads (Miltenyi Biotec).
8. Working medium: X-VIVO 15 medium + 5% pooled AB Human Serum.

9. Anti-CD3 mAb (clone OKT3). Dilute in 0.1 M Tris pH 7.5 at the final concentration of 10 μg/ml and store in ice until needed.
10. Anti-CD28 mAb (clone CD28.2). Dilute in working medium at the final concentration of 10 μg/ml and store in ice until needed.
11. Anti-CD3/CD28 mAb-coated T Cell Expander® beads (Dynal Biotech ASA).
12. Rapamycin: dissolve 1 mg of rapamycin powder in 547 μl Et-OH to have a stock at 2 mM. Aliquot (20 μl) and store at –80°C for up to 3 months. When needed, thaw and dilute 1:100 in working medium (20 μM stock) and store at +4°C for up to 1 week. Use 5 μl of rapamycin Stock Solution for 1 ml of culture (final concentration 100 nM).
13. Rapamune: dilute 25 μl of Rapamune® drug (Whyet) in 1,338 μl PBS 1×. Store this solution (20 μM stock) for 1 week at +4°C (see Note 7). When needed use 5 μl of Rapamune Stock Solution for 1 ml of culture (final concentration 100 nM). Once the Rapamune drug is open, the batch can be stored at +4°C for 3 months.
14. IL-2: dissolve IL-2 powder with RPMI to have a stock at 500 U/μl. Aliquot and store at –20°C. When needed, thaw and dilute 1:10 in working medium (50 U/μl) and store at +4°C for up to 1 week.
15. 5-(and-6)-carboxyfluorescein diacetate succinimidyl ester (CFSE): Dissolve 25 mg CFSE powder (Molecular Probes) in 2.24 ml of DMSO to have a stock of 20 mM. Aliquote (4 μl) and store at –20°C. When needed, thaw and dilute 4 μl of CFSE 20 mM in 80 μl PBS 1× (CFSE 1 mM). Add 50 μl CFSE 1 mM in 10 ml of PBS 1× (CFSE 5 μM) and store in ice until needed. NOTE: Protect CFSE from the light.
16. PBS/HS: D-PBS 1× + pooled AB Human Serum (2:1 volume).
17. PBS-FACS: D-PBS 1× supplemented with 1% Bovine Serum Albumin and 0.1% NaN_3.
18. FIX SOLUTION: 1% Formaldehyde solution in PBS-FACS.
19. Surface staining: anti-CD4 PerCp (clone SK3), anti-CD25 APC (clone 2A3), anti-CD62L PE (clone DREG-56), anti-CD8 APC (clone SK1), and anti-CD127 PE (clone R34.34).
20. Intracellular staining: anti-FOXP3-Alexa 488 mAb (clone 259D), FOXP3 Fix/Perm Buffer and FOXP3 Perm Buffer, all from BioLegend.
21. Freezing solution: 10% dimethylsulfoxide (DMSO) + 90% Fetal Bovine Serum.

3. Methods

The use of rapamycin in cell cultures for the expansion of FOXP3^{+} Treg cells is highly recommended to allow for a selective expansion of T cells with regulatory properties while inhibiting proliferation and survival of non Treg cells. Protocols for the enrichment and expansion of human FOXP3^{+} starting from unfractionated CD4^{+} T lymphocytes and for the expansion of human enriched CD25^{+} Treg cells have been published (5, 6, 8) and they are separately described in detail in this chapter.

3.1. Peripheral Blood Mononuclear Cell (PBMC) Isolation

1. The following procedure is described as working with 20 ml of whole peripheral blood. Adjust volumes and tubes accordingly. Transfer 20 ml whole human blood into a 50-ml tube. Dilute the blood 1:2 with PBS 1× and mix slowly.
2. Centrifuge at 150 × *g*, for 15 min, NO BRAKE at room temperature. Remember to turn off the brake.
3. Gently eliminate the upper phase (diluted PLASMA) and discard.
4. Gently resuspend the remaining cells adjusting the volume to 30 ml with PBS 1×.
5. Slowly stratify the 30 ml of diluted blood in a new 50-ml tube already filled with 15 ml Lymphoprep.
6. Centrifuge at 600 × *g*, for 30 min, NO BRAKE at room temperature. Remember to turn off the brake.
7. Transfer in a new tube the white blood cell ring fraction (PBMC) formed between two phases. Fill the tube with PBS 1× and centrifuge at 300 × *g*, for 10 min, at room temperature. NOTE: this centrifugation is necessary to fully wash out the Lymphoprep, which is toxic to the cells.
8. Discard the supernatant and resuspend well the cell pellet in 3 ml of sterile AC solution. Let the cells rest for 5 min at room temperature. NOTE: this step is necessary to eliminate the eventually contaminant red cells.
9. Fill the tube with PBS 1× and centrifuge at 150 × *g*, for 15 min, at room temperature. NOTE: low speed centrifugation is necessary to maintain platelets and deposits in solution.
10. Discard the supernatant and repeat as per Subheading 3.1, step 9 until the supernatant appears clean.
11. Resuspend the cell pellet at 10×10^6/ml in Store Buffer and proceed to Subheading 3.2 or keep the cells at +4°C for a maximum of 2 hours. NOTE: in case the PBMC will be used the following day, resuspend the cells at 10×10^6/ml in ACD solution. NOTE: always cryopreserve 100×10^6 PBMC for later use (see Subheading 3.8).

3.2. $CD4^+$ T Cell Purification

1. The following procedure is described as working with the $CD4^+$ T cell isolation kit II (Miltenyi Biotec) by negative selection according to the manufacturer's instructions. However, alternative approaches can be used.
2. Centrifuge the PBMC obtained at Subheading 3.1, step 11 at $300 \times g$, for 7 min, at room temperature.
3. Discard supernatant and resuspend the cell pellet in MACS medium. Centrifuge at $300 \times g$, for 7 min, at +4°C.
4. Discard supernatant and resuspend 10×10^6 cells in 40 μl of MACS medium (see Note 1). Add 10 μl of Biotin-Antibody Cocktail. Scale up volume according to cell number. Mix well and incubate for 10 min at +4°C.
5. Add 30 μl of MACS medium per 10×10^6 cells + 20 μl of Anti-biotin MicroBeads. Scale up volume according to cell number. Mix well and incubate for 15 min at +4°C.
6. Wash the cells by adding 2 ml of MACS medium (centrifuge at $300 \times g$, for 7 min, at +4°C) and resuspend the cells in 1 ml MACS medium up to 100×10^6 total cells. Purify the cells on magnetic column accordingly to the manufacture's instruction.
7. Resuspend the purified cells in Working medium and count the cells. NOTE: use 50,000 cells to determine the $CD4^+$ T-cell purity by flow cytometry (see Subheading 3.7); cryopreserve at least 10×10^6 purified $CD4^+$ T cells and all the $CD4^-$ cells (see Subheading 3.8).

3.3. Cell culture for the Enrichment and Expansion of Human $FOXP3^+$ Treg Cells Starting from $CD4^+$ T Cells

1. The purified $CD4^+$ T cells are polyclonally activated with anti-CD3 and anti-CD28 mAbs in the presence or absence of rapamycin for three rounds of stimulation (1 week each round). T_{MED} refers to the cells cultured without rapamycin, T_{RPM} refers to the cells cultured in the presence of rapamycin. The anti-CD3 mAb is used coated on the plate (see Note 2) while the anti-CD28 mAb is used soluble. Based on the amount of cells used, choose the plate accordingly: 48 well/plates for $1–1.5 \times 10^6$ cells (max 1 ml volume); 24 well/plates for $1.6–2.5 \times 10^6$ cells (max 2 ml medium); 12 well/plates for $2.6–4 \times 10^6$ cells (max 4 ml medium); 6 well/plates for up to 4×10^6 cells (max 8 ml medium). A detailed scheme depicting the T cell culture is shown in Fig. 1.
2. Day 0 (first stimulation): Resuspend the purified $CD4^+$ T cells (see Subheading 3.2, step 6) at $1–1.5 \times 10^6$/ml in Working medium containing the anti-CD28 mAb at the final concentration of 1 μg/ml. Plate the cells in previously anti-CD3 coated wells (see Note 2). For T_{RPM} cells, add 100 nM rapamycin (i.e., 100 ng/ml) (see Note 3).
3. Day 7 (second stimulation): Harvest and count the cells. Change completely the medium and resuspend the cells at

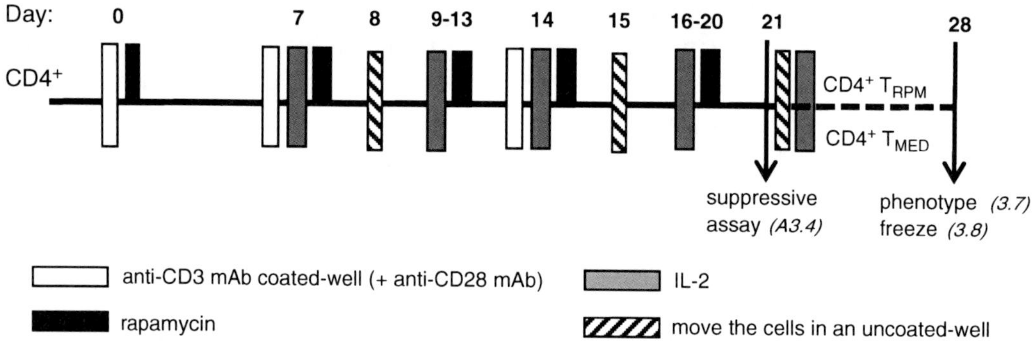

Fig. 1. The diagram outlines the time points (day) and the procedures followed for the ex vivo expansion of CD4⁺ T cells. T_{MED} are the cells expanded in the absence of rapamycin, and T_{RPM} are the cells expanded in the presence of rapamycin.

$1–1.5 \times 10^6$/ml in Working medium containing anti-CD28 mAb at the final concentration of 1 μg/ml and IL-2 at 50 U/ml. Plate the cells as described before in new previously anti-CD3 coated wells (see Note 2). For T_{RPM} cells, add 100 nM rapamycin (i.e., 100 ng/ml).

4. Day 8: Transfer the cells and their medium in a new uncoated well (see Note 4).
5. Carefully monitor the medium and the cells during the culture. When more than 10×10^6/ml of cells are present and/or when the medium turns yellow, split the cells. Add new IL-2 based on the final volume (since IL-2 is rapidly consumed by the cells), while add new rapamycin in T_{RPM} cells based on the new volume added (since rapamycin is not consumed by the cells).
6. Day 14 (third stimulation): Harvest and count the cells. Perform the third stimulation as described in Subheading 3.3, step2.
7. Day 15: Transfer the cells in a new plate as described in Subheading 3.3, step 3 and follow the cell culture as per Subheading 3.3, step 4.
8. Day 21: Harvest and count the cells (see Note 5). Keep at least 2×10^6 cells to perform the suppressive assay as described in Subheading 3.6.
9. Resuspend the remaining T_{MED} and T_{RPM} cells at 10×10^6 cells/ml in Working medium + 50 U/ml IL-2 and plate in uncoated wells (see Note 6).
10. Day 28: At the end of the 7-day resting phase, collect all the T_{MED} and T_{RPM} cells. Perform the phenotype analysis by flow cytometry (see Subheading 3.7) and, if needed, freeze the remaining cells (see Subheading 3.8) for future use.

3.4. CD25⁺ T-Cell Purification

1. The following procedure is described as working with the Miltenyi beads, according to the manufacturer's instructions.

However, alternative approaches can be used. In addition, the procedure assumes the use of the autoMACS™ Separation System, but it can be easily adaptable to the manual isolation with regular columns. Prior to start, select the optimal cell separation approach following the scheme provided by Miltenyi; the correct choice depends on the isolation strategy and on the frequency of magnetically labeled cells.

2. Centrifuge the PBMC obtained in Subheading 3.1, step 11 at 300 × *g*, for 7 min, at room temperature.
3. Resuspend the pellet in autoMACS™ Running Solution and count the cells.
4. *CD8+ T-cell depletion*: $CD8^+$ MicroBeads are used according to the manufacturer's protocol. Resuspend 10×10^6 cells in 80 μl autoMACS™ Running Solution + 20 μl $CD8^+$ MicroBeads. Scale up volumes according to cell number. Mix well and incubate for 15 minutes at +4°C.
5. Centrifuge at 500 × *g*, for 7 min, at +4°C.
6. Discard the supernatant and resuspend the cells in 1 ml autoMACS™ Running Solution for 100×10^6 total cells. In case working with less than 100×10^6 cells, always use 1 ml.
7. Purify the labeled cells on the autoMACS™ Separation System using the "DEPLETES" program.
8. Collect both $CD8^+$ and $CD8^-$ cells. Fill the tube with autoMACS™Running Solution and centrifuge at 300 × *g*, for 7 min, at +4°C. Discard the supernatant.
9. Count the cells and keep 50,000 $CD8^-$ cells to determine the T-cell recovery and the purity (see Subheading 3.4, step 16).
10. *CD25+ T cells positive selection*: $CD25^+$ MicroBeads are used according to the manufacturer's protocol. Resuspend 10×10^6 $CD8^-$ cells (see Subheading 3.4, step 7) in 90 μl autoMACS™ Running Solution + 10 μl $CD25^+$ MicroBeads. Scale up volumes according to cell number. Mix well and incubate for 15 min at +4°C.
11. Centrifuge at 300 × *g*, for 7 min, at +4°C.
12. Discard the supernatant. Resuspend the cells in 1 ml autoMACS™ Running Solution for 100×10^6 total cells. In case working with less than 100×10^6 cells, always use 1 ml.
13. Purify the labeled cells on the autoMACS™ Separation System using the "POSSELD2" program.
14. Collect both $CD8^-CD25^-$ and $CD8^-CD25^+$ cells.
15. Fill the tube with autoMACS™Running Solution and centrifuge at 300 × *g*, for 7 min, at +4°C. Discard the supernatant.
16. Resuspend the purified cells in Working medium and count. Keep the necessary $CD8^-CD25^+$ cell number to perform the expansion (see Subheading 3.5).

17. Take 0.5×10^6 $CD8^-$ (see Subheading 3.4, step 8) and $CD8^-CD25^+$ cells (see Subheading 3.4, step 14) to determine the T-cell recovery (see Note 7) and the purity by flow cytometry (see Subheading 3.7).
18. Cryopreserve the remaining purified $CD8^-CD25^+$ cells (see Subheading 3.4, step 14) and all the $CD8^-CD25^-$ cells (see Subheading 3.4, step 14) and the $CD8^+$ T cells (see Subheading 3.4, step 7) for future use.

3.5. Cell Culture for the Enrichment and Expansion of Human FOXP3⁺ Tregs Starting from CD25⁺ T Cells

1. The purified $CD8^-CD25^+$ cells (see Subheading 3.4, step 15) are activated polyclonally with magnetic beads coated with anti-CD3 and anti-CD28 mAbs in presence or absence of rapamycin for 14 days. $CD25^+$ T_{MED} refers to the cells cultured without rapamycin, and $CD25^+$ T_{RPM} refers to the cells cultured in the presence of rapamycin. Based on the amount of cells used, choose the plate accordingly: 24 well/plates for 0.1×10^6 cells in 1 ml medium; 12 well/plates for 0.25×10^6 cells in 2.5 ml medium; 6 well/plates for 0.5×10^6 cells in 5 ml medium. A detailed scheme depicting the T cell culture is shown in Fig. 2.
2. Day 0: resuspend the purified $CD8^-CD25^+$ cells (see Subheading 3.4, step 15) in working medium containing the anti-CD3/CD28 coated beads at 1:3 ratio [cell:beads]. For T_{RPM} cells, add 100 nM rapamycin.
3. Day 2: add IL-2 at 500 U/ml in each well.
4. Day 5: count and split the cells to keep the cell concentration constant at 0.5×10^6/ml. Add new IL-2 on the final volume (since IL-2 is rapidly consumed by the cells), while add new rapamycin in T_{RPM} cells based on the new volume added (since rapamycin is not consumed by the cells).
5. Day 7: repeat step 4, Subheading 3.5.

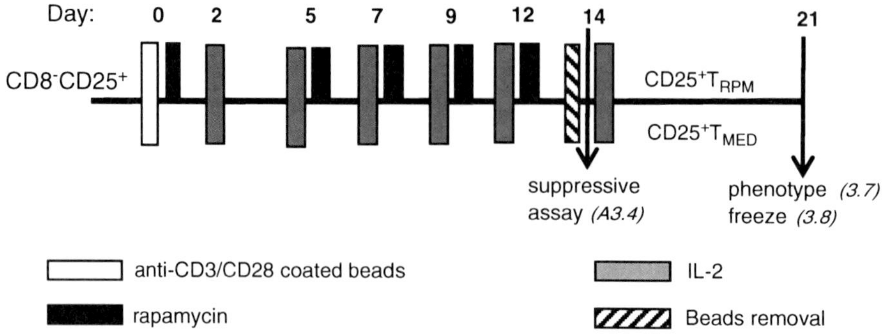

Fig. 2. The diagram outlines the time points (day) and the procedures followed for the ex vivo expansion of $CD8^-CD25^+$ T cells. T_{MED} are the cells expanded in the absence of rapamycin, T_{RPM} are the cells expanded in the presence of rapamycin.

6. Day 9: repeat step 4, Subheading 3.5.
7. Day 12: repeat step 4, Subheading 3.5.
8. Day 14: harvest the CD25$^+$T$_{MED}$ and CD25$^+$T$_{RPM}$ cells. Remove the magnetic beads by adherence. Fill the tubes with Working medium and centrifuge at 300 × *g*, for 7 min, at room temperature. Discard the supernatant, count the cells and resuspend at 1×10^6/ml (see Note 8).
9. Take the necessary amount of cells to perform the suppressive assay as described in Subheading 3.6.
10. Plate the remaining CD25$^+$T$_{MED}$ and CD25$^+$T$_{RPM}$ cells at 10×10^6 cells/ml in new wells in Working medium + 50 U/ml IL-2 (see Note 6) for 7 days.
11. Day 21: at the end of the 7-day resting, collect all the CD25$^+$T$_{MED}$ and CD25$^+$T$_{RPM}$ cells. Keep 0.5×10^6 cells to perform the phenotype analysis by flow cytometry (*see* Subheading 3.7) and, if needed, freeze the remaining cells (*see* Subheading 3.8) for future use.

3.6. Suppressive Assay

1. The following assay is fundamental to determine the suppressive function of the T$_{RPM}$ cells (generated in Subheading 3.3 and 3.5). A typical experiment is performed with responder T cells (R) which are polyclonally activated in the absence or presence of suppressor T cells (S), which in this case are the T$_{RPM}$ cells, generated from the same donor as for the Responder cells or from allogeneic donors. The proliferation of responder T cells in the absence of expanded T cells is compared to that in the presence of expanded T cells.
2. The Responder cells that can be used are the following: total PBMC (see Subheading 3.1, step 11), CD4$^+$ T cells and CD4$^-$ T cells (see Subheading 3.2, step 6) either freshly isolated or previously frozen. Perform the assay in U bottom 96 well/ plates precoated with anti-CD3 mAb (see Note 2).
3. Stain the Responder cells with CFSE:
   (a) Take double the amount of the necessary number of Responder cells (see Fig. 3) and resuspend in PBS 1×. NOTE: this high cell number is required given that during CFSE staining, almost 50% of the cells dye. Centrifuge at 300 × *g*, for 7 min, at room temperature.
   (b) Resuspend the cell pellet at 20×10^6 cell/ml in PBS 1×. Add equal volume of CFSE 5 μM and incubate for 8 min swirling (on the wheel). NOTE: the CFSE is light-sensitive. Protect samples from light.
   (c) Add 3 volumes of PBS/HS and incubate for 5 min, at room temperature, in dark. This step is necessary to stop the CFSE-labeling of the cells. Centrifuge at 300 × *g*, for 7 min, at room temperature.

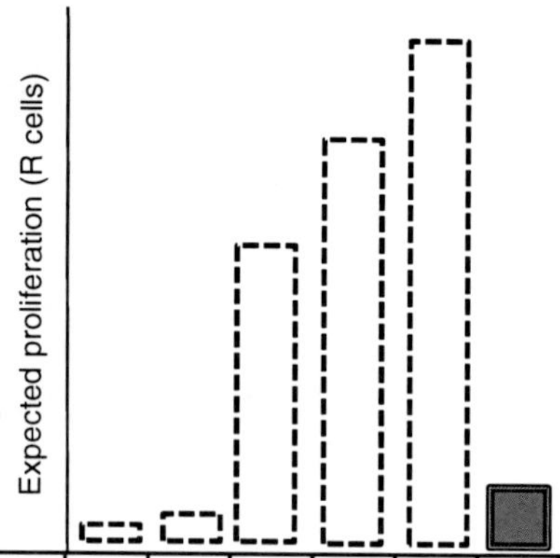

anti-CD3 coated anti-CD28 soluble	-	-	+	+	+	+
Responder cells 100.000	+	-	+	-	+	+
Responder cells 200.000	-	+	-	+	-	-
Suppressor cells $-T_{MED}$ 100.000	-	-	-	-	+	-
Suppressor cells $-T_{RPM}$ 100.000	-	-	-	-	-	+

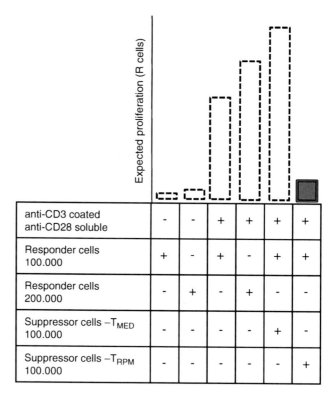

Fig. 3. Representative suppression experiment showing the necessary cell/stimuli combination. The expected cell proliferation in each culture condition is depicted on *top*.

(d) Resuspend the cells in Working medium, incubate for 5 min, at room temperature, in dark. Centrifuge $300 \times g$, 7 min, room temperature.

(e) Resuspend the cells in Working medium at the concentration of 2×10^6/ml. These are the CFSE-labeled Responder cells ready to be used

4. Plate 0.1×10^6 CFSE labeled Responder cells (50 μl), 0.1×10^6 unlabeled suppressor cells (50 μl) according to the scheme shown in Fig. 3, in previously precoated plate.
5. Add 100 μl of Working medium with 2 μg/ml anti-CD28 (to have 1 μg/ml final concentration) according to the scheme shown in Fig. 3.
6. Twenty four hours after initiation of the culture (see Subheading 3.6, step 4) move the cells in new uncoated wells (see Note 4).
7. Seven days after initiation of the culture (see Subheading 3.6, step 4) transfer the cells into a V bottom 96 well/plates and perform the flow-cytometry-based analysis (see Subheading 3.7, step 4).

3.7. Flow Cytometry-Based Analysis

1. The following procedure is fundamental to determine: the phenotype of the starting T cell subset (see Subheading 3.2, step 6 and 3.4, step 16) and the expanded T_{MED} and T_{RPM} cells (see Subheading 3.3, step 10 and 3.5, step 11), and the proliferation of CFSE-labeled Responder cells used in the suppressive assay (see Subheading 3.6, step 7) (see Note 9).
2. Surface phenotype staining (total cells needed: 0.3×10^6):
   (a) Transfer 0.1×10^6 cells in a FACS-tube, fill the tube with 3 ml of PBS-FACS and centrifuge at 300 × *g*, for 7 min, at room temperature.
   (b) Resuspend the cells in 50 μl PBS-FACS containing 1 μl of the following mAbs:
   tube 1: // [negative control]
   tube 2: anti-CD4 PerCP + anti-CD25 APC + anti-CD62L PE [test]
   tube 3: anti-CD4 PerCP + anti-CD25 APC + anti-CD127 PE [test]
   (c) Dark incubation for 30 min on ice.
   (d) Fill the tube with 3 ml of PBS-FACS and centrifuge at 300 × *g*, for 7 min, at +4°C.
   (e) Discard the supernatant, resuspend in 100 μl of FIX solution. The samples are ready for FACS acquisition. In case the samples are not going to be analyzed the same day they can be stored at +4°C for up to 7 days.
3. Intracellular phenotype staining (total cells needed: 0.4×10^6):
   (a) Transfer 0.2×10^6 cells in a FACS-tube, fill the tube with 3 ml of PBS-FACS and centrifuge at 300 × *g*, for 7 min, at room temperature.
   (b) Resuspend the cells in 50 μl of PBS-FACS containing 1 μl of the following mAbs:
   tube 1: // [negative control]
   tube 2: anti-CD4 PerCP + anti-CD25 APC [test]
   (c) Dark incubation for 30 minutes on ice, gently shake every 5 minutes.
   (d) Fill the tube with 3 ml of PBS-FACS and centrifuge at 300 × *g*, for 7 min, at room temperature. Discard the supernatant.
   (e) Resuspend in 50 μl of Fix/Perm Buffer 1×. Vortex and incubate for 20 min, at room temperature, in dark.
   (f) Fill the tube with 3 ml of PBS-FACS and centrifuge at 300 × *g*, 7 min, room temperature. Discard the supernatant.
   (g) Fill the tube with 3 ml of FOXP3 Perm Buffer and centrifuge at 300 × *g*, for 7 min, at room temperature. Discard the supernatant.

(h) Resuspend in 50 μl of FOXP3 Perm Buffer and incubate for 15 min, at room temperature, in dark. Centrifuge at 300×g, 7 min, at room temperature. Discard the supernatant.

(i) Resuspend the cells in 50 μl of FOXP3 Perm Buffer containing 4 μl of the following mAbs:

tube 1: // [negative control]

tube 2: anti-FOXP3 Alexa488 [test]

Vortex and incubate for 30 min, at room temperature, in dark.

(j) Fill the tube with 3 ml of PBS-FACS and centrifuge at 300×g, for 7 min, at room temperature. Discard the supernatant. Repeat once.

(k) Resuspend in 150 μl of PBS-FACS. The samples are ready for FACS acquisition. In case the samples are not analyzed the same day they can be stored at +4°C for up to 7 days.

4. Suppressive-assay staining:

(a) Prepare the mAb-cocktail for the surface cell-labeling: 96 μl PBS-FACS + 2 μl anti-CD4 PerCp + 2 μl anti-CD8 APC per each well used in the suppressive assay.

(b) Centrifuge the plate (see Subheading 3.6, step 7) at 500×g, for 7 min, at room temperature and discard the supernatant.

(c) Resuspend well with 100 μl of the mAb-cocktail (see Subheading 3.7, step 3a) and incubate at +4°C for 30 min, in dark.

(d) Add 100 μl PBS-FACS in each well, centrifuge the plate at 500×g, for 7 min, at room temperature and discard the supernatant.

(e) Resuspend each well in 200 μl PBS-FACS. The samples are ready for FACS acquisition, which has to be performed within few hours. NOTE: keep the plate protected from the light.

5. Data acquisition and analysis:

(a) Acquire at least 10,000 events and perform data acquisition and analysis with FCS Express V3 software (DE Novo Software).

(b) CFSE analysis. The percentages of $CFSE^+$ Responder T cells proliferating *in vitro* is determined by gating on $CFSE^+$ cells depicted in histograms (see Fig. 4). The number of gated cells (# events) in a given cycle (division: n) is divided by 2 raised to the power of n, to calculate the percentage of original precursor cells from which they arise. The sum of original precursors from division 1–6 represents the number of precursor cells which proliferated. The percentage of $CFSE^+$ divided cells is calculated

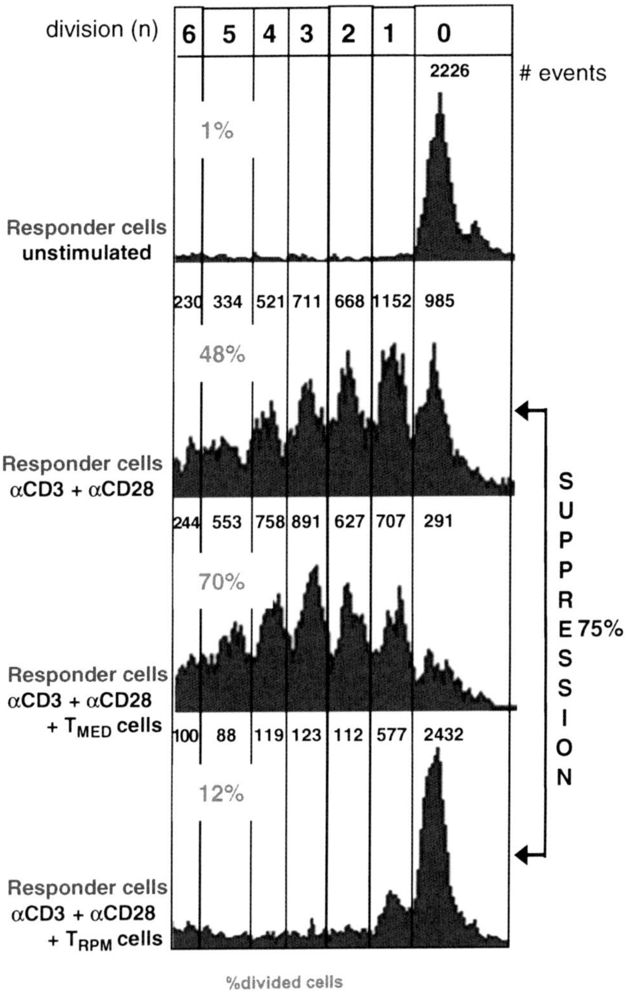

Fig. 4. Responder T cells were stained with CFSE and were not activated or activated with anti-CD3/CD28 mAbs. T_{MED} or T_{RPM} cells were added in equal number to CFSE⁺ Responder cells. After 7 days of culture, cell division was monitored by levels of CFSE dilution. Histograms show the FACS profile of CFSE⁺ Responder cells. The number of events in each cell division (*n*) is indicated on top of each peak. The amount of CFSE⁺ cells proliferating in the absence or presence of cultured T cells was calculated as described Subheading 3.7 step 5b and percentages of divided cells in each culture condition is indicated in *red*. Percentages of suppression in comparison to proliferation of responder cells is shown.

by [(# of precursors that proliferated 1–6/# of total precursors 0–6) × 100]. The percentage of $CFSE^+$ Responder cells divided in the presence of expanded cells is compared to the percentage of $CFSE^+$ divided Responder cells in the absence of any added cells.

3.8. Cryopreservation

1. Centrifuge the cells at 300 × *g*, for 7 min, at +4°C.
2. Discard the supernatant and resuspend gently in cold Freezing solution at no more than 100×10^6 cell/ml (optimal cell

concentration is 20×10^6 cell/ml) and transfer to a 2-ml cryovial kept in ice.

3. Place the cryovials in the Cryo 1° freezing container (Nalgene) and transfer the container at −80°C for 24 h.
4. Store the cryovials in liquid nitrogen until needed.

4. Notes

1. To obtain pure $CD4^+$ T cells always use cold MACS medium during the purification steps and always centrifuge at +4°C. When working with less than 10×10^6 cells use the same volumes as for 10×10^6 cells. When working with more than 10×10^6 cells scale up all volumes accordingly.
2. Add 150 μl/well (in 48 well/plates) or 300 μl/well (in 24 well/plates) or 1.2 ml/well (in 6 well/plates) of 10 μg/ml anti-CD3 mAb (see Subheading 2, item 9). Seal the plate with parafilm and store in the 37°C incubator for 3 h or at +4°C overnight. When needed, just before plating the cells, wash the coated wells twice with PBS 1×.
3. During this first week of culture it is crucial not to change the medium but to keep the same conditioned medium (full of cytokines and growth factors released by the cells themselves). Thus, do not change the medium even if it is yellow (i.e., exhausted) but add new medium (with new rapamycin for T_{RPM} cells) and split the cells in bigger wells if necessary. Please note that during the first week of culture the T_{RPM} cells do not proliferate but they rather dye significantly.
4. T cells repetitively stimulated via their TCR tend to dye by activation induced cell death. After the second round of stimulation it is important to move the cells from the wells coated with the anti-CD3 mAb to uncoated wells to reduce the strength of TCR-mediated stimulation.
5. The fold increase of $CD4^+$ T_{RPM} cells at the end of the three rounds of stimulation is about 1/10 of $CD4^+$ T_{MED} cells. Importantly, the T_{RPM} cells look smaller and less activated than T_{MED} cells and this can be appreciated directly at the microscope (see Fig. 5 as example).
6. The T_{RPM} and T_{MED} cells are left resting for 7 days in the presence of IL-2 and in the absence of any compound (i.e., NO rapamycin) to let all the activation markers (such as CD25, FOXP3, …) to return to their physiological levels. This step is fundamental to really discriminate activation-induced expression of Treg-associated markers versus their real and stable expression.
7. The average of $CD8^-$ and $CD8^-CD25^+$ cells commonly recovered from PBMC is 63% ± 15 and 1.5% ± 0.2, respectively [$n = 8$].

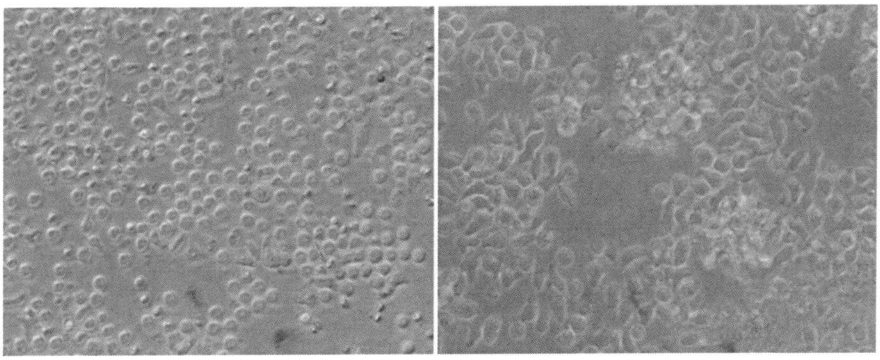

Fig. 5. Pictures of CD25$^+$T$_{RPM}$ (*left*) and CD25$^+$T$_{MED}$ (*right*) cells at the end of the expansion protocol.

8. The fold increase of CD25$^+$ T$_{RPM}$ cells at the end of 14-day culture is about 1/10 of CD25$^+$T$_{MED}$ cells (45-fold versus 450-fold as mean).
9. Fluorescinate mAbs are light-sensitive. Protect the samples from light.

Acknowledgments

This work was supported by the Italian Telethon Foundation and the Juvenile Diabetes Research Foundation (GJT04014). This study was also supported in part by research funding from BD Biosciences.

References

1. Curotto de Lafaille, M. A., and J. J. Lafaille (2009) Natural and adaptive foxp3+ regulatory T cells: more of the same or a division of labor? *Immunity* **30**, 626–635.
2. Wing, K., and S. Sakaguchi (2010) Regulatory T cells exert checks and balances on self tolerance and autoimmunity. *Nat Immunol* **11**, 7–13.
3. Malek, T. R., and I. Castro (2010) Interleukin-2 Receptor Signaling: At the Interface between Tolerance and Immunity. *Immunity* **33**, 153–165.
4. Roncarolo, M. G., and M. Battaglia (2007) Regulatory T-cell immunotherapy for tolerance to self antigens and alloantigens in humans. *Nat Rev Immunol* **7**, 585–598.
5. Battaglia, M., A. Stabilini, and M. G. Roncarolo (2005) Rapamycin selectively expands CD4+CD25+FoxP3+ regulatory T cells. *Blood* **105**, 4743–4748.
6. Battaglia, M., A. Stabilini, B. Migliavacca, J. Horejs-Hoeck, T. Kaupper, and M. G. Roncarolo (2006) Rapamycin promotes expansion of functional CD4+CD25+FOXP3+ regulatory T cells of both healthy subjects and type 1 diabetic patients. *J Immunol* **177**, 8338–8347.
7. Foxp3-mediated induction of pim 2 allows human T regulatory cells to preferentially expand in rapamycin. Basu S, Golovina T, Mikheeva T, June CH, Riley JL. *J Immunol.* 2008 May 1;180(9):5794–8.
8. Stability of human rapamycin-expanded CD4+CD25+ T regulatory cells. Tresoldi E, Dell'albani I, Stabilini A, Jofra T, Valle A, Gagliani N, Bondanza A, Roncarolo MG, Battaglia M. Haematologica. 2011 Sep; 96(9):1357–65.

Chapter 18

Rapamycin-Induced Enhancement of Vaccine Efficacy in Mice

Chinnaswamy Jagannath and Pearl Bakhru

Abstract

Th1 immunity protects against tuberculosis infection in mice and humans. The widely used BCG vaccine primes CD4 and CD8 T cells through signaling mechanisms from dendritic cells and macrophages. The latter express MHC-II and MHC-I molecules through which peptides from BCG vaccine are presented to CD4 and CD8 T cells, respectively. Since BCG sequesters within a phagosome that does not fuse with lysosomes, generation of peptides within antigen-presenting cells infected with BCG occurs with reduced efficiency. We demonstrate that activation of DCs containing BCG vaccine with rapamycin leads to an enhanced ability of DC vaccines to immunize mice against tuberculosis. Coadministration of rapamycin with BCG vaccine also enhanced Th1 immunity. We propose that rapamycin-mediated increase in Th1 responses offers novel models to study mTOR-mediated regulation of immunity.

Key words: Rapamycin, *Mycobacterium tuberculosis*, BCG vaccine, Mouse, mTOR, CD4, CD8 T cells, Th1 immunity

1. Introduction

Rapamycin is a well-known drug with antineoplastic properties. By repressing the mammalian target of rapamycin (mTOR), it triggers a wide variety of biochemical signaling within mammalian cells (1–4). An interesting property is the induction of autophagy (5, 6). Autophagy is a process in which a double-membrane phagophore is initiated which engulfs the cytosol and organelle. The phagophore then closes forming an autophagosome (AP) and subsequently fuses with the lysosome leading to an autophago-lysosome (APL). The degradation of cytosol and other membrane-bound vesicles or organelle within APL generates the much needed metabolites for the cell. Our initial studies have shown that the process of AP

Thomas Weichhart (ed.), *mTOR: Methods and Protocols*, Methods in Molecular Biology, vol. 821,
DOI 10.1007/978-1-61779-430-8_18, © Springer Science+Business Media, LLC 2012

fusion with lysosome can be exploited to enhance the efficacy of vaccines for tuberculosis, where the causative pathogen hides within macrophages and dendritic cells (6). In this study, we illustrate the beneficial effects of rapamycin using a mouse model to evaluate the BCG vaccine which is used to protect against human tuberculosis.

Tuberculosis is the leading cause of death due to a single infectious agent in humans. The causative pathogen *Mycobacterium tuberculosis* has a unique property of sequestering within early endosome like membrane compartments or phagosomes that do not fuse with lysosomes. Interestingly, the BCG vaccine, which is derived from a similar bovine pathogen *M. bovis* also avoids lysosomal fusion and localizes to phagosomes that are sequestered (7–9). It is well-known that phagosome–lysosome (PL) fusion is critical for degradation of mycobacterial peptides leading to the generation of peptides which are usually sorted to the MHC-II-containing compartments (MIIC). Such peptides are ultimately loaded into MHC-II for surface presentation to the CD4 cells initiating one major arm of the Th1 immunity.

Since BCG vaccine avoids PL fusion, it was obvious that peptide presentation to T cells was likely to be defective. Indeed, in our earlier studies, we and others demonstrated such defects using a T-cell hybridoma (10, 11). We have now developed a protocol using rapamycin to induce lysosomal localization for BCG vaccine that in turn leads to enhanced degradation, peptide production, and better priming of T cells and generation of better Th1 immunity (6).

2. Materials

2.1. BCG and Reagents

1. Culture BCG vaccine (Pasteur strain) in 7H9 broth for 7 days and freeze log phase organisms. Before use, thaw aliquots, wash three times in PBS (3000 × *g*; 15 min), sonicate at 4 W using a soniprobe, and disperse suspension to match with McFarland #1 in turbidity (10^8 CFU/mL).
2. Use this as stock BCG. A single-cell (colony-forming unit, CFU) suspension of BCG for infections is made by gentle centrifugation of this suspension at 300 × *g* for 5 min and using supernatants that contain 10^7 CFU/mL.
3. Viability of vaccine is confirmed by using live–dead stain from Invitrogen (USA). More than 90% of organisms should stain green.

2.2. H37Rv Strain of *M. tuberculosis*

1. Stock strain obtained from ATCC is cultured to log phase (7–10 days) in 7H9 broth using rotary shakers and frozen as aliquots as described above for BCG vaccine.

2.3. Rapamycin

1. Dissolve 1 mg of rapamycin in 100 μL of dimethyl sulfoxide and make up to 1 mL in distilled water or as per manufacturer's recommendation.

2.4. Mice and Dendritic Cells for Immunization

1. C57Bl/6 mice (M/F, 6–8 weeks of age) are sacrificed as per standard approved procedures and harvested for bone marrow from femurs and tibia. Centrifuged pellet of marrow cells are treated with a red cell lysing solution, such as ACK buffer, and the washed white cells are plated at a density of 10^6 per mL in 6-well TC plates using Isocove's modification of DMEM (IDM) with 10% fetal bovine serum, penicillin and gentamicin (100 U/mL and 50 μg/mL, respectively), and 50 μM 2-mercaptoethanol. Add 10 ng/mL of GM-CSF (Cell Sciences, USA) to IDM for cultures and replenish GM-CSF medium every 2–3 days.
2. Between days 7 and 10, harvest cells, wash, and use CD11c beads and a bead fractionator (Miltenyi Inc, USA) to obtain >95% pure CD11c-positive, CB11b-negative immature DCs.
3. Count and adjust to 10^8 cell/mL in IDM without GM-CSF. Pass through a 27-gauge needle to disperse cells and store in an ice bath.

2.5. Rapamycin Activation and Preparation of DC Vaccines

1. Use 4×2 mL aliquots of DCs kept cold in an ice bath, each with 10^8 cells/ mL.
2. To two aliquots of DCs, add rapamycin to a final 100 and 10 μg/mL and mix cells at 37°C and 5% CO_2 for 4 h. Mix also two more aliquots of DCs without the addition of rapamycin.
3. While cells are mixing, prepare single-cell BCG suspension as above. Remove DCs, and add 2×10^7 BCG CFU to rapamycin (10 and 100 μg)-treated DCs and to one aliquot of untreated DCs. Do not add BCG to the fourth aliquot. Note: All DC vaccines are at 2×10^8 per 2 ml. Higher infection ratios do not make a difference as long as infection ratio does not exceed 1:5 (DC:BCG).
4. Mix again for 4 h at 37°C and 5% CO_2. Remove DCs and wash three times with cold Hank's balanced salt solution (HBSS) ($1000 \times g \times 5$ min) to remove unbound bacteria. Make up cells in HBSS, gently pass through a 27-gauge needle to obtain single-cell suspension, and then adjust the count of cells to 10^8/mL in HBSS; keep cells cold in an ice bath.
5. Check cell viability using trypan blue; it should be more than 95%. (e) Inject groups of 20–25 mice per vaccine group i.p. with 100 μL suspension of DCs containing 2×10^6 cells per mouse.
6. From each vaccine preparation, pellet 100 μL aliquots, lyse with 0.05% SDS, and plate the lysate dilutions on 7H11 agar media to determine colony counts (CFUs) of BCG organisms contained within the DCs vaccine inoculum.
7. House mice under BSL-3 conditions and provide food and water ad libitum.

3. Methods

3.1. Assay to Determine if Th1 Immune Responses Are Accelerated by Rapamycin

1. Seven or 14 days after vaccination, four mice per group were sacrificed and splenic T cells immunostained for surface receptors (CD4 and CD8) and intracellular staining for IFNγ following standard procedures.
2. T cells from four individual mice per group are acquired using a BD Facscan and analyzed using the Cellquest software. Data are represented as histograms to determine the log shift in fluorescence of T cells.
3. T cells from mice are also stainable using CD8 antigen-specific tetramers (TB10.4-specific tetramer obtained from NIH tetramer facility, Emory Vaccine Center, Atlanta GA, USA). Such cells are as usual analyzed in combination with intracellular IFNγ stain. Note: Use of antigen-specific tetramer stains enable more precise analysis of the CD8 memory T cells. An example of TB10.4 tetramer-stained CD8 T cells is shown in Fig.1b.

3.2. Assay of Vaccine Efficacy After Rapamycin Activation of DCs

1. After 4 weeks of vaccination, mice in groups of 20 are exposed to an aerosol dose of a suspension of *M. tuberculosis* calculated to deliver 100-CFU counts per mouse using a Glas-Col (Indiana, USA) aerosol chamber. Mice are then housed in cages for another 4 weeks.
2. On the 4th week following aerosol challenge, the mice (five per vaccine group) were sacrificed as per standard procedures and lungs and spleens harvested individually as follows.
3. Lungs and spleens are homogenized in 5-mL saline with 0.05% tween 80 and then 100 μL aliquots of ten-fold dilutions are plated in replicates on 7H11 agar plates with 5 μg/mL thiophene 2-carboxylic acid hydrazide (TCH). Plates are read for CFU counts of *M. tuberculosis* and plotted against vaccine groups. Vaccine-induced protection is expressed as log10 reduction in CFU counts compared to controls and unvaccinated or naïve group, and significance was determined using ANOVA.
4. Four additional mice per group are sacrificed and T cells were purified from the organs and analyzed for IFNγ-secreting CD4 and CD8 T cells using antibody staining and flow cytometry. Note: As needed, the same pellet can also be stained for multiple cytokines using appropriate colored antibodies.
5. In order to confirm antigen specificity, cells were restimulated with 100 ng/mL of mycobacterial antigen 85B (available from BEI, NIH repository, MD, USA) and Elispot assay to determine IFNγ-secreting T cells. Elispot kit was obtained from either BD (USA) or Ebiosciences.

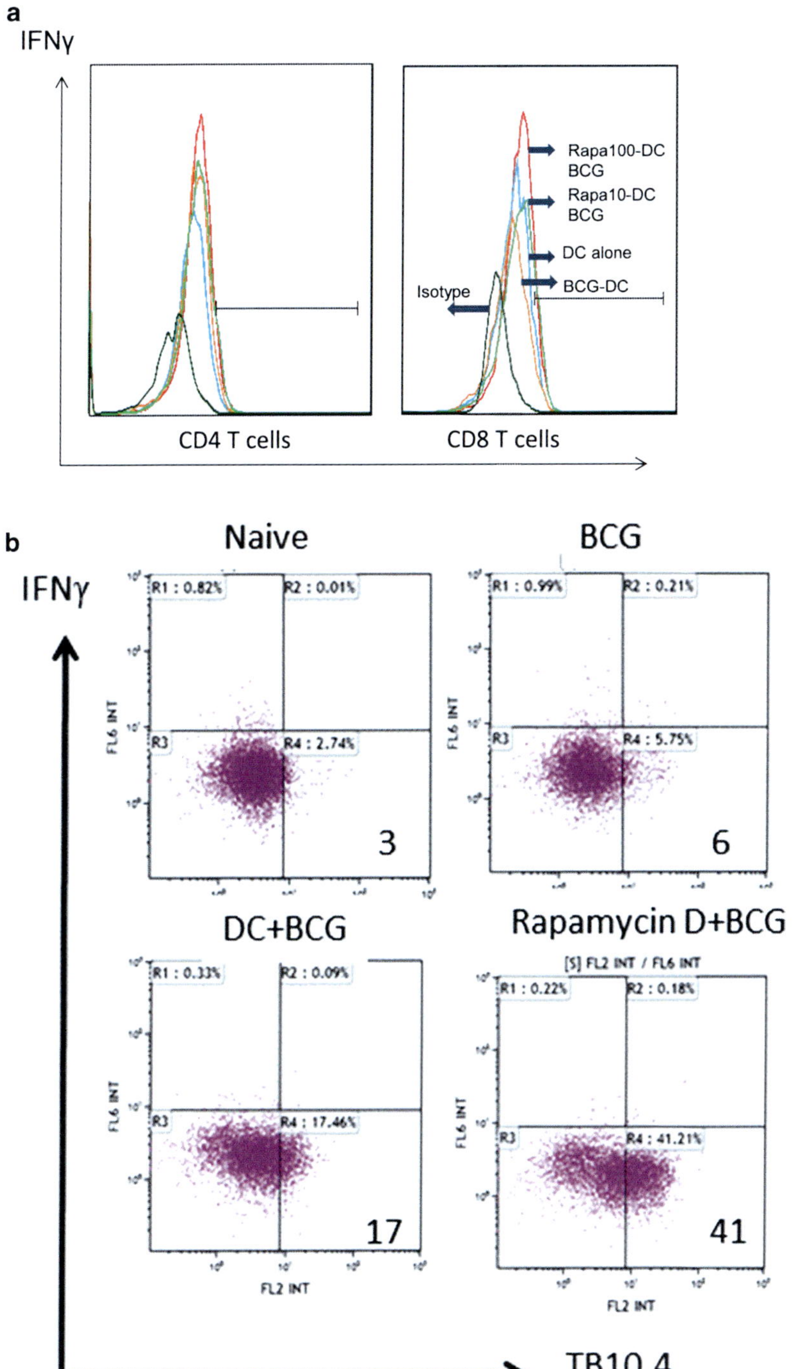

Fig. 1. Rapamycin treatment of dendritic cells with BCG vaccine accelerates the Th1 response in mice. C57Bl/6 mouse bone marrow-derived dendritic cells (DCs) were treated with either 10 or 100 μg of rapamycin followed by infection with *Mycobacterium bovis* Bacillus Calmette Guerin (BCG) bacteria for 4 h and washed. Five mice per group were injected i.p. with, 2×10^6 DCs per mouse per group as follows: (1) rapamycin-activated and BCG-infected DCs (10 or 100 μg doses), (2) untreated DCs infected with BCG, and (3) naïve DCs. An untreated group was an additional control. (**a**) Seven days after injection, splenic T cells were typed for CD4 and CD8 T cells secreting IFNγ after restimulation with antigen 85B, a major antigen of BCG vaccine. Flow cytometric analysis shows increased expression of IFNγ-secreting CD4 and CD8 T cells in mice given rapamycin-activated DCs containing BCG. (**b**) Mice were immunized as in panel (**a**). Two weeks later, the spleen-derived T cells were stained using a tetramer specific for BCG TB10.4 antigen (NIH Tetramer facility Emory Vaccine Center, USA). Rapamycin-activated, BCG-infected DCs induce a stronger expansion of tetramer-specific CD8 T cells compared to BCG-infected DCs alone.

3.3. Variations of a Protocol for Rapamycin Effects In Vivo on Vaccine Efficacy

In a recent study, a low-dose administration of rapamycin given i.p. to mice was found to enhance the longevity of CD8 immune responses to viral pathogens (3). We, therefore, developed a modified protocol to demonstrate an increase in CD8 responses to BCG vaccination and *M. tuberculosis* infection in mice.

1. Mice (C57Bl/6, M/F, 6–8 weeks of age; 20 per group) are immunized with BCG given once subcutaneously at 1×10^6 CFU per mouse.
2. From day one post vaccination, mice are given daily intraperitoneal injections of 75 μg/kg dose of rapamycin for 30 days. Control groups received BCG vaccine alone, rapamycin alone, or left untreated as naïve.
3. On day 32, two days after the last rapamycin dose, mice are aerosol infected as above and protection against tuberculosis measured. Before aerosol infection, spleens and inguinal lymph nodes are evaluated for IFNγ-secreting CD8 T cells using a 4-h restimulation with phorbol myristyl acetate in the presence of brefeldin.

4. Notes

1. We observed earlier that rapamycin enhances the MHC-II-mediated peptide presentation in DCs and macrophages infected with BCG vaccine. Antigen-presenting cells like DCs prime CD4 and CD8 T cells through MHC-II and MHC-I in a cytokine-rich environment. We, therefore, propose that activation of DCs would enhance peptide processing within the DCs and lead to an enhancement in vaccine-induced priming and thereby efficacy.
2. In this study, we noted that rapamycin drove an expansion of antigen-specific T cells. Figure 1b shows the tetramer-specific CD8 T cells in response to BCG vaccination. Figure 2c shows that the lungs and spleens of rapamycin-DC-BCG-vaccinated mice contained larger number of antigen-85B-specific T cells than comparable organs of other groups. This correlates with the hypothesis that Ag85B drives a major immune activation against tuberculosis in the lungs and that rapamycin has an amplifying response on antigen-specific T cells (4).
3. In this study, we have described a protocol that can be used to enhance the efficacy of the BCG vaccine in mice against tuberculosis using rapamycin-activated DCs (Figs. 1 and 2). Rapamycin has multiple effects on immune cells, including DCs, macrophages, and T cells. Therefore, a modified protocol was developed that allowed BCG antigens to be more effective *in vivo* when coadministered with rapamycin (Fig. 3).

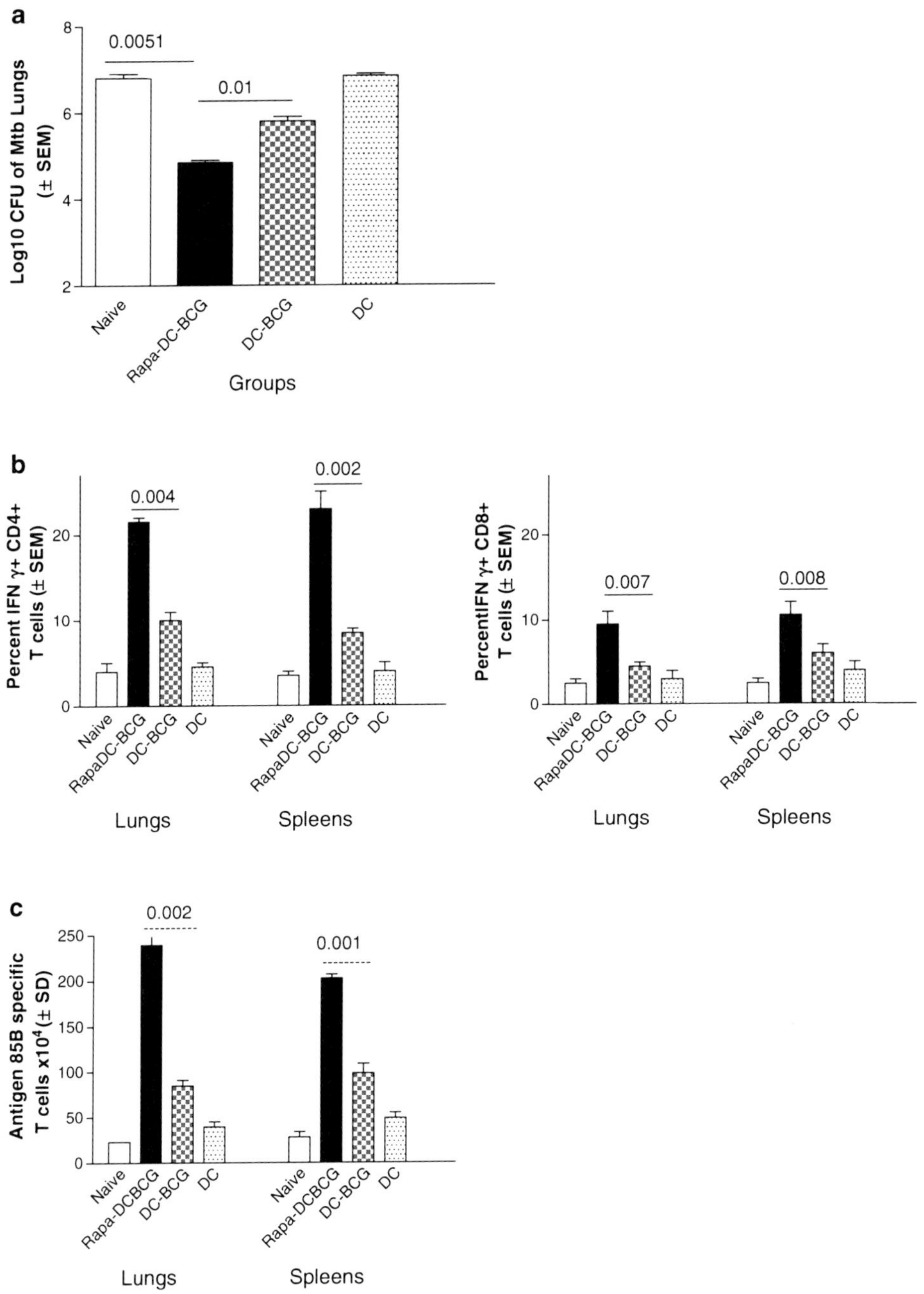

Fig. 2. Rapamycin-activated dendritic cell-BCG vaccine induces a better protection against tuberculosis in mice. C57Bl/6 mice were immunized subcutaneously with one dose of DCs (2×10^6 cells per mouse containing cell containing 1×10^6 BCG CFU) as follows. DCs were (1) preactivated with rapamycin (10 or 100 μg) and infected with BCG, (2) untreated but infected with BCG, and (3) naïve DCs. A group of mice did not receive DCs or BCG as control. Mice were analyzed as follows. (**a**) Mice were aerosol challenged with 100 CFU of *M. tuberculosis* 4 weeks after vaccination and bacterial counts of lungs determined by plating lung homogenates on 7H11 agar. Data show log10 reduction in CFUs of lungs and mice vaccinated with rapamycin-DC-BCG vaccine are protected better than mice given DC-BCG vaccine or no vaccine (five mice per group; three separate experiments; *p*-values determined using ANOVA shown against groups compared). (**b**) Lungs and spleens of mice were analyzed for Th1-type T cells using flow cytometry 4 weeks after aerosol challenge. Rapamycin-DC-BCG-vaccinated mice show higher levels of Th1-type CD4 (*left panel*) and CD8 (*right panel*) T cells secreting IFNγ (*p*-values shown against groups compared, *t* test). (**c**) Four weeks after aerosol challenge, Lungs and spleens of vaccinated mice (four per group) were analyzed for T cells specific for Ag85B using Elispot analysis. Rapamycin-DC-BCG vaccine again induces higher levels of antigen-specific T-cell responses in the lungs and spleens (*p*-values shown for groups compared, *t* test).

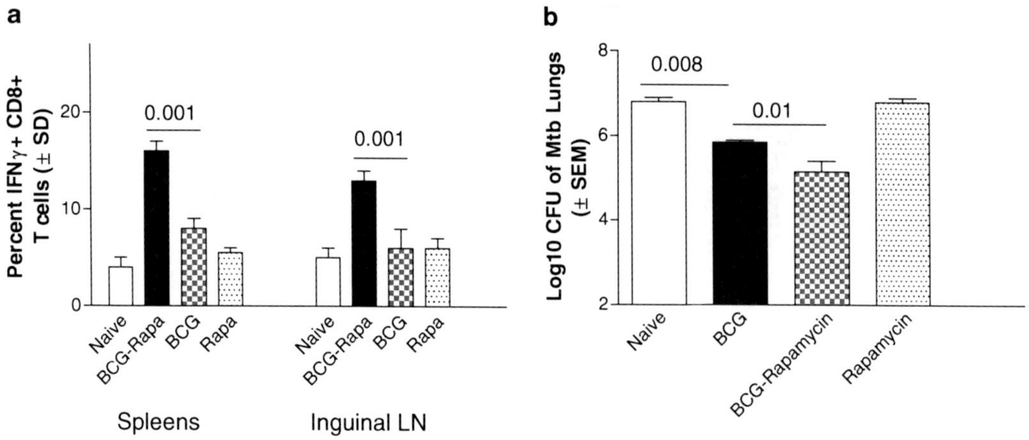

Fig. 3. Coadministration of BCG vaccine and rapamycin induces a better protection against tuberculosis correlating with increased CD8 T-cell responses. C57Bl/6 mice were vaccinated with BCG alone or BCG vaccine followed by 30 daily injections of rapamycin given i.p. On day 32, inguinal lymph nodes and spleens were analyzed for antigen-specific Th1 T-cell responses and then aerosol challenged with *M. tuberculosis* to determine protection. (**a**) On day 32, lymphoid compartments show an increased frequency of CD8 T cells that secrete IFNγ as analyzed by flow cytometry (four mice per group, *t* test). (**b**) Rapamycin combination with BCG vaccine enhances protection against tuberculosis compared to BCG vaccine alone (five mice per group; two separate experiments, *p*-values shown for groups compared, ANOVA).

4. Although enhanced processing of antigens in DCs is a major mechanism of rapamycin, recent studies have found that rapamycin affects mTOR signaling within T cells (4, 12, 13). Thus, we propose that when rapamycin is coadministered with BCG vaccines it can act on multiple cell populations to increase vaccine efficacy. The molecular mechanism of rapamycin *in vivo* needs additional investigation.
5. Note that all infection procedures need to be carried out under BSL-3 conditions and require approved institutional protocols.
6. The protocol describes measures to enhance protection against infectious tuberculosis. However, an identical protocol can also be used to measure the sterilizing effect of BCG vaccine in combination with rapamycin against an infectious challenge with live attenuated BCG vaccine. In this case, the challenge organism is marked with either green fluorescent or red fluorescent protein to enable detection and quantitation of marked BCG in organs. Such studies are useful to analyze the role of various T-cell populations that are modulated by rapamycin. We, therefore, anticipate that the models we described herein will pave the way for dissecting the role of mTOR in the regulation of T cell-mediated immunity.

Acknowledgments

This study was supported by AI49534 and AI78420. The authors acknowledge the technical assistance of Arshad Khan, Kari Herdtner, and Devin R. Lindsey.

References

1. Luo, H., W. Duguid, H. Chen, M. Maheu, and J. Wu. 1994. The effect of rapamycin on T cell development in mice. *Eur J Immunol* 24:692–701.
2. Anderson, K. M., and J. C. Zimring. 2006. Rapamycin prolongs susceptibility of responding T cells to tolerance induction by CD8+ veto cells. *Transplantation* 81:88–94.
3. Araki, K., A. P. Turner, V. O. Shaffer, S. Gangappa, S. A. Keller, M. F. Bachmann, C. P. Larsen, and R. Ahmed. 2009. mTOR regulates memory CD8 T-cell differentiation. *Nature* 460:108–112.
4. Ferrer, I. R., M. E. Wagener, J. M. Robertson, A. P. Turner, K. Araki, R. Ahmed, A. D. Kirk, C. P. Larsen, and M. L. Ford. Cutting edge: Rapamycin augments pathogen-specific but not graft-reactive CD8+ T cell responses. *J Immunol* 185:2004–2008.
5. Saemann, M. D., M. Haidinger, M. Hecking, W. H. Horl, and T. Weichhart. 2009. The multifunctional role of mTOR in innate immunity: implications for transplant immunity. *Am J Transplant* 9:2655–2661.
6. Jagannath, C., D. R. Lindsey, S. Dhandayuthapani, Y. Xu, R. L. Hunter, Jr., and N. T. Eissa. 2009. Autophagy enhances the efficacy of BCG vaccine by increasing peptide presentation in mouse dendritic cells. *Nat Med* 15:267–276.
7. Clemens, D. L., and M. A. Horwitz. 1996. The Mycobacterium tuberculosis phagosome interacts with early endosomes and is accessible to exogenously administered transferrin. *J Exp Med* 184:1349–1355.
8. Deretic, V., and R. A. Fratti. 1999. Mycobacterium tuberculosis phagosome. *Mol Microbiol* 31:1603–1609.
9. Russell, D. G. 2001. Mycobacterium tuberculosis: here today, and here tomorrow. *Nat Rev Mol Cell Biol* 2:569–577.
10. Singh, C. R., R. A. Moulton, L. Y. Armitige, A. Bidani, M. Snuggs, S. Dhandayuthapani, R. L. Hunter, and C. Jagannath. 2006. Processing and presentation of a mycobacterial antigen 85B epitope by murine macrophages is dependent on the phagosomal acquisition of vacuolar proton ATPase and in situ activation of cathepsin D. *J Immunol* 177:3250–3259.
11. Soualhine, H., A. E. Deghmane, J. Sun, K. Mak, A. Talal, Y. Av-Gay, and Z. Hmama. 2007. Mycobacterium bovis bacillus Calmette-Guerin secreting active cathepsin S stimulates expression of mature MHC class II molecules and antigen presentation in human macrophages. *J Immunol* 179:5137–5145.
12. Rao, R. R., Q. Li, K. Odunsi, and P. A. Shrikant. The mTOR kinase determines effector versus memory CD8+ T cell fate by regulating the expression of transcription factors T-bet and Eomesodermin. *Immunity* 32:67–78.
13. Sathaliyawala, T., W. E. O'Gorman, M. Greter, M. Bogunovic, V. Konjufca, Z. E. Hou, G. P. Nolan, M. J. Miller, M. Merad, and B. Reizis. Mammalian target of rapamycin controls dendritic cell development downstream of Flt3 ligand signaling. *Immunity* 33:597–606.

Chapter 19

Utilizing a Retroviral RNAi System to Investigate *In Vivo* mTOR Functions in T Cells

Koichi Araki and Bogumila T. Konieczny

Abstract

RNA interference (RNAi) is an intracellular mechanism for silencing gene expression utilizing short fragments of double-strand RNA that are complementary to the target messenger RNA. This gene silencing technique has now become an invaluable research tool due to its specific and strong repressive effect on a target transcript. We have recently applied a retrovirus-based RNAi system to investigate the *in vivo* role of the mammalian target of rapamycin (mTOR) in antigen-specific CD8 T cells, and have found that mTOR regulates memory CD8 T-cell differentiation. Here, we provide a detailed protocol for knocking down mTOR and its related molecules (raptor and FKBP12) in antigen-specific CD8 T cells. In our protocol, a mouse model of lymphocytic choriomeningitis virus infection is used, but the methods can be extended to other viral and bacterial infections as well as vaccinations. Also, the similar approach can be applied to analysis of CD4 T-cell responses.

Key words: LCMV, T cells, *In vivo*, RNAi, Retrovirus, mTOR, Raptor, FKBP12

1. Introduction

Mammalian target of rapamycin (mTOR) is an essential serine/threonine protein kinase in regulating cell growth, cell proliferation, and metabolism (1). Its specific inhibitor rapamycin is currently used in human transplant recipients as an immunosuppressive drug (2). How mTOR regulates immune responses *in vivo* is still far from understood due to its ubiquitous expression in all cells, and its particular control over a variety of immune and inflammatory responses in various types of immune cells, including dendritic cells, macrophages, and CD4 and CD8 T cells (3, 4). It has recently been demonstrated that the mTOR signal impacts the development of effector and memory T cells. Specifically, inhibition of

Thomas Weichhart (ed.), *mTOR: Methods and Protocols*, Methods in Molecular Biology, vol. 821,
DOI 10.1007/978-1-61779-430-8_19, © Springer Science+Business Media, LLC 2012

mTOR signaling negatively impacts the differentiation of naïve CD4 T cells into effector T-cell subsets, such as Th1, Th2, and Th17, whereas mTOR-deficient CD4 T cells efficiently differentiate into regulatory T cells (5). In contrast, we and others have found that inhibiting the mTOR signal in CD8 T cells paradoxically enhances memory CD8 T-cell differentiation (6–8) so that rapamycin treatment generates higher number of memory CD8 T cells with superior quality.

Using a retrovirus-based RNA interference (RNAi) system combined with adoptive transfer of virus-specific CD8 T cells into recipient mice, we have demonstrated that the mTOR pathway plays a major role in memory CD8 T-cell differentiation *in vivo* (3). This technique is relative easy and cost-effective and does not require special equipment. In addition, gene delivery into primary T cells using retrovirus is generally more efficient compared to transfection methods, and introduced genes that work for a long time because they are incorporated into the host's genome by retrovirus integrase enzyme. In this chapter, we introduce our *in vivo* approach of retrovirus-based RNAi methods in antigen-specific CD8 T cells.

2. Materials

2.1. Retrovirus-Based RNAi Vector and the Sequences for Double-Strand Oligonucleotides for Short Hairpin RNA

1. pMKO.1 GFP retroviral vector (Addgene plasmid 10676) was kindly provided by William Hahn.
2. Double-strand oligonucleotides for short hairpin RNA (shRNA) against mTOR (1), raptor (9, 10), and Fkbp12 (1) are cloned into pMKO.1 GFP between restriction sites AgeI and EcoRI (6) (see Note 1). The sequences for mTOR shRNA are 5′-CCGGGCCAGAATCCATCCATTCATTCTCGAGAATGAATGGATGGATTCTGGCTTTTTG-3′ (sense strand) and 5′-AATTCAAAAAGCCAGAATCCATCCATTCATTCTCGAGAATGAATGGATGGATTCTGGC-3′ (antisense strand). The sequences for raptor shRNA are 5′-CCGGGCCCGAGTCTGTGAATGTAATCTCGAGATTACATTCACAGACTCGGGCTTTTTG-3′ (sense strand) and 5′-AATTCAAAAAGCCCGAGTCTGTGAATGTAATCTCGAGATTACATTCACAGACTCGGGC-3′ (antisense strand). The sequences for Fkbp12 shRNA are 5′-CCGGGCCAAACTGATAATCTCCTCACTCGAGTGAGGAGATTATCAGTTTGGCTTTTTG-3′ (sense strand) and 5′-AATTCAAAAAGCCAAACTGATAATCTCCTCACTCGAGTGAGGAGATTATCAGTTTGGC-3′ (antisense strand).

2.2. Cell Culture and Transfection

1. Complete medium: Dulbecco's modified Eagle's medium (DMEM) supplemented with 10% fetal bovine serum (FBS), penicillin, and streptomycin.
2. Opti-MEM® I Reduced Serum Media (Invitrogen, Carlsbad, CA).
3. TransIT®-293 Transfection Reagent (Mirus Bio LLC, Madison, WI).
4. 293T cells (ATCC, Manassas, VA).
5. Trypsin: 0.05% trypsin with ethylenediaminetetraacetic acid (EDTA), liquid.
6. pCL-Eco retrovirus packaging vector (Imgenex, San Diego, CA).

2.3. Mice, Virus, and Retrovirus Transduction

1. Congenically marked Thy-1.1^{+} P14 transgenic mice bearing the D^{b}GP33-specific TCR were fully backcrossed to C57BL/6 mice in our animal colony (11).
2. C57BL/6j mice (The Jackson Laboratory, Bar Harbor, ME).
3. Red blood cell lysis buffer: 0.84% (w/v) ammonium chloride in water.
4. Washing medium: RPMI1640 medium supplemented with 2% FBS.
5. Lymphocytic choriomeningitis virus (LCMV) Armstrong (11).
6. Polybrene is dissolved at 5 mg/ml in water and stored at 4°C.

2.4. Flow Cytometry

1. FACS buffer: Dulbecco's phosphate-buffered saline (D-PBS) supplemented with 2% FBS and 0.1% (w/v) sodium azide.
2. Cytofix/Cytoperm™ (BD, San Diego, CA) and Perm/Wash™ (BD).
3. Antibodies: Anti-CD8 PerCP, anti-Bcl2 PE, and anti-CD25 PE (BD). Anti-Thy1.1 APC (eBioscience, San Diego, CA).
4. Fixation buffer: 2% (w/v) paraformaldehyde in D-PBS.

3. Methods

Our methods for use of a retrovirus-based RNAi system in antigen-specific CD8 T cells have several important points for successful gene knockdown: (1) making recombinant retrovirus by transfection, (2) preactivation of antigen-specific CD8 T cells, (3) retrovirus transduction, (4) adoptive transfer of antigen-sepecific CD8 T cells, followed by LCMV infection. The time line of this procedure is shown in Fig. 1.

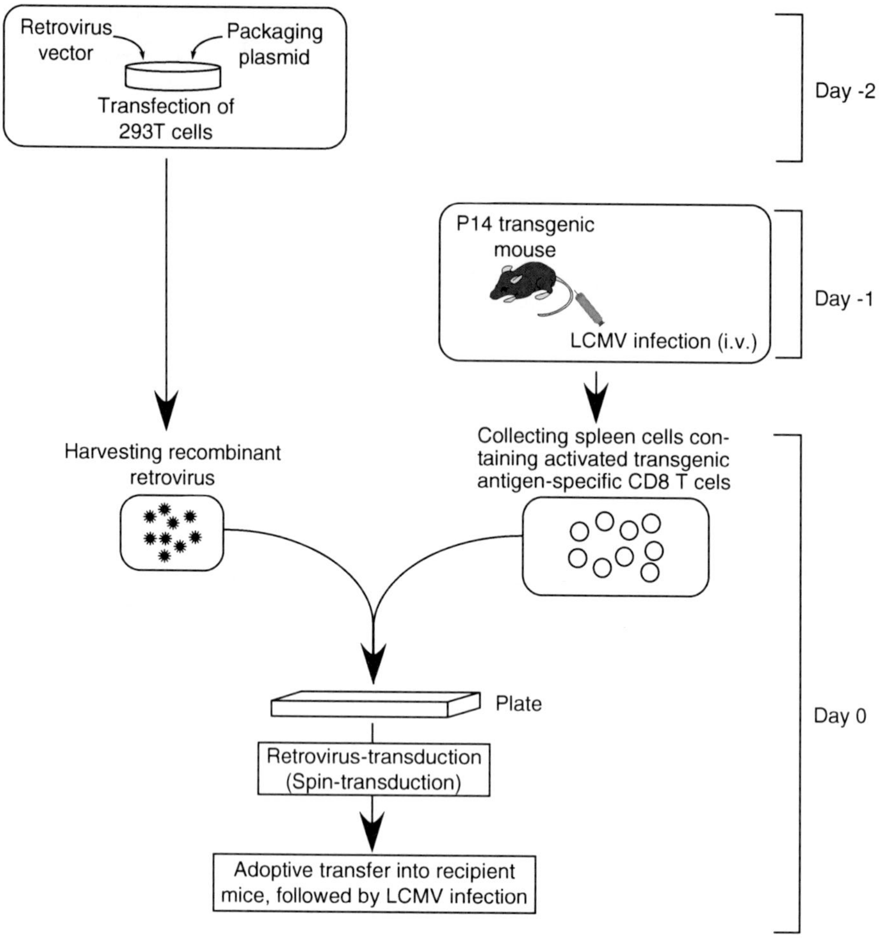

Fig. 1. The time line of a retrovirus-based RNAi system combined with adoptive transfer of virus-specific CD8 T cells.

3.1. Preparation of 293 T Cells for Transfection

1. Trypsinize cultured 293 T cells for 12–18 h prior to transfection and collect the cells. Make a single-cell suspension by gentle pipetting. Plate the 293 T cells at $1–1.5 \times 10^6$ cells per well of a 6-well plate in 2.5 ml of complete medium. Gently rock the plate and distribute the cells evenly in a well. Incubate overnight.
2. Confirm that the confluency of the 293 T cells is 50–80% before transfection (see Note 2).

3.2. Preparation of Transfection Mixture and Transfection

1. Add 9 μl of TransIT-293 transfection reagent into 200 μl of Opti-MEM medium, and mix thoroughly by pipetting or vortexing.
2. Incubate the mixture for 10 min at room temperature.
3. Add 1 μg of pCL-Eco retrovirus packaging vector and 2 μg of pMKO.1 GFP retrovirus RNAi vector or empty control vector into the TransIT-293/Opti-MEM mixture. Mix by gentle pipetting.

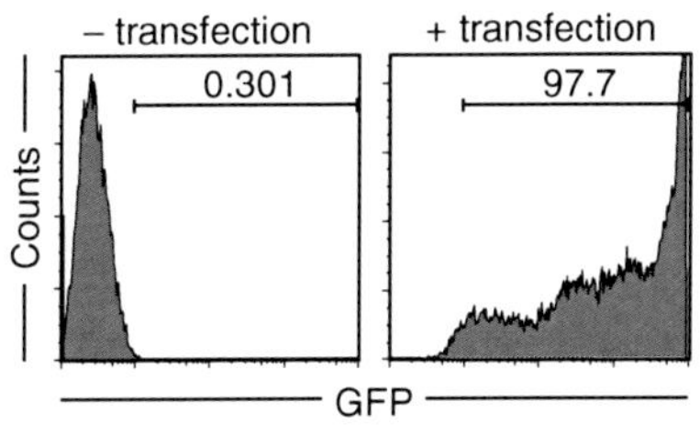

Fig. 2. GFP expression in 293 T cells after transfection. 293 T cells were transfected with retrovirus vector and packaging plasmid. The cells were collected at 48 h after transfection, and GFP expression was analyzed by a flow cytometer. As a negative control, a flow panel of 293 T cells without transfection is shown.

4. Incubate the mixture for 30 min at room temperature.
5. Add the entire DNA/TransIT-293/Opti-MEM mixture dropwise into previously prepared wells described in Subheadings 3.1, step 2, and gently rock the plate. Incubate the plate at 37°C in a 5% CO_2 incubator (see Note 3).
6. Harvest the culture supernatant containing recombinant retrovirus for 48 h after transfection for the following transduction.
7. Collect the remaining 293 T cells using trypsin and resuspend into 1–2 ml FACS buffer. Measure the transfection efficiency using a flow cytometer by analyzing GFP expression (Fig. 2) (see Note 4).

3.3. Activation of Transgenic CD8 T Cells for Retrovirus Transduction

1. Intravenously (i.v.) infect a Thy-1.1$^+$ P14 transgenic mouse (most CD8 T cells in these mice have an LCMV-specific TCR) with 2×10^6 PFU LCMV to activate LCMV-specific CD8 T cells (see Note 5).
2. Twenty-four hours following infection, remove the spleen from the LCMV-infected P14 transgenic mouse with sterile dissecting instruments. Place the spleen in a sterile 15-ml centrifuge tube containing 3 ml of RPMI1640 medium supplemented with 2% FBS. The tube should be kept on ice.
3. Transfer the spleen with medium onto a wire screen in a 35-mm petri dish. Grind the spleen against the wire screen using the plunger of 5-cc syringe. Grind gently to prevent cell damage.
4. Transfer the cell suspension to a 15 ml centrifuge tube. Fill the tube to the top with 2% FBS RPMI1640 medium.
5. Centrifuge the tube at $500 \times g$ for 10 min at 4°C, and then discard supernatant. Resuspend the cell pellet by tapping the tube.
6. Add 1–2 ml of red blood cell lysis buffer into the cell pellet. Mix thoroughly and leave at room temperature for 1–2 min. Caution: Incubation beyond 2 min decreases cell viability.
7. Following the red blood cell lysis, add 2% FBS RMPI1640 medium to top of the tube to neutralize the red blood cell lysis

buffer. Mix and let stand for 3–4 min on ice to allow large clumps of debris to settle.

8. Carefully collect only the supernatant containing cell suspension without disturbing the large clumps that have settled at the bottom of the tube.
9. Centrifuge the tube at 500 × *g* for 10 min at 4°C, and then discard the supernatant. Resuspend the cell pellet by tapping the tube.
10. Add 3 ml of 2% FBS RPMI1640 medium and mix thoroughly. Pass the cell suspension through a cell strainer to eliminate any small clumps of debris. Leave the cell suspension on ice until use.
11. Count the cells in the presence of trypan blue to determine the concentration of live cells.

3.4. Confirmation of Activation Status of Thy1.1⁺ P14 Transgenic CD8 T Cells

1. Transfer 0.5–1 × 10^6 cells/well of the spleen cell suspension obtained in Subheading 3.3, step 10, into a 96-well round-bottom plate.
2. Wash the cells as follows.
   (a) Add 100 μl FACS buffer to each well, and then centrifuge the plate at 1,000 × *g* for 2 min at 4°C. Confirm that a cell pellet is formed, and then discard the wash buffer.
   (b) Add 200 μl FACS buffer and centrifuge as described in part a of step 2.
   (c) Wash the cells once more with 200 μl FACS buffer and centrifuge as described in part a of step 2.
3. While washing the cells, prepare the antibody solution in FACS buffer. Dilute both anti-CD8 PerCP and anti-CD25 PE antibodies 1:100 in FACS buffer.
4. Add 50 μl antibody solution into a well containing the washed spleen cells. Mix thoroughly by gentle pipetting.
5. Incubate the plate on ice in the dark for 30 min.
6. Wash the cells as described in step 2 above.
7. After the final wash, resuspend each well of cells in 200 μl fixation buffer and incubate the plate on ice in the dark for more than 20 min, and then the sample is ready for flow cytometry analysis.
8. Analyze the sample by flow cytometry, and determine the activation status of P14 cells by checking for upregulation of CD25 expression and increased cell size (Fig. 3) (see Note 6).

3.5. Retrovirus Transduction and Adoptive Transfer of Transgenic CD8 T Cells

1. Centrifuge the culture supernatant containing recombinant retrovirus obtained in Subheading 3.2, step 6, at 1,800 × *g* for 10 min at 4°C to remove live and/or dead cells and cell-derived large materials.

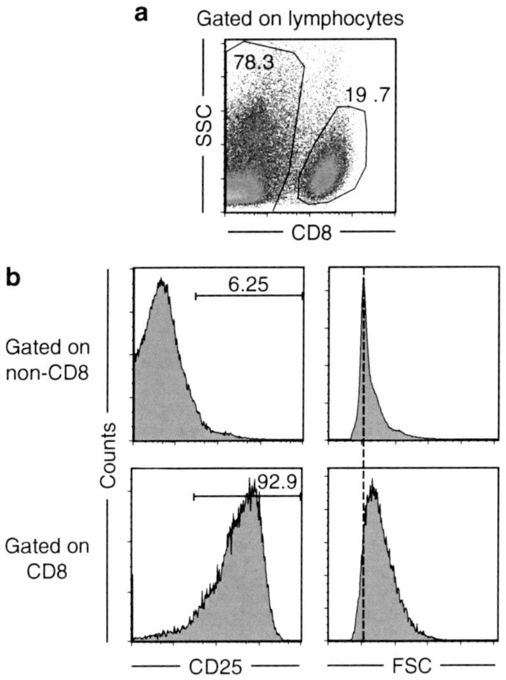

Fig. 3. Activation status of transgenic, antigen-specific CD8 T cells. A P14 mouse was i.v. infected with LCMV, and spleen cells were isolated and analyzed at 24 h post infection. (**a**) CD8 T cells and non-CD8 lymphocytes were separately gated. (**b**) CD25 expression and cell size of both transgenic P14 CD8 T cells and non-CD8 lymphocytes are shown as a measure of the activation status of the P14 CD8 T cells.

2. Collect only the supernatant. Be careful not to disturb the pellet formed on the bottom. This supernatant is used for retrovirus transduction.
3. Transfer 1×10^6 cells/well of the single-cell splenocyte suspension obtained from Subheading 3.3, step 10, into a 96-well round-bottom plate.
4. Centrifuge the plate at $1{,}000 \times g$ for 2 min at 4°C. Confirm that a cell pellet is formed, and then discard the medium. Resuspend the cell pellet by tapping the plate.
5. Add 290 μl of recombinant retrovirus obtained in Subheading 3.5, step 2, into a well containing the cells, and mix thoroughly by gentle pipetting.
6. Add 10 μl polybrene (240 μg/ml: dilute original polybrene solution (5 mg/ml) with complete medium) into the mixture of recombinant retrovirus and spleen cells, and mix thoroughly by gentle pipetting: total volume should be 300 μl per well with the final concentration of polybrene at 8 μg/ml. Polybrene increases retrovirus transduction efficiency.

7. Place the plate into a prewarmed centrifuge. Spin the plate at 1,800 × *g* for 90 min at 37°C to transduce the transgenic CD8 T cells.
8. After centrifugation, discard the supernatant and wash cells as follows.
   (a) Add 200 μl of 2% FBS RPMI1640 medium to each well, and then centrifuge the plate at 1,000 × *g* for 2 min at 4°C. Confirm that a cell pellet is formed, and then discard the medium.
   (b) Wash the cells once more with 200 μl of 2% RPMI1640 medium and centrifuge as described in part a of step 8.
9. Resuspend the cells in plain RPMI1640 medium without FBS at the desired concentration for adoptive transfer.
10. Intravenously inject the desired number of retrovirus-transduced spleen cells into naïve C57BL/6 (Thy-1.2$^+$) recipients: we usually transfer 1×10^4 to 5×10^5 spleen cells containing approximately 2×10^3 to 1×10^5 LCMV-specific P14 transgenic CD8 T cells.
11. Immediately after adoptive transfer, infect the mice intraperitoneally with 2×10^5 PFU LCMV (Fig. 4a).

3.6. Flow Cytometric Analysis of Bcl2 Expression in LCMV-Specific P14 Transgenic CD8 T Cells

1. Here, we show a typical flow cytometric analysis of Bcl2 expression in LCMV-specific P14 transgenic CD8 T cells that were transduced with retrovirus that expresses shRNA targeting mTOR. Bcl2 is an antiapoptotic intracellular molecule, and we found that it was highly expressed in P14 cells with mTOR knocked down (6). In this subsection, we describe how to stain and analyze retrovirus-transduced P14 cells isolated from LCMV-infected mice by using changes in expression of the antiapoptotic protein Bcl2 as an example of a functional consequence on the target cell population (see Note 7).
2. Isolate and make a single-cell suspension (previously described in Subheading 3.3) of spleen cells from a mouse that was infected with LCMV and had adoptively received LCMV-specific transgenic P14 cells transduced with retrovirus.
3. Stain spleen cells with fluorescein-conjugated antibodies as described in Subheading 3.4, steps 1–6. Make the antibody solution in FACS buffer: anti-CD8 PerCP (1:100 dilution) and anti-Thy-1.1 APC (1:1,000 dilution).
4. After washing the cells, thoroughly resuspend the cells in 100 μl of Cytofix/Cytoperm solution.
5. Incubate the plate on ice in the dark for 20 min.
6. Wash the cells as described in Subheading 3.4, step 2, but use 1× Perm/Wash buffer.

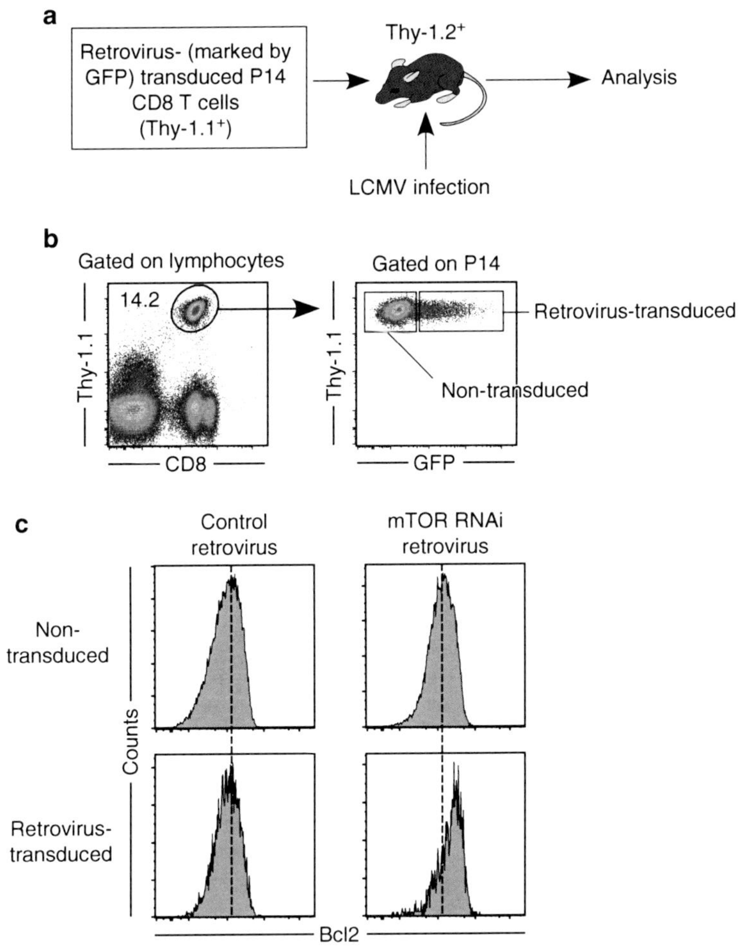

Fig. 4. Bcl2 expression increases in antigen-specific CD8 T cells with mTOR knockdown. (**a**) LCMV-specific P14 transgenic naïve CD8 T cells (Thy-1.1+) were activated *in vivo* and then transduced with retroviruses (marked by GFP expression) expressing RNAi for mTOR or empty control vector. These transduced P14 cells were then transferred into naïve Thy-1.2+ mice and these mice were infected with LCMV. (**b**) P14 cells in spleen cells on day 15 post infection were identified by CD8 and Thy-1.1 staining (*left panel*), and then retrovirus-transduced cells were distinguished from nontransduced cells by GFP expression (*right panel*). (**c**) Bcl2 expression was compared between retrovirus-transduced (GFP^+) and nontransduced (GFP^-) P14 cells. The *left panel* shows an empty control retrovirus vector and the *right panel* shows an mTOR RNAi retrovirus vector.

7. While the cells are being pelleted in the centrifuge, prepare an anti-Bcl2 PE antibody solution in Perm/Wash buffer. For one staining, mix 20 μl of anti-Bcl2 PE and 30 μl of Perm/Wash buffer, and pipette the mixture into a well containing spleen cells treated with Cytofix/Cytoperm. Resuspend the cells thoroughly by gentle pipetting.
8. Incubate the plate on ice in the dark for 30 min.

9. Wash as described in Subheading 3.4, step 2, but use 1× Perm/Wash buffer.
10. Perform two additional washes using FACS buffer. Use 200 μl of FACS buffer per wash and centrifuge the plate at 1,000 × *g* for 2 min at 4°C.
11. After the final centrifugation, resuspend the cells in 200 μl of fixation buffer and wrap the plate with aluminum foil. Keep the wrapped plate at 4°C until analysis.
12. Acquire the sample by flow cytometry, and analyze the data with an adequate software, such as Flowjo (Tree Star, Inc., Ashland, OR).
13. Create a gate to identify LCMV-specific transgenic P14 CD8 T cells ($CD8^+$ and Thy-1.1^+) (Fig. 4b).
14. Determine retrovirus-transduced (GFP^+) and nontransduced (GFP^-) populations in P14 cells (Fig. 4b) (see Note 8).
15. As shown in Fig. 4c, Bcl2 expression in control retrovirus vector-transduced P14 cells is similar to nontransduced P14 cells while Bcl2 expression increases in P14 cells transduced with an mTOR RNAi retrovirus vector compared to nontransduced cells (see Note 9).

4. Notes

1. A detailed method for cloning double-strand DNA into the pMKO.1 retrovirus vector can be obtained from Addgene Web site: http://www.addgene.org/pgvec1?f=c&identifier=8452&atqx=pmKO.1&cmd=findpl.
2. Cell confluency should be 50–80% to maximize transfection efficiency and to make high-titer recombinant retrovirus.
3. Transfection can be performed in complete medium instead of serum-free medium. We usually do not change the medium after transfection.
4. Carefully collect 293 T cells to avoid cell death, and pass the cells through a cell strainer to remove cell clump before flow cytometry analysis. If the procedure is carried out appropriately, you observe more than 90% GFP^+ 293 T cells.
5. Retrovirus transduction requires preactivation of T cells. We and others infected P14 transgenic mice i.v. with high-titer LCMV to activate all transgenic, antigen-specific CD8 T cells *in vivo* (6, 12). This method efficiently activates transgenic CD8 T cells, and provides optimal expansion and survival of the transgenic CD8 T cells following adoptive transfer into recipient mice infected with LCMV. Alternative stimulation

methods, such as epitope peptides and anti-CD3/CD28, may be used for preactivation of T cells.

6. CD25 (IL-2 receptor alpha) is an activation marker of T cells. As shown in Fig. 3b, a majority of transgenic CD8 T cells should be CD25$^+$ for maximal transduction efficiency. Cell size of transgenic CD8 T cells is also a good marker for T-cell activation. The transgenic CD8 T cells are usually larger than non-CD8 T-cell population after *in vivo* activation (Fig. 3b).
7. Here, we only show an example of mTOR RNAi. You can examine the role of mTOR complex 1 by knocking down raptor. Also, FKBP12 RNAi makes T cells rapamycin insensitive.
8. We usually obtain 5–20% GFP$^+$ retrovirus-transduced P14 cells. You can sort GFP$^+$ retrovirus-transduced P14 cells using FACS to measure the knockdown efficiency by western blotting. A knockdown effect of 75–90% on a target molecule (mTOR, raptor, or FKBP12) should be seen in RNAi retrovirus-transduced cells compared to control retrovirus-transduced cells (6).
9. This *in vivo* retrovirus-based RNAi approach is useful to examine the intrinsic effect of a target gene in T cells because only a fraction of transgenic, antigen-specific CD8 T cells adoptively transferred lack the target gene while all the other cells in the mouse retain the gene.

Acknowledgments

The authors would like to thank Professor Rafi Ahmed for providing financial support to establish this retrovirus-based RNAi system; Dr. Ben Youngblood for critical reading of our manuscript; and Dr. William Hahn for providing the pMKO.1 GFP vector.

References

1. Wullschleger S, Loewith R, Hall MN (2006) TOR signaling in growth and metabolism. Cell 124:471–484
2. Halloran PF (2004) Immunosuppressive drugs for kidney transplantation. N Engl J Med 351:2715–2729
3. Araki K, Youngblood B, Ahmed R (2010) The role of mTOR in memory CD8 T-cell differentiation. Immunol Rev 235:234–243
4. Thomson AW, Turnquist HR, Raimondi G (2009) Immunoregulatory functions of mTOR inhibition. Nat Rev Immunol 9:324–337
5. Delgoffe GM, Kole TP, Zheng Y, Zarek PE, Matthews KL, Xiao B, Worley PF, Kozma SC, Powell JD (2009) The mTOR kinase differentially regulates effector and regulatory T cell lineage commitment. Immunity 30:832–844
6. Araki K, Turner AP, Shaffer VO, Gangappa S, Keller SA, Bachmann MF, Larsen CP, Ahmed R (2009) mTOR regulates memory CD8 T-cell differentiation. Nature 460:108–112
7. Pearce EL, Walsh MC, Cejas PJ, Harms GM, Shen H, Wang LS, Jones RG, Choi Y (2009) Enhancing CD8 T-cell memory by modulating fatty acid metabolism. Nature 460: 103–107
8. Rao RR, Li Q, Odunsi K, Shrikant PA (2010) The mTOR Kinase Determines Effector versus

Memory CD8(+) T Cell Fate by Regulating the Expression of Transcription Factors T-bet and Eomesodermin. Immunity 32:67–78

9. Hara K, Maruki Y, Long X, Yoshino K, Oshiro N, Hidayat S, Tokunaga C, Avruch J, Yonezawa K (2002) Raptor, a binding partner of target of rapamycin (TOR), mediates TOR action. Cell 110:177–189

10. Kim DH, Sarbassov DD, Ali SM, King JE, Latek RR, Erdjument-Bromage H, Tempst P, Sabatini DM (2002) mTOR interacts with raptor to form a nutrient-sensitive complex that signals to the cell growth machinery. Cell 110:163–175

11. Wherry EJ, Teichgraber V, Becker TC, Masopust D, Kaech SM, Antia R, von Andrian UH, Ahmed R (2003) Lineage relationship and protective immunity of memory CD8 T cell subsets. Nat Immunol 4:225–234

12. Joshi NS, Cui W, Chandele A, Lee HK, Urso DR, Hagman J, Gapin L, Kaech SM (2007) Inflammation directs memory precursor and short-lived effector CD8(+) T cell fates via the graded expression of T-bet transcription factor. Immunity 27:281–295

Chapter 20

Exploring Functional *In Vivo* Consequences of the Selective Genetic Ablation of mTOR Signaling in T Helper Lymphocytes

Greg M. Delgoffe and Jonathan D. Powell

Abstract

The mammalian Target of Rapamycin (mTOR) defines a crucial link between nutrient sensing and immune function. In CD4+ T cells, mTOR has been shown to play a critical role in regulating effector and regulatory T cell differentiation as well as the decision between full activation versus the induction of anergy. In this chapter, we describe how our group has employed the Cre-lox technology to genetically delete components of the mTOR signaling complex in T cells. This has enabled us to specifically interrogate mTOR function in T cells both *in vitro* and *in vivo*. We also describe techniques used to assay immune function and signaling in mTOR-deficient T cells at the single-cell level.

Key words: T cells, CD4, mTOR

1. Introduction

The initiation of an adaptive immune response requires the integration of many varied signals. The mammalian Target of Rapamycin (mTOR), an evolutionarily conserved serine–threonine protein kinase, is a nutrient sensor which interprets environmental cues (1). T cells utilize mTOR to integrate many immunologic signals and promote T helper cell differentiation (2). In CD8+ T cells, mTOR has been shown to play a role in regulating the generation of memory cells (3, 4). Inhibition of mTOR with the macrolide compound rapamycin or genetic deletion of the mTOR kinase results in failed T helper cell effector differentiation and alternate regulatory T cell generation (2, 5–7).

mTOR signals via two nutrient-sensitive protein complexes: mTORC1 and mTORC2. mTORC1 is characterized by the adaptor

Thomas Weichhart (ed.), *mTOR: Methods and Protocols*, Methods in Molecular Biology, vol. 821,
DOI 10.1007/978-1-61779-430-8_20, © Springer Science+Business Media, LLC 2012

protein raptor and the small GTPase Rheb, and is read out by the phosphorylation of the ribosomal S6 kinase (S6K1) and 4E-BP1. mTORC1 has been implicated in the initiation of translation, inhibition of apoptosis, and initiation of mitochondrial metabolism (8). mTORC2 is characterized by the adaptor protein rictor and the mSIN1 proteins and is read out by the phosphorylation of Akt on its hydrophobic motif, serine 473 (9). While less is known about mTORC2 signaling and function, it has been implicated in actin reorganization as well as cell survival.

Here, we describe T cell-specific deletion of the mTOR kinase resulting in the ablation of total mTOR signaling. We discuss an *in vivo* model of Th1 differentiation (viral infection) and the interrogation of mTOR activity at the single-cell level using flow cytometry.

2. Materials

2.1. Generation and Genotyping of CD4-Cre × Floxed Mice

1. Tail snips from pups.
2. Tail lysis buffer: 100 mM Tris–HCl, pH 8.0, 5 mM EDTA, 0.2% SDS, 200 mM NaCl.
3. Proteinase K (Qiagen).
4. ddH_2O.
5. Platinum PCR Supermix (Invitrogen).
6. Forward and reverse primers (see Table 1).
7. 2% TAE-agarose gels.

2.2. Purification of CD4 T Cells

1. Ack lysing buffer (Quality Biological).
2. MACS CD4 isolation kit (negative selection, Miltenyi-Biotec).
3. LS columns (Miltenyi-biotec).
4. Midi MACS magnet (Miltenyi-biotec).
5. Phosphate-buffered Saline (PBS), pH 7.4 (Quality Biological).
6. Magnetic sorting buffer: PBS supplemented with 2 mM EDTA and 0.5% BSA.

Table 1
Primers and product sizes for genotyping mTOR-deficient mice

Gene	Direction	Sequence	Product length
Cre	FOR REV	CGA TGC AAC GAG TGA GG GCA TTG CTG TCA CTT GGT CGT	~300
Mtor	FOR REV	CCC AGC ACT TGG GAA TCA GAC AG CAG GAC TCA GGA CAC AAC TAG CCC	~550 flox ~350 wt

2.3. Vaccinia-OVA Infection and OT-II Adoptive Transfer

1. *Vaccinia*-OVA (kept at stock solution of PBS at 2×10^7 PFU/mL).
2. C57/BL6 host mice (Jackson Laboratories).
3. OT-II wild-type or mutant cells (bred from stock at Jackson Laboratories).
4. Mouse immobilizer (Braintree Scientific).
5. 28 G ½ insulin syringes (Becton Dickinson).
6. Ceramic heat lamp.

2.4. Direct Ex Vivo Interrogation of T Cells

1. Culture medium: 45% RPMI 1640 medium, 45% EHAA (Click's) medium, 10% fetal bovine serum (FBS) supplemented with L-glutamine, Gentamicin reagent (Quality Biologicals), Ciprofloxacin (Sigma), and antibiotic/mycotic (Mediatech) (10).
2. Anti-CD3 (clone 2 C11) and anti-CD28 (clone 37.51).
3. $OVA_{323\text{-}339}$ (class II-restricted) peptide (AnaSpec), reconstituted in water at 10 mg/mL.
4. GolgiPlug (brefeldin A) (BD Biosciences).

2.5. Intracellular Cytokine Staining

1. PBS, pH 7.4 (Quality Biological).
2. Surface staining buffer: PBS supplemented with 2% FBS and 0.2% sodium azide.
3. PerCP-conjugated antibody to CD4 (L3T4, BD Biosciences).
4. BD Cytofix/Cytoperm (BD Biosciences).
5. BD Permwash (BD Biosciences).
6. FITC-conjugated anti-IFN-γ (XMG1.2).
7. APC-conjugated anti-IL-4 (BD Biosciences).

2.6. Multiparameter Phospho-FACS

1. PBS, pH 7.4 (Quality Biological).
2. Surface staining buffer: PBS supplemented with 2% FBS and 0.2% sodium azide.
3. Biotin-anti-CD4 (L3T4) (BD Biosciences).
4. Fixation buffer (Formalin diluted to 4% in PBS) (Sigma).
5. Ice cold 90% methanol (Sigma).
6. Blocking buffer: PBS supplemented with 10% FBS, and 500-fold dilution of FcBlock (BD Biosciences).
7. Intracellular staining buffer: PBS supplemented with 1% FBS.
8. Monoclonal mouse antibody to pS6K1 (T389) (Cell Signaling Technology).
9. Monoclonal rabbit antibody to pAkt (S473, clone D9E) (Cell Signal Technology).
10. DyLight 649-conjugated anti-rabbit IgG secondary (Jackson ImmunoResearch).

11. Oregon Green 488-anti-mouse IgG secondary (Invitrogen).
12. Strepdavidin-conjugated-PE (BD Biosciences).

3. Methods

The use of the macrolide antibiotic rapamycin has greatly facilitated the discovery and elucidation of mTOR function (11). While it was originally thought that rapamycin only inhibited the mTORC1 signaling pathway, it is clear that rapamycin can affect mTORC2 as well (12). We find that mTORC2 in lymphocytes is exquisitely sensitive to inhibition by rapamycin even at concentrations as low as 20 nM. Further, it is clear that rapamycin has a wide variety of diverse effects on many cells regulating immune responses (13). In order to study the specific role of mTOR function in T cells, we have taken a genetic approach. First, we have taken advantage of the expertise and generosity of other investigators by breeding previously generated floxed mice with CD4-Cre. Since CD4 is expressed at the double-positive stage of T cell development, breeding CD4-Cre mice with mTOR-floxed mice leads to the efficient deletion of mTOR in both CD4 and CD8 T cells. Further, because CD4 comes up relatively late in T cell development, the ultimate elimination of mTOR protein which is even later in development does not appear to significantly affect the generation of single-positive T cells. By breeding CD4-Cre, mTOR-floxed mice to TCR transgenic mice, we can greatly enhance our ability to specifically activate the genetically altered cell of interest. Further, backcrossing the mice to a congenic marker allows for the ability to track antigen-specific, genetically altered T cell *in vivo* in a wild-type host. Proper genotyping and husbandry are absolutely critical to the success of these assays.

To assess the role of mTOR in T cells in regulating CD4+ T cell function in response to infection, we routinely adoptively transfer the genetically altered T cells into a host prior to infection. *Vaccinia*, a potent inducer of an antiviral Th1 response, can be engineered to express a number of model antigens. Here, we report the use of *Vaccinia*-OVA to induce Th1 differentiation of OT-II (OVA specific) CD4+ T cells. These adoptively transferred cells are marked with the congenic marker Thy1.1 and thus are readily distinguished from host T cells by FACS. CD4+ T cells adoptively transferred into vaccinated hosts become IFN-gamma producing Th1 cells that do not express IL-4, a Th2 cytokine. However, T cells deficient in total mTOR signaling fail to differentiate into Th1 or Th2 cells (2). In as much as the frequency of the antigen-specific T cells is relatively low *in vivo*, we have employed FACS as a means of both interrogating cells for cytokine production as well as multiparameter phospho-FACS to detect mTORC1 and mTORC2 activation in T cells at the single-cell level.

3.1. Generation and Genotyping of CD4-Cre × Floxed Mice

1. CD4-Cre mice on a B6 background should be bred to mice homozygous for floxed mTOR (sometimes, known as *Frap1* or *Mtor*).
2. The F1 generation should be bred back to homozygous floxed founders such that some progenies have a CD4-Cre transgene and be homozygous floxed at the locus of choice.
3. These mice should also be bred to a Thy1.1 (or other congenic marker) and, ideally, to a TCR-transgenic background (this chapter uses OT-II) (see Note 1).
4. At 3 weeks of age, separate pups from dams and sterilely snip 1–2 mm of tail for genotyping.
5. Incubate in 100 μL tail lysis buffer containing 2 μL proteinase K overnight at 55°C (see Note 2).
6. Dilute 2 μL lysate into 48 μL ddH_2O and transfer to a PCR tube (this is the template for PCR).
7. Add 28 μL of Platinum PCR Supermix and 1 μL of each specific primer, diluted to 10 μM.
8. Mix well and run PCR with 52°C annealing temperature (see Note 3).
9. Add 10× loading buffer and run on a 2% TAE agarose gel.
10. Banding patterns indicate genotype (see Table 1).

3.2. Purification of CD4 or CD8 T Cells

1. Isolate splenocytes/lymphocytes.
2. Resuspend in 1 mL Ack red blood cell (RBC) lysis solution (see Note 4).
3. Incubate at RT for 2 min.
4. Wash 1× with 10 mL PBS (see Note 5).
5. Count cells in an appropriate volume of PBS and spin down.
6. Aspirate the supernatant as well as possible.
7. Resuspend pellet in 3 μL magnetic sorting buffer and 0.75 μL Antibody-Biotin Cocktail (CD4 isolation kit) per 10^6 cells.
8. Incubate at 4°C for 15 min (see Note 6).
9. Add 2 μL magnetic sorting buffer and 1.5 μL anti-biotin microbeads (CD4 isolation kit) per 10^6 cells.
10. Incubate at 4°C for 30 min (see Note 6).
11. Add 10–20× the labeling volume magnetic sorting buffer and spin at 300×*g* for 5 min.

12. Equilibrate an LS MACS column with 3 mL magnetic sorting buffer, discarding the flow through.
13. Resuspend pellet in 1 mL of magnetic sorting buffer and add to the column, allowing it to enter the column by gravity flow, collecting the flow through (this is the "negatively selected fraction").
14. Wash 3× with 3 mL of magnetic sorting buffer collecting all the flow through ("negatively selected fraction").
15. The positively selected fractions (column bound) are CD4−; they can be discarded.
16. Count cells and resuspend in PBS or culture media, as necessary.

3.3. Vaccinia Infection and Adoptive Transfer

1. Isolate $CD4^+$ T cells (Subheading 2.2) and resuspend in PBS at 10×10^6 cells/mL.
2. Resuspend at 20×10^6 cells/mL in PBS (see Note 7).
3. Thaw Vaccinia virus from frozen (−80°C) stock. Stocks are kept at 2×10^7 PFU/mL.
4. Place host mice (C57/B6 mice for OT-II transfer/Vaccinia-OVA infection) into a cage and heat them slowly using a radiant ceramic heat lamp (see Note 8).
5. After the mice have heated up, slide them into the immobilizer, tail sticking out, and inject 100 μL (2×10^6 CD4+ T cells) intravenously using the insulin syringe (see Note 9).
6. Bring up Vaccinia virus into a syringe. Inject 100 μL intraperitoneally ($1–2 \times 10^6$ PFU).
7. Let the mice harbor the infection for 3–5 days. Sacrifice the mice and remove spleens (see Note 10).

3.4. Direct Ex Vivo Interrogation of T Cells

1. Splenocytes from host mice or directly from mutant mice should be in single-cell suspension.
2. Resuspend in RBC lysis buffer.
3. Incubate at RT for 2 min.
4. Wash 1× in 10 mL of PBS.
5. Resuspend at 20×10^6/mL in culture medium.
6. Make up 2× stimulation medium (6 μg/mL anti-CD3 and 4 μg/mL anti-CD28, or 20 μg/mL OVA peptide) (see Notes 11 and 12).
7. If performing intracellular cytokine staining, also supplement the stimulation medium with Golgi Plug (see Note 13).
8. Stimulate cells by adding equal volumes (1:1) of cells and 2× stimulation medium and move to an appropriate cell culture vessel.
9. Incubate overnight at 37°C.

3.5. Intracellular Cytokine Staining

1. If necessary, transfer the cells to a round-bottomed plate and pellet the cells (1,500 rpm for 5 min).
2. Stain the surface molecules diluted in 50 μL of surface staining buffer.
3. Incubate for 5–15 min at 4°C.
4. Wash with 150 μL of unsupplemented surface staining buffer.
5. Resuspend the cells in 100 μL BD Cytofix/Cytoperm solution.
6. Incubate at RT in the dark for 15 min.
7. Wash 2× with 100 μL BD Permwash.
8. Resuspend in 50 μL BD Permwash supplemented with:
   (a) FITC anti-IFN-g (1:500)
   (b) APC anti-IL-4 (1:100)
9. Incubate at RT in the dark for 30 min.
10. Wash with 150 μL BD Permwash.
11. Resuspend in 200 μL PBS and run samples on a flow cytometer (see Fig. 1).

3.6. Multiparameter Phospho-FACS

1. Harvest stimulated cells by centrifugation (3 min at 1,500 rpm), transferring to a 96-well U-bottom plate if the stimulation was done in a separate vessel.
2. Wash 1× in 200 μL PBS.
3. Resuspend in 50 μL surface staining buffer supplemented with anti-CD4–biotin (1:500 dilution) (see Note 14).
4. Incubate at 4°C for 15 min.
5. Add 150 μL of unsupplemented surface staining buffer and spin down.
6. Wash 1× in 200 μL unsupplemented surface staining buffer and spin down.
7. Resuspend in 100 μL fixation buffer.
8. Incubate at 37°C for 15 min.
9. Wash with 100 μL PBS and spin down (see Note 15).
10. Wash 1× in 200 μL PBS and thoroughly remove all supernatant.
11. Add 100 μL ice-cold 90% methanol and gently pipet up and down twice (see Note 16).
12. Incubate at –20°C for 20 min.
13. Spin down.
14. Carefully remove the methanol.
15. Wash 2× in 200 μL of PBS and spin down.
16. Resuspend cells in 100 μL blocking buffer.
17. Incubate for 15 min at RT.

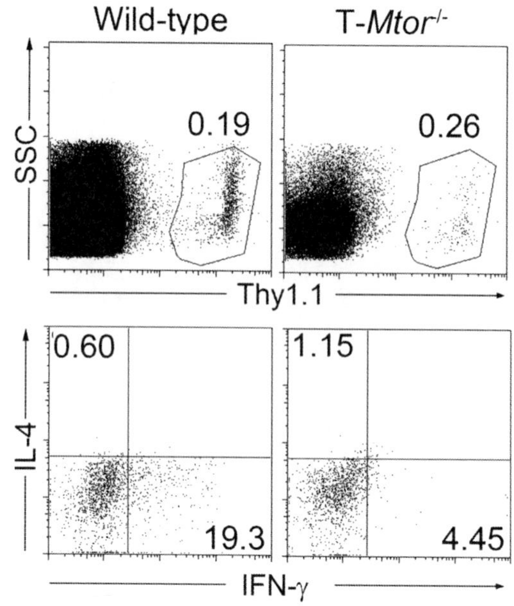

Fig. 1. Wild-type and T-*Mtor*$^{-/-}$ mice were sacrificed. Spleens were harvested and subjected to CD4 purification by magnetic sorting. 2×10^6 CD4 cells were injected intravenously into C57/BL6 host mice previously immunized with 2×10^6 PFU *Vaccinia*-OVA. Four days post transfer, host mice were sacrificed, and splenic cells were rechallenged with 50 μg/mL OVA peptide overnight in the presence of a protein transport inhibitor. Splenocytes were then stained for Thy1.1 (*upper panels*) and intracellularly stained for IL-4 and IFN-γ to assess cytokine production (*lower panels*, gated on Thy1.1+ cells). Wild-type cells produce copious amounts of IFN-γ upon rechallenge, indicating that they have differentiated into Th1 cells. Cells deficient in mTOR fail to produce either cytokines when restimulated.

18. Wash with 100 μL 1% FCS in PBS.
19. Resuspend cells in 50 μL staining buffer supplemented with:
    (a) Mouse anti-phospho-S6K (T389) (1:200)
    (b) Rabbit anti-phospho-Akt (S473) (1:200)
20. Incubate at RT in the dark for 45 min.
21. Wash 1× with 1% FCS in PBS.
22. Resuspend cells in 50 μL intracellular staining solution supplemented with:
    (a) Anti-mouse Oregon Green 488 (1:200)
    (b) Anti-rabbit DyLight 649 (1:200)
    (c) SA-conjugated PE (1:500)
23. Incubate at RT in the dark for 45 min.
24. Wash 2× with 200 μL unsupplemented intracellular staining buffer.
25. Resuspend in 200 μL PBS and run on cytometer (See Note 16, Note 17, and Fig. 2).

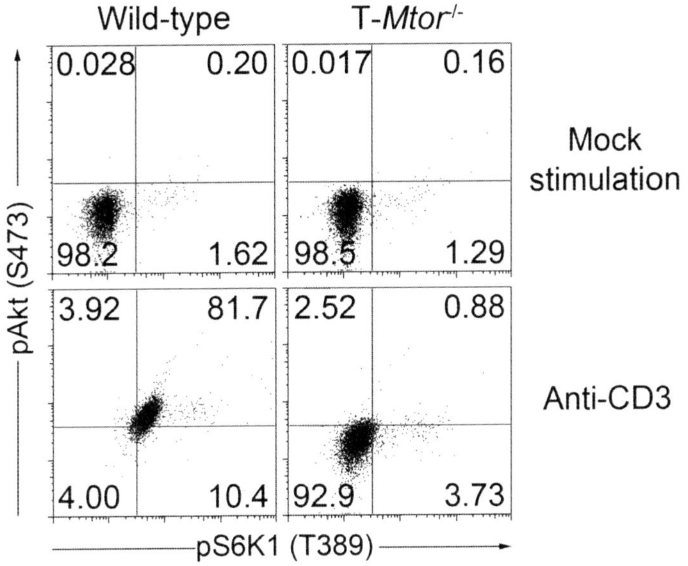

Fig. 2. Spleen and lymph node cells were harvested from wild-type and T-*Mtor*$^{-/-}$ mice. After RBC lysis and extensive washing, splenocytes were stimulated with 3 μg/mL anti-CD3 and 2 μg/mL anti-CD28 overnight. Cells were surface stained with anti-CD4–biotin, fixed in 4% formalin, and permeablized with 100% methanol. After a 10-min block in 10% FCS, cells were stained intracellularly for phosphorylation of S6K1 and Akt. Plots are gated on CD4 cells. Voltage was adjusted based on secondary-only controls and gates were set with an unstimulated control. Wild-type cells activate both mTORC1 (pS6K1) and mTORC2 (pAkt) signaling when stimulated, but mTOR-deficient T cells lack activation of both mTORCs.

4. Notes

1. A congenic marker (like Thy (CD90), CD45, or Ly isoforms) facilitates the ability to track adoptively transferred T cells. Likewise, a TCR transgene recognizing a cognate antigen expressed by *Vaccinia* virus (like OT-II and OVA-expressing *Vaccinia*) can be used to both stimulate and identify such cells *in vivo*. Backcrossing several generations to the host strain (B6, in this case) is highly recommended if the floxed line was made on a different background.
2. Our laboratory uses 96-well PCR plates to help streamline the genotyping by being able to use multichannel pipettors. After an overnight incubation, tails should be almost completely dissolved.
3. Optimizing for individual thermocyclers and primer pairs is highly recommended.
4. Red cell lysis uses a hypotonic solution (Ack lysing buffer, NH_4Cl) to lyse RBCs. It is critical that the incubation during hemolysis is kept as short as possible to prevent damage to lymphocytes.

5. Unless otherwise stated, the term "wash" refers to resuspension of cell pellets in the solution mentioned followed immediately by centrifugation at 300×*g* for 5 min.

6. Incubations at 4°C during magnetic sorting can be extended for up to 60 min with no effects on purity, yield, or viability.

7. Syringes typically have a void volume of 100 μL; it is critical to account for this when preparing your sample so that you have enough cells to inject all the needed host mice.

8. It is critical that your donor mice have a distinguishing cytometric marker (our laboratory uses Thy1.1 and Thy1.2, but CD45.1 and CD45.2 or Ly markers can work).

9. Heating the mice allows easy visualization of the tail veins; they are on either side of the midline of the tail. When properly performed, there should be very little pressure on the plunger of the syringe when depressed. Start the furthest from the base of the tail. If pressure is felt, try injecting the vein again, slightly closer to the base of the tail. Precision generally comes with practice. Retro-orbital injections are easier to perform, but the delivery is not always consistent.

10. Spleens from *Vaccinia*-infected hosts should be generally larger than mice that have not been immunized.

11. Anti-CD3 and anti-CD28 cross-link all TCRs and deliver costimulation. This is done via the Fc receptor on APCs. This stimulation should be used for non-TCR transgenic T cells or when all cells should be activated. OVA peptide stimulates only the OT-II cells. For the *Vaccinia* experiment listed in Subheading 2.3, OVA peptide should be used as the stimulation of choice.

12. Always include a no-stimulation control, containing only medium and Golgi Plug. This is a critical experimental control for setting negative gates on the flow cytometer.

13. Golgi Plug (brefeldin A) is ideal for overnight stimulations and the majority of cytokines. Some cytokine procedures work better with Golgi Stop (monensin), which is ideal for short-term stimulations. This should be optimized based on your particular needs.

14. Some antigens, especially cell surface antigens, become degraded with the harsh methanol permeabilization step in this protocol. To overcome this hurdle, our laboratory stains cell surface molecules with a biotinylated antibody prior to fixation. The marker can then be detected with fluorochrome-conjugated streptavidin during the secondary incubation step.

15. Cells can be stored at 4°C at this point. It is critical for consistency that all the intracellular phospho-staining is done at the same time to compare across conditions.

16. 90% ice-cold methanol should be added drop by drop, very gently.
17. If the samples are not to be run immediately, a second fixation step in fixation buffer is recommended.

Acknowledgments

The authors would like to thank members of the Powell laboratory for technical assistance in optimizing these models. In addition, we would like to thank Dr. Sara C. Kozma (U. Cincinnati) for generating the original floxed mouse lines. This work was supported by R01AI077610-01A2.

References

1. Sabatini, D. M. (2006) mTOR and cancer: insights into a complex relationship, *Nat Rev Cancer 6*, 729–734.
2. Delgoffe, G. M., Kole, T. P., Zheng, Y., Zarek, P. E., Matthews, K. L., Xiao, B., Worley, P. F., Kozma, S. C., and Powell, J. D. (2009) The mTOR kinase differentially regulates effector and regulatory T cell lineage commitment, *Immunity* ***30***, 832–844.
3. Araki, K., Turner, A. P., Shaffer, V. O., Gangappa, S., Keller, S. A., Bachmann, M. F., Larsen, C. P., and Ahmed, R. (2009) mTOR regulates memory CD8 T-cell differentiation, *Nature* ***460***, 108–112.
4. Rao, R. R., Li, Q., Odunsi, K., and Shrikant, P. A. The mTOR kinase determines effector versus memory CD8+ T cell fate by regulating the expression of transcription factors T-bet and Eomesodermin, *Immunity 32*, 67–78.
5. Kopf, H., de la Rosa, G. M., Howard, O. M., and Chen, X. (2007) Rapamycin inhibits differentiation of Th17 cells and promotes generation of FoxP3+ T regulatory cells, *Int Immunopharmacol 7*, 1819–1824.
6. Valmori, D., Tosello, V., Souleimanian, N. E., Godefroy, E., Scotto, L., Wang, Y., and Ayyoub, M. (2006) Rapamycin-mediated enrichment of T cells with regulatory activity in stimulated CD4+ T cell cultures is not due to the selective expansion of naturally occurring regulatory T cells but to the induction of regulatory functions in conventional CD4+ T cells, *J Immunol 177*, 944–949.
7. Battaglia, M., Stabilini, A., and Roncarolo, M. G. (2005) Rapamycin selectively expands CD4+CD25+FoxP3+ regulatory T cells, *Blood* ***105***, 4743–4748.
8. Wang, B., Xiao, Z., Chen, B., Han, J., Gao, Y., Zhang, J., Zhao, W., Wang, X., and Dai, J. (2008) Nogo-66 promotes the differentiation of neural progenitors into astroglial lineage cells through mTOR-STAT3 pathway, *PLoS One 3*, e1856.
9. Guertin, D. A., and Sabatini, D. M. (2007) Defining the role of mTOR in cancer, *Cancer Cell 12*, 9–22.
10. Zheng, Y., Delgoffe, G. M., Meyer, C. F., Chan, W., and Powell, J. D. (2009) Anergic T cells are metabolically anergic, *J Immunol 183*, 6095–6101.
11. Guertin, D. A., and Sabatini, D. M. (2009) The pharmacology of mTOR inhibition, *Sci Signal 2*, pe24.
12. Sarbassov, D. D., Ali, S. M., Sengupta, S., Sheen, J. H., Hsu, P. P., Bagley, A. F., Markhard, A. L., and Sabatini, D. M. (2006) Prolonged rapamycin treatment inhibits mTORC2 assembly and Akt/PKB, *Mol Cell 22*, 159–168.
13. Delgoffe, G. M., and Powell, J. D. (2009) mTOR: taking cues from the immune microenvironment, *Immunology 127*, 459–465.

Chapter 21

Evaluating the Therapeutic Potential of mTOR Inhibitors Using Mouse Genetics

Huawei Li, Jennifer L. Cotton, and David A. Guertin

Abstract

Extensive efforts are underway to develop small-molecule inhibitors of the mammalian target of rapamycin (mTOR) kinase. It is hoped that these inhibitors will have widespread clinical impact in oncology because mTOR is a major downstream effector of PI3K signaling, one of the most frequently activated pathways in cancer. In cells, mTOR is the catalytic core subunit of two distinct complexes, mTORC1 and mTORC2, which are defined by unique mTOR-interacting proteins and have unique functions downstream of PI3K. Two classes of mTOR inhibitors are currently being evaluated as cancer therapeutics: rapamycin and its analogs, which partially inhibit mTORC1 and in some cell types mTORC2, and the recently described ATP-competitive inhibitors, which inhibit the kinase activity of both complexes. Although small molecules that selectively target mTORC2 do not yet exist, experiments using mouse genetics suggest that a theoretical mTORC2 inhibitor may have significant therapeutic value. Here, we discuss an approach to model mTOR complex specific inhibitors using mouse genetics and how it can be applied to other gene products involved in oncogenic signaling to which inhibitors do not exist.

Key words: Drug discovery, Cancer, mTORC2, Mouse model, Conditional knockout, PTEN, Prostate cancer

1. Introduction

Aberrantly active PI3K signaling is a characteristic of many human cancers, and as such, the effort to develop PI3K pathway inhibitors is intense (Reviewed in ref. 1). Much of this effort focuses on targeting the mammalian Target of Rapamycin (mTOR) kinase, which mediates many of the downstream functions of PI3K signaling. In cells, mTOR nucleates two complexes (mTORC1 and mTORC2) with distinct substrates and functions that are dictated in part by unique regulatory subunits that are specific to each complex (Fig. 1) (2).

Thomas Weichhart (ed.), *mTOR: Methods and Protocols*, Methods in Molecular Biology, vol. 821,
DOI 10.1007/978-1-61779-430-8_21, © Springer Science+Business Media, LLC 2012

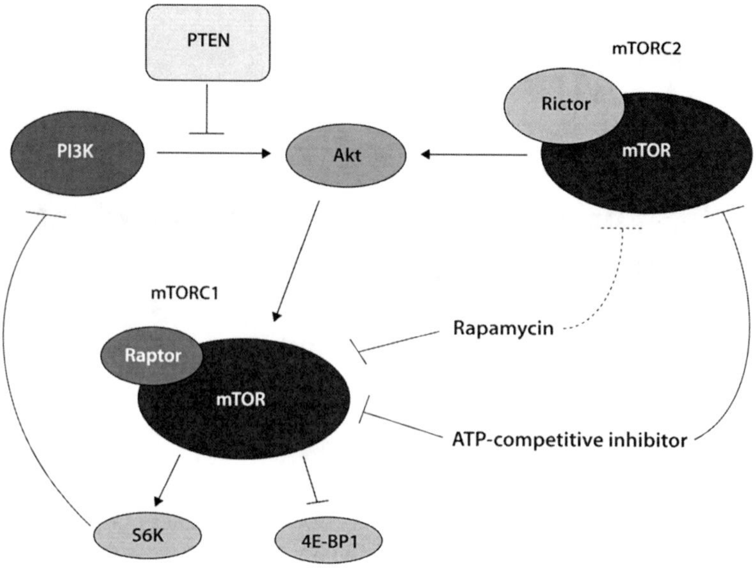

Fig. 1. Simplified model of the PI3K-mTOR pathway. mTOR is the catalytic subunit of two complexes called mTORC1 and mTORC2, both of which function downstream of PI3K. Akt is the most proximal effector of PI3K activity, which is negatively regulated by the PTEN tumor suppressor. Akt activates mTORC1, which contains the essential regulatory subunit Raptor. Active mTORC1 phosphorylates S6K and 4E-BP1, which increases protein and lipid biosynthesis and leads to cell growth. mTORC2, which contains the essential subunit Rictor, directly phosphorylates Akt, resulting in full Akt activation. Rapamycin partially inhibits mTORC1 and additionally inhibits mTORC2 but only in certain cell types. ATP-competitive inhibitors, which target the kinase domain of mTOR, inhibit both mTORC1 and mTORC2. Selective inhibitors that can specifically target either mTORC1 or mTORC2 have not been described.

For example, mTOR associates with an essential regulatory protein called Raptor in mTORC1. By contrast, mTOR associates with a distinct regulatory protein called Rictor in mTORC2 (Fig. 1). Depleting cultured cells of Raptor or Rictor by RNA-interference is commonly used to selectively reduce either mTORC1 or mTORC2 signaling, respectively. Rapamycin analogs were the first mTOR inhibitors evaluated in clinical trials for use in oncology and some are now FDA-approved cancer therapeutics (3). However, rapamycin only partially inhibits mTORC1, which has limited its clinical success as a cancer therapy. This is not entirely unexpected as the FKBP12-rapamycin complex binds mTOR outside the kinase domain, functioning not as a catalytic domain inhibitor, but as an allosteric inhibitor. A second factor limiting rapamycin's efficacy is that active mTORC1 dampens PI3K activity by way of a strong negative feedback loop; rapamycin relieves this feedback inhibition and potentiates PI3K activity and Akt-mediated cell survival. A new class of small molecules targeting the mTOR kinase domain with high specificity recently became available and it is anticipated that this next generation of mTOR inhibitors will have broader clinical success compared to rapamycin (3–5). Because mTOR kinase inhibitors are ATP-competitive inhibitors, they have the advantages of more directly inhibiting mTOR kinase activity while simultaneously targeting both mTORC1 and mTORC2. Interestingly, prolonged rapamycin treatment can additionally inhibit mTORC2 activity by blocking complex assembly, but this is functionally observed only in a subset of cell types (6). The reason behind the cell specificity of this phenomenon is still a mystery. The ATP-competitive mTOR inhibitors have superior inhibitory effects on mTORC1 activity compared to rapamycin (Reviewed in refs. 3–5). For example, ATP-competitive mTOR inhibitors are far more potent inhibitors of protein translation. Evaluation of the mTOR kinase inhibitors as cancer therapeutics is just underway, and it is unknown whether such inhibitors will have toxic side effects or whether loss of mTORC1-dependent feedback loops will be problematic. An unresolved question is whether there are any benefits to developing mTOR inhibitors that selectively target either mTORC1 or mTORC2. Because the mTOR ATP-competitive inhibitors target both complexes, determining the relevance of additionally inhibiting mTORC2 verses more completely inhibiting mTORC1 toward a clinical response is difficult.

Selective mTOR complex inhibitors are not currently available. Therefore, alternative approaches must be taken to evaluate the function of each individual mTOR complex in cancer. One possibility is to use murine cancer models in combination with selective inhibition of either complex by a gene-knockout strategy. We tested this approach by inhibiting mTORC2 in a well-known mouse prostate cancer model by deleting *rictor* (7). In this model, mTORC2 inhibition blocks prostate tumor development, but importantly, normal nontransformed prostate epithelial cells do not require mTORC2 activity. In the following sections, we describe this approach and how it can be useful in evaluating the potential of theoretical inhibitors that target signaling molecules important in cancer. Modeling cancer in mice is a widely used approach to study tumor biology *in vivo*. Crossing conventional deletion alleles of a particular gene into a tumor model is one strategy to study that gene's function in tumor development. Homologous recombination in mice has been used to generate conventional knockout models of mTORC1 and mTORC2 by deleting *raptor*, *rictor*, or other essential regulatory subunits (Table 1); however, disrupting either complex results in embryonic lethality. This precludes the study of tumor development in mice globally lacking mTORC1 or mTORC2 activity. Notably, heterozygous deletions of key mTOR regulators can be useful to study partial inactivation of mTOR signaling in tumor formation (discussed below) (7).

Cre-Lox mediated conditional knockout strategies have been the main recourse in studying essential genes in mice. Conditionally

Table 1
Mouse models of mTORC1 and mTORC2

Gene	Knockout strategy	Complex affected	Phenotype	References
Partial loss-of-function models				
mTOR	Ethyl-nitroso-urea mutagenesis	Both	Embryonic lethal (e12.5)	(22)
Conventional Knockout models				
mTOR	Homologous recombination	Both	Embryonic lethal (preimplantation)	(23)
mTOR	Cre-mediated germ-line deletion	Both	Embryonic lethal (preimplantation)	(24)
mTOR	Gene trap mutagenesis	Both	Embryonic lethal (preimplantation)	(25)
mTOR	Kinase dead knock-in	Both	Embryonic lethal (preimplantation)	(26)
Raptor	Homologous recombination	mTORC1	Embryonic lethal (preimplantation)	(25)
Rictor	Gene trap mutagenesis	mTORC2	Embryonic lethal (e10.5)	(25)
Rictor	Gene trap mutagenesis	mTORC2	Embryonic lethal (e10.5)	(27)
mSIN1	Homologous recombination	mTORC2	Embryonic lethal (not described)	(28)
mLST8	Homologous recombination	mTORC2[a]	Embryonic lethal (e10.5)	(25)
Conditional Knockout models				
mTOR	Promoter through exon 5 floxed	Both	–	(24)
Raptor	Exon 6 floxed	mTORC1	–	(29)
Raptor	Exon 6 floxed	mTORC1	–	Peterson et al. (unpublished)
Rictor	Exon 3 floxed	mTORC2	–	(30)
Rictor	Exons 4–5 floxed	mTORC2	–	(29)

[a]mLST8 tightly associates with both mTOR complexes but is only required for mTORC2 function in development

knocking out the genes encoding *raptor* or *rictor* in a particular tissue allows targeted inactivation of mTORC1 or mTORC2, respectively, bypassing the embryonic lethality of a whole body knockout (Table 1). In theory, this strategy is only limited by the availability and specificity of Cre-recombinases. Most Cre drivers are constitutively active under control of tissue specific promoters although regulatable Cre drivers are increasingly being used (Reviewed in ref. 8). Combining a mouse model of cancer with conditional alleles of either *raptor* or *rictor* can be used to study each individual mTOR complex in tumor formation. For example, we developed a mouse model of mTORC2 function in a *PTEN*-deletion dependent prostate cancer model by deleting both *rictor* and *PTEN* specifically in the prostate epithelium (7). A similar strategy was taken to study *mTOR* and the ER chaperone protein GRP78/BiP in prostate cancer (9, 10). Of note, targeting *mTOR* will not distinguish between the functions of mTORC1 and mTORC2 but may be useful for predicting the effectiveness of mTOR kinase inhibitors. However, deleting *mTOR* destabilizes the mTOR complexes and therefore differs from the effects of a typical kinase inhibitor that would preserve complex integrity and reversibly suppresses mTOR activity, likely in a dose-dependent manner. Thus, expressing an mTOR kinase-dead allele might be a more faithful representation of the effects of an mTOR kinase inhibitor. This is nicely demonstrated in a report comparing the differential effects of knocking out *PIK3CA* versus expressing a kinase dead version of the gene (11).

The *PTEN*-deletion model of prostate cancer has several advantages for studying *rictor*/mTORC2 in tumor development: (1) Human prostate cancer often associates with *PTEN* deficiency; (2) Deleting *PTEN* induces tumorigenesis by activating PI3K signaling; (3) Tumors occur in 100% of the mice with *PTEN* deleted in the prostate epithelium; (4) the latency to tumor formation is relatively short and occurs with well-defined kinetics; (5) and tumors progress in a manner similar to human prostate cancer (Fig. 2) (12). In this model, Cre-Lox mediated deletion of *PTEN* "floxed" (i.e., $PTEN^{loxP/loxP}$) alleles is driven by the ARR_2PB-*Cre* (*PB-Cre4*) transgene, the expression of which is postnatal and mostly restricted to the prostate epithelium (13). Limited Cre activity is also detectable in testes and ovaries and importantly, if females carry the *PB-Cre4* transgene, recombination can occur in maternally derived floxed alleles, generating progeny ubiquitously deleted for the maternal copy of that allele ((13) and unpublished observations).

Generating a mouse harboring the necessary experimental allele combinations can be expensive and labor intensive if not carefully planned because one must combine the *PB-Cre4* transgene, two $PTEN^{loxP}$ alleles, and two $rictor^{loxP}$ alleles (i.e., *PB-Cre4* $PTEN^{loxP/loxP}$ $rictor^{loxP/loxP}$) in a male. Once founder mice are generated, future breeding should only occur between the offspring to maintain strain composition. All possible experimental and control

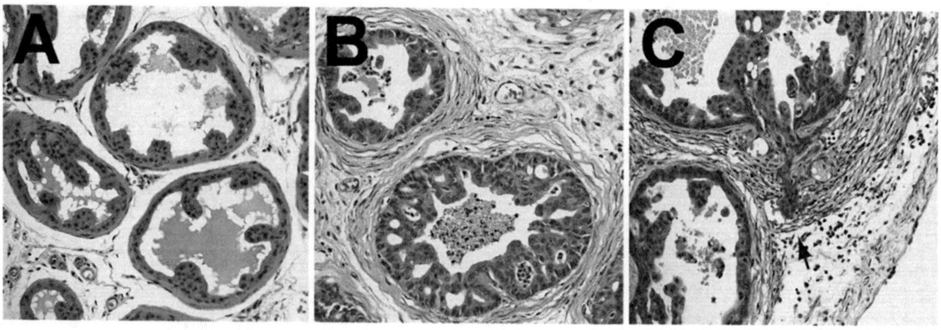

Fig. 2. The *PTEN* prostate cancer model. (**a**) H&E stain of a normal mouse prostate epithelium. (**b**, **c**) Representative images of prostate adenocarcinoma caused by *PTEN* deletion. Note the malignant cells invading the stroma in panel (**c**) (*arrow*). Because PTEN is the main negative regulator of PI3K activity, this mouse model of cancer is useful to study the role of a particular gene in tumorigenesis driven by PI3K pathway activation. Images are with a 20× objective lens (×200 magnification).

mice can be generated from crosses between *PB-Cre4* $PTEN^{loxP/+}$ $rictor^{loxP/+}$ males and $PTEN^{loxP/+}$ $rictor^{loxP/+}$ females. Modifying subsequent crosses (such as crossing to $PTEN^{loxP/loxP}$ $rictor^{loxP/loxP}$ females) can improve the probability of generating certain combinations. Generating a significant number of age and/or litter matched mice is challenging; therefore, it is important to establish a large number of breeding pairs. We use *PB-Cre4* $PTEN^{loxP/loxP}$ $rictor^{+/+}$ males as positive controls (i.e., they develop prostate cancer), and *PB-Cre4* $PTEN^{+/+}$ $rictor^{+/+}$ males, or any male lacking *PB-Cre4* as negative controls (i.e., prostate cancer free). It is very important to characterize the extent of tumor formation in the positive controls because strain composition, which varies depending upon the background of the mice, can affect tumor incidence and severity (14). Heterozygous (*loxP/+*) combinations will also be obtained from the described crosses and can be used to determine haploinsufficiencies. Importantly, in our study we also characterized *PB-Cre4* $PTEN^{+/+}rictor^{loxP/loxP}$ males to determine the effects of *rictor* deletion (i.e., mTORC2 inactivation) alone on prostate cells. This is critical for determining the role of mTORC2 in nontransformed cells.

2. Materials

2.1. Digesting Tail DNA

Tail Lysis Buffer (with Proteinase K).

180 μL Tail Lysis Buffer (DirectPCR (Tail), Viagen, #102-T).

20 μL Proteinase K (Lambda Biotech, #DB0451).

2.2. PCR Primers for Genotyping

PTEN	
P1:	5′-ACT CAA GGC AGG GAT GAG C-3′
P2:	5′-AAT CTA GGG CCT CTT GTG CC-3′
P3:	5′-GCT TGA TAT CGA ATT CCT GCA GC-3′
Rictor	
T41:	5′-ACT GAA TAT GTT CAT GGT TGT G-3′
T45:	5′-GAC ACT GGA TTA CAG TGG CTT G-3′
Pb-Cre	
COO1:	5′-ACC AGC CAG CTA TCA ACT CG-3′
COO2:	5′-TTA CAT TGG TCC AGC CAC C-3′
COO3:	5′-CTA GGC CAC AGA ATT GAA AGA TCT-3′
COO4:	5′-GTA GGT GGA AAT TCT AGC ATC ATC C-3′

2.3. Histology, Immunohistochemistry, and Protein Biochemistry

1. Stains, antibodies and kits useful for characterizing mTOR pathway activity in tumors and cancer cells (see *Table* 2 *for more details*): Hematoxylin & eosin (H&E) stain: For routine histology.
2. Trichrome stain: For visualizing connective tissue surrounding tumor cells.
3. Ki67 and Cleaved Caspase 3 antibodies: For proliferation and apoptosis, respectively, by Immunohistochemistry (IHC).
4. BrdU labeling: Measuring proliferation of tumor cells.
5. TUNEL stain: Measuring apoptosis of tumor cells.
6. Phospho-specific and total protein antibodies to S6K, S6, 4E-BP1 (to measure mTORC1 activity) and Akt (to measure mTORC2 activity) by IHC and Western blot.
7. Phospho-specific and total protein antibodies to GSK3β, FoxO1/3, TSC2, and PRAS40 to evaluate AKT activity by IHC and Western blot.
8. AKT1/PKBα Immunoprecipitation Kinase Assay (Millipore, #P31749): To measure *in vitro* AKT kinase activity.

2.4. Immunohistochemistry

1. 10% Neutral-buffered Formalin (NBF).
2. Tissue Cassettes.
3. Xylenes.
4. 100% Ethanol.
5. Wash Buffer (1× PBS/0.1% Tween)

   10× PBS: 2.03 g Sodium Phosphate Monobasic, 2.17 g Sodium Phosphate Dibasic heptahydrate, 85.0 g NaCl, dH_2O to 1.0 L.

Table 2
Useful antibodies for studying mTOR signaling in cancer

Antibody	Application	Company	Catalog #
mTOR	W	CST	#2983
Tuberin/TSC2 (28A7)	W	CST	#3635
Phospho-tuberin/TSC2 (T1462)	W	CST	#3617
Rictor	W	CST	#2140
Rictor	IHC	Santa Cruz	#50678
Raptor	W	CST	#2280
FoxO3a	W	CST	#2497
Phospho-FoxO1-T24/FoxO3a-T32	W	CST	#9464
Phospho-p70 S6 kinase-T389	W	CST	#9234
p70 S6 kinase α (H-160)	W	Santa Cruz	sc-9027
PTEN	W	CST	#9559
PTEN (D4.3) XP™	IHC	CST	#9188 S
AKT	W	CST	#9272
Phospho-AKT-T308	W	CST	#4056
Phospho-AKT-S473	W	CST	#4058
Phospho-Akt (Ser473) D9E XP™	IHC	CST	#4060
GSK-3β	W	CST	#9315
Phospho-GSK-3β-S9	W	CST	#9323
S6	W	CST	#2217
Phospho-S6-S240/246	W	CST	#4838
pS6 Ribosomal protein (S235/236) (D572.2.2E)	IHC	CST	#4858 S
4EBP1	W	CST	#9644
Phospho-4EBP1-T37/46	W	CST	#2855
Phospho-4EBP1-S65	W	CST	#9456
PRAS40	W	Upstate	#05-988
PRAS40-T246	W	BioSource	#44-1100 G
Ki67	IHC	BD Pharmingen	#550609
Cleaved Caspase-3 (D175)	IHC	CST	#9661

W Western blot, *IHC* immunohistochemistry, *CST* Cell Signaling Technologies

1× PBS/0.1% Tween (1× PBST): Dilute 100 mL 10× PBS + 0.5 mL Tween-20 with enough dH_2O to a final volume of 1.0 L.

6. Citrate Buffer (10 mM Sodium Citrate Buffer)

   Add 2.94 g of sodium citrate trisodium salt dihydrate to 1 L dH_2O, then adjust the buffer's pH to 6.0. Alternatively, one could dilute a 10× stock.

7. 3% Hydrogen Peroxide: Dilute 10 mL of 30% H_2O_2 with 90 mL H_2O.
8. SignalStain® Ab Diluent (#8112L, Cell Signaling Technology).
9. Elite VECTASTAIN ABC Kit (Vector Labs, #PK6101).
10. DAB Substrate Kit for Peroxidase (Vector Labs, #SK-4100).
11. Cytoseal™60 (Richard-Allan Scientific, #8310-4).

2.5. Cell Lysis

1. Cell Lysis Buffer:

   50 mM 4-(2-hydroxyethyl)-1-piperazineethanesulfonic acid–potassium hydroxide (Hepes–KOH) pH 7.4,

   40 mM Sodium Chloride (NaCl),

   2.0 mM Ethylenediaminetetraaceticacid (EDTA),

   1.5 mM Sodium Orthovanadate ($NaVO_4$),

   50 mM Sodium Fluoride (NaF),

   10 mM Sodium Pyrophosphate ($Na_4P_2O_7$),

   10 mM Sodium β-Glycerophosphate ($C_3H_9o_6PNa$),

   1% Triton X-100 (if performing immunopurifications of intact complexes, use 0.3% CHAPS) (see Note 1),

   1 tablet SigmaFast Protease Inhibitor (1 tablet/100 mL) (see Note 2).

2.6. Tissue Lysis

1. Tissue Lysis Buffer (115% strength):

   57.5 mM 4-(2-hydroxyethyl)-1-piperazineethanesulfonic acid–potassium hydroxide (Hepes–KOH) pH 7.4,

   46 mM Sodium Chloride (NaCl),

   2.3 mM Ethylenediaminetetraaceticacid (EDTA),

   1.5 mM Sodium Orthovanadate ($NaVO_4$),

   57.5 mM Sodium Fluoride (NaF),

   11.5 mM Sodium Pyrophosphate ($Na_4P_2O_7$),

   11.5 mM Sodium β-Glycerophosphate ($C_3H_9O_6P.Na$),

   tablet SigmaFast Protease Inhibitor (1 tablet/100 mL).

2. Detergent mix (1:20:10)

   20% Sodium Dodecylsulfate (SDS) (1 part),

   10% Sodium Deoxycholate (20 parts),

   20% Triton (10 parts).

3. Polytron®PT-MR2100 homogenizer (Kinematica AG).
4. 5-mL Polystyrene round-bottom tubes.

2.7. Western Blots

1. PVDF 0.45-μM membranes (Immobilon-P, Millipore, Cat. #IPVH00010).
2. Chromatography paper.
3. Transfer buffer

   Mix 8.85 g CAPS (Sigma, #C2632-250 G), 13 NaOH pellets (Fischer, #S318-500), 400 mL 100% ethanol, and 3.6 L H_2O to bring final volume to 4.0 L.
4. Western Lighting®-ECL chemiluminescence reagent (Perkin-Elmer, # NEL100001EA).

3. Methods

Here, we describe methods useful for characterizing the role of mTORC2 in a mouse prostate cancer model. An approach such as this can provide valuable information about a particular gene product's role in the early stages of tumor formation. (We attempted to generate a model of mTORC1 specific-inhibition in parallel, but the wild type *raptor* allele was linked to the *PB-Cre4* transgene and despite numerous rounds of breeding, we were unable to uncouple them (Guertin, Sabatini unpublished]). In the *rictor* conditional knockout model, the tumor-initiating event – *PTEN* deletion – occurs simultaneously with inactivation of the potential drug target – mTORC2. However, in a clinical setting, patients are unlikely to be treated at the moment of tumor initiation. Thus, strategies investigating more advanced, therapeutically relevant stages of cancer should be considered in follow-up studies to such an approach. The potential negative impact of inhibiting mTORC2 on the normal biology of other tissues (as would be the case with a systemically delivered small-molecule inhibitor) also cannot be determined by this approach. To more completely understand the importance of mTORC2, or any essential gene, in normal versus cancer cells a much wider spectrum of tissue types must be studied. Such studies are important in helping to predict potential side effects caused by inhibiting the target. Addressing *rictor*-dependent processes in cancers driven by other mutations and originating in other tissues, is equally as important. Finally, the physiological difference between mice and humans is always challenging for the cancer biologist to reconcile. Thus, while genetic studies can serve as guides for drug development, it is wise to consider them in the context of other approaches (see Notes 3–5 for complementary approaches).

3.1. Preparing Mouse Tail DNA for Genotyping

1. At time of weaning, use a razor blade to cut 2–3 mm piece off the mouse tail (see Note 6).
2. Incubate tail pieces at 55°C overnight in 200 μL Tail Lysis Buffer (with Proteinase K).
3. The next morning, incubate samples for 45 min at 85°C.
4. Centrifuge the samples at least 5 min at full speed on a bench-top centrifuge.
5. Store the samples at 4°C (short term) or -20°C (long term).
6. Use genomic DNA samples in PCR genotyping reactions (see Fig. 3 for examples of the expected results for *PTEN* and *rictor* conditional alleles).

3.2. PCR Programs

1. *PCR Program for PTEN conditional:*

   Step 1: 94°C for 1 min,

   Step 2: 94°C for 30 s,

   Step 3: 61°C for 90 s,

   Step 4: 72°C for 1 min,

   Step 5: Repeat steps 2–4 (35 cycles),

   Step 6: 72°C for 10 min,

   Step 7: HOLD at 4°C

   WT allele: 900 bp

   Targeted allele: 1,000 bp

   Exon 5 deletion: 300 bp
2. *PCR Program for Rictor conditional:*

   Step 1: 94°C for 1 min,

   Step 2: 94°C for 30 s,

   Step 3: 55°C for 30 s,

   Step 4: 72°C for 1 min,

   Step 5: Repeat steps 2–4 (40 cycles),

   Step 6: 72°C for 10 min,

   Step 7: HOLD at 4°C

   WT allele: 773 bp

   Targeted allele: 890 bp

   Ex3del allele: 281 bp
3. *PCR Program for Pb-Cre:*

   Step 1: 94°C for 1 min,

   Step 2: 94°C for 30 s,

   Step 3: 55°C for 30 s,

   Step 4: 72°C for 45 s,

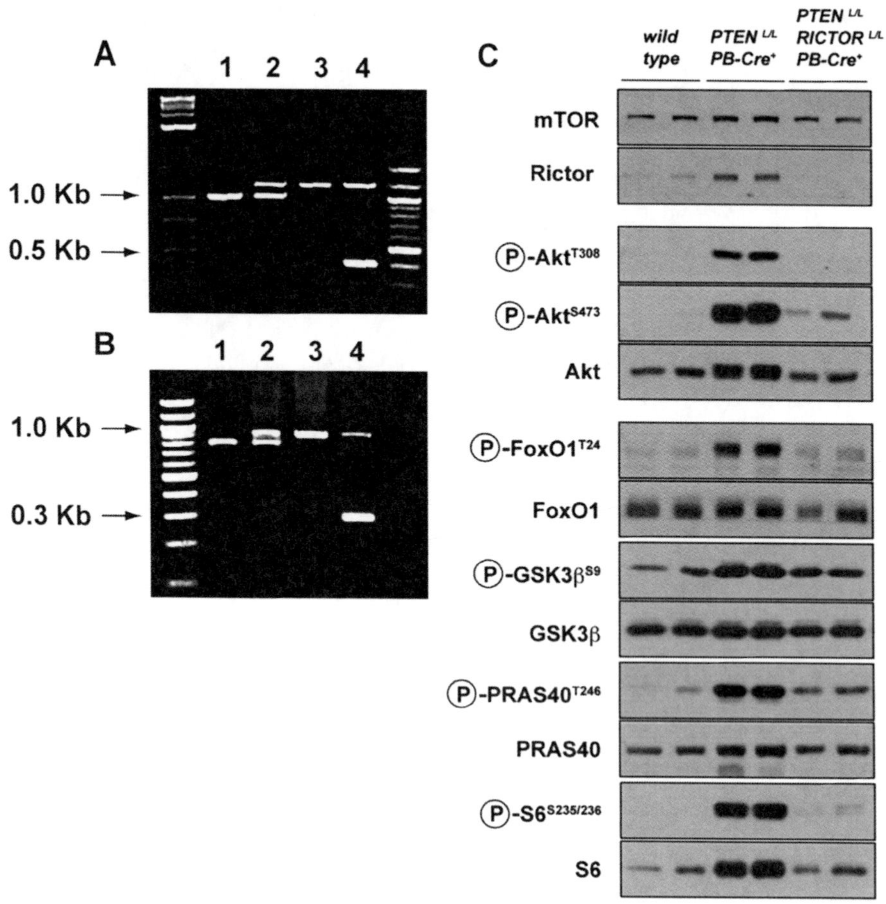

Fig. 3. Example PCR and Western blots for *PTEN* and *rictor* conditional mice. (**a**) PCR for *PTEN* conditional mice. Examples of the expected PCR products using genomic DNA from the tissue of wild type (lane 1), *loxP/+* (lane 2), *loxP/loxP* (lane 3), and *loxP/loxP cre+* (lane 4) mice are shown (originally described in ref. 12). (**b**) PCR for *rictor* conditional mice. Examples from wild type (lane 1), *loxP/+* (lane 2), *loxP/loxP* (lane 3), and *loxP/loxP cre+* (lane 4) mice are shown [originally described in ref. 30). (**c**) Western blots of mTORC2 pathway components. Protein lysates were prepared from wild type, *PTEN*-deficient, and *PTEN rictor* doubly deficient prostate tissue and probed with the indicated antibodies to mTORC2 components, to total and phosphorylated AKT, and to downstream AKT pathway components. Reprinted from *Cancer Cell*, 15, Guertin et al., mTOR Complex 2 Is Required for the Development of Prostate Cancer Induced by Pten Loss in Mice, 148–158, 2009, with permission from Elsevier.

Step 5: Repeat steps 2–4 (35 cycles),
Step 6: 72°C for 10 min,
Step 7: HOLD at 4°C
Internal Control: 324 bp
Transgene: 199 bp

3.3. Prostate Tissue Harvest and Sectioning for Immunohisto-chemistry

1. *PB-Cre4 PTEN$^{loxP/loxP}$* mice develop mPIN (murine prostatic intraepithelial neoplasia) by 6 weeks of age and invasive adenocarcinoma by 9–10 weeks (which can reportedly progress to metastasis by as early as 12 weeks (12)). Examine prostate cancer progression by histology from the earliest signs of hyperplasia

(4–6 weeks) to invasion (9–12 weeks) to determine the age of mice in which the stage of interest presents.

2. Humanely sacrifice mice at desired stage of cancer progression using CO_2 inhalation or other IACUC-approved method.
3. Remove the entire genitourinary (GU) block, which contains four pairs of prostate lobes (ventral prostate, lateral prostate, dorsal prostate, and coagulating gland or anterior prostate) that are characteristic of the mouse prostate (see Note 7).
4. Immediately after the GU block is removed, descriptive features, length, mass, and photographs of the prostate should be recorded.
5. Place tissues in cassettes to prepare for routine immunohistochemistry
6. Fix tissues for 4 h at room temperature in 10% neutral buffered formalin (NBF) then processes for sectioning or store in 70% ethanol at 4°C.
7. Embed tissues in paraffin using the urethra for orientation.
8. Using the urethra as a landmark cut 5 μM transverse sections containing all four pairs of lobes, or the particular lobes of interest. Most institutions have a histology core that can section tissues and provide H&E and other standard stains of tissue sections (see Note 8).

3.4. Immunohistochemistry

1. (Adapted from Cell Signaling Technologies, Danvers, MA). Incubate unstained slides three times for 5 min each in xylene to remove paraffin wax.
2. Rehydrate the slides:
   (a) Two washes in 100% ethanol for 10 min each,
   (b) Two washes in 95% ethanol for 10 min each,
   (c) Two washes in dH_2O for 5 min each,
3. Unmasking antigen:
   (a) Bring the slides to a boil in 10 mM citrate buffer using a microwave,
   (b) Incubate the slides at 95°C for 10 min (see Note 9),
   (c) Allow the slides to cool slowly at room temperature for approximately 30 min,
4. Wash the slides three times in dH_2O for 5 min each
5. Incubate the slides in 3% H_2O_2 for 10 min
6. Wash the slides twice in dH_2O for 5 min each
7. Incubate the slides with wash buffer (1× PBS/0.1% Tween) for 5 min
8. Incubate the slides with 100–400 μL blocking solution (according to VECTASTAIN protocol)

9. Incubate with primary antibody diluted in SignalStain® Ab Diluent or other appropriate buffer (see Note 10)
10. Wash the slides three times for 5 min each in wash buffer
11. Incubate the slides in 100–400 μL secondary antibody (according to VECTASTAIN protocol) for 30 min at room temperature
12. Prepare Elite VECTASTAIN ABC Reagent 30 min before use (according to Elite VECTASTAIN protocol). Incubate 30 min at room temperature.
13. Wash the slides three times in wash buffer for 5 min each
14. Incubate the slides in Elite VECTASTAIN ABC Reagent for 30 min
15. Wash the slides three times in wash buffer for 5 min each
16. Prepare DAB Substrate immediately prior to use
17. Add 100–400 μL DAB Substrate to the slides and monitor while stain develops (usually 5–10 min)
18. Stop DAB reaction by submerging slides in dH_2O
19. Briefly counterstain in hematoxylin for 5–7 s
20. Rinse in dH_2O twice for 5 min
21. Dehydrate the slides:
    (a) Two incubations in 95% ethanol for 10 s each,
    (b) Two incubations in 100% ethanol for 10 s each,
    (c) Two incubations in xylene for 10 s each.
22. Mount coverslips

3.5. Cell Lysis Protocol

1. Aspirate off cell culture media
2. Wash cells with cold 1× PBS
3. Completely aspirate off 1× PBS (see Note 11)
4. Add appropriate amount Cell Lysis Buffer directly to the dish
5. Incubate on ice for 10 min
6. Scrape cells and transfer to a microfuge tube
7. Centrifuge lysate at 4°C for 10 min at full speed
8. Transfer supernatant to a new tube
9. Measure protein concentration
10. Normalize concentrations between samples and boil in 5× SDS sample loading buffer
11. Store lysate at −20°C or resolve by SDS-PAGE.

3.6. Tissue Harvesting and Tissue Lysis Protocol

1. Humanely sacrifice mouse according to IACUC approved method.
2. Using a dissection microscope, carefully dissect individual prostate lobes.

3. For DNA or protein analysis, the tissues should be placed in a microfuge tube or wrapped in foil, snap-frozen in liquid nitrogen, and stored at −80°C.
4. For RNA analysis, place the tissue samples in a tube containing 300 μL RNAlater (Applied Biosystems, AM7021) and store at −80°C.
5. Thaw snap-frozen tissues on ice in 115% strength Tissue Lysis Buffer with no detergent.
6. Grind speed for 8s at mid-full speed using a Kinematica Polytron PT 2100 homogenizer.
7. Immediately after grinding, place sample on ice for 30–60 s, repeat grinding once or twice as necessary.
8. Add detergent mix to bring final strength of Tissue Lysis Buffer to 100%.
9. Centrifuge at 4°C for 10 min at full speed.
10. Transfer supernatant to a new tube.
11. Measure protein concentration.
12. Normalize concentrations between samples and boil in 5× SDS sample buffer.
13. Resolve by SDS-PAGE or store protein lysates at −20°C.

3.7. Western Blotting

1. Charge PVDF membranes in 100% ethanol for 3–5 s
2. Soak the membranes in transfer buffer
3. Assemble transfer apparatus with the membranes and filter paper
4. Transfer for 2–3 h at 35 V
5. Block the membranes 1 h in 5% instant milk in 1× PBST
6. Rinse the membranes briefly with 1× PBST
7. Dilute primary antibodies in either 5% milk or 5% BSA
8. Seal the membranes with primary antibody dilution in plastic bags with Impulse Sealer
9. Incubate overnight at 4°C
10. Wash the membranes three times for 5 min each in 1× PBST
11. Incubate in secondary antibody diluted in 5% milk for 1 h at room temperature
12. Wash the membranes three times for 5 min each in 1× PBST
13. Incubate the membranes for 1 min in chemiluminescence reagent
14. Expose to film (see Fig. 3 for an example of an mTORC2 pathway Western blot using protein lysates derived from prostate tissue)

4. Notes

1. Triton containing buffer will destabilize the mTOR complexes; therefore, the weaker CHAPS buffer should be used if intact complexes are desired, such as for coimmunopurification studies or when purifying mTOR complexes for *in vitro* kinase assays.
2. Lysis buffer supplemented with protease inhibitor tablets can be stored for up to 1 week at 4°C.
3. One complementary strategy to selectively inhibit mTORC1 or mTORC2 in mammalian cultured cancer cells is to use RNA-interference to stably deplete cells of Raptor or Rictor, respectively. For instance, stably expressing shRNAs targeting Rictor for RNAi-mediated silencing in human prostate cancer cell lines revealed that PC-3 cell proliferation is particularly sensitive to mTORC2 activity (7). Moreover, PC-3 cells with decreased mTORC2 activity exhibit reduced tumor formation when injected subcutaneously into nude mice. The sequences and lentiviral delivery protocols for Rictor shRNAs are described in ref. 15. A general limitation of this approach includes the fact that it is often difficult to achieve knockdown robust enough to block downstream signaling. It is important to screen numerous shRNA constructs to identify a minimum of two that are efficient (as a rule of thumb, greater than 90% knockdown of total RNA or protein should be achieved for a target with catalytic activity). Of note, cultured cancer cells lack normal cell interactions as well as a tumor microenvironment, making it difficult to predict *in vivo* effects.
4. Xenograft models offer another highly utilized, rapid, and cost-effective alternative to *in vitro* studies. However, xenograft tumors are often based on the subcutaneous injection of hundreds to thousands of human cancer cells into immunocompromised mice, and thus, tumors do not evolve naturally with the surrounding stroma or in the presence of a normal immune response.
5. Another complementary approach takes advantage of haploinsufficiencies associated with heterozygous gene deletion. *PTEN* heterozygous mice have a shortened life span because they develop tumors in multiple organs including the prostate (16–18). Although mice deleted for *mTOR* or the mTORC2 components *rictor* or *mLST8* die in development, heterozygous mice are viable and can be combined with *PTEN* heterozygous mice to determine the effects of partial mTOR inactivation on tumor development (7). In this case, partial loss of mTORC2 activity extended the life span of $PTEN^{+/-}$ mice and protected them against prostate cancer formation. Similar strategies have been employed to study the role of *AKT* and *PDK1* in tumor development (19, 20).

6. Storing tails in tubes on dry ice until processing improves DNA quality. This strategy can also be used to prepare genomic DNA from prostate tissue, although for a cleaner DNA preparation we recommend using a kit, such as the DNeasy Blood & Tissue Kit from Qiagen.
7. For histological analysis, we find it useful to remove the entire genitourinary (GU bloc) as described in ref. 21.
8. It is often necessary to obtain sections from multiple levels for H&E staining. We find it useful to first analyze the initial 1–2 levels (i.e., surface cuts) by H&E to confirm correct tissue orientation and identify which tissue samples are ideally suited for follow-up IHC. Subsequent sections of multiple levels (typically 25–50 μm apart) can then be prepared. Obtaining unstained sections from each level for IHC or other procedures is necessary. One section from each level should always be stained with H&E.
9. Using a microwave, we find that the citrate buffer can be heated to boiling and then the power level can be lowered, allowing the slides to incubate for the allotted 10 min at subpar boiling. The power level required must be determined for each microwave model.
10. We find that using 75 μL of primary antibody solution under 1.5 cm square of parafilm works well and prevents the slide from drying during the overnight incubation. Incubating the slides in a large tissue culture dish with damp filter paper prevents slides from drying out.
11. It is imperative to tilt the dish on ice for several seconds after aspirating the wash buffer to collect any residual buffer that could unnecessarily dilute the lysis buffer.
12. A major challenge to a successful cancer therapy is developing a drug that blocks the target activity with high specificity and potency, and selectively kills the cancer cells while leaving normal cells functionally intact. Because inhibitors typically target an enzymatic activity (e.g., kinase activity), developing an mTORC2-selective inhibitor will be challenging because it shares the catalytic mTOR subunit with its sibling mTORC1. Thus by current standards, an mTORC2 inhibitor would have to function by a somewhat unconventional mechanism of action. Possible mechanisms include preventing complex assembly, blocking enzyme–substrate interactions, or disrupting normal intracellular localization. Without detailed structural information it is difficult to imagine the molecular basis of such an inhibitor. Nevertheless, genetic studies can provide critical proof of principle and serve as benchmarks from which drug discovery efforts are built.

Acknowledgments

We thank Cynthia Sparks for critical reading of the manuscript. D.A.G. is supported by grants from the NIH (R00CA129613), the Charles Hood Foundation, and the PEW Charitable Trusts.

References

1. Engelman, J.A. (2009). Targeting PI3K signalling in cancer: opportunities, challenges and limitations. Nat Rev Cancer *9*, 550–562.
2. Guertin, D.A., and Sabatini, D.M. (2007). Defining the Role of mTOR in Cancer. Cancer Cell *12*, 9–22.
3. Sparks, C.A., and Guertin, D.A. (2010). Targeting mTOR: prospects for mTOR complex 2 inhibitors in cancer therapy. Oncogene.
4. Shor, B., Gibbons, J.J., Abraham, R.T., and Yu, K. (2009). Targeting mTOR globally in cancer: Thinking beyond rapamycin. Cell Cycle *8*.
5. Feldman, M.E., and Shokat, K.M. New Inhibitors of the PI3K-Akt-mTOR Pathway: Insights into mTOR Signaling from a New Generation of Tor Kinase Domain Inhibitors (TORKinibs). Curr Top Microbiol Immunol.
6. Sarbassov, D., Ali, S.M., Sengupta, S., Sheen, J.H., Hsu, P.P., Bagley, A.F., Markhard, A.L., and Sabatini, D.M. (2006). Prolonged rapamycin treatment inhibits mTORC2 assembly and Akt/PKB. Mol Cell *22*, 159–168.
7. Guertin, D.A., Stevens, D.M., Saitoh, M., Kinkel, S., Crosby, K., Sheen, J.H., Mullholland, D.J., Magnuson, M.A., Wu, H., and Sabatini, D.M. (2009). mTOR complex 2 is required for the development of prostate cancer induced by Pten loss in mice. Cancer Cell *15*, 148–159.
8. Heyer, J., Kwong, L.N., Lowe, S.W., and Chin, L. (2010). Non-germline genetically engineered mouse models for translational cancer research. Nat Rev Cancer *10*, 470–480.
9. Nardella, C., Carracedo, A., Alimonti, A., Hobbs, R.M., Clohessy, J.G., Chen, Z., Egia, A., Fornari, A., Fiorentino, M., Loda, M., et al. (2009). Differential requirement of mTOR in postmitotic tissues and tumorigenesis. Sci Signal *2*, ra2.
10. Fu, Y., Wey, S., Wang, M., Ye, R., Liao, C.P., Roy-Burman, P., and Lee, A.S. (2008). Pten null prostate tumorigenesis and AKT activation are blocked by targeted knockout of ER chaperone GRP78/BiP in prostate epithelium. Proc Natl Acad Sci USA *105*, 19444–19449.
11. Foukas, L.C., Claret, M., Pearce, W., Okkenhaug, K., Meek, S., Peskett, E., Sancho, S., Smith, A.J., Withers, D.J., and Vanhaesebroeck, B. (2006). Critical role for the p110alpha phosphoinositide-3-OH kinase in growth and metabolic regulation. Nature *441*, 366–370.
12. Wang, S., Gao, J., Lei, Q., Rozengurt, N., Pritchard, C., Jiao, J., Thomas, G.V., Li, G., Roy-Burman, P., Nelson, P.S., et al. (2003). Prostate-specific deletion of the murine Pten tumor suppressor gene leads to metastatic prostate cancer. Cancer Cell *4*, 209–221.
13. Wu, X., Wu, J., Huang, J., Powell, W.C., Zhang, J., Matusik, R.J., Sangiorgi, F.O., Maxson, R.E., Sucov, H.M., and Roy-Burman, P. (2001). Generation of a prostate epithelial cell-specific Cre transgenic mouse model for tissue-specific gene ablation. Mech Dev *101*, 61–69.
14. Freeman, D., Lesche, R., Kertesz, N., Wang, S., Li, G., Gao, J., Groszer, M., Martinez-Diaz, H., Rozengurt, N., Thomas, G., et al. (2006). Genetic background controls tumor development in PTEN-deficient mice. Cancer Res *66*, 6492–6496.
15. Sarbassov, D.D., Guertin, D.A., Ali, S.M., and Sabatini, D.M. (2005). Phosphorylation and regulation of Akt/PKB by the rictor-mTOR complex. Science *307*, 1098–1101.
16. Di Cristofano, A., Pesce, B., Cordon-Cardo, C., and Pandolfi, P.P. (1998). Pten is essential for embryonic development and tumour suppression. Nat Genet *19*, 348–355.
17. Podsypanina, K., Ellenson, L.H., Nemes, A., Gu, J., Tamura, M., Yamada, K.M., Cordon-Cardo, C., Catoretti, G., Fisher, P.E., and Parsons, R. (1999). Mutation of Pten/Mmac1 in mice causes neoplasia in multiple organ systems. Proc Natl Acad Sci USA *96*, 1563–1568.
18. Suzuki, A., de la Pompa, J.L., Stambolic, V., Elia, A.J., Sasaki, T., del Barco Barrantes, I., Ho, A., Wakeham, A., Itie, A., Khoo, W., et al. (1998). High cancer susceptibility and embryonic lethality associated with mutation of the PTEN tumor suppressor gene in mice. Curr Biol *8*, 1169–1178.
19. Chen, M.L., Xu, P.Z., Peng, X.D., Chen, W.S., Guzman, G., Yang, X., Di Cristofano, A., Pandolfi, P.P., and Hay, N. (2006). The

deficiency of Akt1 is sufficient to suppress tumor development in Pten$^{+/-}$ mice. Genes Dev *20*, 1569–1574.

20. Bayascas, J.R., Leslie, N.R., Parsons, R., Fleming, S., and Alessi, D.R. (2005). Hypomorphic mutation of PDK1 suppresses tumorigenesis in PTEN(+/−) mice. Curr Biol *15*, 1839–1846.
21. Shappell, S.B., Thomas, G.V., Roberts, R.L., Herbert, R., Ittmann, M.M., Rubin, M.A., Humphrey, P.A., Sundberg, J.P., Rozengurt, N., Barrios, R., et al. (2004). Prostate pathology of genetically engineered mice: definitions and classification. The consensus report from the Bar Harbor meeting of the Mouse Models of Human Cancer Consortium Prostate Pathology Committee. Cancer Res *64*, 2270–2305.
22. Hentges, K., Thompson, K., and Peterson, A. (1999). The flat-top gene is required for the expansion and regionalization of the telencephalic primordium. Development *126*, 1601–1609.
23. Murakami, M., Ichisaka, T., Maeda, M., Oshiro, N., Hara, K., Edenhofer, F., Kiyama, H., Yonezawa, K., and Yamanaka, S. (2004). mTOR is essential for growth and proliferation in early mouse embryos and embryonic stem cells. Mol Cell Biol *24*, 6710–6718.
24. Gangloff, Y.G., Mueller, M., Dann, S.G., Svoboda, P., Sticker, M., Spetz, J.F., Um, S.H., Brown, E.J., Cereghini, S., Thomas, G., et al. (2004). Disruption of the mouse mTOR gene leads to early postimplantation lethality and prohibits embryonic stem cell development. Mol Cell Biol *24*, 9508–9516.
25. Guertin, D.A., Stevens, D.M., Thoreen, C.C., Burds, A.A., Kalaany, N.Y., Moffat, J., Brown, M., Fitzgerald, K.J., and Sabatini, D.M. (2006). Ablation in Mice of the mTORC Components raptor, rictor, or mLST8 Reveals that mTORC2 Is Required for Signaling to Akt-FOXO and PKCalpha, but Not S6K1. Dev Cell *11*, 859–871.
26. Shor, B., Cavender, D., and Harris, C. (2009). A kinase-dead knock-in mutation in mTOR leads to early embryonic lethality and is dispensable for the immune system in heterozygous mice. BMC Immunol *10*, 28.
27. Yang, Q., Inoki, K., Ikenoue, T., and Guan, K.L. (2006). Identification of Sin1 as an essential TORC2 component required for complex formation and kinase activity. Genes Dev *20*, 2820–2832.
28. Jacinto, E., Facchinetti, V., Liu, D., Soto, N., Wei, S., Jung, S.Y., Huang, Q., Qin, J., and Su, B. (2006). SIN1/MIP1 Maintains rictor-mTOR Complex Integrity and Regulates Akt Phosphorylation and Substrate Specificity. 2006 Oct 6;127(1):125–137.
29. Bentzinger, C.F., Romanino, K., Cloetta, D., Lin, S., Mascarenhas, J.B., Oliveri, F., Xia, J., Casanova, E., Costa, C.F., Brink, M., et al. (2008). Skeletal muscle-specific ablation of raptor, but not of rictor, causes metabolic changes and results in muscle dystrophy. Cell Metab *8*, 411–424.
30. Shiota, C., Woo, J.T., Lindner, J., Shelton, K.D., and Magnuson, M.A. (2006). Multiallelic Disruption of the rictor Gene in Mice Reveals that mTOR Complex 2 Is Essential for Fetal Growth and Viability. *Dev Cell* 11(4):583–589.

Chapter 22

Inhibition of PI3K-Akt-mTOR Signaling in Glioblastoma by mTORC1/2 Inhibitors

Qi-Wen Fan and William A. Weiss

Abstract

Amplification of the gene encoding the epidermal growth factor receptor (EGFR) occurs commonly in glioblastoma (GBM), leading to activation of downstream kinases, including phosphatidylinositol 3′-kinase (PI3K), Akt, and mammalian target of rapamycin (mTOR). A serine-threonine kinase, mTOR controls cell growth by regulating mRNA translation, metabolism, and autophagy; acting as both a downstream effector and upstream regulator of PI3K. These signaling functions are distributed between at least two distinct complexes, mTORC1 and mTORC2 with respect to pathway specificity. We have investigated mTOR signaling in glioma cells with the allosteric mTORC1 inhibitor rapamycin, the mTORC1/2 inhibitor Ku-0063794, a dual PI3K/mTORC1/2 kinase inhibitor PI-103, and siRNA against raptor, rictor, or mTOR, and evaluated the value of mTOR inhibitors for the treatment of glioblastoma.

Key words: Glioblastoma, PI3-kinase, Akt:mTORC1, mTORC2, mTOR, EGFR

1. Introduction

Gliomas represent the most common primary brain tumor and are among the most lethal of all cancers. *EGFR* is commonly mutated in GBM, leading to overexpression and activation of downstream signaling pathways. EGFR signals through a complex network of intermediates, including PI3K/AKT/mTOR, MAPK, and PLCγ. Inactivation of *PTEN* and activating mutations in PI3K itself collectively occur in a majority of GBM tumors, effectively uncoupling PI3K from upstream control by EGFR (1).

PI3Ks are lipid kinases activated by a wide range of RTKs to generate the second messenger phosphatidylinositol-3,4,5-trisphosphate (PIP_3). PIP_3 couples PI3K to downstream effectors, such as Akt, a serine-threonine kinase that suppresses apoptosis, promotes

Thomas Weichhart (ed.), *mTOR: Methods and Protocols*, Methods in Molecular Biology, vol. 821,
DOI 10.1007/978-1-61779-430-8_22, © Springer Science+Business Media, LLC 2012

growth, and drives proliferation. PIP_3 also indirectly activates the mammalian target of rapamycin (mTOR), a protein kinase critical for cell growth (1). The mTOR kinase contains a PI3K homology domain (making mTOR a PIK-related kinase – PIKK), although mTOR itself has no lipid kinase activity (2).

Signaling functions of mTOR are distributed between at least two distinct mTOR protein complexes: mTORC1 and mTORC2. In mTORC1, mTOR is associated with a number of proteins, including PRAS40 and the rapamycin-sensitive adapter protein of mTOR (Raptor), whereas in mTORC2, mTOR is associated with a separate protein complex, including the rapamycin-insensitive companion of mTOR (Rictor). Stimulation of PI3K in response to growth factors leads to phosphorylation and activation of Akt. In addition, phospho-Akt phosphorylates and separately inhibits PRAS40 and the Tsc1/2 (hamartin-tuberin) complex. PRAS40 is inhibitory to mTORC1, while tuberin is inhibitory to GTPase RHEB, which in turn is inhibitory to mTORC1. The detailed signaling leading to activation of mTORC2 is less clearly understood (3).

The activated mTORC1 complex phosphorylates substrates, including Thr-389 S6K, Ser-209 eIF4E, and 4EBP1. The mTORC2 complex phosphorylates Akt on Ser-473, and also phosphorylates additional substrates, including serum glucocorticoid-induced protein kinase (SGK) and PKCα. Inhibition of mTORC1/S6K1 by allosteric inhibitors, including rapamycin (Sirolimus) (Fig. 1), CCI-779 (Tensirolimus), RAD001 (4), or other similar agents triggers a negative feedback loop through an IRS-I-dependent mechanism, resulting in increase phosphorylation of Akt. This negative feedback loop is prominent in glioma, however the robustness with which inhibition of mTORC1 activates Akt varies across multiple cancer types (5–7). Unlike rapamycin, ATP-competitive inhibitors of mTORC1/mTORC2 by mTOR inhibitors, including Torin, Ku-0063794 (8, 9), and pp242 (10) blocks

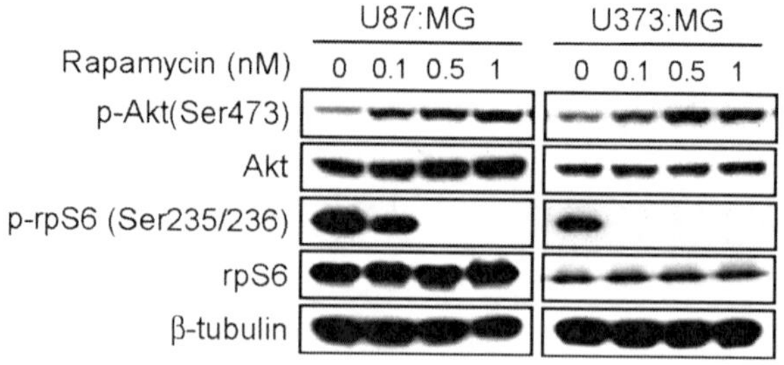

Fig. 1. The allosteric mTORC1 inhibitor rapamycin induced p-Akt in glioma cells. *PTEN* mutant-type U87:MG and U373:MG cells were treated with rapamycin at doses shown and were assayed for total and phospho Akt and rpS6 by immunoblot. Rapamycin blocked p-rpS6 in both cell lines at a dose of 0.5 nM. Activation of Akt was dose-dependent, as indicated by increased levels of the p-Akt at Ser-473. β-tubulin is shown as loading control.

the phosphorylation of Akt at Ser473. As an integrator of cell growth and proliferation, mTOR also regulates autophagy, a program of cellular self-digestion activated during periods of nutrient and growth factor deprivation (11). The signaling linking activation of mTOR signaling to blockade of autophagy in metazoan cells is poorly understood.

Preclinical evaluation of dual PI3K/mTOR inhibitors, such as PI-103 and NVP-BEZ235 have demonstrated efficacy for these agents in blocking the growth of glioblastoma (GBM) cells *in vitro* and *in vivo* (5, 12). NVP-BEZ235 and other dual inhibitors are therefore being evaluated in early clinical trials. Thus, inhibitors of mTOR and of PI3K/mTOR provide a new class of agents and therapeutic for glioma. Pure ATP-competitive inhibitors of mTORC1/2 inhibit both mTOR complexes, and have been evaluated in less detail in glioma. We have directly compared rapamycin, Ku-0063794, PI-103, or siRNA against mTOR in glioma cells. In contrast to rapamycin, both Ku-0063794 and PI-103 blocked the phosphorylation of Akt and prevented its activation. Ku-0063794 and PI-103 also decreased the phosphorylation of the mTORC1 target 4EBP1 and induced autophagy more effectively in comparison with the allosteric mTORC1 inhibitor rapamycin (Figs. 2 and 3).

2. Materials

2.1. Cell Culture and Lysis

1. Dulbecco's modified Eagle's medium (DMEM H-21) supplemented with 10% fetal bovine serum (FBS, Gibco/BRL) and 1% penicillin–streptomycin and stored at +4°C.
2. Solution of trypsin (0.25%) stored at +4°C.
3. Rapamycin (Cell Signaling) is dissolved at 100 μM in methanol and stored in single use aliquots at –20°C.
4. Ku-0063794 is dissolved at 20 mM in DMSO and stored in single use aliquots at –20°C.
5. PIK-90 is dissolved at 20 mM in DMSO and stored in single use aliquots at –20°C.
6. PI-103 is dissolved at 20 mM in DMSO and stored in single use aliquots at –20°C.

2.2. SDS-Polyacrylamide Gel Electrophoresis

1. Tris-buffered saline with Tween 20 (10× TBS-T).
2. NOVEX-NuPAGE LDS Sample Buffer (4×) (Invitrogen Corporation).
3. NOVEX-NuPAGE MOPS Running Buffer (20×) (Invitrogen Corporation).
4. NOVEX-NuPAGE Tris-Acetate Runing Buffer (25×) (Invitrogen Coropration).

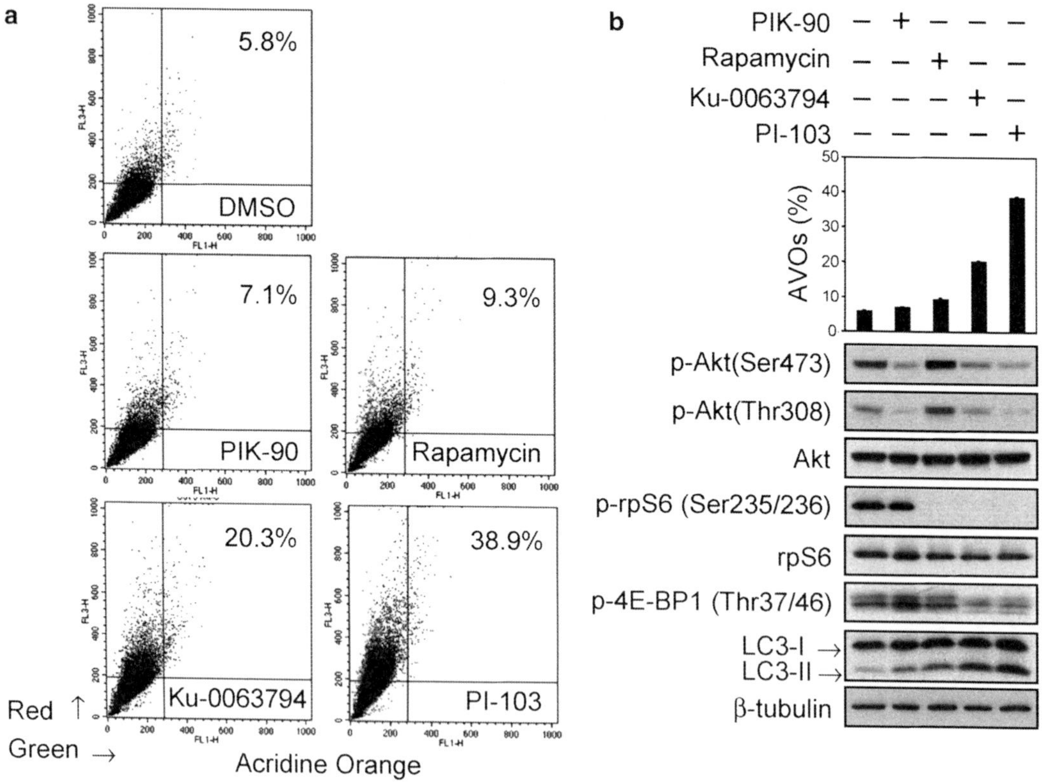

Fig. 2. The mTOR kinase inhibitor KU-0063794 and the dual PI3K/mTOR inhibitor PI-103 both block phosphorylation of Akt and induce autophagy. In comparison, the mTOCR1 inhibitor rapamycin-induced phosphorylation of Akt and was a less potent inducer of autophagy. (**a**) *PTEN* mutant U373:MG cells were treated with DMSO, PI3Kα inhibitor PIK-90 (1 μM), mTORC1 inhibitor rapamycin (100 nM), mTOR inhibitor Ku-0063794 (5 μM), or a dual PI3K/mTOR inhibitor PI-103 (1 μM) for 48 h and stained with acridine orange (1 μg/ml) for 15 min. Cells were analyzed by flow cytometry. Autophagy was quantified by the accumulation of acidic vescular organelles. Percentage of AVOs is indicated. (**b**) Percentages of cells positive for AVOs; mean ± S.E. for triplicate samples (*top panel*). Cells were treated as in (**a**) for 24 h and lysates examined by immunoblot using antibodies shown (*bottom panel*). In contrast to rapamycin, Ku-0063794 and PI-103 block both p-Akt and p-rpS6. Ku-0063794 and PI-103 also decreased the phosphorylation of 4EBP1 and induced autophagy more potent than rapamycin as indicated by increased levels of the autophagy marker LC3-II. The PI3Kα inhibitor PIK-90 blocked phosphorylation of Akt, but did not affect phosphorylation of 4EBP1 or of rpS6, demonstrating that inhibition of PI3K and Akt activation alone is not sufficient to block the phosphorylation of mTORC1 targets 4EBP1 and of rpS6. (Reproduced from ref. 13 with permission from AAAS).

5. NOVEX-NuPAGE Tris-Glycine Running Buffer (10×) (Invitrogen Corporation).
6. NOVEX-NuPAGE Transfer Buffer (20×) (Invitrogen Corporation) and NOVEX-NuPAGE Tris-Glycine Transfer Buffer (25×).
7. NOVEX-NuPAGE 4–12% BT SDS-polyacrylamide gel electrophoresis (SDS-PAGE) gel, 1.0 mm 15 well, NOVEX-NuPAGE 3–8% Tris-Acetate gel, 1.0 mm, 15 well, and NOVEX-NuPAGE 16% Tris-cyclin gel, 1.0 mm, 15 well (Invitrogen Corporation).

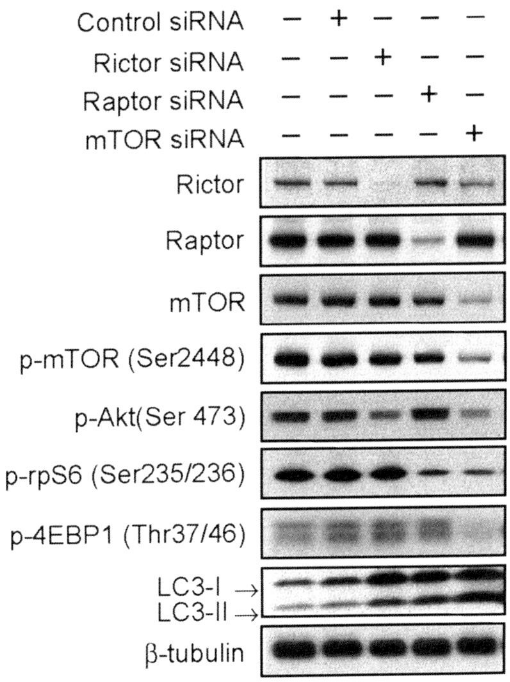

Fig. 3. Knockdown of raptor, but not rictor or mTOR, activates Akt on Ser-473. We transfected U373:MG cells with control siRNA, siRNA against the mTORC2-associated mRNA rictor, the mTORC1-associated mRNA raptor, or to mTOR itself. Lysates were collected at 72 h and examined by immunoblot using antibodies shown. Knockdown of raptor blocked phosphorylation of rpS6 and increased the phosphorylation of Akt. In contrast, knockdown of rictor or mTOR both decreased the phosphorylation of Akt. Knockdown of rictor or raptor also induced autophagy, albeit to a lesser extent than that observed in response to knockdown of mTOR (as indicated by accumulation of LC3-II). (Reproduced from ref. 13 with permission from AAAS).

8. Full-range rainbow molecular weight markers (GE Healthcare).
9. BCA™ Protein Assay Reagent (Thermo Scientific Corporation).
10. 10× Cell lysis buffer (Cell Signaling Technology).
11. Complete protease inhibitor cocktail tables (Roche).

2.3. Western Blotting for PI3K-Akt-mTOR Pathway

1. NOVEX-NuPAGE Transfer Buffer (20×) (Invitrogen Corporation) and NOVEX-NuPAGE Tris-Glycine Transfer Buffer (25×).
2. Methanol (Sigma-Aldrich).
3. PVDF Transfer Membrane (GE Healthcare).
4. Blocking buffer: 5% (w/v) nonfat dry milk in TBS-T.
5. Membrane were blotted with p-Akt (Ser473), p-Akt (Thr308), Akt (pan), p-S6 ribosomal protein (Ser235/236), S6 ribosomal protein, rictor, raptor, p-mTOR (Ser 2448), mTOR, p-4E-BP1 (Thr37/46) (Cell Signaling Technology). LC3 (Novus) or β-tubulin (Millipore).

6. Primary antibody dilution buffer: TBS-T supplemented with 5% (w/v) fraction V bovine serum albumen (Sigma).
7. Secondary antibody: Goat anti-mouse IgG peroxidase conjugate and goat anti-rabbit IgG peroxidase conjugate (EMD Chemicals Inc).
8. Hybond™ ECL™ Western blotting detection Kit (GE Healthcare).

2.4. siRNA Transfection

1. Control small interfering RNA (siRNA) was purchased from Santa Cruz Biotechnology. Raptor siRNA (NM 020761 RAPTOR ON-TARGET plus SMARTpool), Rictor siRAN (NM 152756 RICTOR ON-TARGET plus SMARTpool), mTOR siRNA (NM 004958 FRAP1 ON-TARGET plus SMARTpool) were purchased from Thermo Scientific.
   5× siRNA buffer (300 mM KCL, 30 mM HEPES-pH 7.5, 1.0 mM $MgCL_2$); RNase-free water (molecular grade RNase-free water for dilution of 5× buffer). Resuspend siRNAs to 20 μM stock in 1× siRNA buffer and stored in single use aliquots at −20°C.
2. Lipofectamine 2000 (Invitrogen) stored at +4°C.

2.5. Detection and Quantification of AVOs

1. Acridine orange (Sigma Chemical Co.). Dissolve 1 g of acridine orange to 100 ml of distilled water. Store in a foil wrapped container at 4°C (see Note 1) and bring to room temperature prior to use. 5 ml polystyrene round-bottom tube with cell-strainer cap (BD Biosciences).

3. Methods

3.1. Preparation of Samples for Assay of PI3K Pathway by Western Blotting

1. U87:MG and U373:MG cells are grown in 10% FBS and are passaged every 3 days with trypsin/EDTA to provide new maintenance cultures on 100-mm tissue dishes and experimental culture on 6-well plates. Two 6-well plates are required for each data point.
2. Cells are seeded in 6-well plates at $3\times10^5/cm^2$ and will approach confluence after 72 h. Cells are treated by adding 2 ml fresh media containing inhibitors for desired time periods (typically 24 h in our experiments).
3. Media is aspirated, cells washed twice with ice-cold 1× PBS, and aspirated again. Cells are lysed by adding 1× ice-cold cell lysis buffer (50–100 μl per 6-well plate). Immediately scrape the cells off the plate and transfer the extract to a 1.5 ml microcentrifuge tube. Keep on ice for 30 min.

4. Sonicate for 5 s to shear DNA and reduce sample viscosity. Microcentrifuge at 12,700 rpm (15,000 × *g*) for 5 min at 4°C, and transfer supernatant to new tube.
5. Measure the protein concentration of samples by BCA protein Assay kit. Mix reagent A with Reagent B to generate a working reagent (WR). Load the 96-well microplate with WR (200 μL per well) and then add the protein samples (2 μL each) alone with BSA (bovine serum albumin) protein standard dilutions. Incubate 30 min at 37°C and read at 562 nm with a multimode microplate reader (BioTek Instruments Inc).
6. Heat sample by adding 25% 4× LDS sample buffer to 70°C on the heat block for 10 min. Cool on ice and microcentrifuge for 5 min at room temperature. Samples are now ready for separation by SDS-PAGE.

3.2. SDS-PAGE

1. These instructions assume the use of a Hoefer SE260B gel system.
2. Prepare the running buffer by diluting 20 ml of 20× running buffer (see Note 2) with 380 ml of Milli-Q water in a measuring cylinder. Cover with parafilm invert to mix.
3. Once the stacking gel has set, carefully remove the comb and use a 5 ml syringe fitted with a 22-gauge needle to wash wells with running buffer.
4. Add the running buffer to the upper and lower chambers of the gel unit and load equal amount of each sample in the well. Include one well for 10 μl of full-range rainbow molecular weight markers.
5. Complete the assembly of the gel unit and connect to a power supply. Two gels can be run at 110 mA (see Note 3) for 1 h. 126 V/two gels (start); 200 V/two gels (end).

3.3. Western Blotting for PI3K-Akt-mTOR Pathway

1. Samples that have been separated by SDS-PAGE are transferred electrophoretically to nitrocellulose membranes. These directions assume the use of a Bio-RAD Trans-Blot SD semidry transfer cell system. Prepare the transfer buffer (see Note 4) by diluting 15 ml of 20× transfer buffer with 285 ml of Milli-Q water in a measuring cylinder. Cover with parafilm. Invert to mix.
2. Cut a sheet of the PVDF-transfer membrane just larger than the size of the separating gel. Clip upper right corner to clarify orientation. Wet membrane in 100% methanol for 1 min. Rinse briefly in Milli-Q and then soak for 15 min in transfer buffer before use.
3. Cut filter paper to the dimensions of the gel. Two pieces of extra thick filter paper per gel are needed for each gel/membrane

sandwich. You can substitute four pieces of thick or six pieces of thin filter paper. Completely saturate the filter paper by soaking in transfer buffer.

4. Place a presoaked sheet of extra thick filter paper onto the platinum anode. Roll a pipet over the surface of the filter paper to exclude all air bubbles.
5. Carefully place the equilibrated gel on top of the transfer membrane, aligning the gel on the center of the membrane.
6. Place the other sheet of presoaked filter paper on the top of the gel. Carefully roll a pipet over the filter paper to remove bubbles from between each layer.
7. Complete the assembly of the gel unit and connect to a power supply. Two gels can be run at 110 mA (see Note 5) for 1 h. 3–5 V/two gels (start); 10–12 V/two gels (end).
8. After transfer, wash the membrane with 25 ml TBS for 5 min at room temperature. The membrane is then incubated in 25 ml of TBS-0.5% Tween (TBS-T)/5% w/v nonfat dry milk for 1 h at room temperature. Wash three times with 25 ml of TBS-T, 10 min per wash.
9. Incubate the membrane and primary antibody at the appropriate dilution (see Note 6) in 10 ml primary antibody in TBS-T/5% BSA with gentle agitation on a rocking platform overnight at 4°C.
10. Wash three times for 10 min each with 25 ml of TBS-T. Incubate the membrane with HRP-conjugated secondary antibody (1:2,000) in 10 ml TBS-T/2%BSA with gentle agitation for 1 h at room temperature. Wash with TBS-T six times, 10 min per wash.
11. During the final wash, 5 ml aliquots of each portion of the ECL reagent are warmed separately to room temperature. Once the final wash is removed from the blot, the ECL reagents are mixed together and then immediately added to the blot, which is then rotated by hand for 1 min to ensure even coverage.
12. Remove blot from the ECL reagents. Drain membranes of excess developing solution (do not let dry), wrap in plastic warp and expose to X-ray film. An initial 10-s exposure should indicate the proper exposure time.

3.4. siRNA Transfection

1. Control siRNA was purchased from Santa Cruz Biotechnology. siRNAs against rictor, raptor, and mTOR were purchased from Dharmacon, and transfected using Lipofectamine 2000 (Invitrogen).
2. One day before transfection, cells were seeded in 6-well plates at $3 \times 10^5/cm^2$ in 10% FBS growth medium without antibiotics (see Note 7).

3. Dilute 8 μl of 20 μM siRNA in 250 μl medium without serum (final concentration of RNA when added to cells is 80 nM). Mix gently. Incubate for 5 min at room temperature dilute 8 μl of lipofectamine in 250 μl medium without serum. Mix gently. Incubate for 5 min at room temperature.
4. After the 5 min incubation, combine the diluted siRNA with the diluted lipofectamine 2000. Mix gently and incubate for 20 min at room temperature to allow complex formation to occur.
5. Add 500 μl of the siRNA-lipofectamine 2000 complex to each well of a 6-well plate containing 1.5 ml 10% FBS growth medium without antibiotics. Mix gently by rocking the plate back and forth.
6. After 6 h (see Note 8), wash cells with PBS and adding 1.5 ml of 10% FBS growth media.
7. Incubate the cells at 37°C in a 5% CO_2 for 72 h, then assay for gene knockdown.

3.5. Detection and Quantification of AVOs

1. In acridine orange-stained cells, the cytoplasm and nucleolus fluorescent bright green and dim red, respectively, whereas acidic compartments fluorescent bright red. The intensity of the red fluorescence is proportional to the degree of acidity. Therefore, the volume of the cellular acidic compartment can be quantified.
2. U373:MG cells were seeded in 6 well plates at $3 \times 10^5/\text{cm}^2$ in 10% FBS growth medium. Cells were treated by adding 2 ml fresh media containing inhibitors for desired time periods (Typically 48 h in our experiments).
3. Cells were stained with 1 μg/ml of acridine orange for 15 min, wash cells once with PBS and incubate with 0.25% trypsin-EDTA for 3–5 min, add media with serum to stop trypsin-EDTA, removed from the plate with typsin-EDTA. Microcentrifuge at 500 rpm for 5 min at room temperature, the pellet was resuspended in 500 μl of phenol red-free growth medium, collected with a 5 ml in a foil wrapped polystyrene round-bottom tube with cell-strainer cap (see Note 1), and examined immediately by flow cytometry.
4. Analyze cells on a flow cytometry, green (510–530 nm) and red (650 nm) fluroscence emission from 1×10^4 cells illuminated with blue (488 nm) excitation light was measured with an FACSCalibur from (Becton Dickinson) using CellQuest software.

4. Notes

1. Acridine orange is light sensitive.
2. MOPS SDS running buffer for NOVEX-NuPAGE 4–12% BT SDS-PAGE gel, Tris-acetate running buffer for NOVEX-NuPAGE 3–8% Tris-acetate gel (for high molecular weight protein detection, such as mTOR, p-mTOR (MW 289 kDa)), and Tris-glycine running buffer for NOVEX-NuPAGE 16% Tris-cyclin gel (for low molecular weight protein detection, such as LC3 (MW 18 kDa for LC3I and MW 16 kDa for LC3II)).
3. For NOVEX-NuPAGE 16% Tris-cyclin gel, two gels run at 125 V constant for 1.5 h. 30–40 mA (start); 8–12 mA (end).
4. 1× NOVEX-NuPAGE transfer buffer with 10% methanol for NOVEX-NuPAGE 4–12% gel and 3–8% gel. 1× NOVEX-NuPAGE Tris-glycine transfer buffer with 20% methanol for NOVEX-NuPAGE 16% Tris-cyclin gel.
5. For NOVEX-NuPAGE 16% Tris-cyclin gel, run at 20 V constant for 1.5 h.
6. For the primary antibodies purchased from cell signaling was used at 1:1,000 dilution, for LC3 at 1:500 dilution, and for β-tubulin at 1:2,000 dilution.
7. Transfecting cells at lower density can minimize the loss of cell viability due to cell overgrowth after 3 days.
8. Removing the complexes after 6 h without loss of transfection activity, but can reduce the toxicity to cells.

Acknowledgments

We thank Zachary Knight, Benjamin Houseman, Morri Feldman, and Kevan Shokat for providing PI-103, PIK-90, and Ku-0063794. We acknowledge support from NIH grants PCA133091, NS055750, CA102321, CA097257, CA128583, CA148699 P01 CA081403, Burroughs Wellcome Fund, American Brain Tumor Association, The Brain Tumor Society, Accelerate Brain Cancer Cure; Alex's Lemonade Stand, Children's National Brain Tumor, Katie Dougherty, Pediatric Brain Tumor, Samuel G. Waxman and V Foundations.

References

1. A. Gschwind, O. M. Fischer, A. Ullrich. (2004) The discovery of receptor tyrosine kinases: Targets for cancer therapy. *Nat. Rev. Cancer* 4, 361–370.
2. C. K. Cheng, Q. W. Fan, W. A. Weiss. (2009) PI3K signaling in glioma – animal models and therapeutic challenges. *Brain Pathol* 19, 112–120.
3. X. M. Ma, J. Blenis. (2009) Molecular mechanisms of mTOR-mediated translational control. *Nat Rev Mol Cell Biol* 10, 307–318.
4. Faivre S, Kroemer G and Raymond E. (2006) Current development of mTOR inhibitors as anticancer agents. *Nat Rev drug discov* 5, 671–688.
5. Q. W. Fan, Z. A. Knight, D. D. Goldenberg, W. Yu, K. E. Mostov, D. Stokoe, K. M. Shokat, W. A. Weiss. (2006) A dual PI3 kinase/mTOR inhibitor reveals emergent efficacy in glioma. *Cancer Cell* 9, 341–349.
6. S. Y. Sun, L. M. Rosenberg, X. Wang, Z. Zhou, P. Yue, H. Fu, F. R. Khuri. (2005) Activation of Akt and eIF4E survival pathways by rapamycin-mediated mammalian target of rapamycin inhibition. *Cancer Res* 65, 7052–7058.
7. Q. W. Fan, C. K. Cheng, T. P. Nicolaides, C. S. Hackett, Z. A. Knight, K. M. Shokat, W. A. Weiss. (2007) A Dual Phosphoinositide-3-Kinase {alpha}/mTOR Inhibitor Cooperates with Blockade of Epidermal Growth Factor Receptor in PTEN-Mutant Glioma. *Cancer Res* 67, 7960–7965.
8. C. C. Thoreen, S. A. Kang, J. W. Chang, Q. Liu, J. Zhang, Y. Gao, L. J. Reichling, T. Sim, D. M. Sabatini, N. S. Gray. (2009) An ATP-competitive mTOR inhibitor reveals rapamycin-insensitive functions of mTORC1. *J. Biol. Chem* 284, 8023–8032.
9. J. M. Garcia-Martinez, J. Moran, R. G. Clarke, A. Gray, S. C. Cosulich, C. M. Chresta, D. R. Alessi. (2009) Ku-0063794 is a specific inhibitor of the mammalian target of rapamycin (mTOR). *Biochem J* 421, 29–42.
10. M.E. Feldman, B. Apsel, A. Uotila, R. Loewith, Z.A. Knight, D. Ruggero. (2009) Active-site inhibitors of mTOR target rapamycin-resistant outputs of mTORC1 and mTORC2. *PLoS Biol* 7, 0371–0383.
11. B. Levine, G. Kroemer. (2008) Autophagy in the pathogenesis of disease. *Cell* 132, 27–42.
12. T. J. Liu, D. Koul, T. LaFortune, N. Tiao, R. J. Shen, S. M. Maira, C. Garcia-Echevrria, W. K. Yung. (2009) NVP-BEZ235, a novel dual phosphatidylinositol 3-kinase/mammalian target of rapamycin inhibitor, elicits multifaceted antitumor activities in human gliomas. *Mol Cancer Ther* 8, 2204–2210.
13. Q. W. Fan, C. K. Cheng, C. S. Hackett, M.E. Feldman, B. T. Houseman, T. P. Nicolaides, D. A. Haas-Kogan, C. D. James, S. A. Oakes, J. Debnath, K. M. Shokat, W. A. Weiss. (2010) Akt and autophagy cooperate to promote survival of drug-resistant glioma. *Sci Signal* 3(147), ra81.

Chapter 23

Assessing the Function of mTOR in Human Embryonic Stem Cells

Jiaxi Zhou, Dong Li, and Fei Wang

Abstract

We described a protocol for dissecting the function of an important serine/threonine protein kinase, mammalian target of rapamycin (mTOR), in regulating the long-term undifferentiated growth of human embryonic stem cells (hESCs). The function of mTOR in hESCs was inactivated with a highly specific chemical inhibitor, rapamycin, and gene-specific small-hairpin RNAs, and the effects were evaluated under self-renewal or early differentiation conditions. We found that inactivation of mTOR impairs proliferation and enhances mesoderm and endoderm activities of hESCs. This protocol described a general strategy for studying the function of key genes and signaling events during hESC long-term self-renewal and early lineage specifications with pharmacological and genetic approaches.

Key words: Human embryonic stem cell, mTOR, Long-term self-renewal, Differentiation

1. Introduction

Human embryonic stem cells (hESCs), derived from blastocyst-stage embryos, can undergo long-term self-renewal and have the remarkable ability to differentiate into multiple cell types in the human body (1). They, therefore, represent a unique system to study early human development and a theoretically inexhaustible source for producing differentiated cells for drug screening and cell replacement therapy (2, 3). Acquisition of various types of differentiated cells with high purity in large-scale and cost-effective ways is a prerequisite for achieving those purposes. However, derivation of large amounts of functional differentiated cells from hESCs remains as a major challenge in the field of stem cell biology. A more detailed delineation of the signaling mechanism and pathways underlying hESC self-renewal

Thomas Weichhart (ed.), *mTOR: Methods and Protocols*, Methods in Molecular Biology, vol. 821,
DOI 10.1007/978-1-61779-430-8_23, © Springer Science+Business Media, LLC 2012

and differentiation facilitates the design of strategies for efficient differentiation and the applications of differentiated cells.

Mammalian target of rapamycin (mTOR), a protein which in human is encoded by *FRAP1* gene, is a serine/threonine protein kinase that regulates cell proliferation, differentiation, and growth (4). Loss of *Frap1* (which encodes mouse mTOR) in mice causes early embryonic lethality (5, 6). As a result, mouse ESCs cannot be established from *Frap1*-knockout embryos, making it difficult to study mTOR in ESC self-renewal and early differentiation. Here, we studied the roles of mTOR in regulating hESC long-term undifferentiated growth (7). We provided an example of how to dissect the functions of gene(s) in hESCs during self-renewal and early differentiation. First, we applied rapamycin, a highly specific chemical inhibitor of mTOR (8), to hESCs cultured on Matrigel in medium containing basic FGF (bFGF) and mouse embryonic fibroblast (MEF)-conditioned medium (CM), a culture condition known to support hESC self-renewal and pluripotency (9). We found that rapamycin impaired the compact morphology of hESC colonies, in which cells flattened and were spread out in a time-dependent manner. Expression of pluripotency markers, OCT-4, SOX2, and NANOG proteins, was further analyzed with western blotting and immunofluorescence. We found that the expression of OCT-4, SOX2, and NANOG proteins was markedly decreased 6 days after rapamycin treatment. The decrease was confirmed at the transcription level by experiments with real-time PCR. Small hairpin (sh)RNAs that depleted mTOR in hESCs recapitulated the phenotype induced by rapamycin. By using real-time PCR, we assessed the expression of differentiation markers (*SOX1* and *NEUROD1* for ectoderm, *Brachyury* (*T gene*) and *MESP1* for mesoderm, *GATA4*, *GATA6*, and *SOX17* for endoderm, and *CGB7* for trophoectoderm) in hESCs cultured as monolayers or induced to differentiate as embryoid bodies (EBs). Mesoderm and endoderm activities were significantly enhanced in cells treated with rapamycin. In contrast, differentiation toward trophoblasts and neuroectoderm was suppressed.

In summary, we revealed a critical role for mTOR in supporting hESC long-term self-renewal and in suppressing mesoderm and endoderm differentiation and provided general strategies for studying the functions of key genes and signaling pathways in hESCs (7).

2. Materials

2.1. Cell Lines

1. H1 (WA01) hESC line (WiCell Institute, Wisconsin) (1).
2. H9 (WA09) hESC lines (WiCell Institute, Wisconsin) (1).
3. HEK293T cells (ATCC).

2.2. Medium and Supplements

1. Medium for MEF, DMEM, 10% heat-inactivated fetal bovine serum (FBS), 1 mM nonessential amino acid, 1× penicillin/streptomycin. Stable for 1 month at 4°C.
2. MEF cryopreservation medium, DMEM, 10% FBS, 1 mM nonessential amino acid, 1× penicillin/streptomycin, 10% DMSO. Prepare freshly.
3. hESC growth medium, DMEM/F12, 20% knock-out serum replacement (Invitrogen), 1 mM glutamine, 1% nonessential amino acid, 0.1 mM β-mercaptoethanol, 4 ng/mL bFGF (Invitrogen). Stable for 2 weeks at 4°C.
4. Medium for 293T cells, DMEM, 10% heat-inactivated FBS, 2 mM glutamine, 1 mM sodium pyruvate, 1× penicillin/streptomycin. Stable for 1 month at 4°C.
5. Protein lysis buffer, 0.3% Triton X-100, 20 mM Tris/HCl, pH 7.5, 0.1 mM Na_3VO_4, 25 mM NaF, 25 mM β-glycerophosphate, 2 mM EGTA, 2 mM EDTA, 1 mM DTT, 0.5 mM PMSF.
6. EB induction medium, DMEM/F12 with 20% heat-inactivated FBS, 1 mM glutamine, 1% nonessential amino acid, 0.1 mM β-mercaptoethanol. Prepare freshly.
7. Virus packaging medium, DMEM, 30% FBS, 4 mM L-glutamine, 1 mM sodium pyruvate. Stable for 1 month at 4°C.

2.3. Inhibitor and Antibodies

1. Rapamycin (EMD Bioscience).
2. OCT-4 (Santa Cruz), SOX-2 (Millipore).
3. NANOG (R&D System).
4. mTOR (Cell Signaling).
5. S6K1 (Cell Signaling).
6. p-S6K1 (Cell Signaling).

2.4. Others

1. Reverse transcription kit (Promega).
2. Real-time PCR kit (Promega).
3. Cryo 1 freezing container (NALGENE).
4. Ultra-low attachment culture plate (Corning).
5. TRIZOL reagent (Invitrogen).
6. RNase-free DNase set (QIAGEN).
7. RNeasy Mini kit (QIAGEN).

3. Methods

3.1. Derivation and Cultivation of MEFs

1. Sacrifice CF-1 mice on day 13.5 of gestation (Charles River, USA), remove the whole uterine horn, and place it onto a petri dish containing PBS. The dish is subsequently transferred to a sterile tissue culture hood.

2. Wash the uterine horn with sterile PBS for three times, cut out embryonic sacs, and move the embryos into a new petri dish containing sterile PBS.
3. Wash the embryos with sterile PBS for three times and remove the visceral tissue from the embryos. The visceral tissue is the dark tissue surrounding the embryo in the middle.
4. Collect the embryos without visceral tissue and wash two times with sterile PBS.
5. Place the embryos into a new petri dish, add 5–10 ml 0.05% trypsin–EDTA solution, and cut the embryo with sterile scissors into small pieces.
6. Add more 0.05% trypsin–EDTA solution. The volume should be 2 ml/embryo. Incubate for 30–45 min at 37°C until cells are well-dissociated.
7. Add MEF growth medium to stop the trypsin digestion. Pipet up and down vigorously to dissociate the tissue into single cells.
8. Place the cells in 10-cm^2 cell culture dishes. Use one dish for cells from two embryos. Place dishes overnight at 37°C. Cells typically reach confluency the next day and should be subcultured (see Subheading 3.1, step 9) or frozen down (see Subheading 3.1, step 12).
9. For continuous subculture, 24 h after the initial plating, wash the cells once with 5 ml PBS, add 1 ml 0.05% trypsin–EDTA, tilt the plate to ensure that all surface area is covered, and incubate for 3 min at 37°C.
10. After incubation, tap the plate three times to dislodge the cells from the plate and add 5 ml complete growth medium to stop the digestion. Pipet up and down repeatedly to generate single-cell suspension.
11. Cells are further plated onto 10-cm^2 plates by the addition of 1 ml cell suspension and 9 ml complete growth medium. Cells are typically subcultured further at 1:6 every 3 days. Cells can be subcultured twice before irradiation (see Subheading 3.2).
12. Primary MEFs can be kept as frozen stocks if not needed for irradiation treatment immediately. To establish frozen stocks, cells are dissociated with 0.05% trypsin–EDTA as described above.
13. After the addition of growth medium, remove the cell suspension from the plate, place the cells in a 15 ml centrifuge tube, and centrifuge at 120 × *g* for 5 min.
14. Resuspend the cells in ice-cold 2 ml cryopreservation medium, and aliquot cell suspension to cryovials to 1 ml medium/vial.
15. Place the cryovials in a freezing container and transfer them to a –80°C freezer immediately. Store overnight at –80°C.

16. Transfer all the cells to liquid nitrogen storage the next day. The stocks can be stored for at least 2 years without drastic loss of viability.

3.2. Preparation of Irradiated Mouse Embryonic Fibroblast

1. Passage primary MEFs twice as described above to get sufficient cells for irradiation. Typically, cells from one 10-cm^2 plate can be amplified 20 times after two passages.
2. To collect the cells, wash with 5 ml PBS once, add 1 ml 0.05% trypsin–EDTA, and incubate for 3 min at 37°C.
3. Stop the digestion with 4 ml complete growth medium, pipet up and down to dissociate the cells to single-cell suspension, and place the cells in a 50 ml centrifuge tube. Cells from four 10-cm^2 plates can be pooled in one tube. The total volume is 20 ml.
4. Prepare four more tubes of cells as described above.
5. Mix the cells well before irradiation and place the tubes in a γ-irradiator. The dosages are set as 60, 80, 100, 120, and 140 Gy. The optimal dosage(s) can be selected (see Subheading 3.2, step 8).
6. Centrifuge irradiated cells at $120 \times g$ for 5 min.
7. Remove supernatant, resuspend cells with growth medium, and seed the cells in 6-well plates or T-175 flasks precoated with 0.2% gelatin for cultivation of hESCs or production of MEF-CM. The plating density is 1×10^5 cells/cm^2.
8. Check the viability and the growth of irradiated MEFs in the following days. High-quality irradiated MEFs should be viable for 2 weeks without cell proliferation. The optimal dosage for irradiation in our experiments is 100–120 Gy. Irradiated MEFs can also be stored in liquid nitrogen for future utilization.

3.3. Cultivation of hESCs on Irradiated MEF

1. Add 0.2% gelatin solution to each well (2 ml/well) of a 6-well plate. Incubate overnight at 37°C.
2. Remove the gelatin solution, plate irradiated MEFs at the density of 1×10^5 cells/cm^2, and grow at 37°C for 24–48 h.
3. Remove the medium and wash cells once with sterile PBS (2 ml/well) before plating of hESCs. Thaw hESCs from liquid nitrogen by placing cryovials in a 37°C water bath, gently transfer the cells from cryovials to 15 ml centrifuge tubes, add 5 ml prewarmed hESC growth medium, and centrifuge at $120 \times g$ for 5 min.
4. Remove supernatant, gently resuspend cells in 2.5 ml hESC growth medium, place them in one well of the 6-well plate coated with irradiated MEFs, and shake the plate to distribute the cells uniformly.
5. Change medium 24 h after plating and replace medium daily in the next 4–5 days until the colonies reach 200–400 μm in

diameter and are ready for subculture. hESCs should be subcultured before the colonies contact one another.

6. Prepare 1 mg/ml type IV collagenase solution using DMEM/F12 medium and store at 37°C before use.
7. Remove the medium, add 1 ml prewarmed type IV collagenase solution (1 mg/ml) to each well in a 6-well plate, and incubate at 37°C for 5 min.
8. Aspirate the collagenase solution and wash the residual collagenase in each well with 2 ml sterile PBS.
9. Add 2 ml hESC growth medium to each well, scrape cells from the plate, transfer them to a 15 ml centrifuge tube, wash the well with another 2 ml growth medium to collect residual cells, and put the medium in the same centrifuge tube.
10. Centrifuge at 120 × *g* for 5 min.
11. Resuspend cells in 5 ml growth medium, and pipet up and down to break the colonies to smaller cell clumps. Place the cells in four to five separate wells, add additional medium to make the total volume to 2.5 ml for each well, and continue to grow the cells on MEF feeders at 37°C, 5% CO_2.

3.4. Preparation of MEF-CM and Cultivation of hESCs Under Feeder-Free Condition

1. To collect MEF-CM, plate irradiated MEFs on gelatin-coated T-175 tissue culture flasks. Gelatin coating has been described earlier (see Subheading 3.3, steps 2 and 3).
2. Twenty-four hours after plating, wash the cells once with 20 ml sterile PBS, add 25 ml hESC growth medium lacking bFGF, and incubate at 37°C, overnight.
3. Collect MEF-CM and store at 4°C. Add another 25 ml fresh medium. MEFs can be used for up to 2 weeks, with medium collected daily. MEFs plated on one T-175 flask can produce up to 350 ml CM in 2 weeks.
4. Filter the CM with 0.2 μm filters and store at 4°C for future experiments. The CM can be stored at 4°C for up to 1 month without drastic loss of activity.
5. Thaw hESC-qualified Matrigel on ice, dilute it with ice-cold DMEM/F12 medium at the ratio of 1:40, coat 6-well plates with diluted Matrigel solution (1 ml/well), and incubate at RT for 1 h.
6. Dissociate and centrifuge hESCs grown on irradiated MEFs as described above (see Subheading 3.3, steps 7–10).
7. Resuspend hESCs in medium containing MEF-conditioned hESC medium with bFGF (8 ng/mL), aspirate Matrigel solution from the 6-well plate, plate the cells on the plate (2 ml medium/well), and cultivate the cells at 37°C, 5% CO_2.
8. Replace medium daily. Cells can be subcultured for 5–6 days if split at 1:5 to 1:6.

3.5. Treatment of hESCs with Rapamycin

1. Plate H9 hESCs on Matrigel-coated plates and cultivate in medium containing MEF-CM and bFGF as described above (see Subheading 3.4, steps 6 and 7).
2. Change medium the next day, add 2 μl rapamycin stock solutions (100 mM: 1,000×) to make the final concentration 100 nM. DMSO (vehicle) is used as a control. Mark a few colonies on the bottom of the plate using the colony marker attached to the microscope for both rapamycin-treated and control wells to track the morphological changes after the treatment. Lyse cells in one well before rapamycin treatment with TRIZOL (for RNA analysis) or protein lysis buffer (for western blotting analysis) and designate as the day 0 samples.
3. Change medium daily and add rapamycin or DMSO when replacing the medium. Lyse cells on day 2, day 4, and day 6 for RNA or protein analysis.
4. For OCT-4 and SOX2 protein detection, cells (5×10^6) were lysed directly in 200 μl Laemmli sample buffer (Bio-Rad). For analysis of other proteins, protein lysis buffer is used. 25 μl of each sample is used for western blotting. The blots are developed using SuperSignal West Pico Chemiluminescent Substrate (Pierce).
5. For real-time RT-PCR and microarray analyses, total RNA is isolated using TRIZOL according to the manufacturer's instructions.
6. Total RNA is treated with DNase I to eliminate DNA contamination. 2.5 μl DNase I and 10 μl buffer are added to 87.5 μl RNA (total RNA are diluted to this volume at the final step of extraction). The final solution is incubated at RT for 10 min.
7. Purify RNA with the RNeasy mini kit.
8. cDNA is synthesized from the purified RNA using Reverse Transcription System (Promega) and random primers. Total RNA (1 μg), 10× buffer (2 μl), $MgCl_2$ (4 μl) dNTP mixture (10 mM, 2 μl), recombinant RNasin ribonuclease inhibitor (0.5 μl), random primer (2 μl), and AMV reverse transcriptase (0.5 μl) are mixed, and nuclease-free water is added to make the final volume 20 μl. The mixture is incubated at RT for 10 min and then at 42°C for 1 h. The reaction is terminated at 95°C (5 min) and then on ice for 2 min. cDNA is diluted to 1:10 prior to real-time PCR analysis.
9. Real-time PCR is performed by using the QuantiTech SYBR Green PCR kit (QIAGEN). The reaction mixture includes the SYBR Green mix (2×, 12.5 μl), 10× primer assay (2.5 μl), cDNA (1 μl), and nuclease-free water with the total volume of 25 μl. The following program suggested by QIAGEN is used: 95°C for 15 min followed by 40 cycles of 94°C for 15 s, 56°C

for 15 s, and 72°C for 34 s. *ACTB* gene is used as internal control to equalize cDNA loading.

10. Cells cultured under feeder-free conditions are washed once with cold PBS, fixed with 4% paraformaldehyde for 20 min, permeabilized with 0.1% Triton X-100 in PBS, and blocked with PBS containing 3% skim milk. Samples are incubated with primary antibody overnight at 4°C followed by incubation with FITC (or Alexa)-conjugated secondary antibody. Nuclei are counterstained with DAPI (Sigma). The images are collected with an Axiovert 200 M microscope (Zeiss).

3.6. Rapamycin Treatment During EB Induction

1. Prior to EB induction, culture H9 hESCs on irradiated MEFs as described in Subheading 3.3.
2. Aspirate the hESC growth medium, add type IV collagenase (1 ml/well in 6-well plates), and incubate at 37°C for 10 min.
3. Remove collagenase and wash the cells gently with PBS once (2 ml/well).
4. Add 2 ml hESC growth medium in each well and flush the well to make hESCs detach as colonies. Make sure to keep the colonies intact and transfer them to a 15 ml centrifuge tube. Wash the plate with 2 ml hESC growth medium to collect remaining hESC colonies and place them in the same centrifuge tube.
5. Centrifuge at 120 × *g* for 5 min.
6. Gently resuspend the cells in 2 ml EB induction medium and add the cell suspension to 6-well plates with flat and ultralow attachment surface. Take an aliquote of cells as the day 0 sample, lyse the cells by adding TRIZOL, and store the sample at −80°C.
7. To change medium, gently transfer EB-containing medium to a 15 ml centrifuge tube using a 10-ml pipette. Keep the tube still vertically at RT (for 10 min) to allow EBs to settle down to the bottom of the tube.
8. Carefully aspirate the medium, add fresh prewarmed EB growth medium, place EBs back to the plate, and incubate at 37°C for continuous culture.
9. Collect EBs on day 2, day 4, and day 6. Lyse cells by the addition of TRIZOL.
10. Extract total RNA, perform reverse transcription, and conduct real-time PCR as described in Subheading 3.5, step 5.

3.7. Design and Validation of shRNAs for mTOR Knockdown

1. Obtain the full-length sequence of human mTOR cDNA (NCBI, Gene ID: 2475).
2. Design candidate shRNAs with the Oligomaker 1.5 software. Briefly, input the human mTOR cDNA and select four output candidate oligos with the score higher than 8 as final candidates (designated as shRNA1, shRNA2, shRNA3, and shRNA4) (Table 1).

Table 1
Sequences of mTOR-targeting shRNAs

hmTOR-1R	5′-/5Phos/TCG AGA AAA AAG CTG GCT CTT GCT CAT AAA TCT CTT GAA TTT ATG AGC AAG AGC CAG CA-3′
hmTOR-1 F	5′-/5Phos/TGC TGG CTC TTG CTC ATA AAT TCA AGA GAT TTA TGA GCA AGA GCC AGC TTT TTT C-3′
hmTOR-2R	5′-/5Phos/TCG AGA AAA AAG GCC TAT GGT CGA GAT TTA TCT CTT GAA TAA ATC TCG ACC ATA GGC CA-3′
hmTOR-2 F	5′-/5Phos/TGG CCT ATG GTC GAG ATT TAT TCA AGA GAT AAA TCT CGA CCA TAG GCC TTT TTT C-3′
hmTOR-3R	5′-/5Phos/TCG AGA AAA AAG GAA GAT CCT GCA CAT TGA TCT CTT GAA TCA ATG TGC AGG ATC TTC CA-3′
hmTOR-3 F	5′-/5Phos/TGG AAG ATC CTG CAC ATT GAT TCA AGA GAT CAA TGT GCA GGA TCT TCC TTT TTT C-3′
hmTOR-4R	5′-/5Phos/TCG AGA AAA AAG GAG GCT GAT GGA CAC AAA TCT CTT GAA TTT GTG TCC ATC AGC CTC CA-3′
hmTOR-4 F	5′-/5phos/TGG AGG CTG ATG GAC ACA AAT TCA AGA GAT TTG TGT CCA TCA GCC TCC TTT TTT C-3′

3. Oligos are synthesized at Integrated DNA Technology (IDT, USA).
4. Modify pSicoRGFP lentivector by replacing the CMV promoter with the EF-1α promoter. Briefly, the CMV promoter is removed from pSicoRGFP backbone by using AgeI/BamHI double digestion. The EF-1α promoter is amplified from the 2 K7 lentivector (7) and then inserted into pSicoRGFP. The correct insert is confirmed by sequencing.
5. Generate pSicoRGFP shRNA constructs as described in 10. Briefly, dissolve the oligos in DNase-free water to the final concentration of 100 nM, prepare annealing buffer containing nuclease-free water (23 μl), 2× annealing buffer (25 μl), sense oligo (1 μl), and anti-sense oligo (1 μl), and incubate at 95°C (4 min) and 70°C (10 min). The reaction is cooled down slowly in a 70°C water bath. Annealed oligos (5 μl) are mixed with nuclease-free water (12 μl), T4 ligase buffer (10×, 2 μl), and T4 kinase (2 μl), and phosphorylation is carried out at 37°C (30 min) and then 70°C (10 min). The phosphorylated double-strand oligos are then ligated with the pSicoRGFP vector. The correct plasmids are confirmed by sequencing and purified using the PureYield™ plasmid midiprep system.
6. Culture HEK293T cells by following the suggested protocol (ATCC).

7. Passage HEK293T cells and plate them on 6-well plates (0.5×10^5 cells/cm^2). Prepare two 6-well plates for each shRNA plasmid.
8. Twenty-four hours after plating, package lentivirus by mixing the following in one well of a 6-well plate: the shRNA plasmid (0.5 μg), super-mixture plasmids (Invitrogen, 1.5 μg), and Fugene 6 reagent (6 μl). The plasmids and Fugene 6 reagent are further mixed with Opti-MEM (Invitrogen), which, after a 25-min incubation at RT, is added to HEK293T cells dropwise. Incubate at 37°C overnight.
9. Replace medium with virus-packaging medium (2.5 ml/well).
10. Collect supernatant 48 h after the addition of virus-packaging medium and centrifuge at 1,500 × *g*, 4°C, for 15 min. Discard residual cells and cell debris at the bottom. Keep the supernatant, which contains the lentivirus.
11. Dissociate HEK293T cells with 0.05% trypsin–EDTA and generate single-cell suspension in growth medium (1×10^5 cells/ml). Place 1-ml cell suspension in 1 well of a 6-well plate.
12. Add virus-containing supernatant (1 ml) to each well containing HEK293T cell suspension. Add 2 μl polybrene (final concentration 6 μg/ml from 1,000× stock (6 mg/mL)) to the cell suspension, mix well, and incubate at 37°C overnight.
13. Twenty to twenty-four hours after incubation, replace with fresh HEK293 cell growth medium (2.5 ml for each well in a 6-well plate).
14. Four to five days after lentiviral infection, determine the number of GFP^+ cells under a fluorescence microscope (Zeiss). GFP^+ cells should be nearly 100% if the virus is well-prepared. Collect cells for western blotting analysis using mTOR-specific antibody.
15. Among the four shRNAs selected, shRNA3 and shRNA4 give rise to more than 90% knockdown efficiency (Fig. 1). They are selected further for infection of hESCs.

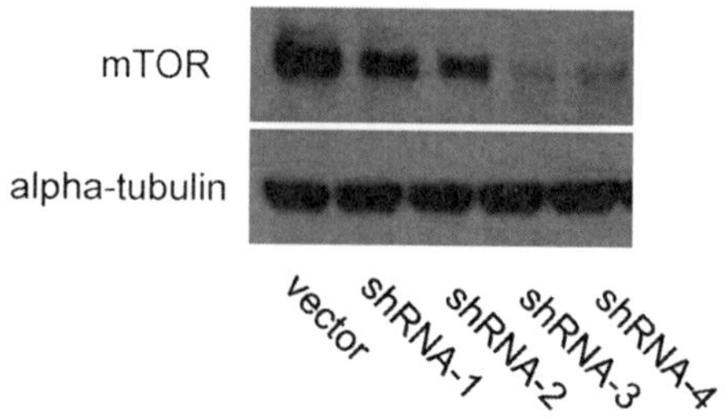

Fig. 1. Western blot analysis of mTOR in HEK293 cells with or without mTOR-targeting shRNAs. Cells infected with lentivirus containing shRNA1, shRNA2, shRNA3, or shRNA4. shRNA3 and shRNA4 effectively depleted mTOR. Virus containing the pSicoRGFP vector was used as a negative control. Alpha-tubulin was a loading control.

3.8. Knockdown of mTOR in hESCs

1. For effective knockdown in hESCs, lentivirus should be concentrated before infection. Prior to concentration, package lentivirus containing shRNA3 and shRNA4 and centrifuge as described in Subheading 3.7, steps 5–10.
2. Transfer 30 ml supernatant to ultracentrifuge tubes and centrifuge for 2 h at 52,000×*g* and 4°C using SW28 ROTOR (Beckman).
3. Remove supernatant. A small pellet can be seen at the bottom of the centrifuge tubes if virus is generated properly. Resuspend the pellet with 300 μl PBS. Virus is concentrated ~100 times. Concentrated virus can be stored as 50-μl aliquots at −80°C for at least 1 year. Avoid repeated freeze/thaw cycles.
4. Before infection, hESCs are cultured on irradiated MEFs for 5 days as descried in Subheading 3.3. Prepare 6-well plates coated with Matrigel as described in Subheading 3.4, step 5.
5. hESCs in one well of a 6-well plate are dissociated as clusters with type IV collagenase and resuspended in 15 ml MEF-CM supplemented with bFGF.
6. Add 15 μl polybrene (6 mg/ml stock) to a final concentration of 6 μg/ml. Plate hESCs on Matrigel in a 6-well plate. Put 2.5 ml medium in each well.
7. Add concentrated virus (50 μl) containing pSicoRGFP (vector only), mTOR shRNA3, and shRNA4 to each well. Shake the plate to mix the virus and to distribute the hESC colonies equally. Transfer the plate back to a 37°C incubator with 5% CO_2. Incubate overnight.
8. Twenty to twenty-four hours after infection, remove virus-containing medium and replace with fresh CM medium supplemented with bFGF. Medium is replaced daily.
9. Forty-eight to seventy-two hours after infection, more than 95% cells should become GFP positive when observed under the fluorescence microscopy (Zeiss).
10. Collect cells for further analysis 5–6 days after infection.

4. Notes

1. Derivation of high-quality MEFs is one of the most critical steps for successful cultivation of hESCs. The quality of MEF cells should be tested before being used in large scale to culture hESCs. We find that some batches of MEFs cannot support the undifferentiated growth of hESCs with unknown reasons. It appears important to obtain mouse embryos between E13 and E14. MEFs from different mouse stains can also exhibit variations. CF-1 mice work well in our experiments.

2. Preparation of high-quality hESCs before EB induction and rapamycin treatment is extremely important. Cell survival becomes an issue if hESCs are not cultivated properly. EBs might not form properly and instead might form small and loose EBs due to poor cell viability. Rapamycin treatment can cause substantial amount of cell death if hESCs are not grown properly before treatment.
3. CMV promoter shows extremely low activity in hESCs in our experiments with lentiviral infection. The EF-1α promoter is a good choice for ectopic expression of genes in hESCs.
4. Lentivirus-mediated gene delivery in hESCs works significantly better than other approaches in our experiments. Addition of lentivirus after hESCs are attached to Matrigel renders the cells at the center of the colonies difficult to infect. We find that adding virus at the time of cell plating improves the infection efficiency, which can reach >90% after a single round of infection with the MOI around 50.

Acknowledgments

The authors would like to thank the Wang lab for helpful discussion. The authors' work is supported by grants from the National Institute of Health (GM083812 and GM083601), the Illinois Regenerative Medicine Institute (IDPH 2006-05516), NSF CAREER award (0953267), NSF-EBICS award, the Beckman award from the University of Illinois, and the National Natural Science Foundation of China (30728022).

References

1. Thomson JA, *et al.* (1998) Embryonic stem cell lines derived from human blastocysts. *Science* 282(5391):1145–1147.
2. Daley GQ & Scadden DT (2008) Prospects for stem cell-based therapy. *Cell* 132(4):544–548.
3. Rossant J (2008) Stem cells and early lineage development. *Cell* 132(4):527–531.
4. Sarbassov DD, Ali SM, & Sabatini DM (2005) Growing roles for the mTOR pathway. *Curr Opin Cell Biol* 17(6):596–603.
5. Gangloff YG, *et al.* (2004) Disruption of the mouse mTOR gene leads to early postimplantation lethality and prohibits embryonic stem cell development. *Mol Cell Biol* 24(21):9508–9516.
6. Murakami M, *et al.* (2004) mTOR is essential for growth and proliferation in early mouse embryos and embryonic stem cells. *Mol Cell Biol* 24(15):6710–6718.
7. Zhou J, *et al.* (2009) mTOR supports long-term self-renewal and suppresses mesoderm and endoderm activities of human embryonic stem cells. *Proc Natl Acad Sci USA* 106(19): 7840–7845.
8. Bain J, *et al.* (2007) The selectivity of protein kinase inhibitors: a further update. *Biochem J* 408(3):297–315.
9. Xu C, *et al.* (2001) Feeder-free growth of undifferentiated human embryonic stem cells. *Nat Biotechnol* 19(10):971–974.
10. Ventura A, *et al.* (2004) Cre-lox-regulated conditional RNA interference from transgenes. *Proc Natl Acad Sci USA* 101(28):10380–10385.

Chapter 24

Video-EEG Monitoring Methods for Characterizing Rodent Models of Tuberous Sclerosis and Epilepsy

Nicholas R. Rensing, Dongjun Guo, and Michael Wong

Abstract

Tuberous Sclerosis Complex (TSC) is a genetic disease involving dysregulation of the mTOR pathway and resulting in disabling neurological manifestations, such as epilepsy. Animal models may recapitulate epilepsy and other behavioral features of TSC and are useful tools for investigating mechanisms of epileptogenesis and other neurological deficits in TSC. In this chapter, methods for performing video-electroencephalography (video-EEG) to characterize epilepsy and neurological dysfunction in rodent models are reviewed. In particular, technical aspects of surgical implantation of EEG electrodes, video-EEG recording, and analysis and interpretation of EEG data are detailed. These methodological approaches should be helpful in characterizing seizures and background EEG abnormalities not only in animal models of TSC but also in many rodent epilepsy models in general.

Key words: Electroencephalography, Video-EEG, Telemetry, Tuberous sclerosis complex, Rat, Mice, Seizure, Epileptiform

1. Introduction

Tuberous Sclerosis Complex (TSC) is the prototypical genetic disease involving abnormal signaling of the mammalian target of rapamycin (mTOR) pathway (1, 2). TSC is caused by mutation of either the *TSC1* or *TSC2* gene, which results in hyperactivation of the mTOR pathway. In turn, abnormal mTOR signaling has been implicated in activating numerous downstream pathophysiological mechanisms that likely mediate many of the phenotypic features of this disease, including tumor or hamartoma formation in multiple organs and neurological symptoms, such as epilepsy, cognitive deficits, and autism. Epilepsy in TSC is an especially common neurological manifestation of TSC, with the seizures usually being very disabling

Thomas Weichhart (ed.), *mTOR: Methods and Protocols*, Methods in Molecular Biology, vol. 821, DOI 10.1007/978-1-61779-430-8_24, © Springer Science+Business Media, LLC 2012 and intractable to available treatments (3, 4). Even in patients whose seizures are well-controlled on medication, it is generally accepted that current antiepileptic drugs simply suppress seizures symptomatically, but do not possess true disease-modifying or "antiepileptogenic" properties for correcting the underlying brain abnormalities that cause epilepsy in the first place. Thus, novel drugs with different mechanisms of action are needed for epilepsy in patients with TSC, as well as people with other types of epilepsy.

Development of more effective treatments for epilepsy in TSC depends on having a better understanding of the underlying cellular and molecular mechanisms of epileptogenesis. Animal models can greatly facilitate the mechanistic investigation of epilepsy and other neurological symptoms of human disease. Several rodent models of TSC, based on spontaneous or induced mutation of one of the two *Tsc* genes, have been developed and characterized, some of which appear to have seizures or other neurological abnormalities (5–8). In one knockout mouse model of TSC, involving conditional inactivation of the *Tsc1* gene predominantly in glial cells (*Tsc1*^{GFAP}CKO mice), progressive epilepsy has been particularly well characterized and underlying cellular and molecular mechanisms of epileptogenesis mediated by the mTOR pathway have been identified, aided greatly by the use of video-EEG monitoring techniques (8–10). Although EEG is an old, established electrophysiological method with a long history of clinical use and applications in people, relatively few sources exist in the literature detailing video-EEG methods and analysis in rodents in a comprehensive fashion (11–13). In this chapter, methods for performing video-EEG to characterize epilepsy and neurological dysfunction in rodent models are described, including protocols for surgical implantation of EEG electrodes, EEG recording, and analysis and interpretation of EEG data. These methodological approaches should be helpful in characterizing seizures and background EEG abnormalities not only in animal models of TSC, but also many rodent epilepsy models in general.

2. Materials

2.1. Surgical Implantation of EEG Electrodes

1. Anesthesia and surgical setup: A typical anesthesia device, such as a rodent anesthesia mask (Kopf, Tunjunga, CA) connected to an isoflurane mixer (Matrx, Orchard Park, NY, 5% for induction, 1.5–2% for surgery), or rodent anesthesia cocktail (ketamine: 100 mg/kg; xylazine: 15 mg/kg, i.p.) is needed to anesthetize the rodent in a stereotactic frame (Kopf, Tujunga, CA). A high-speed micro drill (Fine Science Tools, Foster City, CA) with appropriately sized bits is attached to the stereotactic arm to allow precision drilling of holes in the skull. Standard sterile surgical supplies (e.g., gauze, cotton swabs, betadine ointment)

and instruments (e.g., forceps, scalpel, small screwdriver) are needed to expose the skull and secure the electrodes to the rodent.

2. Epidural screw electrodes: Wired EEG electrodes consist of small stainless steel screws (Small Parts, Miami Lakes, FL) connected to Teflon insulated stainless steel wire (AM Systems, Carlsborg, WA) approximately 2–3 cm in length. Screw size should be matched to the age and size of rodent used (see Note 1). A razor blade is used to strip the Teflon coating approximately 4 mm from both ends. A stainless steel screw is held with a banana clip and the exposed portion of one end of the wire looped and tightened around the neck of the screw. A small drop of solder flux may be added to stabilize the connection between the wire and the screw. Construct the desired number of screw electrodes with wires using the same technique, then solder the opposite end of the wires to individual pins of a customized electric pin header (Newark Electronics, Chicago, IL) (see Note 2). Use dental cement (Parkell, Farmingdale, NY) to cover the soldering points on the header and build a base for stability. Check impedance between each electrode and pin contact to confirm electrical continuity. The final electrode assembly consists of the pin header with multiple screw electrodes attached by insulated wires (Fig. 1).

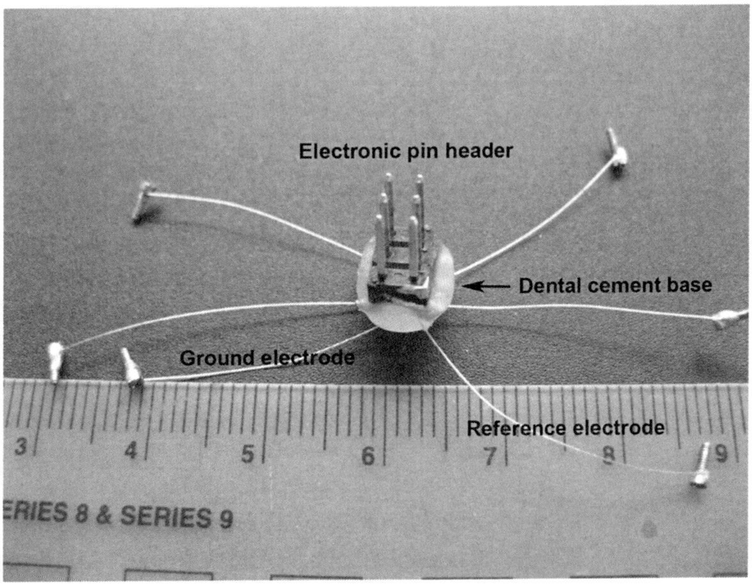

Fig. 1. An EEG electrode assembly with six electrodes and an electronic pin header which connects to a cable plug. The reference, ground, and "active" electrodes each consist of an insulated steel wire soldered to an individual pin on one end (covered with dental cement at the base of the pin header) and a stainless steel screw on the other end, to be anchored in the skull of the rodent.

3. Depth electrodes: Subcortical or hippocampal depth electrodes are constructed in a similar fashion using a wire inserted into silica tubing (Agilent Technologies, Santa Clara, CA). The wire tip is cut at an angle and extended approximately 2 mm longer than the silica tubing. The opposite end is soldered to an electronic header pin. A drop of epoxy is placed on the wires entering the tubing to prevent movement. In some applications, the silica tubing can be glued to the sides of the header with the appropriate distance between the two electrodes.
4. Wireless telemetry electrodes: Wireless EEG electrodes have epidural screws soldered to the leads extending from the telemetry module. Pull the coiled wire from the insulation and solder to a stainless steel screw.

2.2. Video-EEG Recording and Data Analysis

1. Custom cages for video-EEG monitoring (Fig. 2): Individual animal cages for video-EEG monitoring may be constructed using 8–12 in. OD cast acrylic tubes with wire mesh glued to the base. Cage heights range from 12 to 20 in. depending on rodent being monitored. The wire mesh allows minimal rodent contact during bedding changes. Upright water bottles and a food container are fastened to the exterior of the cage. A Plexiglas sheet with drilled holes provides a cage-top to prevent escape. Multiple animal cages may be grouped together within a custom sound-insulated structure that also allows mounting of preamplifiers, cameras, and lights and switches producing a 12-h light–dark cycle.
2. Cable systems: Cables with swivel commentators are essential for stress-free movement during EEG acquisition (Fig. 2). Flexible six wire cables (Cooner Wire, Chatsworth, CA) are inserted into stainless steel springs (Pastics1, Roanoke, VA) to prevent destruction while monitoring adult rats. Highly flexible cable (4–6 leads) without the spring covering is suitable for mouse applications. Solder the wires of the cable to a female header plug (matching the male pin header on the electrode assembly), and on the opposite end, the wire leads are soldered to a male header with the same pin positions as the lower plug. Another female header plug to connect to the cable is soldered to the swiveling leads of the commentator and head-box/amplifier plugs (Grass, West Wawick, RI) are soldered to stationary wires of the commentator unit. All soldered joints are covered with silicon glue or dental cement and then taped, and impedance tested for electrical continuity.
3. Wireless EEG acquisition: Preamplifiers, telemetry devices, and corresponding receiver pads are available with 2–4 EEG/EMG leads (Data Science International, St. Paul, MN).
4. Commercial time-locked video-EEG monitoring system: Several commercial clinical-grade video-EEG systems are available for

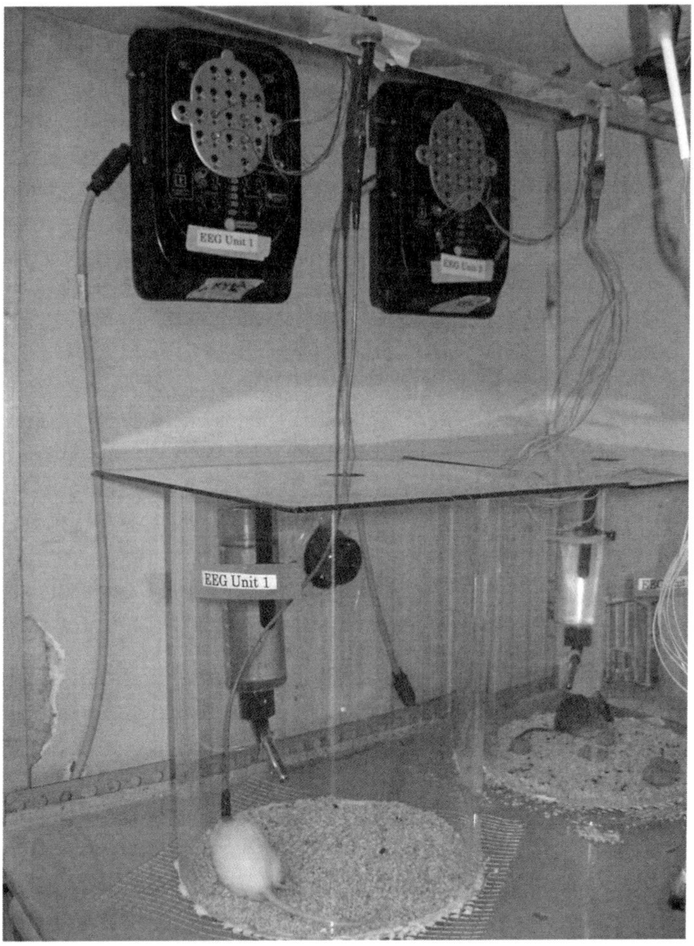

Fig. 2. Custom cages for rodents consist of an acrylic tube with a wire mesh bottom. The wire cable is plugged into the electronic pin header protruding from the head. The opposite end is plugged into a swivel commentator. Wires extending from the stationary output of the commentator are plugged into the appropriate head-box channels and connected to a computer. A mounted wide-angle IR camera records animal activity (lens protruding from rear wall). Multiple animal monitoring units are contained in a custom sound-insulating structure that provides a 12-h light–dark cycle.

data acquisition and analysis: e.g., Stellate Systems (Natus, Montreal, Canada), XLTEK (Optima, London, UK).

5. Custom video and EEG acquisition systems: Alternatively, customized video-EEG systems can be built from individual components: analog or digital video camera, digital video capture card (e.g., MPEGator, Darim, Livermore, CA), personal computer, AC amplifiers (e.g., P511, Grass Instruments, West Warwick, RI), separate video- (e.g., MG100, Darim, Livermore, CA) and EEG-acquisition software (e.g., Axoscope, Molecular Devices, Sunnyvale, CA).

3. Methods

Video-EEG monitoring on rodents may be performed for a number of reasons and applications. Most commonly, video-EEG is utilized to evaluate possible seizures. In particular, EEG can be useful for confirming that an observed behavioral spell is an epileptic seizure, and not some other paroxysmal episode (e.g., movement disorder, normal behavior), based on finding a correlated ictal epileptiform discharge on EEG during the spell. Furthermore, video-EEG can potentially help determine the localization and classification of seizures (e.g., partial vs. generalized, hippocampal vs. neocortical), determine the frequency and temporal progression of seizures, and assess response of seizures to treatment. In addition to characterizing seizures, assessment of the interictal EEG background activity may also provide an objective, physiological assay of overall, general brain function, such as evidence of encephalopathy, although EEG may not be very sensitive or specific in this respect.

There are a variety of different techniques and approaches that can be used to perform video-EEG. Important factors in deciding the specific approach to use include age/size of the animal, anatomical sites of interest (e.g., hippocampus, neocortex, subcortical structures), duration of recording, and availability of equipment. In this section, we focus primarily on our approach for recording long-term/chronic neocortical and/or hippocampal EEG in unanesthetized/freely behaving juvenile (postweaning) or adult mice or rats. We also describe methods for performing video-EEG monitoring using either traditional cable or wireless telemetry connections to the recording equipment. Finally, we present methods for analysis of EEG data, focusing on characterizing seizures and associated interictal EEG background abnormalities, such as those that occur in *Tsc1*^{GFAP}CKO mice. Potential pitfalls and opportunities for adapting these methods for other situations and applications are also noted.

3.1. Surgical Implantation of EEG Electrodes

1. General Considerations: A variety of surgical methods can be used to implant EEG electrodes into rodents. The specific approach to use depends on such factors as the size/age of the animal, the anatomical site(s) of interest to be recorded, and the duration of the recording, which dictate the number of electrodes, the electrode type (epidural screw electrodes or wire depth electrodes), and transmission system (cable tethering vs. wireless).
2. Electrode Surgery: All electrode and EEG implantation techniques are performed using aseptic techniques with the animal anesthetized and secured in a stereotaxic frame over a warming pad (36–37°C). Ensure that the head is mounted stably to

prevent movement while drilling or screwing. Pinch the tail and/or toe to ensure that the animal is fully anesthetized. Apply eye ointment to prevent drying out and irritation from the disinfectant. Shave the hair from the neck up to the eyes. Sterilize the head skin with betadine. Create a vertical midline incision to the skin of the rodent's head. The incision starts at eye level and proceeds posterior to the neck muscles. Spread the skin, hold open with forceps, and clean the skull of blood and connective tissue using forceps, hydrate with saline and clean with cotton swabs. After bleeding is controlled, dry the exposed skull with 3% hydrogen peroxide. Using a high-speed micro drill attached to the sterotaxic arm, drill appropriate size holes for securing screws. The size of the drill bit must match the size of the screw and the hole should avoid major blood vessels. Slowly lower the drill and allow the drill bit to pierce the skull, but take great caution not to disrupt the dura and damage the brain. Minor bleeding may occur on most electrode sites, but particular attention must be taken when drilling near the sutures not to disrupt any major blood vessels.

Generally, adult rat EEG epidural screw electrodes consist of four bilateral recording electrodes (−1.60 mm bregma, 1.80 mm lateral and −4.0 mm bregma, 3.0 mm lateral), a reference electrode (+2.0 mm bregma, 1.0 mm lateral), and a ground electrode (−10.0 mm bregma, 1.0 mm lateral), although more complex configurations with additional electrodes can be designed. For hippocampal depth electrodes, the two posterior recording screws may be replaced with insulated wire electrodes (14). A modified header with fastened depth electrodes is held with a sterotaxic arm and the electrodes lowered into the hippocampus according to the desired stereotactic coordinates based on a rodent atlas. When a modified header is not possible each depth electrode is held with a sterotaxic arm and lowered into place. The depth electrode is glued in place and the sterotaxic arm is removed.

Adult mouse EEG electrodes are similar to the rat. However, due to the smaller head size, usually one set of bilateral recording electrodes (−1.5 mm bregma, 1.5 mm lateral), a reference electrode, (+0.5 mm bregma, 0.5 mm lateral), and a ground electrode (−4.5 mm bregma, 0.5 mm lateral) are sufficient to monitor EEG activity (9). Depth electrodes and four recording screws can also be implanted if more EEG detail is desired or warranted.

Drill all holes in the skull and screw electrodes starting with the anterior-most electrodes. Fasten the screw approximately three full rotations to secure in the skull for most adult rats. Again, use caution not to disturb the dura, and if working with mice or younger rats bear in mind they have thinner skulls and fewer rotations may be warranted. Use a small drop of

Krazy glue to stabilize the screw to the skull. With all the screws secure, grab the pin header and twist the wires close to the head. Keep the pin header on the top of screws and on the center of the skull. Prepare dental cement using equal parts resin and monomer and mix until cement is a thick liquid. Apply cement over screws and wires, allowing the cement to cure only on the exposed dry skull. Keep dental cement away from the muscles of the neck and head, as well as the skin. Multiple applications of dental cement may be required to "build up" to the base of the electronic header. With the last application of dental cement partially cured, mold cement to a smooth surface and place skin around the "cap". Use cotton swabs to stop minor bleeding and dry muscle before suturing the skin anterior and posterior of the exposed header pins. Clean entire head around pins and allow animal to slowly recover in an individual clean cage (see Note 3 for methods for improving long-term stability of recordings).

3. Wireless EEG Surgery: In wireless telemetry EEG applications, a second incision is needed. Shave and clean the skin between the scapulae and make a 1.5–2.0 cm lateral incision. Use blunt forceps or haemostats to create a subcutaneous pocket posterior the incision. Place the telemetry device within the pocket and drill holes for the EEG electrodes in the skull. Using blunt forceps create a subcutaneous channel from the back of the head down to the transmitter wires and gently pull the wires to the exposed skull. Screw the electrodes into the skull and suture the incision used to imbed the telemetry device. Adjust wires so they do not push against the skin or impede movement. Apply dental cement to exposed dry skull, let cure, then clean and suture skin over the electrodes. Allow animal to recover in an individual clean cage.

3.2. Video-EEG Recording

1. General Considerations: A number of approaches can be used to record video-EEG data. Commercially available systems are convenient and pre-packaged with a number of easy-to-use, advanced features, but are relatively expensive. Custom-built systems require more effort to build and may lack some key properties (e.g., automated synchronization between the video and EEG data), but are relatively inexpensive and may allow more customization.

2. Animal Setup: Rodents should have at least 3 days postoperative care and cage acclimation before monitoring. Each electrode (epidural or depth) has a corresponding header pin, needed to plug into the proper head-box channel for monitoring. Predetermined and set pin orientation allows for efficient connection from electrodes, to cable, to commentator, to head-box (see Note 4). If a standardized orientation is followed, plug the

cable connector to the pin header in the same order for each animal. Large rats may require a small amount of dental cement at the connection point on the head to prevent unwanted disconnection between the cable and pin header. On commercially available monitoring systems, check the impedance ($<10\,\Omega$) of connected channels to ensure minimal noise in the EEG channels. The rodent being monitored should be freely moving and the commentator should easily turn with movement (see Note 5). Allow the animal to acclimate to the surroundings for best results.

3. Wireless Recording: Again, rodents should have postoperative care and acclimate to surroundings before starting recording. Rodents with implanted transmitters are placed on the receiver pad, with minimal distance between the receiver pad and the corresponding implanted radio transmitter. A magnet is swiped over the radio transmitter and signal strength checked with the operating program.
4. Video-EEG Monitoring with commercial acquisition systems: A number of commercial video-EEG systems with integrated time-locked digital video and EEG data acquisition are available that are used clinically for video-EEG monitoring in people and may also be adapted for animal research, such as Stellate Systems (Natus, Montreal, Canada) and XLTEK (Optima, London, UK) (see Note 6).
5. Video-EEG monitoring with custom video and EEG acquisition systems: A relatively inexpensive video-EEG system can be custom-made, with the video and EEG acquisition involving separate systems that can be time-locked by manual methods. Video data can be acquired through multiple methods: (1) an analog video camera recorded on tape with a video-cassette recorder, (2) an analog video camera connected to and recorded digitally on a personal computer via a digital video capture card (e.g., MPEGator, Darim, Livermore, CA), or (3) digital video camera connected to a computer with direct storage to the hard drive. The latter two methods are preferable for long-term monitoring, although large hard drives are required for storage (e.g., standard MPEG video files require approximately 10 GB of hard drive space per 24 h of video data). EEG data can be acquired through standard research-grade AC amplifiers (e.g., P511, Grass Instruments, West Warwick, RI) and stored simultaneously on the same computer used for video acquisition with freely available electrophysiology software (e.g., Axoscope, Molecular Devices, Sunnyvale, CA). Although acquired separately, the video and EEG data can be manually synchronized by initiating the two acquisition programs simultaneously and tracking either cumulative or real time.

6. EEG Filtering: Set low and high frequency filters. Filtering can be performed either through the amplifiers during acquisition or post hoc during digital analysis. The most relevant EEG frequencies of interest usually occur between 1 and 100 Hz, with frequencies outside this range often representing artifact (non-EEG) sources. Thus, common filter settings include low frequency (high-pass) filter at 1 Hz and high frequency (low-pass) filter at 100 Hz. A 60 Hz notch filter may also be useful to eliminate common AC electrical noise.
7. EEG Sampling: Set the digital sampling rate. The conversion of analog (i.e., EEG signals) to digital format requires appropriate digital sampling. In order to prevent insufficient, inaccurate sampling of the data ("aliasing"), the digital sampling rate must be at least twice the frequency of the high frequency filter. Thus, with a high frequency filter of 100 Hz, the digital sampling rate should be at least 200 Hz.

3.3. Video-EEG Data Analysis

1. General Considerations: Video-EEG studies usually result in the collection of a voluminous amount of data that can be complicated and laborious to analyze and interpret. Both qualitative and quantitative analysis can be performed on EEG data. The most appropriate analysis method depends on the nature of the studies and questions to be addressed. Most video-EEG experiments focus on evaluating seizures and epilepsy, including confirmation that a behavioral spell represents a seizure, characterization of the type and localization of the seizures, and quantification of seizure frequency and response of seizures to treatment. In addition to directly documenting seizures, assessment of the "interictal" EEG background activity can provide additional evidence for possible epilepsy, encephalopathy or other neurological abnormalities in rodents.
2. Qualitative analysis of EEG background: Descriptive analyses of the normal background EEG in control rodents during the awake and asleep states have been reported previously (15, 16). Briefly, the awake active state is characterized primarily by a sinusoidal ~5–8 Hz theta rhythm, recorded throughout neocortex, as well as in hippocampus (Fig. 3). During drowsiness, the theta rhythm is interrupted by bursts of higher amplitude polymorphic slow delta (~1–4 Hz) waves. During nonrapid eye movement (non-REM) sleep, periodic higher frequency (~6–10 Hz) spindle-like activity intersperses with the background slow waves. Finally, in REM sleep, there is often a reemergence of a ~6 Hz theta rhythm similar to the awake state, but the animals are behaviorally asleep and simultaneously recorded electromyogram (EMG) shows absence of muscle activity.

   Epileptiform spikes are brief, abnormal electrical discharges that are in themselves relatively specific interictal markers or

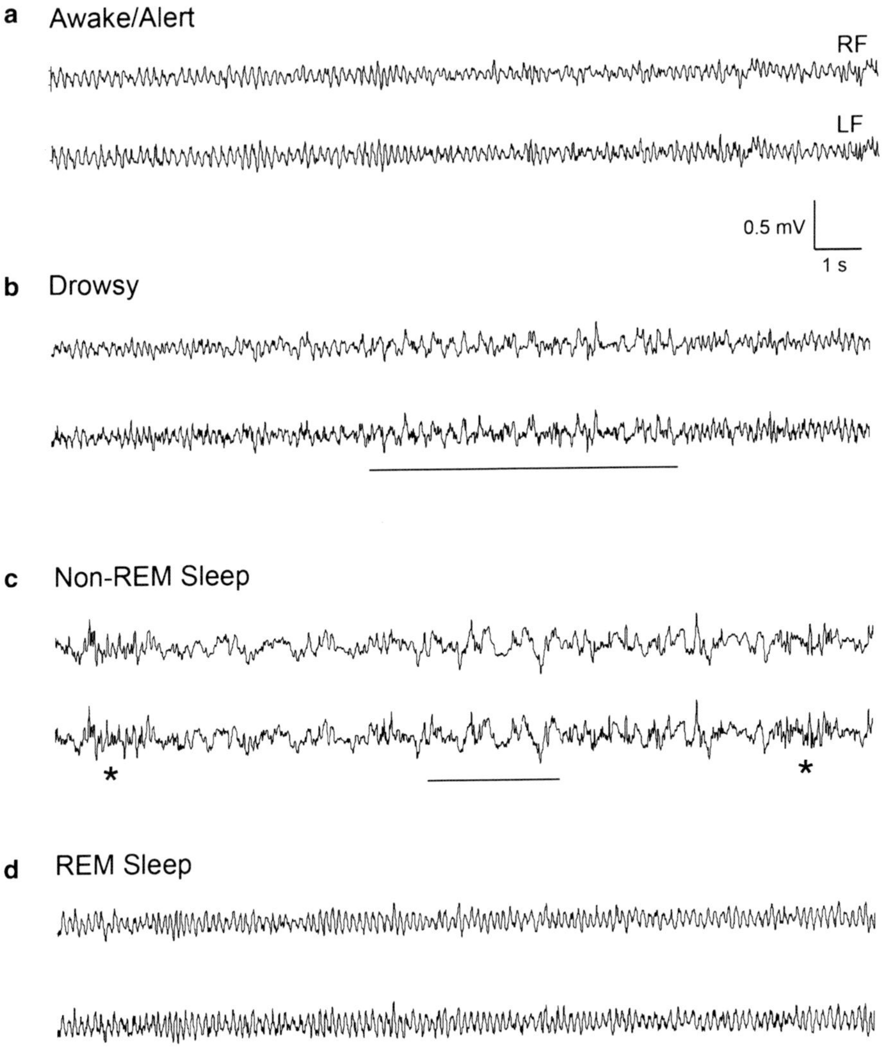

Fig. 3. Normal rodent EEG during the sleep–wake cycle. Typical background EEG activity varies with the sleep–wake cycle. (**a**) In the awake/alert state, a predominant EEG finding is a 5–8 Hz sinusoidal theta rhythm. (**b**) During drowsiness, the theta rhythm is often interrupted by periods of higher amplitude polymorphic slower (1–4 Hz) delta waves (*underlined section*). (**c**) During non-REM sleep, slow delta waves predominate (*underlined section*) and are intermixed with bursts of higher frequency (~6–10 Hz) spindle-like activity (*asterisks*). (**d**) During REM sleep, a sinusoidal theta rhythm, similar to the awake state, may reemerge, but the animals would appear behaviorally asleep and simultaneous electromyogram (EMG) recordings would document absence of muscle activity (not shown).

indicators of animals that have epilepsy or seizures. Interictal spikes are defined as fast (<200 ms) sharp or "pointy" waveforms that stand out from and are usually at least twice the amplitude of the surrounding background activity (Fig. 4a). Many electrical artifacts and other physiological activity (e.g., EMG potentials) can mimic and be mistaken for epileptiform spikes. Review of the concurrent video may help determine whether an apparent epileptiform spike is actually due to an artifact. As true epileptiform

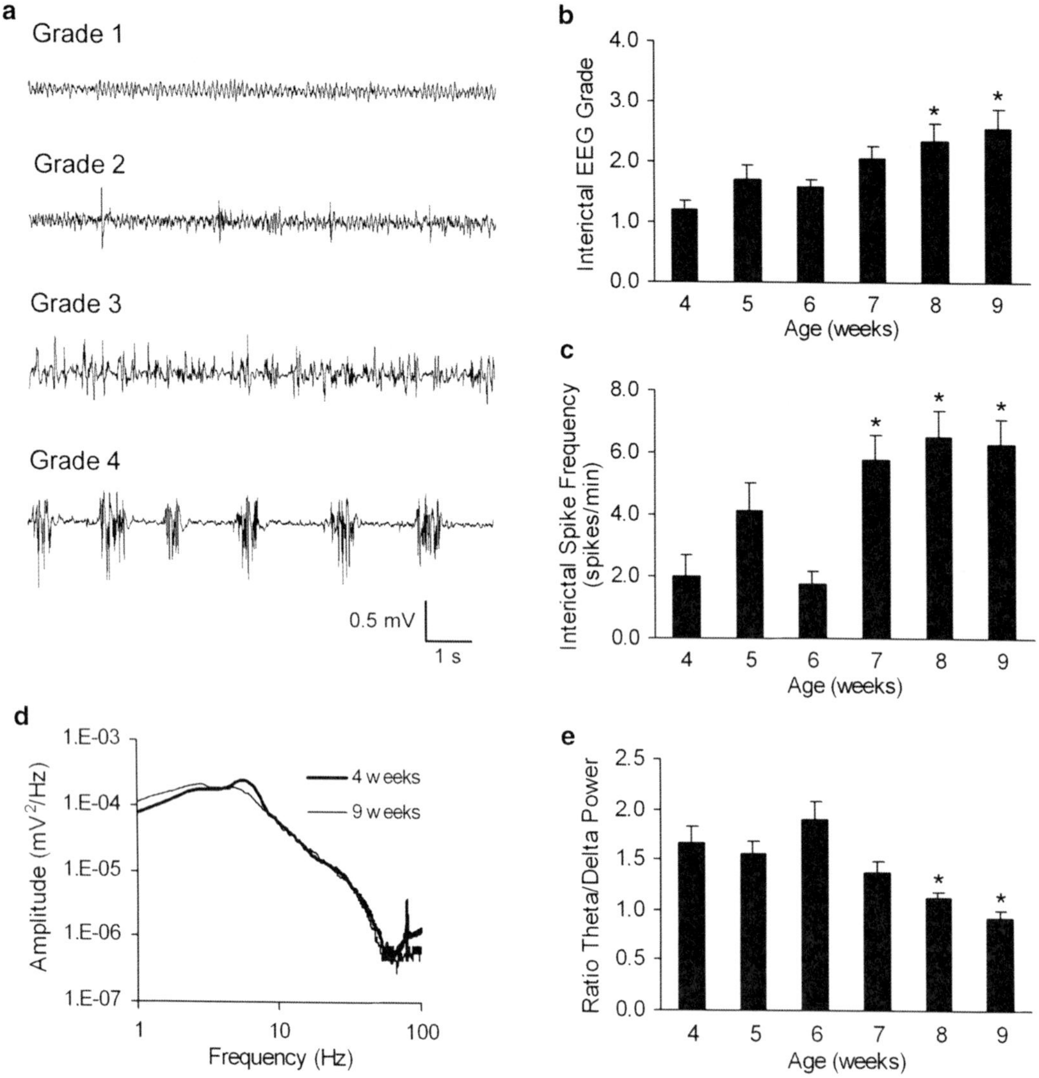

Fig. 4. Representative analysis of interictal EEG background abnormalities. Interictal EEG background activity can exhibit abnormalities in background frequencies and epileptiform spike discharges, as exemplified in *Tsc1*^{GFAP}CKO mice. (**a**) A qualitative grading system has been developed to describe progressive abnormalities in interictal EEG background activity that occur in mouse models of neurological diseases, including *Tsc1*^{GFAP}CKO mice (9, 10, 17, 18). Representative examples of the four interictal EEG background grades are shown: Grade 1 – normal background activity (±5–8 Hz sinusoidal theta rhythm), no epileptiform spikes; 2 – mostly normal background activity, few epileptiform spikes; 3 – mostly abnormal background activity, many spikes; 4 – burst-suppression pattern. (**b**) Average interictal EEG background grade of *Tsc1*^{GFAP}CKO mice shows progressive worsening with age. (**c**) Average interictal spike frequency of *Tsc1*^{GFAP}CKO mice also increases with age. (**d**) Fast Fourier Transform of interictal EEG background of *Tsc1*^{GFAP}CKO mice reveals different frequency distributions dependent on age. Average power spectra for 4 week and 9 week old mice are shown. Theta (5–8 Hz) frequencies predominated in 4 week old mice and slower delta (1–4 Hz) frequencies were more common in older 9 week old mice. (**e**) The ratio of theta/delta power decreased with age. Reproduced with permission from Wiley-Blackwell (9).

spikes usually occur relatively frequently, it is generally advisable not to overinterpret rare, isolated spikes as necessarily being a sign of epilepsy (see Note 7). Commercial video-EEG acquisition programs typically include automated spike detection features.

In some mouse models that exhibit progressive epilepsy and other neurological deficits, such as *Tsc1*^{GFAP}CKO mice, a stereotypical progression of interictal epileptiform spikes and other EEG background abnormalities have been documented (Fig. 4). A qualitative grading system for these abnormal EEG background stages has been established that correlates approximately with seizure frequency and general neurological status of the mice, as well as response to treatment (9, 10, 17, 18): Grade 1: normal background (e.g., normal theta rhythm while awake), no epileptiform spikes; 2 – mostly normal background activity, few epileptiform spikes; 3 – mostly abnormal background activity (absent theta rhythm, excessive slowing), many epileptiform spikes; 4 – burst-suppression pattern (Fig. 4a).

3. Quantitative analysis of EEG background: In addition to qualitative approaches of interpreting EEG, more quantitative methods are available for analyzing EEG data. As EEG basically consists of a composite of sinusoidal waveforms of varying frequencies and amplitudes, most quantitative methods utilize some form of frequency or spectral analysis (e.g., Fourier analysis), which separates out the relative contribution of different underlying frequencies (Fig. 4d, e). The relative power and distribution of different EEG wave frequencies varies depending on sleep state (19, 20), presence of seizures and other neurological abnormalities (9, 10), and response to treatment. Technical details of these quantitative methods are beyond the scope of this chapter, but standard spectral analyses can be performed by a number of commercially available software programs (e.g., pCLAMP, Molecular Devices, Sunnyvale, CA).

4. Qualitative analysis of seizures: One of the most common applications of video-EEG monitoring is to identify and confirm whether an observed behavioral event represents an epileptic seizure and to characterize the behavioral and electrographic features of seizures. The standard definition of a seizure is an abnormal, paroxysmal, stereotypical behavioral event that is caused by or correlated with an abnormal, paroxysmal electrical discharge from cortex. Although theoretically the repertoire of abnormal behaviors that can manifest as seizures is extremely large, especially in people, the number of reliably observable behavioral manifestations of seizures in rodents is relatively limited. A standardized staging system (Racine scale) is commonly utilized to describe different limbic seizures in rodents, such as those that occur with progressive severity in the popular

kainate, pilocarpine, and kindling models of epilepsy (21): stage 1 – behavioral arrest with mouth/facial movements, stage 2 – head nodding, stage 3 – forelimb clonus, stage 4 – rearing, stage 5 – rearing and falling, stage 6 – loss of posture and generalized convulsive activity. In addition, other, often more subtle, types of seizures can be demonstrated in rodents, such as the brief freezing/behavioral arrest of absence seizures, rapid isolated jerks of the body or limbs with myoclonic seizures, or a jackknife flexion of the body with epileptic spasms. Each seizure type tends to have a stereotypic ictal electrographic correlate on EEG.

Video-EEG analysis allows precise correlation between an observed, paroxysmal behavioral event and the EEG to determine whether the event meets the definition of a seizure. By definition, the behavioral manifestations of a seizure should be correlated with an abnormal electrical discharge, although there are some exceptions, such as very focal seizures that originate at a site distant from the electrodes utilized. There are no universal criteria or electrographic patterns for what constitutes a seizure on EEG. Typically, seizures on EEG appear as self-limited (i.e., having a clear start and end that stands out from the background EEG activity) electrical pattern often involving an initial low voltage, high frequency rhythmic discharge or repetitive epileptiform spikes (tonic phase) that evolve in amplitude and frequency into a bursting pattern (clonic phase) (Fig. 5a). After the seizure is over, there is often a period of voltage suppression (postictal phase). While this stereotypical “tonic-clonic” pattern is most widely recognized, other EEG abnormalities may also represent ictal patterns, such as single spike and wave discharges correlating with brief myoclonic seizures or isolated periods of electrodecrement/voltage suppression of the background activity correlating with epileptic spasms. Similar to the pitfalls of interpreting apparent interictal epileptiform spikes, many electrical and physiological artifacts can mimic seizure-like patterns, such as scratching behavior or respiratory artifact. Thus, time-locked video analysis can again be very helpful in distinguishing true seizure activity from artifact.

When the spells of concern are relatively frequent (several times a day), it is relatively easy to identify and capture them by concurrent behavioral observation or retrospective review of the video, and then determine whether an ictal EEG correlate exists on EEG. Infrequent spells (less than once a day) may be difficult or laborious to document. In this case, long periods of EEG data can be initially screened relatively quickly either visually or with a commercial automated seizure detection program; any seizure-like patterns detected on EEG can then be correlated with the video to determine whether the

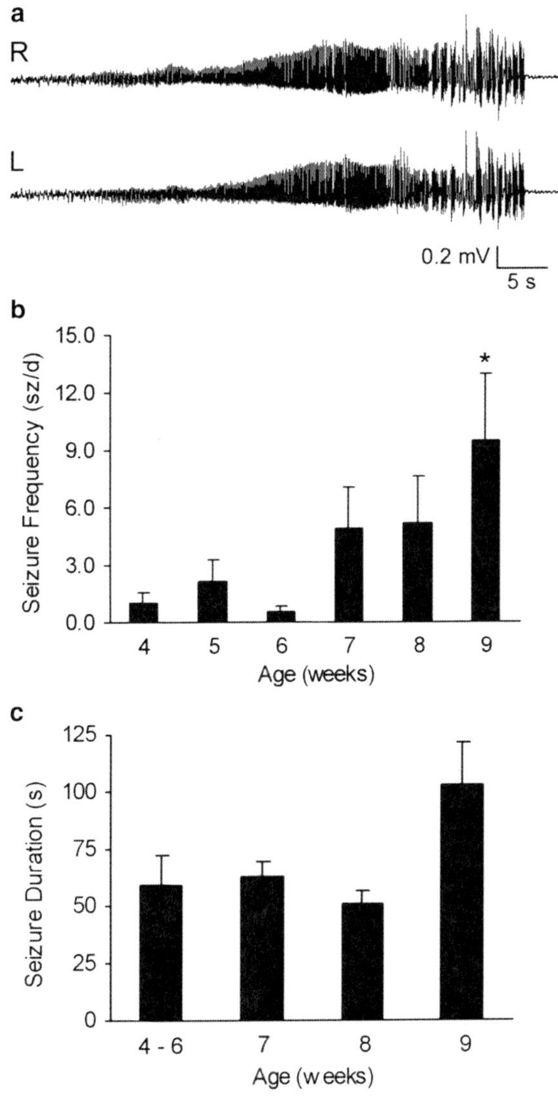

Fig. 5. Representative analysis of seizures. Seizures can be identified and characterized by video-EEG in various rodent models of epilepsy, such as in *Tsc1*^{GFAP}CKO mice. (**a**) A representative example of ictal EEG activity recorded by bilateral neocortical epidural electrodes from a *Tsc1*^{GFAP}CKO mouse. Seizures usually start with bilateral fast tonic activity that evolves into higher amplitude spikes. The end of the seizure typically culminates in bursting spike discharges, followed by postictal suppression. (**b**) Average seizure frequency of *Tsc1*^{GFAP}CKO mice increases with age. (**c**) Average seizure duration of *Tsc1*^{GFAP}CKO mice does not change significantly with age. Reproduced with permission from Wiley-Blackwell (9).

spell of interest was captured and thus represents a seizure. Unfortunately, if the spell of concern is not a seizure (does not have an ictal EEG correlate), then screening the EEG will not help in determining whether the spell of interest was captured, and thus lengthy, laborious cage-side observation or retrospective video review becomes necessary to identify if/when a spell has been recorded on the EEG (see Note 8).

5. Quantitative analysis of seizures: In addition to identification/confirmation of seizures and qualitative characterization of their behavioral and electrographic features, more quantitative analysis of seizures can also be performed. Seizure frequency, such as the number of seizures occurring within a 24 h period, and average seizure duration are easily quantified (Fig. 5b, c). When a reliable, stereotypical ictal EEG pattern can be identified, assessment of seizure frequency can be accomplished relatively quickly by scrolling through the EEG record to find seizures within the given time period or by using an automated seizure detection program (although the positive detections should always be confirmed visually, as false positives are common). Accurate assessment of the duration of status epilepticus (prolonged or repetitive seizures lasting greater than 30 min), such as occurring initially with the kainate or pilocarpine models, often requires EEG, as the behavioral manifestations of seizures can become subtle and "subclinical" electrographic seizures may be frequent, especially during the latter stages of status epilepticus. Quantification of seizure parameters with EEG allows the most accurate assessment of the effect of treatments on seizures (9, 10).

4. Notes

1. EEG screw electrodes: A variety of screw sizes and lengths can be used, but should be selected based on skull thickness which is dependent on the age and size of the rodent. Larger screws (e.g., #000 = 0.034 in. diameter, 1/8 in. length) may be used for older rats. Smaller screws (e.g., #0000 = 0.021 in. diameter, 1/16 in. length) may be used for mice or young rats.
2. Electronic pin headers: Customized pin headers may be constructed from standardized electronic circuit components (Newark Electronics, Chicago, IL), which typically are manufactured in long paired rows of pins (e.g., 2 × 6, 2 × 32). Cut the pin header to match the number of electrodes to be used (e.g., 2 × 3 for 6 electrodes).
3. Improving stability of long-term recording: A common pitfall of long-term EEG recording is the tendency for electrodes to become mechanically or electrically unstable, leading to noisy, uninterpretable recordings or separation of the electrode cap from the animal. A clean/dry skull, aseptic surgery technique, and at least 3 days postoperative care prior to recording are imperative for stable, long-term EEG monitoring. Any inflammation or infection of the surgical site weakens the cement bond to the skull. Trim the nails of the rodent to prevent skin

infection and irritation around the surgical site, and apply a small amount of antibiotic ointment if needed. Also, "anchor" screws (non-electrode screws) placed over the cerebellum or in other areas under the electrode cap are helpful in keeping the electrode cap on adult rats.

4. Electrode orientation: Header pins and electrodes should have a prearranged position and this orientation kept throughout all experimental applications. Distinguish the reference electrode pin, its corresponding plug position on cable, commentator, and channel on head-box. Set a position for remaining ground and "active" electrodes and construct all cables, plugs, commentators and head-box connections reflecting these positions. Standardized pin positions and coordinating head-box channels avoid confusion and simplify tethering of animals for recording.
5. Cable systems: Proper cable length is important to minimize strain to the rodent head and allow free-movement in cages. Wire used in construction of cable should be of best pliability, particularly in smaller rodents. Springs protecting cables should have smallest diameter needed to protect wires and only used with adult rats. For proper commentator swivel and to reduce torque at the surgical site, all electrode pins, cables, and commentators need to function as a single unit. Use electrical tape over all plug-in connections especially at the commentator connection. If possible, animals should be tethered with cables for the duration of experiment with minimal handling. If connected animals need to be exchanged and monitored again, great care should be taken when disconnecting/connecting cables to minimize strain on the electrode cap. Once an animal loses the electrode cap, it typically cannot be replaced and the animal usually must be sacrificed.
6. Commercial video-EEG systems: The primary advantage of purchasing a commercial clinical video-EEG system is the integrated time-locked acquisition and playback of both digital video and EEG data within the same software program. The system software usually also includes a number of advanced features for data acquisition and analysis, including automated spike and seizure detection. The main disadvantage of commercial systems is the expense.
7. EEG analysis: When analyzing and comparing EEG data between different conditions or groups of animals, it is generally advisable for the reviewer to be blinded to the experimental condition to minimize bias and subjectivity that is inherent to EEG interpretation.
8. Seizure analysis: The documentation of a behavioral spell that has no abnormal, time-locked EEG correlate usually means that the spell does not represent an epileptic seizure. This is

especially true for behavioral spells that involve bilateral motor activity of the extremities, because rodent seizures involving bilateral movements are usually detectable by standard cortical EEG electrodes. However, it is always difficult to rule out the possibility that a more subtle, behavioral event represents a focal seizure (e.g., hippocampal) that has not spread to the available electrodes. If the spell appears to have characteristics consistent with a seizure, but no abnormal EEG discharge is detected, more detailed EEG recording strategies, including a larger number of electrodes or the use of depth electrodes, may be warranted.

Acknowledgments

The authors acknowledge grant support from the National Institutes of Health (K02NS045583 and R01NS056872), the Tuberous Sclerosis Alliance, and Citizens United for Research in Epilepsy.

References

1. Crino PB, Nathanson KL, Henske EP (2006) The tuberous sclerosis complex. N Engl J Med 355:1345–1356.
2. Kwiatkowski DJ (2003) Tuberous Sclerosis: from tubers to mTOR. Ann Hum Genet 67:87–96.
3. Holmes GL, Stafstrom CE, and the Tuberous Sclerosis Study Group (2007) Tuberous Sclerosis Complex and epilepsy: recent developments and future challenges. Epilepsia 48:617–630.
4. Chu-Shore CJ, Major P, Camposano S, Muzykewicz D, Thiele EA (2010) The natural history of epilepsy in tuberous sclerosis complex. Epilepsia 51:1236–1241.
5. Waltereit R, Welzl H, Dichgans J, Lipp HP, Schmidt WJ, Weller M (2006) Enhanced episodic-like memory and kindling epilepsy in a rat model of tuberous sclerosis. J Neurochem 96:407–413.
6. Meikle L, Talos DM, Onda H, Pollizzi K, Rotenberg A, Sahin M, Jensen FE, Kwiatkowski DJ (2007) A mouse model of tuberous sclerosis: neuronal loss of Tsc1 causes dysplastic and ectopic neurons, reduced myelination, seizure activity, and limited survival. J Neurosci 27:5546–5558.
7. Way SW, McKenna J 3rd, Mietsch U, Reith RM, Wu HC, Gambello MJ (2009) Loss of Tsc2 in radial glia models the brain pathology of tuberous sclerosis complex in the mouse. Hum Mol Genet 18:1252–1265.
8. Uhlmann EJ, Wong M, Baldwin RL, Bajenaru ML, Onda H, Kwiatkowski DJ, Yamada KA, Gutmann DH (2002) Astrocyte-specific TSC1 conditional knockout mice exhibit abnormal neuronal organization and seizures. Ann Neurol 52:285–296.
9. Erbayat-Altay E, Zeng LH, Xu L, Gutmann D, Wong M (2007) The natural history and treatment of epilepsy in a murine model of tuberous sclerosis. Epilepsia 48:1470–1476.
10. Zeng LH, Xu L, Gutmann DH, Wong M (2008) Rapamycin prevents epilepsy in a mouse model of tuberous sclerosis complex. Ann Neurol 63:444–453.
11. Bertram EH, Williamson JM, Cornett JF, Spradlin S, Chen ZF (1997) Design and construction of a long-term continuous video-EEG monitoring unit for simultaneous recording of multiple small animals. Brain Res Protocols 2:85–97.
12. Williams P, White A, Ferraro D, Clark S, Staley K, Dudek FE (2006) The use of radiotelemetry to evaluate electrographic seizures in rats with kainate-induced epilepsy. J Neurosci Methods 155:39–48.
13. Lapray D, Bergeler J, Dupont E, Thews O, Luhmann HJ (2008) A novel miniature telemetric system for recording EEG activity in freely moving rats. J Neurosci Methods 168: 119–126.

14. Zeng LH, Rensing NR, Wong M (2009) The mammalian target of rapamycin signaling pathway mediates epileptogenesis in a model of temporal lobe epilepsy. J Neurosci 21:6964–2672.
15. Tsubone H, Sawazaki H (1979) Electroencephalographic patterns and their alternating process in mice. Jap J Vet Sci 41:495–504.
16. Eleftheriou BE, Zolovick AJ, Elias MF (1975) Electrocephalographic changes with age in male mice. Gerontologia 21:21–30.
17. Griffey MA, Wozniak D, Wong M, Bible E, Johnson K, Rothman SM, Wentz A, Cooper JD, Sands M (2006) CNS-directed AAV2-mediated gene therapy ameliorates functional deficits in a murine model of infantile neuronal ceroid lipofuscinosis. Molec Therap 13:538–547.
18. Kielar C, Maddox L, Bible E, Pontikis CC, Macauley SL, Griffey MA, Wong M, Sands MS, Cooper JD (2007) Successive neuron loss in the thalamus and cortex in a mouse model of infantile neuronal ceroid lipofuscinosis. Neurobiol Dis 25:150–162.
19. Rosenberg RS, Bergmann BM, Rechtschaffen A (1976) Variations in slow wave activity during sleep in the rat. Physiol Behav 17:931–938.
20. Trachsel L, Tobler I, Borbely A (1988) Electroencephalogram analysis of non-rapid eye movement sleep in rats. Am J Physiol 255: R27–R37.
21. Racine RJ (1972) Modification of seizure activity by electrical stimulation: II. Motor seizure. Electroenceph Clin Neurophysiol 32:281–294.

Chapter 25

A Genetic Model to Dissect the Role of Tsc-mTORC1 in Neuronal Cultures

Duyu Nie and Mustafa Sahin

Abstract

Tuberous sclerosis complex (TSC) is an autosomal dominant disease caused by mutations in either of two genes, *TSC1* or *TSC2*, whose protein products form a complex that is essential in the regulation of mammalian target of rapamycin (mTOR) activity. TSC is characterized by the presence of benign tumors called hamartomas, which within the brain are known as cortical tubers. Neurological manifestations in TSC patients include epilepsy, mental retardation, and autistic features. In response to hormones, growth factors, or nutrients, the phosphatidylinositol 3-kinase or extracellular signal-regulated kinase-Tsc-mTOR pathways activate the translation machinery and regulate cell growth and/or size. Loss of TSC1 or TSC2 function results in constitutive activation of mTOR leading to tumor formation. Nevertheless, regulation of mTOR activity in nondividing neuronal cells and roles of mTOR hyperactivation in the neurological aspects of TSC remain elusive. Here, we have established a genetic model of mTOR complex 1 (mTORC1) activation in culture by using lentiviral vector-mediated TSC2 knockdown, which offers a reliable tool for analyzing the TSC-mTORC1 signaling in neurons.

Key words: Cortical tuber, TSC, mTORC1, Autism, Epilepsy, PI3-K, ERK, Neuron

1. Introduction

Tuberous sclerosis complex (TSC) is caused by mutations in either of two genes, *TSC1* or *TSC2*, whose protein products form a complex that is critical in the phosphatidylinositol 3-kinase (PI3K)-Akt-mTOR pathway. Binding of a growth factor, such as insulin, to its cell surface receptor leads to activation of PI3K, which in turn activates the Akt kinase. Phosphorylation of TSC2 by Akt at Thr1462 and Ser939 releases the inhibitory effect of the TSC1/TSC2 complex toward the GTPase Rheb (Ras homolog enriched in brain) (1, 2).

Thomas Weichhart (ed.), *mTOR: Methods and Protocols*, Methods in Molecular Biology, vol. 821,
DOI 10.1007/978-1-61779-430-8_25, © Springer Science+Business Media, LLC 2012

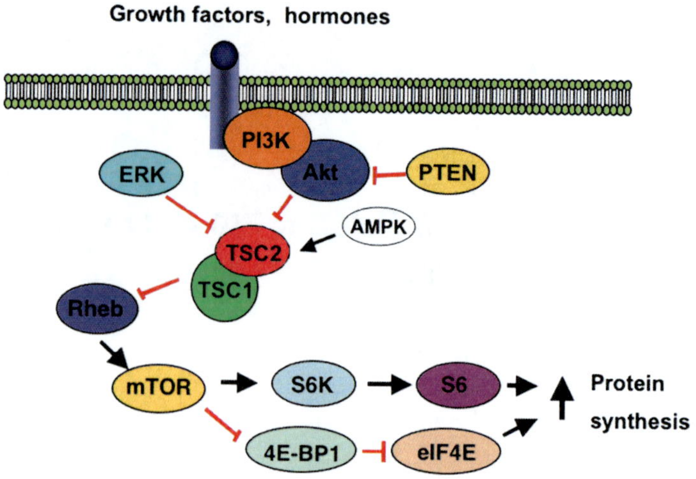

Fig. 1. A simplified model to depict the canonical TSC-mTORC1 pathway. The TSC1/2 complex plays a central role in controlling protein synthesis and cell size via mTORC1 to regulate the S6K and 4E-BP1 phosphorylation status. Also note that TSC2 is modulated by a myriad of cytoplasmic-signaling kinases, such as ERK, PI3-K/Akt, and AMPK.

Phosphorylation at Ser644 by ERK (extracellular signal-regulated kinase, ERK) also modulates the effect of TSC2 on mTOR activation (Fig. 1) (3, 4).

Rheb and its downstream effector mammalian target of rapamycin (mTOR) are master regulators of cell growth. There are two distinct mTOR complexes: mTOR complex 1 (mTORC1) and mTORC2. mTORC1 consists of mTOR, Raptor, and mLST8 and regulates the translational machinery. mTORC2 consists of mTOR, Rictor, and mLST8 and regulates the actin cytoskeleton. When Rheb is activated, the protein synthesis machinery is turned on, most likely via mTORC1, and cell growth programs are initiated. Therefore, the TSC1/TSC2 complex keeps cell size in check by inhibiting mTOR-mediated mRNA translation. Cells with insufficient TSC1 or TSC2 function grow beyond their normal size and form hamartomas.

mTORC1 is acutely sensitive to rapamycin, whereas mTORC2 is not, although mTORC2 can be inhibited by long-term rapamycin treatment. Rheb can activate mTORC1 by disrupting the binding of the inhibitory protein FKBP38, but reportedly inhibits mTORC2 (5). There are at least two mechanisms by which mTORC1 regulates the translational machinery: phosphorylation of p70 S6K, which results in elevated phosphorylation of ribosomal S6 protein at both Ser235/236 and Ser 240/244 leading to increased translation of mRNA transcripts containing a tract of pyrimidine (TOP) motif; and phosphorylation of the eIF-4E-binding proteins (4E-BP), which relieves the inhibitory effect of 4E-BP on cap-dependent translation initiation (Fig. 1). Commercial availability of antibodies that recognize site-specific phosphorylations on p70S6K and S6 proteins has allowed quick and sensitive readout of mTORC1 activation or suppression. Knocking down *Tsc2* gene in neurons with a lentiviral vector has provided a convenient and reliable model of mTORC1 activation and facilitated looking into roles of TSC-mTOR pathway in the central nervous system (3, 6, 7).

2. Materials

2.1. Equipment

1. 4°C microcentrifuge.
2. Dissecting microscope.
3. Upright fluorescent microscope.
4. Minigel system.
5. Power station.
6. Digital dry bath.
7. Vortex mixer.
8. P10, P200, and P1000 pipetman.

2.2. Lentiviral Packaging

1. The lentiviral constructs encoding for Tsc2 short hairpin (sh) RNA were custom made by Cellogenetics. The target sequence used is as follows: 5′-GGTGAAGAGAGCCGTATCACA-3′.
2. EndoFree Plasmid Maxi Kit (Qiagen, Valencia, CA).
3. OPTI-MEM reduced serum medium (GIBCO).
4. Lipofectamine 2000 (Invitrogen).
5. Dulbecco's modified Eagle's medium (DMEM).
6. Human embryonic kidney (HEK) 293 T culture medium: DMEM with 10% fetal bovine serum (FBS), 100 U/ml penicillin plus 100 μg/ml streptomycin (GIBCO), L-glutamine (Invitrogen, 200 mM).
7. Low-serum DMEM medium: DMEM with 1% FBS, 100 μg/ml penicillin plus 100 U/ml streptomycin.
8. Syringe PVDF membrane filters (Millipore, 0.22 and 0.45 μm).

2.3. Cortical and Hippocampal Neurons Isolation, Viral Infection

1. E18.5 pregnant CD1 female rats (Charles River).
2. Autoclave-sterilized dissection tools: Spatula, iris scissors, fine forceps two pairs.
3. 15 and 50-ml Falcon tubes.
4. Papain (Worthington, Lakewood, NJ).
5. Trypsin inhibitors (Sigma–Aldrich, St. Louis, MO).
6. Dissociation medium (DM): Make 500 ml 100 mM $MgCl_2$, 10 mM kynurenic acid (Sigma), 100 mM HEPES in 1× Hank's

Balanced Salt Solution (HBSS) without Ca^{2+} and Mg^{2+} (GIBCO). Heat to dissolve kynurenic acid and then cool it down to room temperature. Adjust pH value to 7.2 and filter through 0.22-μm PVDF filter unit. Make 50-ml aliquots and store at −20°C. This is the 10× stock. To make final 1× dissociation medium, thaw and add one 50-ml aliquot to 450 ml HBSS without Ca^{2+} and Mg^{2+}, and then add penicillin–streptomycin (GIBCO) at 100× dilutions.

7. Neurobasal-B27 medium: Add 10 ml B27 supplement (Invitrogen, 50×), 5 ml of penicillin–streptomycin (100×), 5 ml of L-glutamine (Invitrogen, 200 mM) to 480 ml of Neurobasal medium (Invitrogen). Filter through 0.22-μm pore-sized PVDF membrane filter.
8. Polybrene (Sigma–Aldrich) is dissolved in DMSO at 6 mg/ml and used as 10,000× stock.
9. Rapamycin (Sigma–Aldrich) is dissolved in DMSO at 20 μM (1,000× stock), stored at −20°C.

2.4. Cell Lysis and Western Blotting Analysis

1. RIPA lysis buffer: Tris–HCl, 50 mM, pH 7.4, NP-40, 1%, Na-deoxycholate, 0.25%, NaCl, 150 mM, 1 mM EDTA. Right before use, add PMSF to 1 mM, Na_3VO_4, 1 mM, NaF, 1 mM, and protease inhibitor cocktail (Sigma–Aldrich, 100×).
2. Plastic cell scraper.
3. 3× sample loading buffer: 1 M Tris–HCl, pH 6.8, 2.4 ml, 20% SDS, 3 ml, glycerol 3 ml, ß-mercaptoethanol, 1.6 ml, bromophenol blue, 6 mg, add H_2O to 10 ml. Store a 1-ml aliquot at −20°C.
4. 30% ProtoGel: 37.5:1 acrylamide to bis-acrylamide stabilized solution (National Diagnostics, Atlanta, GA).
5. SDS-PAGE running buffer: To make 10× stock, dissolve 30.2 g Trizma Base, 144 g glycine in 700 ml dH_2O, add 100 ml of 20% SDS, bring the volume up to 1 L by adding more dH_2O. Combine 100 ml of 10× stock and 900 ml of dH_2O for 1× running buffer.
6. SDS-PAGE transfer buffer: To make 10× stock, dissolve 30 g Trizma Base, 140 g Glycine in 700 ml dH_2O with stirring. Add more dH_2O to 1 L. Combine 100 ml of 10× stock, 200 ml of methanol, and 700 ml of dH_2O to make 1× transfer buffer (1 L).
7. TBST: To make 10× stock, stir to dissolve 24.2 g Trizma Base, 80 g NaCl in 700 ml dH_2O. Adjust pH to 7.4–7.6 using HCl. Add 10 ml Tween-20, add enough dH_2O to bring the volume to 1 L. Combine 100 ml of 10× stock and 900 ml of dH_2O to make 1× TBST.
8. Immobilon-P PVDF membrane (0.45 μm, Millipore, Billerica, MA).

9. Antibodies: Phospho-S6K1 (T389), total S6K1, phospho-S6 (S235/236), total S6, ERK1/2 (Cell Signaling, Danvers, MA), anti-tubulin antibody (Sigma–Aldrich). Peroxidase-conjugated goat anti-mouse (Santa Cruz Biotechnology), goat anti-rabbit secondary antibodies (Rockland, Gilbertsville, PA).

3. Methods

3.1. Packaging Lentiviral Vectors Encoding for TSC2 shRNA or Luciferase

It is extremely important to contact the biosafety office at your institution for permission and specific instructions prior to use of lentiviral vectors.

The instructions below are based on the regulatory requirements at our own institution. The biosafety level II (BL2) conditions should be used at all times when handling the virus. All decontamination steps should be performed using 30% bleach or 1% SDS and 70% ethanol (see also Note 1).

The Tsc2 shRNA lentiviral vector, control lentiviral vector encoding for shRNA against firefly luciferase (hereafter referred to as LV-T2sh1 and LV-GL3, respectively) and four packaging helper vectors (rev, tat, gag/pol, vsv-g) (8) are prepared with EndoFree Maxiprep kit as instructed by the manufacturer's manual. All plasmids are prepared 1 μg/μl concentration in water before use.

1. The day before transfection, plate HEK293T cells in 100-mm culture dishes at 10×10^6 per dish (see Note 2). However, other sized dishes (e.g., 150 mm) can be used with adjustment in number of cells and amounts of plasmids accordingly.
2. On the day of transfection, replace HEK293T plating medium with 10 ml of fresh DMEM growth medium, 2–3 h prior to transfection. Dispense 1.5 ml of serum-reduced OPTI-MEM medium to two sets of 15-ml polypropylene Falcon tubes. In one tube, combine the following plasmids: packaging helper plasmids Tat 1 μg, Rev 1 μg, Gag/pol 1 μg, and Vsv-g 2 μg with the shRNA vector (LV-T2sh1 or GL3) 10 μg. Mix gently. To the other tube, add 45–50 μl of Lipofectamine 2000 and mix gently.
3. Add the contents of the DNA/OPTI-MEM tube dropwise to the Lipofectamine/OPTI-MEM tube, mix by propelling air bubbles from the bottom with P1000 pipetman. Incubate the solution for 20 min at room temperature in the tissue culture hood.
4. Aspirate 3 ml of medium from each dish and add back the DNA/Lipofectamine solution in a total volume of 3 ml. Swirl the dishes horizontally and put them back to incubator (37°C, 5% CO_2). After a 5–6-h incubation, replace the DNA/Lipofectamine

solution with at least 6 ml of prewarmed low-serum DMEM medium (see Note 3). Put the dishes back in the incubator.

5. Forty-eight hours post transfection, the medium turns yellowish which is expected. Collect the entire volume of viral supernatant into 15-ml Falcon tube and replace another 5 ml of fresh low-serum DMEM medium. Store the supernatant at 4°C at this moment. Seventy-two hours post transfection, harvest a second time and pool with the previous volume for the same virus.
6. Spin the viral collections at 100 × *g* in a tabletop centrifuge for 5 min to remove large cell debris. The resultant supernatant is then filtered through 0.45-μm pore-sized filter units into a fresh 15-ml Falcon tube. Generally, we make 0.5–2-ml viral aliquots and store them at −80°C. Importantly, the viral tubes, pipettes, and anything in contact with virus must be decontaminated thoroughly with 30% bleach. The working surface in the hood should be wiped with bleach and then cleaned with 70% ethanol.

3.2. Isolation of Cortical and Hippocampal Neurons and Viral Infection

Cortical and hippocampal neurons are isolated from E18.5 rat embryos (CD1, Charles River). There is no difference between cortical and hippocampal neurons in terms of the viral infection efficiency. Therefore, neuronal-type selection depends solely on the purpose of experiments and cell number required.

1. The day before dissection, wash glass coverslips (prewashed with nitric acid and stored in 70% ethanol) three times with autoclaved milli-Q water. Coverslips are then coated with 10 μg/ml poly-D-lysine (PDL) plus 1 μg/ml mouse laminin overnight at 37°C incubator. Coat plastic dishes/plates with only PDL overnight.
2. The day of dissection, wash PDL-coated plastic dishes/plates two times with milli-Q water and one more with plain Neurobasal medium. Wash PDL/laminin-coated coverslips two times with milli-Q water and two more times with plain Neurobasal medium and then transfer coverslips to 24-well plates. Leave the washed coverslips and plates with Neurobasal-B27 medium in 37°C, 5% CO_2 incubator to equilibrate.
3. Weigh out 3.5 mg of L-cysteine into a 15-ml conical Falcon tube, add 10 ml DM, and adjust pH value by adding 1 N NaOH until solution turns the same color as the dissociation medium. Weigh out 0.25 g of trypsin inhibitor to a 50-ml Falcon tube with 25 ml of DM; adjust pH value with 1 N NaOH. Place both tubes into 37°C water bath. Also place Neurobasal-B27 medium into water bath to warm up.
4. Put dissection tools into 95% ethanol. In the meantime, set up dissection area in a dedicated tissue culture room next to the dissecting microscope. Wipe the area and microscopic stage

with 70% ethanol. Set up two ice buckets: one situated in common bench area with a 150-mm dish for harvesting uteri; the other in the dissection area to put 100- and 60-mm dishes for dissecting brains, cortices, and hippocampi.

5. When the preparation is ready, euthanize the rat with CO_2 inhalation. Quickly open the abdomen with a scissors and dissect out all embryos quickly into a 150-mm dish on ice. Transfer embryos to another 100-mm dish with fresh DM. Now, one can transport the embryos to the tissue culture room and finish the rest of dissection procedure with the dissecting microscope.
6. Remove heads from embryos and transfer heads to another 100-mm dish containing DM. In general, one does not need a microscope for this step. Transfer one head each time to a 60-mm dish on ice containing DM, peel away the skin and skull, and scoop out the brain. One by one, transfer the brains to another 60-mm dish with DM.
7. Set up two fresh 60-mm dishes with DM labeled with "cortices" and "hippocampi." Carefully remove blood vessels and meninges surrounding the brain, separate hemispheres from midbrain and hindbrain, dissect out hippocampus from the middle side of each hemisphere, and then harvest cerebral cortices. Transfer isolated cortices and hippocampi to their collection dishes, respectively. Repeat this step for all brains.
8. Once done with the dissection, shake papain bottle to mix the suspension. Aspirate the volume equivalent to 300 units of papain and add to prewarmed 10 ml of L-cysteine solution from step 3. Place it back to 37°C water bath for a couple of minutes. During this waiting time, transfer all cortices and hippocampi to new 15-ml conical tubes. Filter the papain/L-cysteine solution into a fresh 15-ml Falcon tube. Further remove excess DM from the cortices and hippocampi tubes, and then distribute 5 ml of papain/L-cysteine to each tube. Invert the tubes and put them in 37°C water bath for 4–5 min. Filter the trypsin inhibitor solution from step 3 into a fresh 50-ml conical tube.
9. When enzymatic dissociation is over, one should remove the papain/L-cysteine quickly and thoroughly from the cortices and hippocampi, add 5 ml trypsin inhibitor immediately, and invert the tubes (see Note 4).
10. Repeat wash as in step 9, and then place the cortices/hippocampi back to 37°C water bath for 2–3 min. Repeat the same to have a total of three washes.
11. Remove trypsin inhibitor and add 5 ml prewarmed Neurobasal-B27 medium to cortices and hippocampi Falcon tubes. Triturate the tissues with a 5-ml pipette until all tissue clumps have disappeared. Triturate 10–20 more times to ensure a complete

single-cell suspension, and then add another 5 ml warmed Neurobasal-B27 medium.

12. Take a small drop (20 μl) of cell suspension to spread over a hemocytometer under a thin coverslip. Count for total cell yield and adjust to an appropriate density by adding more Neurobasal-B27 medium. In general, we plate 5×10^4 per 13-mm glass coverslip, 1×10^6 per well of 6-well plates, or 10×10^6 per 100-mm dish. Add more medium to an appropriate volume for each culture vessel: 0.5 ml per coverslip, 2 ml per well of 6-well plate, and 10 ml per 100-mm dish.
13. Replace plating medium with prewarmed fresh Neurobasal-B27 medium 2 h after plating. One should not shake the plates/dishes during medium change at this moment.
14. We generally infect isolated cortical or hippocampal neurons the day after plating. Calculate the volume of viral stock needed, which depends on a predetermined dilution factor (see Note 5) and number of plates/dishes. Thaw the virus quickly by putting it to 37°C incubator and dilute into Neurobasal-B27 medium. We dilute a small aliquot of 6 mg/ml stock polybrene with Neurobasal-B27 medium to a final concentration of 0.6 mg/ml, which is then added to cultures at 1,000× dilutions. The role of polybrene is to aid in viral attachment to target cell membrane.
15. Add diluted virus/polybrene to culture dishes/plates, the minimal volume is 250–300 μl per coverslip, 1 ml per well of 6-well plate, 5 ml per 100-mm dish. Return dishes/plates to incubator for 6 h. All viral-contaminated materials are then thrown away to a 30% bleach container. The working area in the hood should also be wiped with 30% bleach and then 70% ethanol.
16. Discard infection medium into a bleach container. Wash cells twice with Neurobasal-B27 medium and return them to incubator. We generally replace half the old medium with fresh Neurobasal-B27 medium every 3 days for the extended period of culture. In general, infection with LV-T2sh1 virus results in >95% TSC2 knockdown within 7–10 days after infection as quantified on the protein levels in western blot analysis.
17. To inhibit mTOR activity, infected or noninfected neurons are treated with 20 ng/ml rapamycin overnight the day before harvesting the cell lysates.

3.3. SDS-PAGE Gel Electrophoresis and Western Blotting Analyses of TSC2 Expression and mTOR Activity

The lentiviral shRNA vectors LV-T2sh1 and LV-GL3 used with this method contain separate GFP expression cassette, which offers a rough but quick estimate of the viral expression efficiency by looking at GFP fluorescence 3 days post infection (DPI). At this time, one may see more GFP-positive cells and/or brighter GFP signals in LV-GL3 than LV-T2sh1 infections (see Note 6). As expected, knocking down TSC2 expression results in enlarged

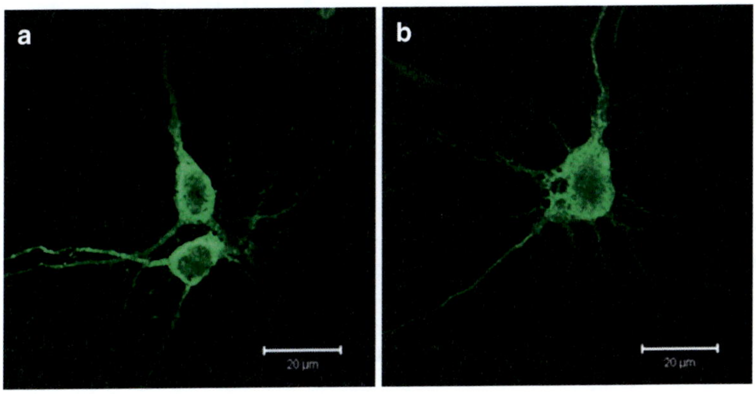

Fig. 2. Representative images of hippocampal neurons infected by control lentivirus LV-GL3 (**a**) or Tsc2 knockdown lentivirus LV-T2sh1 (**b**) at 10 days post infection. Green fluorescence indicates GFP expression. Note that the Tsc2 knockdown neuron display increased soma size. Scale bars: 20 μm.

soma compared to LV-GL3-infected neurons (Fig. 2). We usually harvest neuronal lysates in RIPA buffer at 10 DPI.

1. RIPA buffer is used for preparing neuronal lysates. Immediately before use, orthovanadate, NaF, PMSF, and protease inhibitor cocktail are added to RIPA buffer. The completed RIPA buffer is left on ice.
2. The neurons to be lysed must be washed once with prechilled PBS (pH 7.6) followed by immediately adding RIPA lysis buffer. Typically, we use 100 μl of RIPA buffer per well of 6-well plate, 500 μl per 100-mm dish. Upon addition of lysis buffer, the plates/dished are transferred to ice bucket and swirled to spread lysis buffer over the entire surface. The plates/dishes can then be left on ice for a few minutes.
3. Prechilled plastic scrapers are used to scrape cells mechanically from bottom of plates/dishes. Lysates are then transferred to 1.5-ml Eppendorf tubes that have been prelabeled and prechilled on ice.
4. Now, the lysates/Eppendorf tubes are put on end-over-end shaker in a cold room for 10 min, followed by spinning down at 16,000 × *g* for 10 min to remove large cell debris.
5. Supernatants are then transferred to fresh prechilled 1.5-ml Eppendorf tubes which are then mixed with appropriate amount of 3× sample loading buffer (i.e., 50 μl sample buffer add to 100 μl lysate). Mix lysates well by vortex, put in 95°C heat block for 5 min, and store at −20°C.
6. We pour own SDS-PAGE gels with minigel casters (Hoefer, Holliston, MA) by following a recipe table for different gel

percentages. To make five pieces of 1.0-mm-thick 10% gels, for example, we mix 20 ml H_2O, 16.6 ml 30% ProtoGel, 12.5 ml 1.5 M Tris (pH 8.8), 0.5 ml 10% SDS, 220 μl 10% ammonium persulfate (APS), and 40 μl TEMED in a glass flask for resolving gel. Upon addition of APS and TEMED, one should pour the solution immediately into the casting apparatus. Overlay the acrylamide gel with water-saturated isobutanol.

7. After approximately 30 min, rinse off the isobutanol from the resolving gel and prepare the stack gel solution: 17 ml H_2O, 4.2 ml ProtoGel, 3.3 ml 1.0 M Tris–HCl (pH 6.8), 0.25 ml 10% SDS, 100 μl 10% APS, and 20 μl TEMED. Add the stack gel and put combs immediately. The stack gel should polymerize within about 30 min.
8. Assemble gel running system and pour 1× SDS running buffer into the central loading chamber as well as base container. Mark the lanes and remove the combs. The system is ready for sample loading.
9. Remove the protein samples from −20°C freezer; once thawed, vortex and spin them down briefly. We load 3.5–4 μl prestained protein ladder and desired amounts of protein samples (generally, to start with 10 μl). The gel running chamber is then connected to a power supply station and start running with 65 V for a few minutes and then 90 V when ladder and samples pass through the stack gel.
10. Keep monitoring the gel migration and stop the electrophoresis when the blue-colored front line reaches the bottom end of gel. During the gel running time, gel-sized PVDF membranes are immersed into absolute methanol for 1 min and then transferred to a container with 1× transfer buffer. Also prepare two sponge cushions and two pieces of Whatman filter paper per gel, and wet them in the same container with 1× gel transfer buffer.
11. Unpack the gel running system and remove the gel carefully. One can leave the gel into a plastic container with 1× gel transfer buffer. Set up the "sandwich" layers on a transfer clamp in this order (from the black side to red side): sponge pad, Whatman filter paper, gel, PVDF membrane, Whatman filter paper, sponge pad. One can use a glass pipette or pencil to roll over the "sandwich" layers in order to push out air bubbles between the gel and PVDF membrane, which would otherwise prevent protein transfer to the membrane. Close the transfer clamp tightly, position it to the transfer apparatus, and make sure that the black side of each clamp faces the black wall of the apparatus. One sometimes also marks the clamps to identify different samples in transfer units.

12. Put ice block into the gel transfer box and fill up the box with 1× transfer buffer. If high-molecular-weight proteins (>150 kDa, for example) are of interest, one can add 0.1% SDS into the transfer buffer. Connect the gel transfer box to a power station and run at 90 V for 90–120 min at room temperature. Alternatively, one uses 30–35 V for overnight transfer in cold room.
13. Weigh out skim milk powder or BSA and dissolve it into 1× TBST to make 5% solution. When transfer is finished, unpack the transfer apparatus and remove the PVDF membrane into an appropriate sized container with 1× TBST. One should be able to see the prestained ladders on the membrane. Following a brief rinse, the PVDF membrane is blocked in 5% milk for 1 h with gentle shaking on a swinging platform.
14. Discard the block buffer, rinse blots quickly with TBST, and replace primary antibody. As recommended, we dilute rabbit anti-TSC2 (Santa Cruz) 1:500 in 5% skim milk, and anti-phospho-S6K1 and phospho-S6 in 5% BSA (see Note 7). Transfer containers onto a shaking platform in cold room and incubate with the primary antibody overnight.
15. Save primary antibodies into 15-ml Falcon tubes to be reused, add sodium azide (final concentration: 0.025%) to prevent bacterial growth. Wash the membrane blots three to four times with 1× TBST and replace HRP-conjugated secondary antibody against the species from which the primary antibody is raised. Secondary antibody is diluted in the same medium as the primary antibody and specific dilution factor for each secondary antibody should be determined empirically.
16. Incubate the blots in secondary antibodies with shaking for 1–2 h at room temperature. Discard secondary antibodies, and rinse the blots three to four times with 1× TBST, 5-min-each rinse. During the last rinse, prepare the chemiluminescent solution by mixing solutions A and B at 1:1. Add the appropriate volume of chemiluminescent solution to each blot and incubate for 1 min.
17. Spread a plastic wrap on bench, cut a piece of transparency sheet on top of the plastic wrap, and then transfer all blots onto the transparency sheet. One then folds front of the plastic wrap from one end to cover the blots. Quickly place this wrapped sheet into a developing cassette and tape it firmly.
18. Cut a corner of films to mark the orientation. Develop a couple of films with different exposures. An example of western blots to show TSC2 knockdown by LV-T2sh1 infection and regulation of phospho-S6K and phospho-S6 levels, with or without rapamycin treatments, has been provided in Fig. 3.

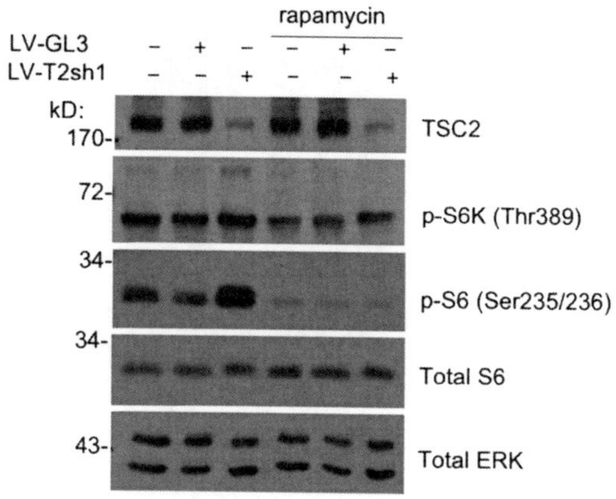

Fig. 3. Western blot analysis of cortical neurons that are left noninfected (lanes 1, 4), infected by control lentivirus LV-GL3 (lanes 2, 5), or infected by Tsc2 knockdown lentivirus LV-T2sh1 (lanes 3, 6) for 10 days in culture. Neurons of lanes 4–6 are also treated with 20 nM rapamycin overnight before harvesting the lysates. Phospho-S6K (Thr389) and phospho-S6 (Ser235/236) are dramatically upregulated in Tsc2 knockdown neurons yet completely suppressed by rapamycin treatment. Total S6 and total ERK staining are used as loading controls.

4. Notes

1. Although the viral vectors and packaging procedure have been modified in such a way that accidental generation of a replication competent virus is completely prevented, some necessary precautions have to be taken in preparing the pseudotyped virus: (1) Always treat the virus solutions and the cells transfected or infected with the virus as a potential biohazard. (2) Do not open the tubes outside of the tissue culture hood. (3) Do not leave virus-containing solutions unattended. (4) Wipe extensively the surfaces of the TC hood that you have used. (5) Clean thoroughly the buckets for the ultracentrifuge. (6) Discard the TC plates and other virus-containing plasticware into dedicated biohazard bags. (7) Absolutely always wear lab coats and gloves, and change gloves before you touch common laboratory areas.
2. If cell detachment is a concern, one can precoat culture dishes with 0.2% gelatin solution (i.e., add 1 g of gelatin to 500 ml of water and autoclaved). Gelatin-coated dishes can be stored at 4°C for later use.
3. Collecting virus with low-serum DMEM would minimize the expansion of glial population during infection of neuronal cultures.
4. One should use pipette to remove the supernatant. Avoid suction use to prevent the possibility of suctioning away the entire tissue pellet.

5. We have routinely diluted viral stock 1:4 and achieved knockdown more efficiently than 90% in TSC2 protein level as determined by western blot assay. However, viral titers may vary from batch to batch, of which the dilution factor must be determined in a pilot study by trying a serial dilutions in the culture. However, if one has followed the packaging steps with everything the same, the same dilution factor would reproduce comparable efficiency in a certain culture system.
6. One even sees weaker GFP signal at later time points during the extended culture after infection. This fade-out may be due to GFP protein degradation but would not affect the knockdown efficacy.
7. Since milk may contain phosphoantigens that could neutralize a phosphorylation site-specific antibody, 5% BSA is recommended to dilute such antibodies.

Acknowledgments

We are grateful to Alessia Di Nardo and Emily Greene-Colozzi for critical reading of the manuscript. Work in the Sahin laboratory is supported in part by grants from the NIH (R01 NS58956), the John Merck Scholars Fund, Tuberous Sclerosis Alliance, and Children's Hospital Boston Translational Research Program to M.S., and the Children's Hospital Boston Mental Retardation and Developmental Disabilities Research Center (P30 HD18655). D.N. was supported by a Mentor Based Postdoctoral Fellowship from Autism Speaks.

References

1. Manning, B.D., Tee A.R., Logsdon M.N., Blenis J., Cantley L.C. (2002) Identification of the tuberous sclerosis complex-2 tumor suppressor gene product tuberin as a target of the phosphoinositide 3-kinase/akt pathway. *Mol Cell* **10**,151–162.
2. Inoki, K., Li Y., Zhu T., Wu J., Guan K.L. (2002) TSC2 is phosphorylated and inhibited by Akt and suppresses mTOR signalling. *Nat Cell Biol* **4**,648–657.
3. Nie, D., Di Nardo A., Han J.M., Baharanyi H., Kramvis I., Huynh T., Dabora S., Codeluppi S., Pandolfi P.P., Pasquale E.B., Sahin M. (2010) Tsc2-Rheb signaling regulates EphA-mediated axon guidance. *Nat Neurosci* **13**,163–172.
4. Ma, L., Chen Z., Erdjument-Bromage H., Tempst P., Pandolfi P.P. (2005) Phosphorylation and functional inactivation of TSC2 by Erk implications for tuberous sclerosis and cancer pathogenesis. *Cell* **121**,179–193.
5. Yang, Q., Inoki K., Kim E., Guan K.L. (2006) TSC1/TSC2 and Rheb have different effects on TORC1 and TORC2 activity. *Proc Natl Acad Sci USA* **103**,6811–6816.
6. Di Nardo, A., Kramvis I., Cho N., Sadowski A., Meikle L., Kwiatkowski D.J., Sahin M. (2009) Tuberous sclerosis complex activity is required to control neuronal stress responses in an mTOR-dependent manner. *J Neurosci* **29**,5926–5937.
7. Choi, Y.J., Di Nardo A., Kramvis I., Meikle L., Kwiatkowski D.J., Sahin M., He X. (2008) Tuberous sclerosis complex proteins control axon formation. *Genes Dev* **22**,2485–2495.
8. Mostoslavsky, G., Kotton D.N., Fabian A.J., Gray J.T., Lee J.S., Mulligan R.C. (2005) Efficiency of transduction of highly purified murine hematopoietic stem cells by lentiviral and oncoretroviral vectors under conditions of minimal in vitro manipulation. *Mol Ther* **11**,932–940.

Chapter 26

Tissue-Specific Ablation of Tsc1 in Pancreatic Beta-Cells

Hiroyuki Mori and Kun-Liang Guan

Abstract

Tuberous sclerosis complex 1 (TSC1) is a tumor suppressor that associates with TSC2 to inactivate Rheb, thereby inhibiting signaling by the mammalian target of rapamycin (mTOR) complex 1 (mTORC1). mTORC1 stimulates cell growth by promoting anabolic cellular processes, such as translation, in response to growth factors and nutrient signals. In order to test roles for TSC1 and mTORC1 in β-cell function, we utilized *Rip2/Cre* to generate mice lacking *Tsc1* in pancreatic β cells (*Rip-Tsc1*cKO mice). While obesity developed due to hypothalamic *Tsc1* excision in older *Rip-Tsc1*cKO animals, young animals displayed a prominent gain-of-function β-cell phenotype prior to the onset of obesity. The young *Rip-Tsc1*cKO animals displayed improved glycemic control due to mTOR-mediated enhancement of β-cell size and insulin production, but not β-cell number consistent with an important anabolic role for mTOR in β-cell function. Thus, mTOR promulgates a dominant signal to promote β-cell/islet size and insulin production, and this pathway is crucial for β-cell function and glycemic control.

Here, we describe the methods of analyzing tissue-specific ablation of Tsc1 in pancreatic β cells.

Key words: Tsc1, Pancreatic β cell, Rapamycin, Conditional knockout mice, Rat insulin promoter 2

1. Introduction

Tuberous sclerosis complex (TSC) is an autosomal dominant genetic disorder characterized by benign tumor growth in a wide range of tissues. TSC is exclusively associated with mutations in either the *TSC1* or *TSC2* tumor-suppressor gene. TSC1 and TSC2 proteins form a tight physical complex with GTPase-activating protein (GAP) activity toward Rheb, a small GTPase of the Ras family. In the TSC1/TSC2 complex, TSC2 contains the catalytic Rheb-GAP activity while TSC1 functions as a regulatory subunit that controls the stability and cellular localization of the complex. Active, GTP-bound Rheb directly binds to the mammalian target of rapamycin (mTOR) complex 1 (mTORC1) to promote

Thomas Weichhart (ed.), *mTOR: Methods and Protocols*, Methods in Molecular Biology, vol. 821,
DOI 10.1007/978-1-61779-430-8_26, © Springer Science+Business Media, LLC 2012

mTORC1 kinase activity. Thus, loss of either TSC1 or TSC2 leads to high Rheb-GTP and constitutive activation of mTORC1.

The mTOR is an evolutionarily conserved serine–threonine kinase and forms distinct functional complexes, the rapamycin-sensitive mTORC1 and the rapamycin-insensitive mTORC2 (1). mTORC1 consists of mTOR, mLST8, raptor, and PRAS40. In addition to the shared mTOR and mLST8, mTORC2 also contains two unique subunits, rictor and sin1. mTORC1 regulates a wide array of cellular processes, including cell growth, proliferation, and autophagy, in response to nutrients (1). Furthermore, mTORC1 phosphorylates S6 kinase (S6K) and 4EBP1 while mTORC2 is required for phosphorylation of AKT, serum, glucocorticoid-inducible kinase (SGK1), and conventional protein kinase C (PKC). Therefore, the two TORC complexes differ in their subunit compositions, physiological functions, and regulations. For the purposes of this manuscript, mTOR refers to mTORC1.

Several studies using genetically engineered mouse models have established an important role of insulin-IR-IRSs-PI3K-Akt cascade not only in glucose metabolism in peripheral tissue but also in pancreatic β-cell development and function. For instance, inactivation of mediators of insulin signaling (such as IR, IRSs, PDK1, Akt, and GLUT4) not only produces peripheral insulin resistance (2–6), but also attenuates pancreatic β-cell function (7–10). The use of rapamycin to inhibit mTOR has suggested crucial roles for mTOR in β-cell function. Not only does rapamycin inhibit β-cell proliferation *in vitro* (11), it also blocks the effects of glucose and Akt activation on β-cell mass and proliferation (12). Recent studies have also begun to explore roles for increased mTOR signaling in β-cell function (13, 14). We utilized genetic ablation of the mTOR inhibitor, TSC1, to examine mTOR function in β cells. Since the conventional knockout of *Tsc1* leads to embryonic lethality (15–17), we generated mice conditionally lacking *Tsc1* in pancreatic β cells using *Rip2/Cre* (*Rip-Tsc1cKO* mice) (18). While older *Rip-Tsc1*cKO animals displayed hyperphagia and obesity due to the activation of mTOR in *Rip/Cre*-expressing hypothalamic neurons (19), we examined β-cell function in younger mice prior to the onset of obesity and insulin resistance (18).

Here, we describe the materials, method, and technique to analyze tissue-specific ablation of Tsc1 in pancreatic β cells that we have reported previously (18).

2. Materials

2.1. Animals and Genotype

1. *Tsc1*$^{lox/lox}$ mice, with exons 17 and 18 of *Tsc1* flanked by loxP sites by homologous recombination, have been described (20, 21). This line is now commercially available (The Jackson Laboratory; strain name: STOCK *Tsc1*$^{tm1Djk/J}$).

2. The *cre* transgenic line driven by the rat insulin promoter 2 (Rip) (7, 9, 22–24). (The Jackson Laboratory; strain name: B6.Cg-Tg (Ins2-cre) 25Mgn/J).
3. Primers to genotype the *Tsc1*alleles, F4536: 5′-AGGAGGC CTCTTCTGCTACC-3′ and R4830: 5′-CAGCTCCGACCA TGAAGTG-3′.
4. Primers to genotype the *Cre* transgene, forward 5′-GCGGT CTGGCAGTAAAAACTATC-3′, reverse 5′-GTGAAACAGC ATTGCTGTCACTT-3′.
5. Tail was used to prepare DNA for PCR analysis.

2.2. Islets Isolation

1. Collagenase-P (Roche Diagnostics, Indianapolis, IN).
2. Hanks' Balanced Salt Solutions (HBSS, GIBCO).
3. Dissecting microscope for both cannulation to common bile duct and hand-pick islets.
4. 60 mm Petri-dishes (Corning).
5. Dissecting equipment (scissors, forceps, Bulldog Clamp to clamp the ampulla).
6. Water bath.
7. Mesh pore size: 1.0 mm to remove undigested tissue.
8. 70 μm cell strainer (BD Falcon, cat. no. 352350).
9. 30 G needle, extension tube, and syringe (1 and 5 ml).
10. 20 and 200 μl pipette to hand pick islets.

2.3. Western Blot Analysis

1. Cell lysis buffer: 10 mM Tris–HCl at pH 7.5, 100 mM sodium chloride, 1% NP-40, 1% Triton X-100, 50 mM sodium fluoride, 2 mM EDTA, 1 mM phenyl methylsulphonyl fluoride, 10 μg/ml leupeptin, and 10 μg/ml aprotinine.
2. BCA Protein Assay Reagent (Thermo Scientific).
3. 4× SDS page sample buffer: 40% glycerol, 8% SDS, 200 mM Tris–HCl, pH 6.8, 0.004% bromophenol blue, 200 mM dithiothreitol.
4. Running buffer: Tris–glycine: 25 mM Tris base, 190 mM glycine, 0.1% SDS.
5. Transfer buffer: 25 mM Tris base, 192 mM glycine, 10% methanol.
6. PVDF membrane: Immobilon-P (Millipore IPVH00010).
7. Blocking buffer and secondary antibody dilution buffer: 5% fat-free milk in TBST (0.05% Tween20).
8. Primary antibody dilution buffer: 5% BSA in TBST (0.05% Tween20).
9. Antibody: (a) Tsc1(#4906, Cell Signaling), (b) Tsc2(#3990, Cell Signaling), (c) IRS1(#2382, Cell Signaling), (d) IRS2(#4502,

Cell Signaling), (e) Phospho-S6 Ribosomal Protein (Ser240/244) (#2215, Cell Signaling), (f) S6 Ribosomal Protein(#2317, Cell Signaling), and (g) α-Tubulin (Sigma–Aldrich).

2.4. Animal Perfusion

1. Nembutal (50 mg/ml solution).
2. Heparin in PBS.
3. 4% paraformaldehyde in PBS.
4. Dissecting equipment (scissors, forceps).
5. 27-G needle, extension tube, and syringe (10 and 20 ml).
6. Perfusion pump (or infusion bag about 150 cm above the operating area).

2.5. Immunohisto-chemistry

1. Paraffin processor.
2. Xylene for deparaffinization.
3. TE: 10 mM Tris/1 mM EDTA, pH 9.0, for antigen unmasking.
4. Microwave oven for antigen unmasking.
5. 3% hydrogen peroxide (dilute 30% hydrogen peroxide by methanol).
6. Antibody: (a) Insulin and glucagon (DakoCytomation Co.), (b) Phospho-S6 (Ser235/236,IHC preferred, Cell Signaling Technology), (c) Phospho-S6 (Ser235/236 Alexa Fluor 488 conjugate; for double staining of pancreas, #4803, Cell Signaling Technology), and (d) β-catenin (Cell Signaling Technology).
7. NI Vision Assistant (version 7.1.0; National Instruments) or Image J (NIH) for analysis of pancreatic morphometry.

2.6. In Vivo Physiological Studies

1. Glucose solution for glucose tolerance tests (GTTs).
2. Human insulin (Novolin R, Novo Nordisk).
3. Blood glucose reader (Accu-Check, Roche).
4. Balance to measure body weight.
5. Insulin ELISA (Crystal Chem, Downers Grove, IL).
6. 30-G needle and 1 ml syringe.

2.7. Rapamycin Treatment Studies

1. Rapamycin (LC laboratories).
2. 100% ethanol.
3. Tween 80 and PEG 400.
4. 30-G needle,1 ml syringe.
5. Balance to measure body weight.

2.8. Insulin Content Measurement

1. Liquid nitrogen.
2. 100% ethanol and HCl.
3. Insulin ELISA Kit same as above.

3. Methods

3.1. Regeneration of β Cell-Specific Tsc1 Knockout Mice

1. Mice containing homozygous floxed *Tsc1* alleles (*Tsc1*$^{lox/lox}$) (20, 21) were mated with the *cre* transgenic line driven by the rat insulin promoter 2 (Rip) (7, 9, 22–24). All subsequent experiments were performed using *Rip-cre/Tsc1*$^{lox/lox}$ (*Rip-Tsc1*cKO) mice and littermate Tsc1$^{lox/lox}$ (control) mice.
2. Primers to genotype the *Tsc1* alleles, F4536: 5′-AGGAGGCC TCTTCTGCTACC-3′ and R4830: 5′-CAGCTCCGACCA TGAAGTG-3′ were used to detect the *Wt* allele (295 bp) and the *lox* allele (486 bp). PCR cycling was performed by the following conditions: step 1: 94°C 3 min; step 2: 94°C 30 s; step 3: 55°C 30 s; step 4: 68°C 40 s; repeat steps 2–4 for 35 cycles; step 5: 72°C 5 min; step 6: 4°C hold.
3. Primers to genotype the *Cre* transgene, forward 5′-GCGGTC TGGCAGTAAAAACTATC-3′, reverse 5′-GTGAAACAGCA TTGCTGTCACTT-3′. PCR condition is same as *Tsc1* PCR.
4. Mice were housed on a 12-h light/12-h dark cycle in the Unit for Laboratory Animal Medicine (ULAM) at the University of Michigan, with free access to water and standard mouse chow.
5. Animal experiments were conducted following protocols approved by the University Committee on the Use and Care of Animals (UCUCA).

3.2. Islet Isolation

We modified collagenase-based method, which was described by Lacy's group (25) and referred to as the Edmonton protocol (26).

1. Mice were sacrificed and skin was sterilized with 70% ethanol.
2. After making an incision around the upper abdomen, remove the bowel to the left side. Then, clamp the ampulla with a Bulldog Clamp.
3. Cannulate 30-G needle to the common bile duct, and 1.5 mg/ml of collagenase-P in HBSS was injected into the duct to distend the pancreas (2 ml injection finally) (Fig. 1a).
4. The pancreas was excised and incubated at 37°C for 20 min. Then, mechanically disrupt the tissue by pipetting in 10 ml of HBSS.
5. Pour pancreas solution from step 4 through the mesh (pore size: 1.0 mm) to remove undigested tissue. Pour 10 ml HBSS to wash the mesh.
6. Pour pancreas solution from step 5 through the 70 μm cell strainer and wash cell strainer with HBSS. Turn the strainer upside down over a new petri dish and rinse the captured islets into a dish with HBSS.
7. Islets were hand picked using a pipette under a microscope (Fig. 1b).

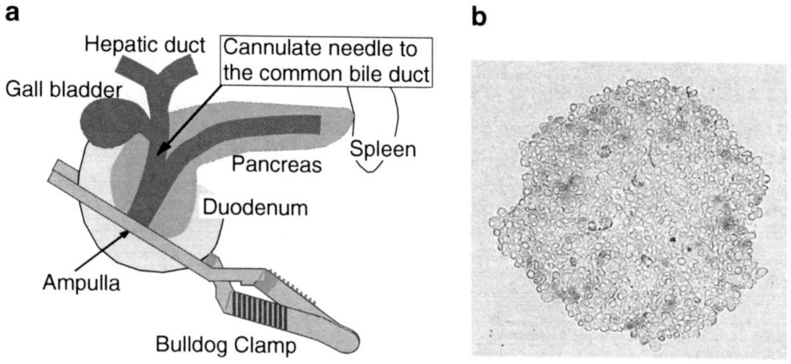

Fig. 1. Islets isolation. (**a**) Schematic presentation of pancreas perfusion. After clamping the ampulla with a Bulldog Clamp, cannulate 30-G needle to the common bile duct, and then 1.5 mg/ml of collagenase-P in Hanks' Balanced Salt Solutions was injected into the duct to distend the pancreas (2 ml injection finally). (**b**) Purified islets form *Tsc1*cKO mouse after hand picking.

3.3. Western Blot Analysis

1. Isolated islets were lysed in lysis buffer (40 islets/15 μl).
2. After centrifugation for 15 min at 20,000 × *g* at 4°C in a microcentrifuge, transfer the supernatant to a new tube.
3. Next, determine protein concentration by a BCA assay. After adjusting the protein concentration and adding 4× sample buffer, samples are boiled at 95–100°C for 5 min.
4. Lysates were subjected to SDS-PAGE (10 μg total protein per well) and proteins were transferred to PVDF membranes.
5. Incubate with blocking buffer, and then samples are immunoblotted with desired antibodies.

3.4. To Confirm Tsc1 Knockout in the Beta Cell

1. To confirm Tsc1deletion, we determined Tsc1 protein levels of islets isolated from *Rip-Tsc1*cKO mice by Western blotting (Fig. 2). Tsc1 protein was dramatically decreased. Tsc2 content was also decreased in *Rip-Tsc1*cKO mice, consistent with the stabilization of TSC2 by TSC1 (Fig. 2a).
2. Immunoblotting for phosphorylated S6 (pS6), which serves as a convenient downstream readout for mTOR, was performed to indirectly measure mTOR activity. Loss of Tsc1 was associated with increased S6 phosphorylation (pS6) in *Tsc1*cKO islets (Fig. 2a).
3. Immunostaining for pS6 showed that, within the pancreas, mTOR was activated only in the β cells of *Rip-Tsc1*cKO mice, but not the neighboring cells (Fig. 2b, c).
4. In addition to confirm *Tsc1* deletion in islets, we tested pS6 immunostaining for the hypothalamus since hypothalamic disruption is well-known in *Rip2/Cre* lines. We found that pS6 immunoreactivity was increased in the hypothalamus, suggesting hypothalamic disruption of *Tsc1* (19) (Fig. 2d).

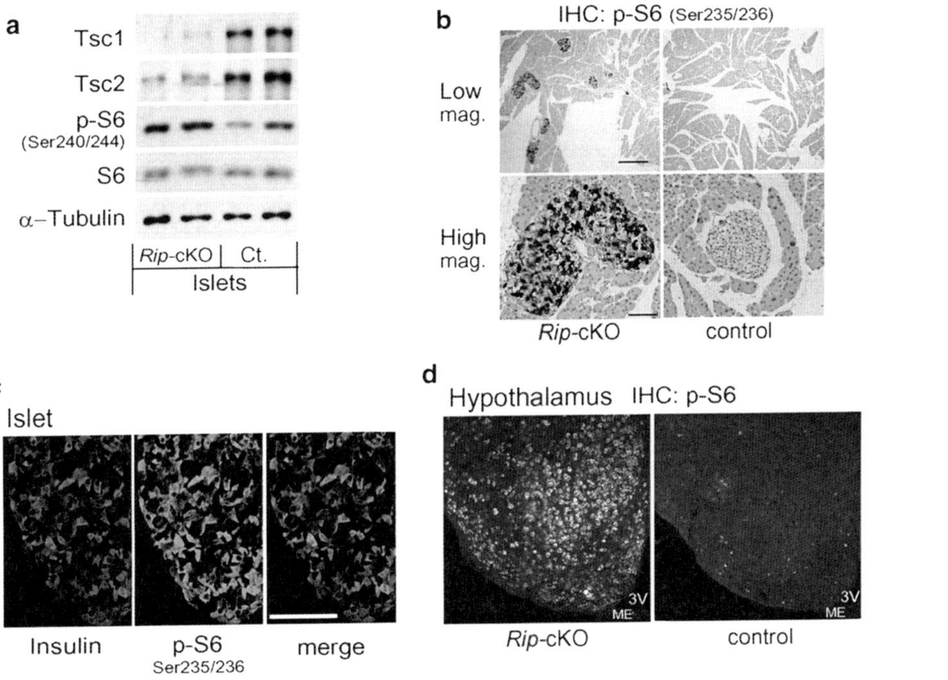

Fig. 2. *Tsc1* knockout by *Rip-Cre* activates mTORC1 in islets. (**a**) Elevated S6 phosphorylation in *Tsc1* knockout islets. Immunoblot analysis of Tsc1, Tsc2, phospho-S6 (pS6), S6, and α-Tubulin is indicated. Islets were isolated and extracts were prepared from independent animals. (**b**) S6 phosphorylation in *Tsc1*cKO β cells. Immunostaining of the pancreatic sections from 4-week-old *Rip-Tsc1*cKO and control mice were performed with pS6 (Ser235/236) antibody. The scale bar represents 500 μm (Low mag.: low magnification) and 100 μm (High mag.: high magnification). (**c**) S6 phosphorylation was observed in *Tsc1*cKO β cells. Double labeling of the pancreatic sections from 4-week-old *Rip-Tsc1*cKO and control mice was performed with pS6 (Ser235/236) antibody and insulin antibody. The scale bar represents 100 μm. (**d**) *Rip-Tsc1*cKO increases S6 phosphorylation in hypothalamus. Representative p-S6 (Ser235/236) staining of hypothalamus from ad libitum fed *Rip-Tsc1*cKO and control mice at 4-week ages. *ME*: median eminence, 3 V: third ventricle. For panels (**a–d**), *Ct.*: control; *Rip*-cKO: *Rip-Tsc1*cKO which is Tsc1 tissue-specific knockout using *Rip/Cre* (modified from ref. 18. Am Physiol Soc, with permission).

3.5. Animal Perfusion and Immunohistochemistry

While energy depletion or hypoxia inhibits mTOR, we chose 4% paraformaldehyde perfusion method which is possible to fix tissues rapidly.

1. Inject the mouse with Nembutal.
2. Once the animal is fully anesthetized, open the abdomen and cut the diaphragm and ribs. Next, insert needle directly into left ventricle and make incision with right atrium.
3. Following the first run about 5–10 ml of 1 U/ml heparin PBS, ice-cold 4% paraformaldehyde perfusion was done (20–30 ml). Since air embolism blocks perfusion flow to peripheral, take care not to introduce air bubbles.
4. Pancreas tissue was harvested and postfixed in 4% paraformaldehyde overnight.

5. Make paraffin block and sections paraffin processor, and then perform deparaffinization and rehydration.
6. Antigen unmasking: Sections were boiled in Tris–EDTA buffer for 10 min using microwave oven, and then cooled on the bench for 20 min.
7. For immunohistochemistry using DAB, incubate sections in 3% hydrogen peroxide for 10 min in order to block endogenous peroxidase.
8. Pancreatic sections were stained with the following antibodies: insulin and glucagon (DakoCytomation Co.), Phospho-S6 (Ser235/236, IHC preferred, Cell Signaling Technology), Phospho-S6 (Ser235/236 Alexa Fluor 488 conjugate; for double staining of pancreas, Cell Signaling Technology), and β-catenin (Cell Signaling Technology) (Fig. 3).

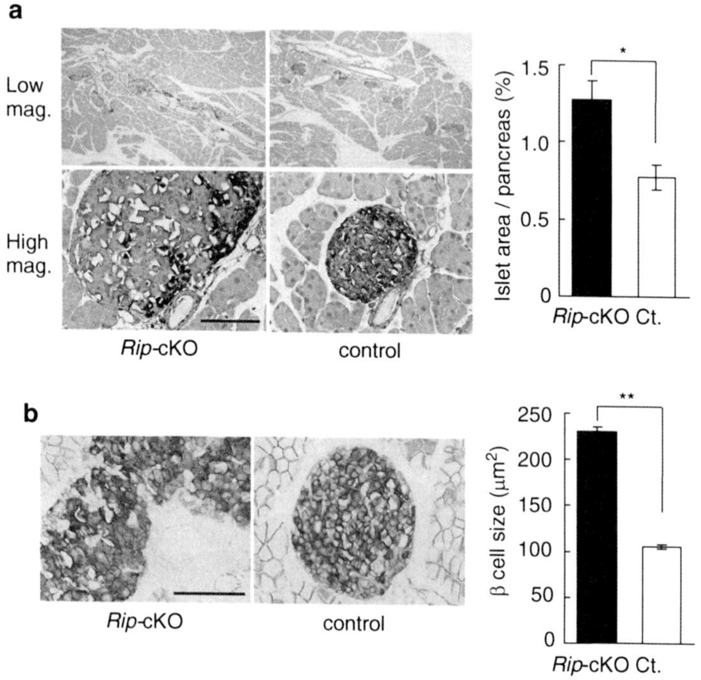

Fig. 3. Tsc1 deletion improves β-cell function. (**a**) Islet size increases in *Rip-Tsc1*cKO. Immunostaining was performed with antibodies against insulin (*original color was brown, now strong gray*) and glucagon (*original color was red, now black*). Scale bar represents 500 μm (Low mag.: low magnification) and 100 μm (High mag.: high magnification). Total insulin- and glucagon-positive areas were expressed as the percent of the total pancreas area. We scored six animals per genotype and nine sections per animal spaced 100 μm apart. (**b**) β-cell size increases in *Rip-Tsc1*cKO. Pancreatic sections were stained with antibodies against insulin (*original color was brown, now strong gray*) and β-catenin (*original color was red, now black*) to determine the size of individual β cells. The scale bar represents 100 μm. The graph shows an average size of at least 250 β cells from each of four mice of each genotype (modified from ref. 18. Am Physiol Soc, with permission).

9. For analysis of pancreatic morphometry, we scored six animals per genotype and nine sections per animal spaced 100 μm apart using NI Vision Assistant (version 7.1.0; National Instruments). Image J is also useful to this analysis (Fig. 3).

3.6. In Vivo *Physiological Studies*

1. Blood glucose levels were determined using an automated blood glucose reader (Accu-Check, Roche) every other week (Fig. 4a).
2. Serum insulin levels were measured by ELISA (Crystal Chem, Downers Grove, IL) following manufacturer's protocol (Fig. 4b).
3. For insulin tolerance tests, mice were fasted 3 h and injected with human insulin (Novolin R, Novo Nordis, 0.75 U/kg body weight) (Fig. 4c).
4. GTTs were performed on mice that were fasted overnight (16 h). Blood was collected immediately before as well as 15, 30, 60, and 120 min after the intraperitoneal injection of glucose (2 g/kg body weight) (Fig. 5b).

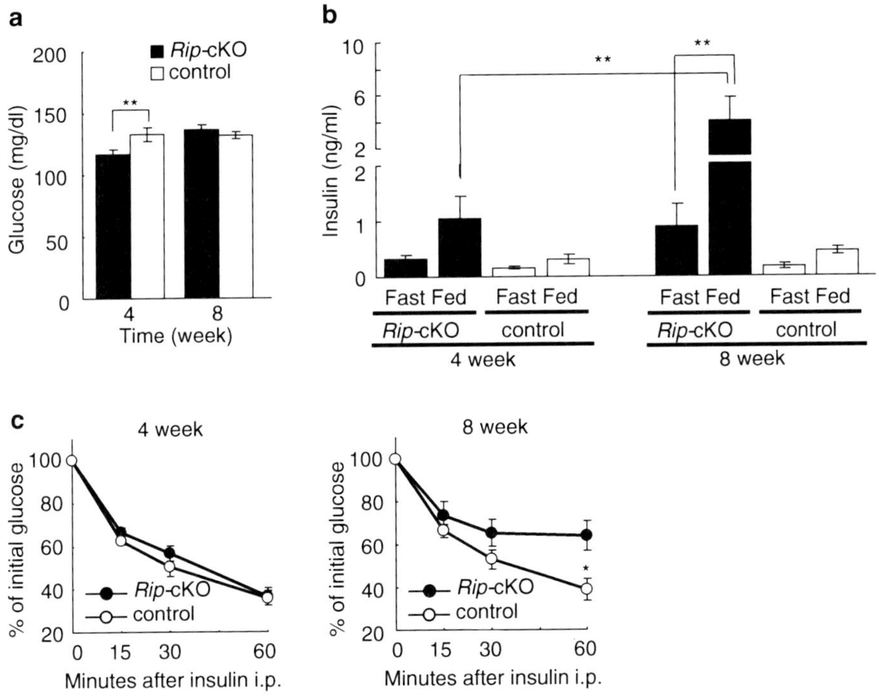

Fig. 4. Normal insulin responsiveness in 4-week-old *Rip-Tsc1*cKO mice. (**a**) Blood glucose levels in random fed *Rip-Tsc1*cKO and control male mice were determined (n=16–17). (**b**) Age-dependent increase of insulin levels in *Rip-Tsc1*cKO mice. Serum insulin levels were determined in overnight-fasted or random-fed mice of each genotype (4, 8 week, n=6–11; 24 week, n=12–15). (**c**) Impaired insulin tolerance in *Rip-Tsc1*cKO mice. Insulin tolerance tests in 3-h-fasted mice at the age of 4 weeks (left, n=9–10) and 8 weeks (right, n=6–10). (**a**–**c**), Values are mean ± S.E.M. $^*p<0.05$, $^{**}p<0.01$. *Rip*-cKO: *Rip-Tsc1*cKO (modified from ref. 18. Am Physiol Soc, with permission).

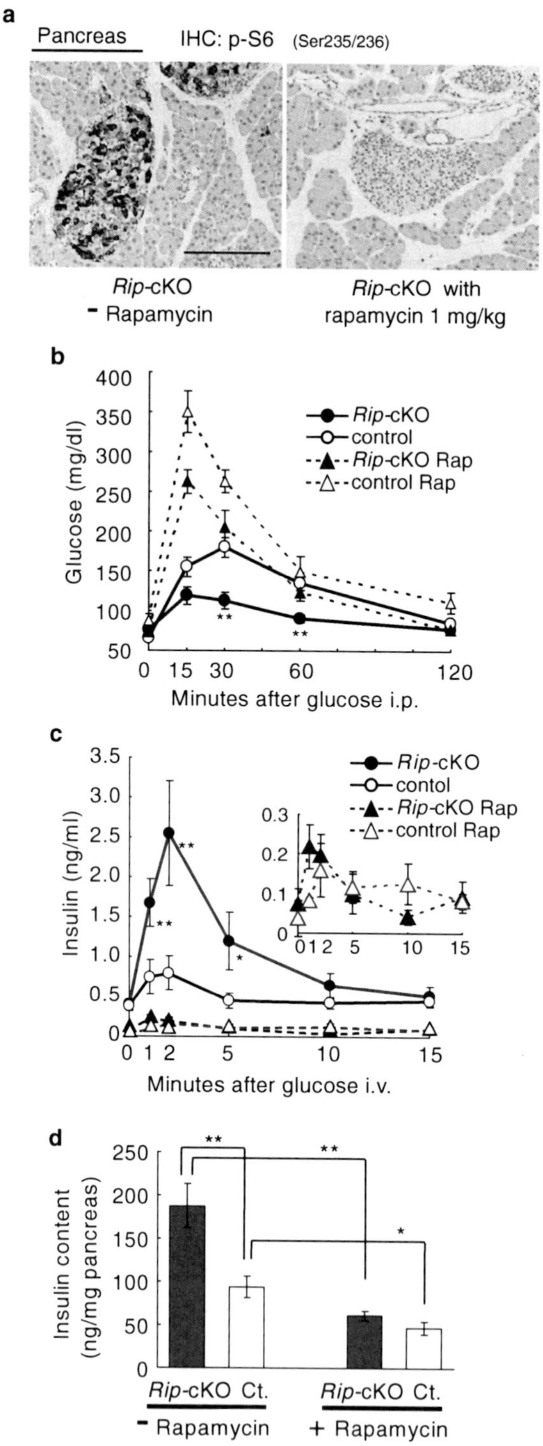

Fig. 5. Tsc1 deletion improves β-cell function. (**a**) p-S6 (Ser235/236) staining of pancreatic sections from *Rip-Tsc1*cKO with or without rapamycin treatment. Scale bar: 100 μm. (**b**) Blood glucose levels after intraperitoneal (i.p.) injection of glucose (2 g/kg body weight) in 4-week-old mice with or without rapamycin treatment ($n=12$–19). * indicates statistical significance between control and *Rip-Tsc1*cKO. (**c**) Serum insulin levels after an intravenous (i.v.) glucose injection. To evaluate acute-phase insulin secretion, 4-week-old mice were injected with glucose (1 g/kg body weight), $n=7$–10 for no treatment, $n=4$ for rapamycin treatment. * indicate same as (**c**). (**d**) Insulin contents of the pancreas from mice of each genotype with or without rapamycin treatment at 4 weeks of age ($n=5$–7). 2-week-old *Rip-Tsc1*cKO and control mice were injected rapamycin (1 mg/kg body weight) every other day for up to 4 weeks. For panels (**a–d**), graphs show mean ± S.E.M. $^{*}P<0.05$, $^{**}P<0.01$. *Rip*-cKO: *Rip-Tsc1*cKO; *Rap* Rapamycin treatment (modified from ref. 18. Am Physiol Soc, with permission).

5. For glucose-stimulated insulin secretion experiments, mice were deprived of food for 16 h and then anesthetized with pentobarbital. Blood samples were taken from tail vein at 0, 1, 2, 5, 10, and 15 min after the intravenous glucose injection (1 g/kg body weight). Insulin levels were measured as described above (Fig. 5c).

3.7. Rapamycin Treatment Studies

1. Rapamycin (LC laboratories) was initially dissolved in 100% ethanol, stored at −20°C, and further diluted in an aqueous solution of 5.2% Tween 80 and 5.2% PEG 400 (final ethanol concentration, 2%) immediately before use (27).
2. Mice were injected rapamycin intraperitoneally (1 mg/kg body weight, every other day).
3. Both pancreas and brain tissues for WB, RT-PCR, and IHC were harvested 12 h after the last rapamycin treatment (Fig. 5a).

3.8. Insulin Content Measurement

1. For measurement of insulin content of whole pancreas, excised pancreata were frozen in liquid nitrogen and disrupted.
2. The pancreas powder was suspended in cold acid–ethanol, and insulin was extracted overnight at 4°C.
3. The supernatants were diluted and subjected to insulin ELISA as described above (Fig. 5d).

4. Notes

About islet isolation using this protocol, we chose hand-pick method which is different from other protocol using Ficoll. The method to use Ficoll takes more than 30–40 min to make different layer, centrifuge, and transfer to new buffer. As reported, long-term Ficoll exposure is toxic for islets. We normally pick islets to a new petri dish with media, and then hand pick again to transfer islets to another new dish. Since it is impossible to pick islets only (little amount of acinar cells are also picked from the first dish), repeating the manual transfer of islets ensures the purity of islet preparation.

Acknowledgments

This work was supported by grants from NIH and DOD (K.L.G), and H.M was supported by a mentor-based postdoctoral fellowship from ADA.

We thank Marta Dzaman, Chris Edwards, Naoko Wanibuchi, and Michael A Reid for technical advice and assistance.

References

1. Loewith, R., Jacinto, E., Wullschleger, S., Lorberg, A., Crespo, J. L., Bonenfant, D., Oppliger, W., Jenoe, P., and Hall, M. N. (2002) Two TOR complexes, only one of which is rapamycin sensitive, have distinct roles in cell growth control, *Molecular cell 10*, 457–468.
2. Withers, D. J., Gutierrez, J. S., Towery, H., Burks, D. J., Ren, J. M., Previs, S., Zhang, Y., Bernal, D., Pons, S., Shulman, G. I., Bonner-Weir, S., and White, M. F. (1998) Disruption of IRS-2 causes type 2 diabetes in mice, *Nature 391*, 900–904.
3. Niswender, K. D., Morton, G. J., Stearns, W. H., Rhodes, C. J., Myers, M. G., Jr., and Schwartz, M. W. (2001) Intracellular signalling. Key enzyme in leptin-induced anorexia, *Nature 413*, 794–795.
4. Araki, E., Lipes, M. A., Patti, M. E., Bruning, J. C., Haag, B., 3rd, Johnson, R. S., and Kahn, C. R. (1994) Alternative pathway of insulin signalling in mice with targeted disruption of the IRS-1 gene, *Nature 372*, 186–190.
5. Tamemoto, H., Kadowaki, T., Tobe, K., Yagi, T., Sakura, H., Hayakawa, T., Terauchi, Y., Ueki, K., Kaburagi, Y., Satoh, S., and et al. (1994) Insulin resistance and growth retardation in mice lacking insulin receptor substrate-1, *Nature 372*, 182–186.
6. Katz, E. B., Stenbit, A. E., Hatton, K., DePinho, R., and Charron, M. J. (1995) Cardiac and adipose tissue abnormalities but not diabetes in mice deficient in GLUT4, *Nature 377*, 151–155.
7. Lin, X., Taguchi, A., Park, S., Kushner, J. A., Li, F., Li, Y., and White, M. F. (2004) Dysregulation of insulin receptor substrate 2 in beta cells and brain causes obesity and diabetes, *J Clin Invest 114*, 908–916.
8. Kulkarni, R. N., Bruning, J. C., Winnay, J. N., Postic, C., Magnuson, M. A., and Kahn, C. R. (1999) Tissue-specific knockout of the insulin receptor in pancreatic beta cells creates an insulin secretory defect similar to that in type 2 diabetes, *Cell 96*, 329–339.
9. Kubota, N., Terauchi, Y., Tobe, K., Yano, W., Suzuki, R., Ueki, K., Takamoto, I., Satoh, H., Maki, T., Kubota, T., Moroi, M., Okada-Iwabu, M., Ezaki, O., Nagai, R., Ueta, Y., Kadowaki, T., and Noda, T. (2004) Insulin receptor substrate 2 plays a crucial role in beta cells and the hypothalamus, *J Clin Invest 114*, 917–927.
10. Hashimoto, N., Kido, Y., Uchida, T., Asahara, S., Shigeyama, Y., Matsuda, T., Takeda, A., Tsuchihashi, D., Nishizawa, A., Ogawa, W., Fujimoto, Y., Okamura, H., Arden, K. C., Herrera, P. L., Noda, T., and Kasuga, M. (2006) Ablation of PDK1 in pancreatic beta cells induces diabetes as a result of loss of beta cell mass, *Nature genetics 38*, 589–593.
11. Liu, H., Remedi, M. S., Pappan, K. L., Kwon, G., Rohatgi, N., Marshall, C. A., and McDaniel, M. L. (2009) Glycogen synthase kinase-3 and mammalian target of rapamycin pathways contribute to DNA synthesis, cell cycle progression, and proliferation in human islets, *Diabetes 58*, 663–672.
12. Kwon, G., Marshall, C. A., Liu, H., Pappan, K. L., Remedi, M. S., and McDaniel, M. L. (2006) Glucose-stimulated DNA synthesis through mammalian target of rapamycin (mTOR) is regulated by KATP channels: effects on cell cycle progression in rodent islets, *The Journal of biological chemistry 281*, 3261–3267.
13. Shigeyama, Y., Kobayashi, T., Kido, Y., Hashimoto, N., Asahara, S., Matsuda, T., Takeda, A., Inoue, T., Shibutani, Y., Koyanagi, M., Uchida, T., Inoue, M., Hino, O., Kasuga, M., and Noda, T. (2008) Biphasic response of pancreatic beta-cell mass to ablation of tuberous sclerosis complex 2 in mice, *Molecular and cellular biology 28*, 2971–2979.
14. Hamada, S., Hara, K., Hamada, T., Yasuda, H., Moriyama, H., Nakayama, R., Nagata, M., and Yokono, K. (2009) Upregulation of the mammalian target of rapamycin complex 1 pathway by Ras homolog enriched in brain in pancreatic beta-cells leads to increased beta-cell mass and prevention of hyperglycemia, *Diabetes 58*, 1321–1332.
15. Murakami, M., Ichisaka, T., Maeda, M., Oshiro, N., Hara, K., Edenhofer, F., Kiyama, H., Yonezawa, K., and Yamanaka, S. (2004) mTOR is essential for growth and proliferation in early mouse embryos and embryonic stem cells, *Molecular and cellular biology 24*, 6710–6718.
16. Kobayashi, T., Minowa, O., Sugitani, Y., Takai, S., Mitani, H., Kobayashi, E., Noda, T., and Hino, O. (2001) A germ-line Tsc1 mutation causes tumor development and embryonic lethality that are similar, but not identical to, those caused by Tsc2 mutation in mice, *Proceedings of the National Academy of Sciences of the United States of America 98*, 8762–8767.
17. Kwiatkowski, D. J., Zhang, H., Bandura, J. L., Heiberger, K. M., Glogauer, M., el-Hashemite, N., and Onda, H. (2002) A mouse model of TSC1 reveals sex-dependent lethality from liver hemangiomas, and up-regulation of p70S6 kinase activity in Tsc1 null cells, *Human molecular genetics 11*, 525–534.

18. Mori, H., Inoki, K., Opland, D., Muenzberg, H., Villanueva, E. C., Faouzi, M., Ikenoue, T., Kwiatkowski, D., Macdougald, O. A., Myers Jr, M. G., and Guan, K. L. (2009) Critical roles for the TSC-mTOR pathway in {beta}-cell function, *American journal of physiology Endocrinol Metab 297*, E1013–E1022.
19. Mori, H., Inoki, K., Munzberg, H., Opland, D., Faouzi, M., Villanueva, E. C., Ikenoue, T., Kwiatkowski, D., MacDougald, O. A., Myers, M. G., Jr., and Guan, K. L. (2009) Critical role for hypothalamic mTOR activity in energy balance, *Cell metabolism 9*, 362–374.
20. Uhlmann, E. J., Wong, M., Baldwin, R. L., Bajenaru, M. L., Onda, H., Kwiatkowski, D. J., Yamada, K., and Gutmann, D. H. (2002) Astrocyte-specific TSC1 conditional knockout mice exhibit abnormal neuronal organization and seizures, *Annals of neurology 52*, 285–296.
21. Meikle, L., McMullen, J. R., Sherwood, M. C., Lader, A. S., Walker, V., Chan, J. A., and Kwiatkowski, D. J. (2005) A mouse model of cardiac rhabdomyoma generated by loss of Tsc1 in ventricular myocytes, *Human molecular genetics 14*, 429–435.
22. Choudhury, A. I., Heffron, H., Smith, M. A., Al-Qassab, H., Xu, A. W., Selman, C., Simmgen, M., Clements, M., Claret, M., Maccoll, G., Bedford, D. C., Hisadome, K., Diakonov, I., Moosajee, V., Bell, J. D., Speakman, J. R., Batterham, R. L., Barsh, G. S., Ashford, M. L., and Withers, D. J. (2005) The role of insulin receptor substrate 2 in hypothalamic and beta cell function, *J Clin Invest 115*, 940–950.
23. Stiles, B. L., Kuralwalla-Martinez, C., Guo, W., Gregorian, C., Wang, Y., Tian, J., Magnuson, M. A., and Wu, H. (2006) Selective deletion of Pten in pancreatic beta cells leads to increased islet mass and resistance to STZ-induced diabetes, *Molecular and cellular biology 26*, 2772–2781.
24. Nguyen, K. T., Tajmir, P., Lin, C. H., Liadis, N., Zhu, X. D., Eweida, M., Tolasa-Karaman, G., Cai, F., Wang, R., Kitamura, T., Belsham, D. D., Wheeler, M. B., Suzuki, A., Mak, T. W., and Woo, M. (2006) Essential role of Pten in body size determination and pancreatic beta-cell homeostasis in vivo, *Molecular and cellular biology 26*, 4511–4518.
25. Lacy, P. E., and Kostianovsky, M. (1967) Method for the isolation of intact islets of Langerhans from the rat pancreas, *Diabetes 16*, 35–39.
26. Shapiro, A. M., Lakey, J. R., Ryan, E. A., Korbutt, G. S., Toth, E., Warnock, G. L., Kneteman, N. M., and Rajotte, R. V. (2000) Islet transplantation in seven patients with type 1 diabetes mellitus using a glucocorticoid-free immunosuppressive regimen, *The New England journal of medicine 343*, 230–238.
27. Wendel, H. G., De Stanchina, E., Fridman, J. S., Malina, A., Ray, S., Kogan, S., Cordon-Cardo, C., Pelletier, J., and Lowe, S. W. (2004) Survival signalling by Akt and eIF4E in oncogenesis and cancer therapy, *Nature 428*, 332–337.

Chapter 27

A Mouse Model of Diet-Induced Obesity and Insulin Resistance

Chao-Yung Wang and James K. Liao

Abstract

Obesity is reaching pandemic proportions in Western society. It has resulted in increasing health care burden and decreasing life expectancy. Obesity is a complex, chronic disease, involving decades of pathophysiological changes and adaptation. Therefore, it is difficult ascertain the exact mechanisms for this long-term process in humans. To circumvent some of these issues, several surrogate models are available, including murine genetic loss-of-function mutations, transgenic gain-of-function mutations, polygenic models, and different environmental exposure models. The mouse model of diet-induced obesity has become one of the most important tools for understanding the interplay of high-fat Western diets and the development of obesity. The diet-induced obesity model closely mimics the increasingly availability of the high-fat/high-density foods in modern society over the past two decades, which are main contributors to the obesity trend in human. This model has lead to many discoveries of the important signalings in obesity, such as Akt and mTOR. The chapter describes protocols for diet induced-obesity model in mice and protocols for measuring insulin resistance and sensitivity.

Key words: Obesity, High-fat diet, Insulin resistance, Body weight, Metabolism

1. Introduction

Obesity results from an imbalance of food intake, basal metabolism, and energy expenditure. At an individual level, multiple endogenous or environmental causes could lead to obesity (1). However, in most cases, a combination of excessive caloric intake and availability of energy-dense meals is thought to be the main contributor to obesity (2). Because complications from obesity such as diabetes and cardiovascular disease usually require decades, surrogate animal models are important for studying the molecular aspects of obesity and its pathophysiological effects. One of these models that are gaining increasing attention is the diet-induced obesity model in mouse.

Thomas Weichhart (ed.), *mTOR: Methods and Protocols*, Methods in Molecular Biology, vol. 821, DOI 10.1007/978-1-61779-430-8_27,

Foods that are rich in fats have been shown to produce increased body weight and diabetes in various strains of mice and rats (3). In the past 20–30 years, there have been many studies characterizing the responses of animals exposed to high-fat diets (4). Some animals show profound increases in their body fat content while some are resistant to weight gain with high-fat diet (5). For example, the outbred Spraque-Dawley rats when fed with high-fat diet have different responses with regard to the development of obesity. In the mice, the A/J mouse and C57BL/KsJ mouse are relatively resistant to high-fat diet when compared to C57BL/6J mouse (6). The B6 mouse is a particularly good model mimicking human metabolic derangements that are observed in obesity because when fed ad libitum with a high-fat diet, these mice develop obesity, hyperinsulinemia, hyperglycemia, and hypertension, but when fed ad libitum to chow diet, they remain lean without metabolic abnormalities (8).

The metabolic abnormalities of B6 mouse closely parallel that of human obesity progression pattern. Although an increase in body weight can be noticed after 2 weeks, the increase is gradual and becomes apparent after 4 weeks. After 16–20 weeks of high-fat diet feeding, mouse will typically exhibit 20–30% increase in body weight when compared to chow-fed mouse (9). The high-fat diet's effects on blood glucose are more discrepant and depend on the type of dietary regimen. Hyperglycemia usually develops within 4 weeks of a high-fat diet (11). The elevation of fasting glucose is usually accompanied by increases in fasting insulin levels. At that time, hyperinsulinemic-euglycemic clamp experiments will demonstrate whole-body insulin resistance. However, there are no reliable predictors for diabetes development or onset and the reported development of overt diabetes is controversial. Nevertheless, the full manifested picture of obesity develops after 16 weeks of high-fat diet with adipocyte hyperplasia, fat deposition in mesentery, increased fat mass, diabetes, and hypertension (12).

Akt and mTOR pathyway integrates several important signals that regulate cell growth and metabolism. Activation of Akt and mTOR signalings by food intake has great implications in obesity and insulin resistance (9). Inhibition of Akt and mTOR pathway by rapamycin has effects in longevity (13), adipocyte differentiation, and obesity (14). Recent studies have shown that S6K1-deficient mice and Akt1 knockout mice exhibit are prevented from diet-induced obesity through model of murine high-fat diet induced obesity described below (15).

Several factors are also important when using this model of diet-induced obesity. First, the genetic nature and degree of model characterization as well as differences in mouse background could lead to different phenotype (16). For examples, AKR/J and DBA/2J mice are very responsive to high-fat diet, while A/J and Balb/cJ mice are more resistant. C57BL/6J mouse are better characterized than some of the other models. Second, gender usually plays an important role. Generally, male mice are more affected by diabetes than are female mice and thus are used more often in diet-induced obesity studies (17). Third, environmental factors are important and should be considered. Obese mice are sensitive to stress and their environmental settings will affect experimental results (18). Therefore, cage placement, mice density, food quality, mice handling, beddings and mice-check frequency will all results in disturbance of the development of obesity in experimental mice.

2. Materials

2.1. Mice

1. C57BL/6J males, 4–6 weeks old (The Jackson Laboratory, stock 000664).
2. Inventoried (DIO) C57BL/6J males (The Jackson Laboratory) (see Note 1).

2.2. Diet

1. 10 kcal% fat diet (Standard diet) (see Note 2).
2. 45 kcal% fat diet (Research Diets, D12451i).
3. 60 kcal% fat diet (Research Diets, D12492i) (see Note 3).

2.3. Glucose and Insulin Test

1. Dulbecco's phosphate-buffered saline (D-PBS), 1× Liquid (Gibco 14190-136).
2. NaCl 0.9% sterilized, Baxter IV solutions (Baxter 6E1322).
3. Disposable sterile animal feeding needles, Popper & Sons, 20 G×1 1/2 in. (VWR 20068-666).
4. Tuberculin syringe with 27 G×1/2 in. needle (Becton-Dickinson, 309623).
5. Heated hard pad.
6. Clean cages with water bottles.
7. D-(+)-glucose 99.5% (Sigma, G7528).
8. Human insulin: Novolin R, 100 U/ml, 10 ml (NovoNordisk).
9. Scalpel.
10. Mineral oil (Sigma M3516).
11. Paper towels.
12. Ascensia ELITE XL Bayer Glucometer Blood (Bayer).
13. Ascensia ELITE XL test strips (Bayer).
14. Batteries, 3 V Lithium cell (Sigma, B0653).
15. SurePrep™ capillary tubes, 75 mm, self-sealing (Becton Dickinson 420315).
16. MicroWell 96-well polystyrene plates, round bottom (non-treated), sterile (Nunc, Sigma, P4241).

17. 20–200-μl pipette with tips.
18. Insulin (Mouse) Ultrasensitive (Alpco Diagnostics, EIA 80- INSMSU-E10).
19. Titer Plate Shaker (Thermo Scientific EW-51402-10).
20. Multilabel Readers, VICTOR (6) (PerkinElmer).

2.4. Blood Collection and Tissue Collection

1. 1-ml syringes with 25 G×5/8 needle (Becton-Dickinson 305122).
2. Microtainer serum separator tubes (BD Diagnostics, 365956).
3. MicroWell 96-well polystyrene plates, round bottom (non-treated), sterile (Nunc, Sigma, P4241).
4. Formalin solution, neutral buffered 10% (Sigma HT501128).
5. 50-ml conical tubes (VWR 21008-178).
6. Rodent restrainer 115 VAC, animal weight range 70–125 g (Harvard apparatus).

3. Methods

Diet induced obesity model usually takes more than 16–20 weeks for completing the experiments (Fig. 1) (12). Moreover, obesity is a multiple organ system disease, involving derangements of body-weight, adipose tissue, glucose metabolism, cholesterol level, and inflammatory markers (19). Therefore, preexperimental design is crucial for the success of the experiments. While multiple organ

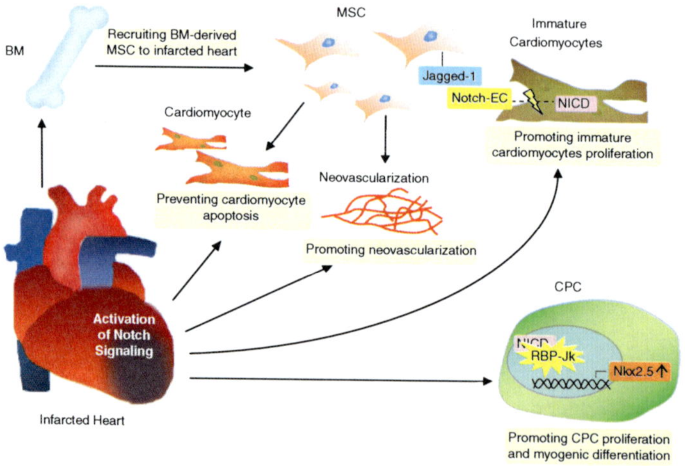

Fig. 1. (**a**) Schematic diagram showing how we design the high-fat and chow diets feeding with and without treatment. In this experiment, we use rapamycin to test if body weight is affected by 4 weeks treatment of rapamycin. (**b**) The typical body weight chart when mice fed with high-fat diet or chow diet (Adapted from Wang et al. Science Signaling (2009) with permission).

systems could be tested and many parameters could be examined, it is important to have upfront experimental hypothesis and goals. During obesity development, one could examine and measure a wide variety of parameters. However, different hypotheses will required different measurements and experimental methods, which in some cases, could affect physiologic parameter. These differences are detailed below.

3.1. Animal Housing

1. Keep mice under standard pathogen-free conditions with food and water ad libitum and regular 12:12 light–dark cycle. Make sure that the light–dark cycle is strictly followed, as checking mice after dark cycle started will disrupt the regular circadian rhythm of mice and will have impact on the metabolism of mice (20).
2. Mice should be housed in the same number per cages (see Note 4).
3. Check cage and bedding conditions daily but avoid check mice too frequently. Although not as frequent as other genetic modified diabetic mice, C57BL/6J mice after becoming obese and diabetic will have polyuria and mice cages may need to be changed more frequently.
4. Perform mice identification with ear notching or ear tagging at the start of the experiments, as these procedures will also affect mice body weight and glucose.
5. Pay attention to the environment nearby the mice cages. Noise, traffic, or other disturbances will affect body weight and cause bias in the diet induced obesity experiments.

3.2. Mice Randomizations and Experimental Design

As there is also individual variation even in the inbred strains of the mice, it is important to collect the baseline data of the mice before experiments. We measure body weight, tibial length, fasting glucose for mice of the same gender and age. The baseline fasting glucose of C57BL/6J male mice ranges from 40–190 mg/dl at age of 6 weeks in our laboratory; we only recruit mice with their fasting glucose between 70 and 130 mg/dl. We also exclude the male mice with body weight below 15 g or over 25 g at 6 weeks of age. Mice with ill looking, teeth problem or hair diseases are also excluded from study.

1. At age of 5 weeks, measure mouse body weight and tibial length.
2. Measure baseline glucose: insert glucometer strip into glucometer. Pick up mouse with gloves. Restrain the mice with either the other hand or the mice restrainer. We usually use unheated version of rodent restrainer. It s cylindrical and have series of holes on each side to facilitate injections or blood testing. Angle type restrainers are simpler if with experiences.

3. Randomize mice by body weight and exclude the mice with extreme body weight.
4. List glucose level in each experimental group.
5. Exclude mice with extreme glucose levels.
6. Perform statistic checkup to make sure that there is no grouping bias.

3.3. High-Fat Diet Feeding and Measurement of Food Intake

1. Select the different diet regimen. For example, for C57BL/6J diet induced obesity model, we usually use 60 kcal% fat diet for the experimental group and 10 kcal% chow diet for the control group, depending on the purpose of the experiments (see Note 5).
2. Measure the diet weight before change the diet. Select the diet with good particles and eliminate the fragmented food particle or powder.
3. Pour the diet into the cages and avoid pouring into the fragments or powder into the cages as they sometimes contaminated the beddings with polyuric mice. These fragments or powder will also cause error in food intake measurement.
4. Observe daily for the food color. These 60% high-fat diet tends to become moisture and oily after 2–3 days exposing to air. However, they usually are good for 1 week.
5. Change the diet every week and measure the changes in food weight for measuring the food intake of the mice. Before measurement, pick up the food particle in the beddings.
6. Usually put more food as needed. We do not add fresh food between each measurement as it could result in differences between groups. As experiments continue, high-fat diet groups usually will have more food intake comparing to control groups.
7. When obese mice develop polyuria, it may require changing bedding more than two to three times per week. Make sure that changing bedding will not cause errors in measuring food intake.
8. Perform body weight measurement at the same time while changing bedding (see Note 6)

3.4. Blood Measurement from Tail

1. Insert glucometer strip into glucometer.
2. Pick up mouse with gloves. Restrain the mice with either the other hand or the mice restrainer. We usually use unheated version of rodent restrainer. It s cylindrical and have series of holes on each side to facilitate injections or blood testing. Angle type restrainers are simpler if with experiences.
3. Cut the tail as less as possible with scalpel.
4. Put one drop of the blood from tail onto glucometer strip and reads.

5. For serum or more blood other than glucose, use capillary tube. If blood stops during capillary tube collection, massage the tail toward the tips.
6. If more blood is needed, collect blood from suborbital area.

3.5. Glucose Tolerance Test

Accurate glucose tolerance test depends on the preparation of the mice. To closely mimic the human part of the glucose tolerance test, we perform mice fasting from 7:00 AM to 15:00 PM. Mice are active during the light–dark phase and this fasting period are more physiological when comparing to human. Without proper fasting, the baseline glucose readings and glucose level after tolerance test will be difficult to interpret.

In our laboratory, we usually administer glucose by intraperitoneal injection. Injection is more consistent with dosage and the results. It also saves time if many mice are going to be tested. However, it is not physiological and cannot totally compare to human glucose tolerance test and omit the effects of glucose on the gastrointestinal tract.

1. Clean the bedding and the food from the cage. Make sure that there is no food left in the beddings. Keep the water bottle.
2. Weigh mice.
3. Prepare a sterile 10% (w/v) glucose solution in PBS. Glucose will be administered at a concentration of 1 g/kg (glucose/body weight) in a 10 ml/kg volume. Occasionally, we choose 2 g/kg concentration earlier in the experiments before mice become diabetic. 2 g/kg concentration will result in unreadable glucose value in the glucometer if mice become diabetic.
4. Preload 1 ml insulin syringes with the glucose solution. Remove air from the syringes.
5. After fasting period is complete. Measure baseline glucose level. Pick up mouse with gloves. Restrain the mice with either the other hand or the mice restrainer. Cut the tail as less as possible with scalpel. Put one drop of the blood from tail onto glucometer strip and reads. Record glucose value. The range of glucose values is 20–600 mg/dl. If the glucometer indicates the value "Hi", we usually take the blood with capillary and diluted 2× with PBS. The actual glucose level seldom reaches 1,200 mg/dl and dilution 2× usually is enough to show the exact level.
6. After adequate fasting period, inject mice with intraperitoneal glucose. If too much mice precludes from injection within 15 min, we usually work with two to three people to avoid prolonged injection time.
7. Repeat blood glucose measurements at 15, 30, 60, and 90 min.
8. Return the mice into cages with food.

3.6. Insulin Level Measurement

After serum collection, store the samples at −20°C. Freezing–thawing of the sample, storing the sample at 4°C, or failure to make sure that the kit reaches room temperature will result in experimental error.

1. Prepare reagent. Dilute the conjugate with ten parts of conjugate buffer. Reconstitute insulin control with 0.6 ml distilled water. Dilute 20 ml wash buffer with 400 ml distilled water.
2. Ensure that microplates are at room temperature prior to opening foil pouch. Designate microplate strips for the standards, controls, and mouse samples.
3. Add 10 μl of each standard, control or mouse sample into its respective wells.
4. Add 75 μl of enzyme conjugate into each well. Seal the plate with film provided.
5. Place the microplate in room temperature and shaking at 700–900 rpm for 2 h.
6. Wash the microplate for six times with wash buffer. After the final wash with wash bottle, remove any residual buffer and bubbles from the wells by inverting and firmly tapping the microplate on absorbent paper towels
7. Add 100 μl of substrate into each well.
8. Incubate for 15 min at room temperature on a horizontal microplate shaker.
9. Pipette 100 μl of stop solution into each well. Gently shake the microplate to stop the reaction.
10. Read the absorbance at 450 nm with a reference wavelength of 620–650 nm. Read the plate within 30 min following the addition of the stop solution.

3.7. Glucose Stimulated Insulin Secretion

We usually perform glucose-stimulated insulin secretion test alone with glucose tolerance test. However, glucose stimulated insulin secretion test is more cumbersome and will need at least two people together to perform glucose tolerance test and glucose stimulated insulin secretion test at the same time.

The peak insulin secretion after glucose intraperitoneal injection is around 9–10 min. We typically collected sample at 0, 5, 10, 15, and 30 min after glucose injection. Glucose stimulated insulin secretion test impairment is a key factor in beta-cell failure and some genetic manipulation mice may only show phenotype in this test other than glucose tolerance test.

1. Prepare glucose and measure the body weight of the mice.
2. Collect the baseline blood into a capillary tube.
3. Administer glucose to mice (1 mg/kg in 10 ml/kg volume).
4. Collect the blood at 5, 10, 15, and 30 min after glucose injection.

5. Spin the capillary tubes into a microhematocrit centrifuge at 12,000 g for 10 min. Cut the capillary tubes at the level between serum and red blood cells.
6. Add the serum into the microplates. Samples could be frozen at –20°C for later measurements.
7. Measure the insulin level as previously described.

3.8. Insulin Tolerance Test

The dose of insulin tolerance test may vary from 1 to 5 U/kg and should take into consideration of the body weight, diabetic status and genetic background of the mice. In the C57BL/6J diet induced obesity model, we usually use 2 U/kg insulin for injection. We usually perform insulin tolerance test in fasting condition. However, it sometimes could be performed in the fed condition if target molecules of different models are considered.

1. Perform 7:00 AM to 15:00 PM fasting.
2. Measure the body weight of the mice.
3. Prepare insulin and glucose. Dilute insulin into 2 U/10 ml by iced sterile PBS. Preload the injection syringes with insulin and keep them on ice. Prepare one or two glucose sterile solution to prevent hypoglycemia of the mice.
4. Collect the blood at 0, 15, 30, 60, 90, and 120 min after insulin injection.
5. Measure glucose level and observe the mice. If the mice are ill-looking, inject glucose immediately to avoid mortality.

3.9. Body Composition Measurement

With the advances of the technology, magnetic resonance imaging (MRI) and dual energy X-ray absorptiometry scanning (DEXA) has gradually replaced the chemical carcass analysis for measuring the fat component (21). Chemical carcass used to be considered as the gold standard for whole body composition analysis. It requires to sacrifice the mice and time-consuming. Dissecting and weighing of fat tissues in individual mice can also measure the total body fat. However, this method is less accurate. The collection of visceral fat can be challenging, as it is often integrated in the internal organs and cannot be accurately dissected.

Body composition measurements by DEXA consist of three components: fat, bone mineral, and lean soft tissue. It shows strong correlation between the percentage of fat area and total body fat as compared with chemical extraction method. There is no need for sacrifice the mice and can be used for serial measurement.

1. Measure the animal's body weight and anesthetized the mice with ketamine/xylazine solution, intraperitoneally.
2. Calibration for the bone mineral density and percent fat content using the phantom mouse unit.

3. Place the mice ventral side down on disposable plastic trays so that the whole mouse can be measured.
4. Record bone mineral density, bone mineral content, lean mass, fat mass, and percentage body fat.
5. Analyze the data using PIXImus2 software.

3.10. Rapamycin Treatment of Mice

1. Dissolve 1 mg Rapamycin in 20 μl ethanol for stack solution.
2. Further dilute dissolved Rapamycin with Ringer's saline solution to a final concentration of 1 mg/ml directly immediately before use.
3. Restrain the mouse and expose the abdomen.
4. Prepare the 25 gauge needle and fill the syringe with diluted Rapamycin.
5. Disinfect the injection site and insert the needle cranially at 30° angle caudal to the umbilicus and lateral to the midline.
6. Before each injection, make sure to aspirate before injection to avoid penetration to intestine or gall bladder.

3.11. Mice Sacrifice and Collection of Tissue and Blood

After the diet feeding experiments, mice are sacrifice to collect their blood, internal organ and analyze the whole body composition. Tissue also will be used for further analyze, including protein or RNA or histology section.

Diet induced obesity model will result in multiple organ changes when comparing to its control mice. High-fat diet will lead to hepatic steatosis and fatty liver changes could be obviously observed after dissection. Adipocyte number and size are increased. Pancreatic islets will become hypertrophic and then degenerate. Skeletal muscle may have decreased insulin sensitivity. Brown fat may show atrophy or white fat deposition. Hence, obesity is a whole body system derangement. Although some organ systems are affected more than other systems, we usually collect involved organs in sequence to avoid loss of data.

1. Prepare the sacrifice plates, syringes, eppendorf tube, dissection tools, perfusion tools, and iced PBS solution before sacrifice.
2. Mark every eppendorf tubes for serum ample, tissue collection or histological fixation. Fill the 3.7% formaldehyde or tissue lysis buffer or RNA extraction solution (Trizol) before sacrifice.
3. Euthanize the mice by CO_2 according to the animal protocol and ethical regulation.
4. Fix the extremity of the mice with needle to dissection plate.
5. Make an incision through the skin horizontally near the umbilical area.
6. Remove the skin and subcutaneous layer.

7. Make a vertical incision through the midline abdomen to open the abdomen.
8. Cut through the chest cavity and reveal the heart and lung.
9. Insert a needle at the left ventricle apex of the heart and draw the blood slowly. Avoid drawing the blood too soon. 300–1,000 μl blood can be obtained in one mouse.
10. Collect the blood into blood collection eppendorf tubes.
11. Collect pancreas first, as their RNA and protein are more prone to damage by pancreatic enzyme.
12. Collect heart, lung, liver, spleen, kidney, and intestine if needed.
13. Collect subcutaneous, inguinal, and mesenteric white fat pad.
14. Collect brown fat at the back.
15. Collect the brain if indicated.
16. Proceed with histology or tissue analysis (see Note 7).

4. Notes

1. Jackson Laboratory provides various diets, which could be used to induce obesity. For example, C57BL/6J mice could be fed with 60% high-fat diet starting at 6 weeks of age while control mice are fed with 10% fat diet. The age availability is from 6 to 26 weeks. They have established physiological database and inventoried mice that could save much time for experiments. Certain care should be managed when using these mice. First, effect of shipping should be taken into account as mice tend to lose weight in transit. We usually acclimate the mice for 1 week in their experimental environment before using them. Second, we usually do not mix these mice cohort with each other. Male mice tend to fight, which will have some impact on their body weight.
2. Research diet D12450B. Its caloric composition is 20 kcal% protein, 70 kcal% carbohydrate and 10 kcal% fat. Every gram of its ingredient contains 3.85 kcal calorie, including 19 mg cholesterol from Lard and yellow dye.
3. Research diet D12492. Its caloric composition is 20 kcal% protein, 20 kcal% carbohydrate, and 60 kcal% fat. Every gram of its ingredient contains 5.24 kcal calorie, including 232 mg cholesterol from lard and blue dye. It should be stored frozen and could last for 6 months. 60 kcal% fat is rarely practical when considering the human diet. However, to induce obesity in mice in shorter time and facilitate the research, 60 kcal% fat is commonly used. However, when screening the effects of drug or genetic manipulation, sometimes it may require

45 kcal% fat diet as it may be more difficult to prevent or reverse the effects of this extreme 60% kcal% diet.

4. As in many mammalians, C57BL/6J mice in cages also establish social hierarchies that dictate the distribution of food, water, resting places, and mating. Different mice numbers in cages will result in different social behavior and will have impact on the development of obesity.
5. Regular chow diet from different vendors could have different ingredients. The calories composition is usually protein 20%, fat 10%, and carbohydrates 70%. However, their cholesterol, vitamin, fiber, and fat components might vary. It is important to be consistent between experiments.
6. Mice are social animals. There are usually some mice that are dominant in the cages and some are not. When measuring body weight, we usually mark which mice are dominant and which are not and compare between groups to get a better insight into their body weight changes.
7. For comparing white adipose tissue with histological section, it requires to comparing multiple location of the white fat with serial section for more than 20 slides to get accurate data. We usually collect subcutaneous, inguinal, mesenteric, gonadal, and white fat nearby brown fat to get overall view for white fat differences between mice.

Acknowledgments

The author would like to thank Professor Ming-Shien Wen for his advice and encouragement. This work was supported by grants from National Institutes of Health (HL052233, DK085006, NS070001), National Health Research Institutes (NHRI-EX99-9925SC) and National Science Council (98-2314-B-182A-082-MY3).

References

1. Flier, J.S. Obesity wars: molecular progress confronts an expanding epidemic. *Cell* **116**, 337–350 (2004).
2. Wisse, B.E., Kim, F. & Schwartz, M.W. Physiology. An integrative view of obesity. *Science (New York, N.Y)* **318**, 928–929 (2007).
3. Sclafani, A. Animal models of obesity: classification and characterization. *Int J Obes* **8**, 491–508 (1984).
4. Zhang, D., *et al.* Resistance to high-fat diet-induced obesity and insulin resistance in mice with very long-chain acyl-CoA dehydrogenase deficiency. *Cell Metab* **11**, 402–411 (2010).
5. Johnson, P.R., Greenwood, M.R., Horwitz, B.A. & Stern, J.S. Animal models of obesity: genetic aspects. *Annu Rev Nutr* **11**, 325–353 (1991).
6. Rossmeisl, M., Rim, J.S., Koza, R.A. & Kozak, L.P. Variation in type 2 diabetes – related traits in mouse strains susceptible to diet-induced obesity. *Diabetes* **52**, 1958–1966 (2003).
7. York, D.A. Lessons from animal models of obesity. *Endocrinol Metab Clin North Am* **25**, 781–800 (1996).
8. Collins, S., Martin, T.L., Surwit, R.S. & Robidoux, J. Genetic vulnerability to diet-induced obesity in the C57BL/6J mouse:

physiological and molecular characteristics. *Physiol Behav* **81**, 243–248 (2004).

9. Speakman, J., Hambly, C., Mitchell, S. & Krol, E. Animal models of obesity. *Obes Rev* **8 Suppl 1**, 55–61 (2007).
10. Wang, C.Y., *et al.* Obesity increases vascular senescence and susceptibility to ischemic injury through chronic activation of Akt and mTOR. *Science signaling* **2**, ra11 (2009).
11. Sato, A., *et al.* Anti-obesity Effect of Eicosapentaenoic Acid in High-fat/High-sucrose Diet-induced Obesity: Importance of Hepatic Lipogenesis. *Diabetes* (2010).
12. Inui, A. Obesity – a chronic health problem in cloned mice? *Trends Pharmacol Sci* **24**, 77–80 (2003).
13. Cox, L.S. & Mattison, J.A. Increasing longevity through caloric restriction or rapamycin feeding in mammals: common mechanisms for common outcomes? *Aging Cell* **8**, 607–613 (2009).
14. Chang, G.R., *et al.* Rapamycin protects against high fat diet-induced obesity in C57BL/6J mice. *J Pharmacol Sci* **109**, 496–503 (2009).
15. Um, S.H., *et al.* Absence of S6K1 protects against age- and diet-induced obesity while enhancing insulin sensitivity. *Nature* **431**, 200–205 (2004).
16. Friedman, J.M., Leibel, R.L., Bahary, N., Siegel, D.A. & Truett, G. Genetic analysis of complex disorders. Molecular mapping of obesity genes in mice and humans. *Ann N Y Acad Sci* **630**, 100–115 (1991).
17. Begin-Heick, N. Of mice and women: the beta 3-adrenergic receptor leptin and obesity. *Biochem Cell Biol* **74**, 615–622 (1996).
18. Coleman, D.L. Obese and diabetes: two mutant genes causing diabetes-obesity syndromes in mice. *Diabetologia* **14**, 141–148 (1978).
19. Walley, A.J., Asher, J.E. & Froguel, P. The genetic contribution to non-syndromic human obesity. *Nat Rev Genet* **10**, 431–442 (2009).
20. Turek, F.W., *et al.* Obesity and metabolic syndrome in circadian Clock mutant mice. *Science* **308**, 1043–1045 (2005).
21. Sjogren, K., *et al.* Body fat content can be predicted in vivo in mice using a modified dual-energy X-ray absorptiometry technique. *J Nutr* **131**, 2963–2966 (2001).

Chapter 28

Rapamycin as Immunosuppressant in Murine Transplantation Model

Louis-Marie Charbonnier and Alain Le Moine

Abstract

The potent immunosuppressive action of rapamycin has been described in many different mouse models of transplantation. In these models, rapamycin prevent or delay allograft rejection. In several models, rapamycin allowed mixed donor–recipient hematopoietic chimerism to develop facilitating tolerance induction. In our own experience, we observed that rapamycin synergized with $CD8^+$ T cell depletion and coreceptor/costimulation blockade to induce long-term survival of Balb/C to C57Bl/6 heterotopic limb allograft. Herein, we describe immunosuppression cocktails containing rapamycin and methods to evaluate several read outs associated with tolerance induction such as mixed donor–recipient hematopoietic chimerism and *in vitro* or *in vivo* recipient alloreactivity.

Key words: Rapamycin, Mouse Transplantation model, Mixed chimerism, Skin graft, Mixed lymphocyte culture

1. Introduction

Rapamycin (RAPA) or sirolimus corresponds to a macrolide antibiotic, initially produced by bacteria such as *Streptomyces hygroscopicus*. The immunosuppressive activity of RAPA was first described in the context of allogeneic transplantation and mediated by the inhibition of mammalian Target of Rapamycin (mTOR), an intracytoplasmic protein kinase implicated in cell T-cell proliferation and in antigen processing by antigen presenting cells (1, 2). *In vivo* administration of RAPA has been described in many mouse models of transplantation to synergize with different immunosuppressive drugs or cell therapy (3, 4). Importantly, RAPA favors *in vivo* and *in vitro* proliferation of regulatory T cells (Tregs) (5, 6) and facilitates tolerance induction through mixed chimerism (7–9).

Thomas Weichhart (ed.), *mTOR: Methods and Protocols*, Methods in Molecular Biology, vol. 821,
DOI 10.1007/978-1-61779-430-8_28, © Springer Science+Business Media, LLC 2012

Herein, we describe the immunosuppressive protocol we used for inducing tolerance of heterotopic limb allograft. We also describe that a single injection of RAPA delays the rejection of skin graft bearing a single MHC class II disparity. Skin grafting, the evaluation of multilineage mixed chimerism and performing mixed lymphocyte cultures are common methods to evaluate recipients that have been made tolerant by different protocols including (or not) RAPA.

2 Materials

2.1 Animals

BALB/c, C57BL/6 (B6), and C57BL/6.C-H-2^{bm12} (bm12) mice were obtained from the Jackson Laboratory. Animals were housed in our specific pathogen-free animal facility. Eight and 12 weeks old female animals were used for experiments. All animals received humane care in compliance with the Principles of Laboratory Animal Care formulated by the National Institute of Health (NIH publication No. 86-23, revised 1985) and protocols were approved by the local committee for animal welfare.

2.2. In Vivo Administration of Rapamycin and Antibodies

1. Rapamycin (Wyeth-Ayerst, Princeton, NJ) was diluted in sterilized medium consisting of deionized water containing 0.2% of sodium carboxymethylcellulose (Sigma-Aldrich, Saint Louis, MO) and 0.25% of Tween 80 (Sigma), using an ultrasound bath (10).

   Alternative method: Rapamycin powder was dissolved in methanol at a concentration of 1 mg/mL and stored in aliquots at −80 °C. A working stock solution was prepared each day from the methanol stock solutions at a concentration of 1 mg/mL.
2. 1 mL syringe with 26 G needle.
3. Sterile phosphate-buffered saline (PBS).
4. Anti-CD4 YTS 177, anti-CD40 L mAb (MR1), and either the anti-CD8 depleting antibodies, YTS 156 and YTS 169 are stored at −20 °C and filtered on 0.22-μm filter just before use at a final concentration of 10 mg/mL in PBS.

2.3. Skin Grafting

1. Anesthetic solution: 10% ketamine/ 5% xylazine diluted in sterile PBS.
2. Hospital Antiseptic Concentrate (H.A.C.) (3.5% in distilled water, stored at room temperature).
3. Balance for weighing animals.
4. Sterile gauze pads.
5. Razor blades.
6. 100 × 20-mm plastic petri dishes.

7. Straight forceps with 1 × 2 teeth.
8. Microdissecting forceps (half-curved, serrated).
9. Microdissecting scissors (curved with sharp points).
10. Blunt-end scissors.
11. Disposable scalpels with surgical stell blades no. 24 (Swann-Morton®).
12. Vaseline petrolatum gauze (10 cm × 10 cm) (Jelonet, Smith and Nephew).
13. Elastic bandage (8 cm long × 2 cm large) (Hansaplast, BSN medical GmbH).
14. Latex-free adhesive fabric tape (2.5 cm large) (Leukoplast® pro LF, BSN medical).
15. Operating board.

2.4. Cell Culture and Cytokine Detection by ELISA

1. Red blood cell lysing buffer (ACK) (0.15 M NH_4Cl, 10 mM $NaHCO_3$, 0.1 mM Ethylene diamine tetra acetic acid (EDTA) in distilled water) (filtered and stored at 4–8°C).
2. Heat-inactivated fetal calf serum (FCS) (25 mL Aliquots are stored at −20°C). Inactivation of complement present in the serum consists in exposition during 20 min at 56°C.
3. Cell culture medium (stored at 4–8°C and prepared on the day of experiment): Roswell Memorial Park Institute medium 1640 (RPMI), 5 FCS, 100 U/ml penicillin, 100 µg/ml streptomycin, nonessential amino acids, and 1 mM Na pyruvate.
4. 48-Well flat-bottom cell culture plates.
5. 96-Well U-bottom cell culture plates.
6. 96-Well flat-bottom immunosorbent plates.

2.5. Flow Cytometry

1. PBS/0.01% bovine serum albumin/2 mM EDTA/0.001% NaN_3 (PBE) stored at 4–8°C.
2. Lysing solution (10× stock solution stored at 4–8°C) (BD FACS™ Ref 349202).
3. BD Cytofix/Cytopem (1× solution stored at 4–8°C) (BD Bioscience Ref 554722).
4. 5 mL polystyrene round-bottom tube (BD Falcon Ref 352052).
5. Fluorochrome conjugated antibodies (stored at 4–8°C protected to light exposure).
   - Purified anti-CD16/CD32 (clone 2.4 G2) (BD Fc Block) (BD Bioscience Ref 553142).
   - Anti-CD45R/B220 fluoresceine isothiocyanate (FITC) (clone RA3-6B2) (BD Bioscience Ref 553087).
   - Anti-H2k^d phycoerythrin (PE) (clone SF1-1.1) (BD Bioscience Ref 553566).

- Anti-CD8 peridinin chlorophyll protein complex (PercP) (clone 53–6.7) (BD Bioscience Ref 553036).
- Anti-CD11b allophycocyanin (APC) (clone M1/70) (BD Bioscience Ref 553312).
- Anti-CD4 APC-Cy7 (clone GK1.5) (BD Bioscience Ref 552051).
- Anti-CD3 Pacific Blue (PB) (clone 500A2) (BD Bioscience Ref 558214).

6. Dual laser flow cytometer (e.g., Cyan LX cytometer, Dako).
7. Data acquisition and analysis software (e.g., Summit 4.1, Dako Cytomation).

3. Methods

As shown in our laboratory, in combination with coreceptor/costimulation blockade, RAPA and $CD8^+$ T-cell depletion synergized to induce long-term acceptance of heterotopic limb allograft in a stringent allogeneic strain (Balb/C to B6) combination. More than 100 days after limb transplantation, recipients tolerated a fresh donor-specific skin allograft while rejecting a third party and displayed a stable mixed hematopoietic chimerism. Recipient allospecific unresponsiveness was demonstrated in mixed lymphocyte reaction with donor-specific or third party stimulation. In addition, a specific (donor-dependent) Vβ deletion of CD4 and CD8 T-cells was noticed (8).

Skin allograft is relevant for testing the recipient capacity to reject alloantigens or to become tolerant after immunomodulation. Moreover, in recipients displaying stable mixed hematopoietic chimerism, the acceptance of a donor type allograft (expressing the same alloantigens than hematopoietic cells) concomitantly with the rejection of a third party skin allograft reflects a status of robust tolerance. Skin grafting is performed according to an adaptation of the method of Billingham and Medawar (11). Establishment of long-term and multilineage mixed chimerism reflects bone marrow engraftment and allospecific tolerance. Here, we describe a method to monitor chimerism with peripheral blood mononuclear cells (PBMCs) by flow cytometry and allospecific tolerance by mixed lymphocyte culture.

3.1. Rapamycin Administration and Immunosuppressive Protocols

First, the potential of RAPA to delay a single MHC class II disparate skin allograft after a single i.p. injection was evaluated. At a dose of 75 µg per mouse (25 g), RAPA is able to delay single MHC class II (bm12 into B6) disparate skin allograft (Fig. 1). Secondly, we describe the immunosuppressive regimen associated with RAPA to

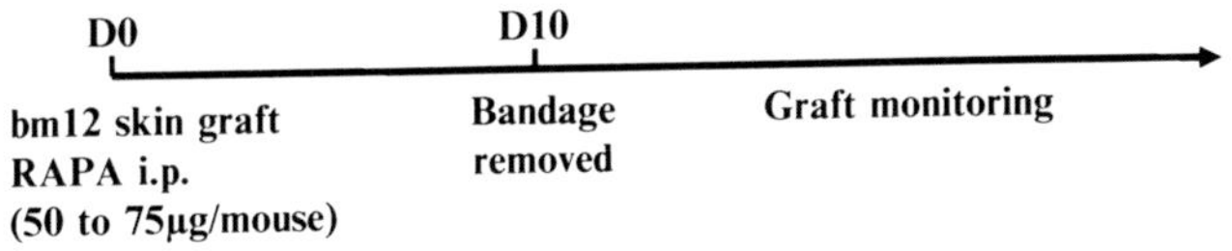

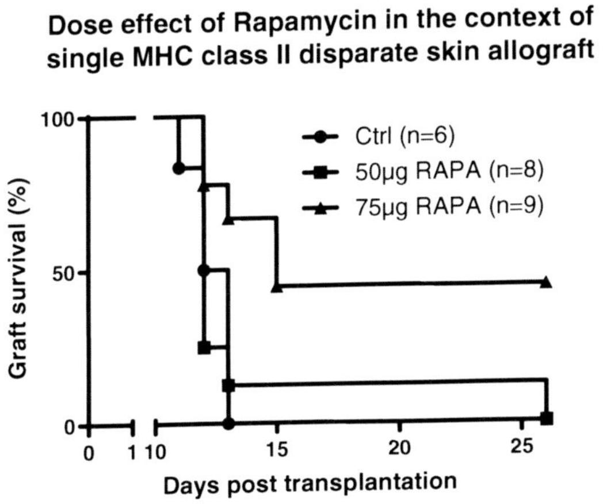

Fig. 1. Rapamycin delays MHC class II disparate skin graft. MHC class II disparate (bm12) skins were grafted on C57Bl/6 recipients treated with a single dose of 50 or 75 μg of rapamycin. Graft survival of RAPA treated mice is compared with untreated controls. A statistical difference was observed using the log-rank test for the dose of 75 μg/mouse (=0.015).

induce long-term acceptance of heterotopic limb allograft in a stringent allogeneic strain (Balb/C to B6) combination (8).

Bm12 skin graft into B6 recipient treated with RAPA:

1. RAPA is administrated intraperitoneally at a dose of 50 or 75 μg/mouse the day of bm12 skin graft.

Immunosuppressive regimen to induce acceptance of heterotopic limb allograft:

2. Recipient mice were treated i.p. with three consecutive injections (days 0, 2, and 4 posttransplantation) of 1 mg of each of the following monoclonal antibodies: rat antimouse nondepleting anti-CD4 YTS 177, anti-CD40 L mAb (MR1) and either the anti-CD8 depleting antibodies, YTS 156 and YTS 169 (1 mg of each). Two doses of 12 mg/kg of RAPA are administered i.p. at days 6 and 30 posttransplantation. RAPA synergized with CD8+ T-cell depletion to induce long-term acceptance of heterotopic limb allograft in a stringent allogeneic strain (Balb/C to B6) combination (Table 1).

3.2. Skin Grafting

Briefly, skin grafting is conducted by grafting full-thickness tail skin (1 cm^2) on the lateral flank. Grafts are monitored daily after the removal of the bandage on day 10 and considered rejected when more than 75% of epithelial breakdown had occurred.

Table 1
RAPA synergizes with immunosuppressive regiment to induce long term acceptance of heterotopic limb allograft

	Graft survival (days)		
Protocol	Heterotopic limb allograft	Donor type second skin graft	Third party donor skin graft
Without RAPA	74, 38, 54	ND	ND
With RAPA	112, 1 CR, >200 × 7	5/7 Accepted	0/7 Accepted

ND not done, *CR* chronic rejection

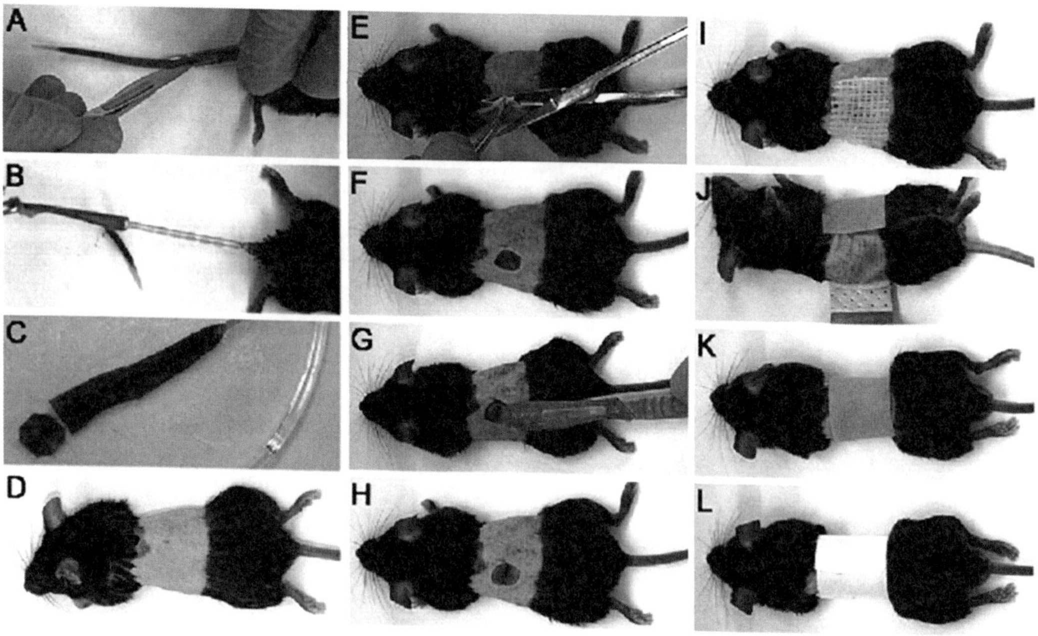

Fig. 2. Skin graft procedures. (**a**) Donor tail skin longitudinal incision after sacrifice. (**b**) Removing tail skin. (**c**) Piece of skin ready for grafting. (**d**) Anesthetized and shaved recipient. (**e**, **f**) Preparing the graft bed. (**g**, **h**) Insertion of the graft into the bed with a scalpel. (**i**) Applying vaseline petroleum gauze. (**j**, **k**) Applying elastic bandage. (**l**) Applying adhesive fabric tape.

1. Recipient mice were anesthetized with a mixture of xylazine (Rompun) 5% and ketamine 10% in PBS and 0.22 μm filtered. After weighing, a total of 5 μL/g of body weight was injected intraperitoneally. Wait approximately 5 min and check for anesthesia by lack of spontaneous movement, slow breathing rate, and lack of response to stimuli (such as pinching a paw).
2. In order to remove tail skin of the donor, after sacrifice, the tail is washed with sterile gauze pad humidified with 3.5% H.A.C. solution. After incising the base circumferentially, the ventral

skin is incised over the entire length of the tail with a scalpel blade (Fig. 2a) and stripped off the base using straight forceps with 1 × 2 teeth (Fig. 2b).

3. The tail skin is placed raw-side down in a petri dish and moisten with cold isotonic solution (PBS or 0.9% NaCl). The tail is immobilized with a pair of forceps and a square piece of skin (around 1 cm^2) is cut. The four corners are removed using a scalpel blade in order to fit with the graft bed (Fig. 2c).
4. The back and the thorax of anesthetized recipient are shampooed with 3.5% H.A.C. solution and the hairs are shaved using a razor blade (Fig. 2d).
5. On the lateral flank, the skin is pinched with a pair of curved serrated forceps (Fig. 2e), and the graft bed is made with a pair of scissors (Fig. 2f).
6. The skin graft is deposited on the scalpel blade and after being dried on sterile gauze pad, inserted into the graft bed (Fig. 2g). The transplant should be slightly smaller than the graft bed, allowed the piece of tail skin to be displaced during bandaging (Fig. 2h).
7. A piece of double-thickness vaseline petrolatum gauze (2 × 2 cm) is applied on the graft (Fig. 2i). It is important to smooth down the gauze so that it is adherent to the surrounding normal skin of the recipient in order to allow the graft to be displaced.
8. A piece of elastic bandage (2 cm large, 8 cm long) is applied on the gauze all around the trunk of the recipient (Fig. 2j, k) (see Note 1).
9. The bandaging is achieved by rolling up 10 cm of 2.5 cm large latex-free adhesive fabric tape on the elastic bandage (Fig. 2l).
10. Following surgery, animal is placed in a warm environment and monitored until it begins to move about freely. Be sure that the bandage has not been applied too tightly and does not prevent motility of paws.
11. The bandage is removed after 10 days by slipping one blade of a pair of blunt-end scissors under the entire bandage and by gently cutting it. Technical failure is characterized by scabbing and contraction of the graft. Tail skin well engrafted should be adherent to the recipient and the surrounding skin (Fig. 3a).
12. Skin grafts are scored daily after the removal of the bandage and until rejection or long-term acceptance. Transplant is considered rejected when more than 75% of epithelial breakdown had occurred. Two types of rejection are described in the literature. Acute rejection is characterized by swelling and erythema followed by graft desiccation, signaling the necrosis of graft tissue (Fig. 3b). Chronic rejection is characterized by

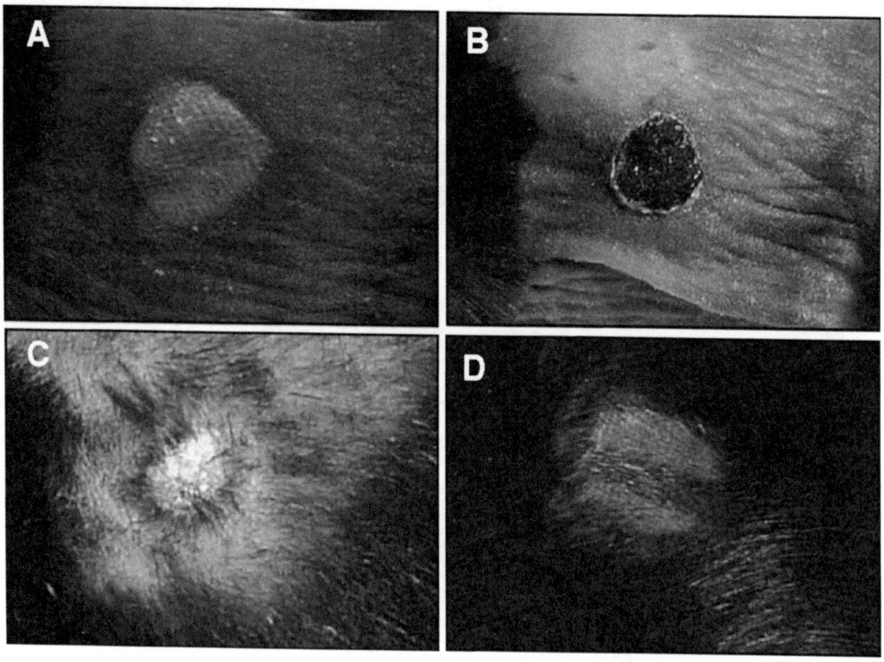

Fig. 3. Features of skin graft. (**a**) skin graft without signs of technical failure just after bandage removal. (**b**) Acute graft rejection with massive necrosis. (**c**) Chronic graft rejection characterized by shrinking and hair loss. (**d**) Long-term accepted skin graft.

loss of hair, pigmentation, and dermal ridges and graft shrinking, reflecting the fibrosis of graft tissue (Fig. 3c). Long-term accepted graft is characterized by absence of erythema, normal thickness (compared to surrounding skin), and presence of a large band of hairs (Fig. 3d).

3.3. Multilineage Mixed Chimerism Evaluation by Flow Cytometry

Multilineage mixed chimerism can be evaluated by co-staining PBMCs for donor-specific MHC class I ($H2k^d$ of BALB/c) and for CD3, CD4, CD8, CD11b, and B220 (CD3, CD4, and CD8 for T cells, CD11b for myeloid cells, and B220 for B cells). Working with PBMC allows paired data on the same animals during time after allogeneic cell graft.

1. Blood drawn from the tail vein can be performed each week. Warm the mouse and place it in a restraining tube. Using a scalpel blade, the tail skin is incised slightly on the tail vein and blood is collected in a capillary tube as drops appear. Apply pressure to stop the bleeding. Fifty to 100 μL of total blood in a 1.5-mL tube with 5 μL of heparin can be collected.
2. Transfer 30 μL of collected total blood in a 5-mL polystyrene round-bottom tube for each sample. The rest of collected blood from every mouse is mixed together and 30 μL of this pool are distributed in eight tubes for simple fluorochrome labeling.

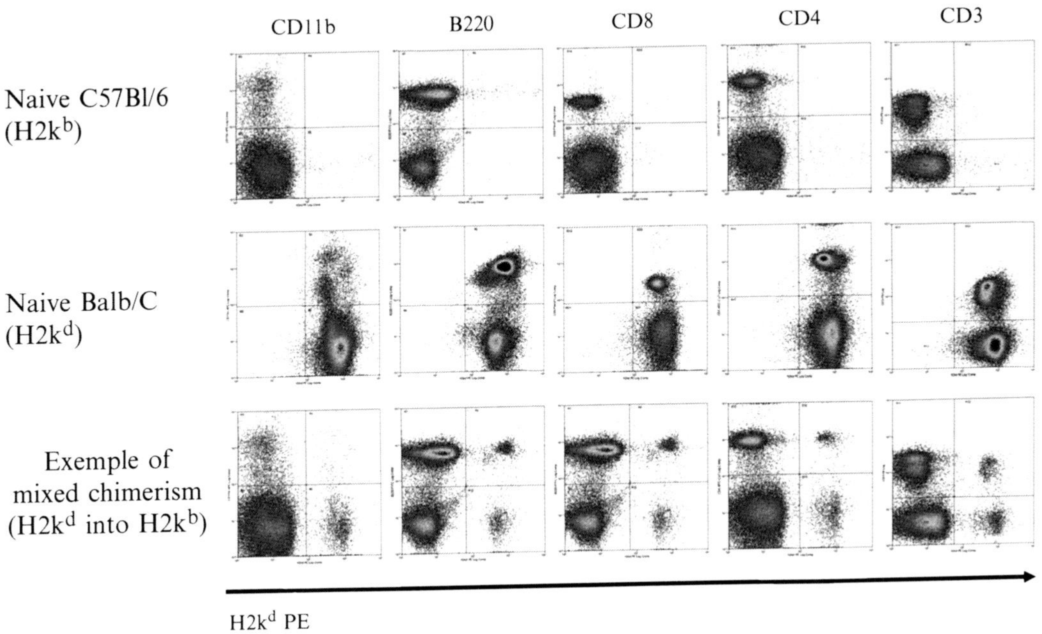

Fig. 4. Multilineage mixed chimerism evaluation. The presence of donor cell is assessed by flow cytometry analysis of PBMCs. Naive C57Bl/6 and Naive Balb/C serve, respectively, as negative and positive controls. H2kd expression reflect chimerism of different cell type discriminated by the expression of specific markers.

3. The cells are incubated 10 min with BD Fcblock® at 1/50 in 50 μL of PBE on ice or at 4–8°C.
4. Without wash, fluorochrome-conjugated Abs are added in 50 μL of PBE (1/50 of every fluorochrome except 1/200 for APC) and incubated 20 min on ice or at 4–8°C in the dark (Antibody cocktail : CD45R/B220 FITC, H2k^d PE, CD8 PercP, CD11b APC, CD4 APC-Cy7, CD3PB) (see Note 2). For this step, it is recommended to make a mix for all samples in order to stain the cells with the same concentration of fluorochrome-conjugated Abs (e.g., for ten samples, 50 μL of a mix constituted with 500 μL of PBE, 10 μL of each Abs of interest noncoupled with APC, and 2.5 μL for APC conjugated one) (see Note 3).
5. Then, the cells are washed by the addition of 2 mL of PBE and discarding the supernatant after centrifugation 5 min at 300 × *g*.
6. PBMC are fixed and red blood cells are eliminated by the addition of 1 mL of lysing solution. After 5 min at 4–8°C, the hypotonic solution is neutralized by the addition of 2 mL of PBE. PBMC are then centrifuged for 5 min at 300 × *g* and the pellet is reconstituted in 400 μL of PBE. The PBMC are ready to be analyzed by flow cytometry (Fig. 4: Exemple of multilineage mixed chimerism).

3.4. Mixed Lymphocyte Culture

Cells isolated from spleen or draining lymph nodes (inguinal and axillary) are used as responders (2.5×10^6 cells/mL) and stimulated with 2.5×10^6 cells/mL irradiated splenocytes (2,000 rad) in 48-well flat-bottom plates for cytokine detection on supernatants by ELISA and in 96-well U-bottom plates for proliferation assay by [^{3}H]-thymidine incorporation.

1. After dissection, the spleen of recipient is crushed in culture medium with a sterile piston of a 5-mL syringe. Then, fragments are eliminated upon 100-μm nylon filter.
2. After centrifugation at 300 × *g* during 5 min, lysis of red blood cells is performed by reconstitution of the pellet with 1 mL of ACK. After 1 min, addition of large volume of culture medium permits the inhibition of hypotonic ACK solution (the next step for spleen cells correspond to point 4).
3. Isolation of draining lymph nodes cells consists in crushing lymph nodes after dissection between two sterile microscope glass slides using the abrasive part in culture medium. Then, fragments are eliminated upon 100-μm nylon filter.
4. Spleen or draining lymph node cells are then centrifuged at 300 × *g* during 5 min and the pellet is reconstituted with 1 mL of culture medium and the cells are used as responders.
5. Number of living cells is evaluated by using Thomas Counting chamber and culture medium is added in order to obtain a final concentration of 4×10^6/mL of living cells.
6. As stimulators, allogeneic or third party spleen are isolated like recipient spleen cells and concentrated at 5×10^6/mL in culture medium. The cells are then irradiated at 2,000 rad.
7. In 48-well flat-bottom plate, 500 μL of responder cells are mixed with 500 μL of stimulators in order to have a final concentration of 2×10^6 and 2.5×10^6/mL for responder and stimulator cells, respectively. If possible, duplicate every condition.
8. The plates are then incubated in a humidified incubator at 37°C with 5% CO_2. Supernatants are collected after 48 h for IL-2 and IL-13 and IFN-γ productions and 72 h for TNF-α and IL-17 production. Cytokine productions were measured in culture supernatants using commercially available enzyme-linked immunosorbent assay (ELISA) kits (Duoset, R&D systems, Minneapolis, MN) following manufacturer's instructions (see Note 4).
9. For proliferation assay, 100 μL of responder (4×10^5 cells) and stimulator cells (5×10^5 cells) are mixed in 96-well U-bottom cell culture plates in triplicates and incubated at 37°C with 5% CO_2. One μCurrie of [^{3}H]-thymidine is added per well at day 3 for 16 h. The plate is harvested on a filter-bottomed 96-well plate and dried for 2–12 h in a dry incubator at 40°C. Thirty microliters of scintillation liquid is added before measuring

count per minute (cpm) in β scintillation counter, evaluating the amount of [^{3}H]-thymidine incorporated into the DNA so the proliferation of responder cells.

4. Notes

1. Be sure that the mouse has mobile anterior and posterior paws.
2. We have found these combinations of antibodies to be particularly interesting for multilineage mixed chimerism and cytokine production evaluations. Numerous competitive reagents are available from other commercial sources.
3. In order to avoid overlaps of the different fluorochromes, simple labeling of each antibody is needed to performed compensations.
4. For cytokine detection by ELISA, it is recommended to dilute supernatant for IFN-γ (1/4) in order to have samples in the same range of values than the standard curve.

References

1. Thomson, A. W., H. R. Turnquist, and G. Raimondi. 2009. Immunoregulatory functions of mTOR inhibition. Nat. Rev. Immunol. 9: 324–337.
2. Weichhart, T., and M. D. Saemann. 2009. The multiple facets of mTOR in immunity. Trends Immunol. 30: 218–226.
3. Raimondi, G., T. L. Sumpter, B. M. Matta, M. Pillai, N. Corbitt, Y. Vodovotz, Z. Wang, and A. W. Thomson. 2010. Mammalian target of rapamycin inhibition and alloantigen-specific regulatory T cells synergize to promote long-term graft survival in immunocompetent recipients. J. Immunol. 184: 624–636.
4. Pilat, N., and T. Wekerle. 2010. Mechanistic and therapeutic role of regulatory T cells in tolerance through mixed chimerism. Curr. Opin. Organ Transplant.
5. Lim, D. G., S. K. Koo, Y. H. Park, Y. Kim, H. M. Kim, C. S. Park, S. C. Kim, and D. J. Han. 2010. Impact of immunosuppressants on the therapeutic efficacyofinvitro-expandedCD4+CD25+Foxp3+ regulatory T cells in allotransplantation. Transplantation 89: 928–936.
6. Daniel, C., K. Wennhold, H. J. Kim, and B. H. von. 2010. Enhancement of antigen-specific Treg vaccination in vivo. Proc. Natl. Acad. Sci. U. S. A 107: 16246–16251.
7. Metzler, B., P. Gfeller, G. Wieczorek, and A. Katopodis. 2008. Differential promotion of hematopoietic chimerism and inhibition of alloreactive T cell proliferation by combinations of anti-CD40Ligand, anti-LFA-1, everolimus, and deoxyspergualin. Transpl. Immunol. 20: 106–112.
8. Li, Z., F. S. Benghiat, L. M. Charbonnier, C. Kubjak, M. N. Rivas, S. P. Cobbold, H. Waldmann, W.De, V, M. Petein, F. Schuind, M. Goldman, and M. A. Le. 2008. CD8+ T-Cell depletion and rapamycin synergize with combined coreceptor/stimulation blockade to induce robust limb allograft tolerance in mice. Am. J. Transplant. 8: 2527–2536.
9. Pilat, N., U. Baranyi, C. Klaus, E. Jaeckel, N. Mpofu, F. Wrba, D. Golshayan, F. Muehlbacher, and T. Wekerle. 2010. Treg-therapy allows mixed chimerism and transplantation tolerance without cytoreductive conditioning. Am. J. Transplant. 10: 751–762.
10. Blaha, P., S. Bigenzahn, Z. Koporc, M. Schmid, F. Langer, E. Selzer, H. Bergmeister, F. Wrba, J. Kurtz, C. Kiss, E. Roth, F. Muehlbacher, M. Sykes, and T. Wekerle. 2003. The influence of immunosuppressive drugs on tolerance induction through bone marrow transplantation with costimulation blockade. Blood 101: 2886–2893.
11. McFarland, H. I., and A. S. Rosenberg. 2009. Skin allograft rejection. Curr. Protoc. Immunol. Chapter 4: Unit.

Chapter 29

Development of ATP-Competitive mTOR Inhibitors

Qingsong Liu, Seong A. Kang, Carson C. Thoreen, Wooyoung Hur, Jinhua Wang, Jae Won Chang, Andrew Markhard, Jianming Zhang, Taebo Sim, David M. Sabatini, and Nathanael S. Gray

Abstract

The mammalian Target of Rapamycin (mTOR)-mediated signaling transduction pathway has been observed to be deregulated in a wide variety of cancer and metabolic diseases. Despite extensive clinical development efforts, the well-known allosteric mTOR inhibitor rapamycin and structurally related rapalogs have failed to show significant single-agent antitumor efficacy in most types of cancer. This limited clinical success may be due to the inability of the rapalogs to maintain a complete blockade mTOR-mediated signaling. Therefore, numerous efforts have been initiated to develop ATP-competitive mTOR inhibitors that would block both mTORC1 and mTORC2 complex activity. Here, we describe our experimental approaches to develop Torin1 using a medium throughput cell-based screening assay and structure-guided drug design.

Key words: mTOR, mTORC1, mTORC2, PI3K, PIKK, Akt, Rapamycin, Torin1

1. Introduction

mTOR (mammalian Target of Rapamycin) is a serine/threonine protein kinase that is highly conserved across eukaryotes. It plays a critical role in the regulation of fundamental cellular physiological processes in response to the upstream cellular signals such as growth factors, energy, stress, and nutrient to control cell growth, proliferation, and metabolism through two known complexes, namely, mTORC1 and mTORC2 (1). mTORC1 is a primary regulator of protein translation through the phosphorylation of the S6 kinase and the inhibitory 4E-binding proteins (4EBPs). mTORC2 phosphorylates and regulates Akt, a downstream effector of the PI3K

Thomas Weichhart (ed.), *mTOR: Methods and Protocols*, Methods in Molecular Biology, vol. 821, DOI 10.1007/978-1-61779-430-8_29, signaling pathway that mediates growth and survival signals. Rapamycin, an allosteric inhibitor of mTOR that was originally identified as an immunosuppressant and later as an anticancer agent, has greatly facilitated the study of mTOR signaling (2, 3). As a key node in the oncogenic phosphoinositide 3-kinase signaling pathway, deregulation of the mTOR kinase has been often observed in human cancers (4–6). Inspired by the promising *in vitro* anticancer effect, it is widely believed that rapamycin or related analogs (rapalogs) would be therapeutically effective by blocking the mTORC1's phosphorylation activity of S6K and 4EBPs, which are key upstream regulators of protein synthesis. However, extensive clinical evaluation has found rapamycin's *in vivo* efficacy to be limited to a few rare cancers, such as renal cell carcinoma and mantle cell lymphoma (7). This may be partially explained by the recent discovery that rapamycin could not block the function of mTORC1 completely and has little effect on mTORC2 complex in the majority of cell types (4, 8). Moreover, the existence of a feedback loop that hyperactivates PI3K when mTORC1 is inhibited (9) and the recent discovery that rapamycin fails to completely inhibit mTORC1 indicates that ATP-competitive small molecule mTOR inhibitors might show broader efficacy.

Structurally, the mTOR catalytic domain exhibits high sequence identity with PI3Ks and PI3K-related kinases (PIKK) such as ATR, ATM, DNA-PK, and SMG-1. The co-crystal structure of LY294002 with PI3Kγ provides valuable insights that enable structure-guided design of numerous PI3K inhibitors (10). The dual PI3K/mTOR inhibitor PI-103, similar to LY294002, which uses the morpholine oxygen as the hinge binding moiety, provides a molecular-entry point for the development of selective mTOR inhibitors (11, 12). For example, PI-103 served as a "lead" compound for the structure-guided development of selective mTOR inhibitors such as KU-0063794 (13) and WYE-354 (14) and other related structures (15) using a rational designed approach (Fig. 1). Preliminary biochemical analyses indicate that these compounds are, indeed, promising candidates for replacing rapamycin as future anticancer agents. Not only do these pharmacological inhibitors have enhanced efficacy in suppressing hyperactive mTOR, but also they are highly

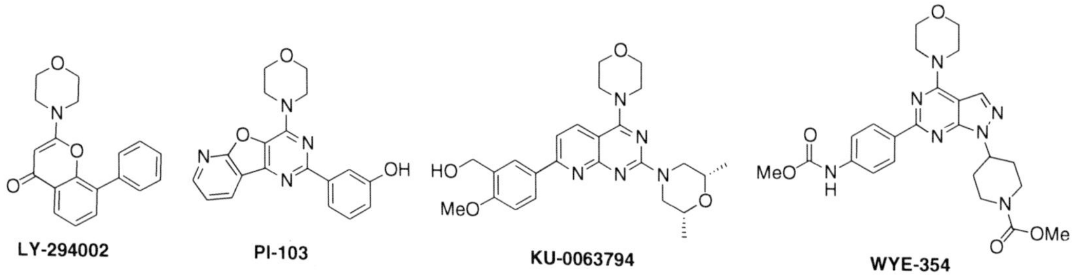

Fig. 1. Structures of PI3K and mTOR inhibitors.

specific in targeting mTOR in a large panel of cancer cell lines. More interestingly, these compounds inhibit the proliferation of diseased cells to a greater extent than rapamycin by targeting the rapamycin-insensitive mTOR complex in addition to the rapamycin-sensitive mTOR complex. We utilized a complementary approach to develop mTOR selective inhibitors which consisted of using medium-throughput biochemical screening of an ATP-site directed heterocyclic compound library followed by iterative-chemical synthesis guided by cell-based SAR analysis and molecular modeling .

2. Materials

2.1. Chemical Synthesis Materials

1. All commercially available chemical materials were used directly without further purification.

2.2. Cell Culture and Lysis

1. Cell culture media: DMEM, 10% fetal bovine serum, penicillin/streptomycin (Invitrogen, Carlsbad, CA).
2. Centrifuge: RC-5 C plus (Sorvall).

2.3. SDS-Polyacrylamide Gel Electrophoresis

1. SDS-Polyacrylamide gel: NuPAGE 4–12% Bis–Tris Gel 1.5 mm, 15 well (Invitrogen, Carlsbad, CA).
2. Running buffer (20×): NuPAGE MES SDS running buffer or NuPAGE MOPS SDS running buffer (Invitrogen, Carlsbad, CA).

2.4. Western Blotting for Active S6K1 and Akt

1. Transfer buffer: 10 mM CAPS, 10% (v/v) ethanol.
2. Immobilon-P transfer membrane from Millipore, Bedford, MA, and 3MM Chr chromatography paper from Whatman, Maidstone, UK.
3. Phosphate-buffered saline with Tween (PBS-T): 0.137 M NaCl, 2.7 mM KCl, 10.0 mM Na_2HPO_4, 1.76 mM KH_2PO_4, pH 7.4, 0.1% Tween-20.
4. Blocking buffer: 5% (w/v) nonfat dry milk in PBS-T.
5. Primary antibody dilution buffer: PBS-T supplemented with 5% (w/v) bovine serum albumin (BSA).
6. Primary antibodies: Phospho-S6K1 (Thr-389) and phospho-Akt (Ser-473) (Cell Signaling Technology, Danvers, MA).
7. Secondary antibody: Anti-mouse IgG (for phospho-S6K1) and anti-rabbit IgG (phospho-Akt) conjugated to horse radish peroxidase (Santa Cruz Biotechnology, Santa Cruz, CA).
8. Enhanced chemiluminescent (ECL) reagent: Western Lighting–ECL (Perkin Elmer, Waltham, MA).

2.5. Material in U87MG Model Anti-Tumor Study

1. Immunodeficient mice: NCR nude, nu/nu (Taconic Laboratories).
2. Drug vehicle: 20%:40%:40% (v/v) *N*-methyl-2-pyrrolidone: PEG400:water.

3. Methods

3.1. Overview of the Process

As illustrated in Fig. 2, the workflow for developing potent and selective mTOR inhibitors commenced with a medium-throughput biochemical screening of a focused ATP-site directed heterocycle library. Compounds that exhibited activity in this assay were then profiled for selectivity across a panel of approximately 400 kinases using the Ambit KinomeScan methodology from which a class of quinoline-derived compounds appeared to exhibit the greatest selectivity for mTOR. Compounds were then evaluated for their ability to inhibit the downstream target of mTORC1 S6K T389 in a cellular assay utilizing mouse embryonic fibroblasts (MEFs). A key goal of this project was to identify compounds that exhibited selectivity for inhibition of mTOR relative to structurally homologous PIKK family members such as the PI3Ks. The ability of the compounds to inhibit PI3K was evaluated by measuring effects on the activation loop phosphorylation site of Akt T308 in the context of the S473D mutant form of Akt in PC-3 cells. The S473D mutant of Akt was employed because this "phospho-mimetic" mutation abolishes the effects that would result from mTORC2-mediated phosphorylation

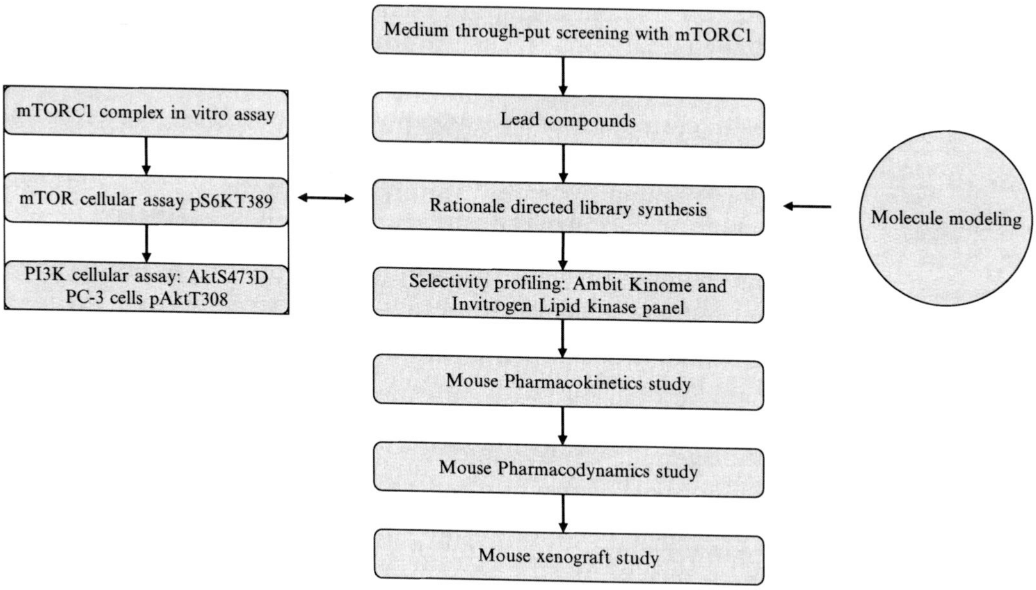

Fig. 2. Flowchart for developing mTOR inhibitors.

of this site (8). This is a necessary precaution because most inhibitors in our series are very potent inhibitors of mTORC2 and this would confound our ability to evaluate how potent they are as inhibitors of PI3K. The structure–activity relationships derived from these cellular assays was interpreted in the context of a molecular model for the presumed binding mode. A homology model of the three-dimensional structure of mTOR was constructed using Autodock Vina. This information was then used to inform the design of iterative rounds of compound synthesis. Given the fairly challenging nature of the chemistry required to prepare these compounds, we were never able to prepare focused libraries of derivatives. The most potent and selective compounds were subjected to broader selectivity profiling across a panel of approximately 450 kinases using the Ambit KinomeScan™ technology. The selected compounds chosen from this profiling were then tested in Invitrogen SelectScreen® lipid kinase panel including all of the PI3K isomers and DNA-PK. The most selective compounds were subject to metabolic stability studies using isolated mouse microsomes and then evaluated in mice for their pharmacokinetic and pharmacodynamic properties. Candidate compounds were then tested for antitumor efficacy using mouse tumor xenograft models.

3.2. Chemical Synthesis Methodology

1. Combinatorial kinase-directed libraries were built from a variety of commercially available heterocycles using both solution and solid phase synthesis (16). Simple and robust chemistries such as nucleophilic aromatic substitution and palladium-mediated couplings were used to install diverse elements. For example, disubstituted quinazolines were constructed starting from compound **1** (4,6-dichloroquinazoline) using a nucleophilic substitution reaction to install a particular aniline to the *C4* position followed by a Suzuki coupling reaction to install an aromatic side chain to the *C6* position of the quinazoline (Scheme 1).

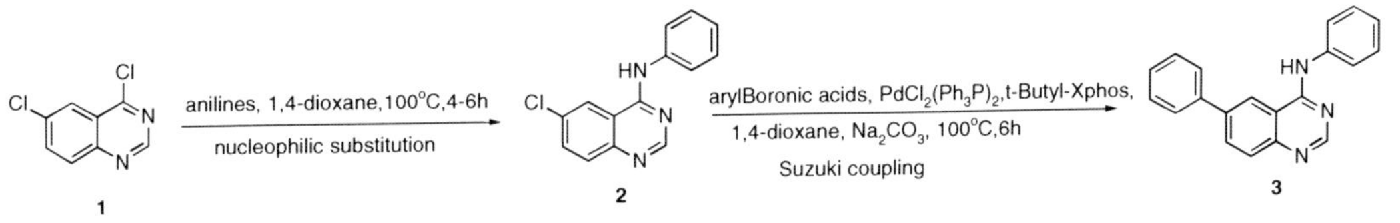

Scheme 1. Synthesis of the high-throughput library.

2. Rationally designed focused library synthesis was accomplished using the synthetic scheme exemplified in Scheme 2 (17). Compound **4** (Ethyl-4,6-dichloroquinoline-3-carboxylate) was subjected to nucleophilic addition by particular aryl anilines to afford compound **5** (Scheme 2). Lithium aluminum hydride (LAH)-mediated reduction of the ethyl ester furnished the corresponding benzyl alcohol compound **6**. Benzylic oxidation

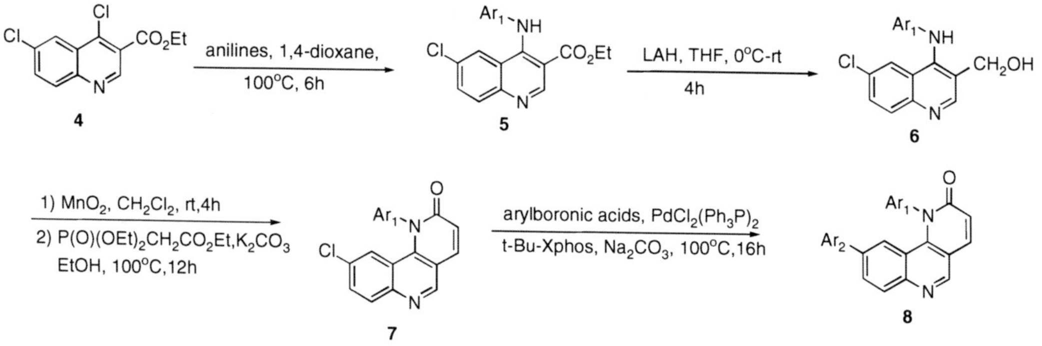

Scheme 2. Synthesis of the focused library.

with MnO_2 followed by Horner–Emmons–Wardsworth olefination generated the cyclized compound 7. Finally, a Suzuki coupling reaction was used to install an additional aryl side chain at *C6* position of the quinoline.

A detailed synthetic protocol is provided below:

(a) To a solution of compound **4** (1 equiv.) in 1,4-dioxane was added aniline (1–2 equiv.) at room temperature. The reaction mixture was then heated to 100°C for 4–6 h and then allowed to cool to room temperature. An aqueous NaOH (1N) solution was added to neutralize the reaction mixture. The resultant solution was diluted with water and extracted with ethyl acetate. After the removal of the solvents under vacuum, the residue was purified by flash column chromatography (hexanes/EtOAc) to afford compound **5** (see Note 1).

(b) To a solution of compound **5** (1 equiv.) in THF at 0°C was added LAH (3–5 equiv.) dropwise. After 15–20 min, the solution was warmed to room temperature and stirred for 1–4 h before carefully quenching with methanol and water. Dilution of the mixture with EtOAc and filtration through celite furnished crude **6**, which was used in the next step without further purification (see Note 2).

(c) To a solution of compound **6** in CH_2Cl_2 (1 equiv.) at room temperature was added MnO_2 (10 equiv. mass). After 1–4 h, the reaction mixture was filtered through celite. The filtrate was concentrated and placed in a sealed tube and dissolved in dry EtOH. K_2CO_3 (3 equiv.) and triethyl phosphonoacetate were then added sequentially. The resulting mixture was heated to 100°C for 12–16 h before cooling to room temperature. Upon removal of the solvents under vacuum, the residue was diluted with water followed by extraction with ethyl acetate (3×). Purification of the residue by flash column chromatography (hexanes/EtOAc) provided compound **7** (see Note 3).

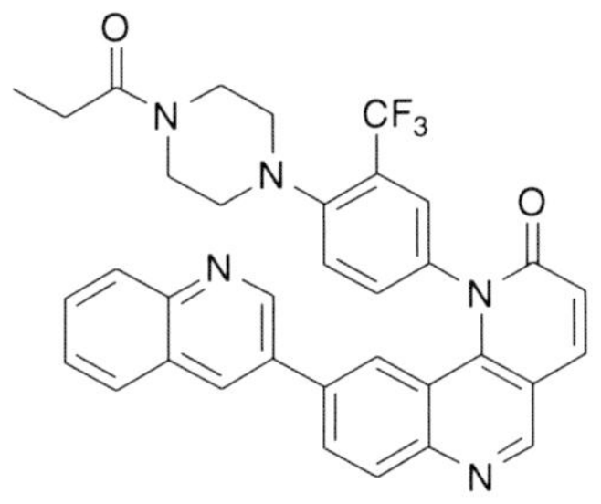

Fig. 3. Structure of Torin1.

(d) To a solution of compound 7 in 1,4-dioxane at room temperature was added subsequently $PdCl_2(Ph_3P)_2$ (0.1 equiv.), *t*-Bu-Xphos (0.1 equiv.), Na_2CO_3 (3 equiv., 1N) and boronic acids or pinacol boronic esters. After degassing, the resultant mixture was heated to 100°C for 6 h before cooling to room temperature and filtering through celite. Upon removal of the solvents, the residue was subjected to column chromatography purification (hexanes/EtOAc) to furnish the desired compounds as exemplified by compound **8** (see Note 4). The structure of Torin1 generated following above procedures was shown in Fig. 3.

3.3. Biochemical Assay of mTORC1 Complex

3.3.1. Preparation of Human Soluble mTORC1 Complex

1. HEK-293T cells that stably expressed N-terminally FLAG-tagged raptor using vesicular stomatitis virus G-pseudotyped MSCV retrovirus was used as a source of mTORC1 complex.
2. The HEK293T cells were lysed in 50 mM HEPES, pH 7.4, 150 mM NaCl, and 0.4% CHAPS at 4°C for 30 min, and the insoluble fraction was removed by centrifugation at 39,000 G for 30 min. Supernatants were incubated with FLAG-M2 monoclonal antibody-conjugated agarose for 1 h and then washed with two column volumes of wash buffer 1 (50 mM HEPES, pH 7.4, 150 mM NaCl, 2 mM DTT and 2 mM ATP, and 0.1% CHAPS) and another two column volumes of wash buffer 2 (50 mM HEPES, pH 7.4, 200 mM NaCl, and 0.1% CHAPS).
3. mTORC1 complex was eluted with 100 μg/ml 3× FLAG peptide in 50 mM HEPES, pH 7.4, 500 mM NaCl, and 0.1% CHAPS, then concentrated by centrifugation. Pure mTORC1 was isolated after gel filtration using a tandem Superose 6 10/300 GL column (GE Healthcare) in 50 mM HEPES, pH 7.4 and 150 mM NaCl on an AKTA purifier (GE Healthcare).

3.3.2. In Vitro mTORC1 Activity Assay (Lanthascreen™ Time-Resolved FRET Assay)

1. mTORC1 (0.1 μg each) was incubated with serially diluted inhibitors (threefold, 11 points) for 30 min in 5 μl kinase buffer (25 mM HEPES, pH 7.4, 10 mM $MgCl_2$, 4 mM $MnCl_2$, 50 mM KCl) in a 384-well low volume plate (Corning).

2. The kinase reaction was initiated by the addition of an equal volume of kinase buffer containing 0.8 μM GFP-labeled 4E-BP1 and 100 μM ATP at room temperature. After the kinase reaction for 1 h, the reaction was stopped by the addition of 10 μl of solution containing 20 mM EDTA and 4 nM Tb-labeled antiphospho 4E-BP1 (T46) antibody (Invitrogen).
3. After incubation for 30 min, the FRET signal between Tb and GFP within the immune complex was read using Envision plate reader (PerkinElmer). Each data point was duplicated and IC_{50} values were calculated using Prism4 software (GraphPad).

3.4. Cellular Assays of mTOR and PI3K Activities

1. Cell Lysis: Cells were rinsed once with ice-cold PBS and lysed in an ice-cold lysis buffer (40 mM HEPES, pH 7.4, 2 mM EDTA, 10 mM pyrophosphate, 10 mM glycerophosphate, and 0.3% CHAPS or 1% Triton X-100, and one tablet of EDTA-free protease inhibitors per 25 ml). The soluble fractions of cell lysates were isolated by centrifugation at 13,000 rpm for 10 min in a microcentrifuge (18).
2. Cellular IC_{50} values for mTOR were determined using $p53^{-/-}$ MEFs. Cells were treated with vehicle or increasing concentrations of compound for 1 h and then lysed. Phosphorylation of S6K1 Thr-389 was monitored by immunoblotting using a phospho-specific antibody.
3. Cellular IC_{50} values for PI3K were determined using $p53^{-/-}$/$mLST8^{-/-}$ MEFs. Cells were treated with vehicle or increasing concentrations of compound for 1 h and then lysed. Phosphorylation of Akt Thr-308 was monitored by immunoblotting using a phospho-specific antibody. The summary of Torin1's biochemical and cellular IC_{50}s was shown in Table 1.

3.5. Molecular Modeling

An mTOR homology model was built with Modeller(9v6) based on a published PI3Kγ crystal structure complexed with GDC-0941 (PDB: 3DBS). Compounds were docked into the model with Autodock Vina (19). The presumed proper binding conformation was selected based on the binding modes of structurally related inhibitors whose co-structures had been determined crystallographically. An example of the docking result of Torin1 with mTOR complex was shown in Fig. 4.

Table 1
Biological evaluation summary of Torin1

Recombinant mTOR biochemical assay IC_{50} (nM)	mTORC1 *in vitro* assay IC_{50} (nM)	mTOR in cellular assay IC_{50} (nM)	PI3K in cellular assay IC_{50} (nM)
4	0.29	2	1800

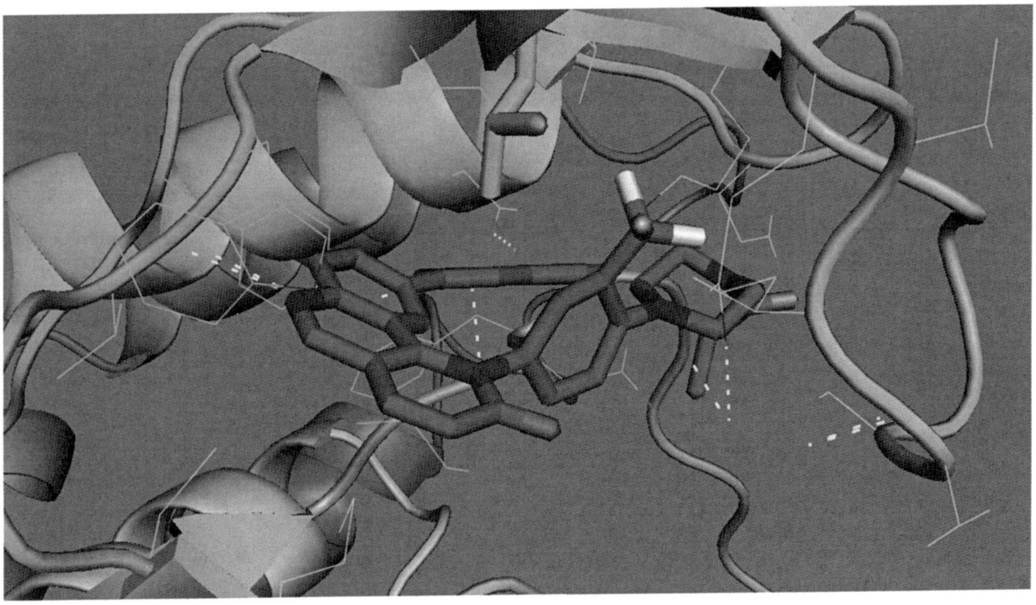

Fig. 4. Torin1 was docked into homology modeled mTOR 3D structure.

3.6. Ambit Kinome Wide and Invitrogen Lipid Kinase Panel Selectivity Profiling

1. Ambit kinome wide selectivity profile was performed in Ambit Bioscience with KinomeScan™ technology. Interesting compounds were profiled at a concentration of 10 μM against a diverse panel of 442 kinases by Ambit Biosciences. Scores for primary screen hits are reported as a percent of the DMSO control (% control). For kinases where no score is shown, no measurable binding was detected. The lower the score, the lower the Kd is likely to be, such that scores of zero represent strong hits. Scores are related to the probability of a hit, but are not strictly an affinity measurement. At a screening concentration of 10 μM, a score of less than 10% implies that the false-positive probability is less than 20% and the Kd is most likely less than 1 μM. A score between 1 and 10% implies that the false-positive probability is less than 10%, although it is impossible to assign a quantitative affinity from a single-point primary screen. A score of less than 1% implies that the false-positive probability is less than 5% and the Kd is most likely less than 1 μM.
2. Invitrogen lipid kinase panel profile was conducted in LifeScience Inc. using SelectScreen® technology by testing IC_{50} on individual kinase with ten testing points from a starting 1 μM drug concentration. PIKK family kinase profiling result of Torin1 in Ambit and Invitrogen was shown in Table 2.

3.7. Pharmacokinetic Studies

1. The study was performed in Sai Advantium Pharma Limited Company (India) with male Swiss Albino mice following single intravenous bolus, oral administration, and intraperitoneal injection.

Table 2
Summary of Ambit and Invitrogen selectivity profiles of PIKK-family kinases

Kinase entry	Ambit score	Invitrogen IC_{50} (nM)	Kinase entry	Ambit score	Invitrogen IC_{50} (nM)
PIK3C2B	25	549	PIK3CA(I800L)	0	ND
PIK3C2G	11	ND	PIK3CA(M1043I)	8.6	ND
PIK3CA	1.3	250	PIK3CA(Q546K)	6.8	ND
PIK3CA(C420R)	0.7	ND	PIK3CB	65	ND
PIK3CA(E542K)	1	ND	PIK3CD	51	564
PIK3CA(E545A)	1	ND	PIK3CG	1.2	171
PIK3CA(E545K)	0.6	ND	PIK4CB	96	6,680
PIK3CA(H1047L)	2.6	ND	DNA-PK	ND	6.34
PIK3CA(H1047Y)	12	ND	mTOR	0	4.32

2. Nine mice were injected 1 mg/kg of Torin1 solution (10% v/v *N*-methyl pyrrolidone and 50% v/v polyethylene glycol-200 in normal saline) via tail vein. Blood samples were collected at 0, 0.08, 0.25, 0.5, 1, 2, 4, and 6 h (see Note 5).
3. Nine mice were dosed orally 10 mg/kg of Torin1 suspension (0.5% w/v Na CMC with 0.1% v/v Tween-80 in water). Blood samples were collected at 0, 0.08, 0.25, 0.5, 1, 2, 4, 6, 8, 12, and 24 h (see Note 5).
4. Nine mice were injected 10 mg/kg of Torin1 solution (10% v/v *N*-methyl pyrrolidone and 20% v/v polyethylene glycol-200 in normal saline and 20% PG in water) via peritoneum. Blood samples were collected at 0, 0.08, 0.25, 0.5, 1, 2, 4, 8, and 24 h (see Note 6).
5. All samples were processed for analysis by precipitation using acetonitrile and analyzed with a partially validated LC/MS/MS method (LLOQ – 1.138 ng/ml). Pharmacokinetic parameters were calculated using the noncompartmental analysis tool of WinNonlin® Enterprise software (version 5.2). The I.V., P.O., and I.P. studies of Torin1 were summarized in Table 3.

3.8. Pharmacodynamic Studies

1. Torin1 was formulated by dissolving the compound at a concentration of 25 mg/ml in 100% *N*-methyl-2-pyrrolidone followed by dilution (1:4) with sterile 50% PEG-400 prior to injection.
2. Six-week-old male C57BL/6 mice were fasted overnight prior to drug treatment. The mice were treated with vehicle (for 10 h) or Torin1 (20 mg/kg for 2, 6, or 10 h) by IP injection, and then refed 1 h prior to sacrifice (CO_2 asphyxiation).

Table 3
Summary of pharmacokinetic test (reproduced from ref. 17)

Route	C_{max} (ng/ml)	T_{max} (h)	AUC (h ng/ml)	$T_{1/2}$ (h)	MRT (h)	CL (ml/min/kg)	V_{ss} (l/kg)	*F* (%)
I.V.	2,757	ND	720	0.5	0.43	23.0	0.59	ND
P.O.	223	0.25	396	0.79	1.51	ND	ND	5.49
I.P.	5,121	0.08	5,718	4.52	ND	ND	ND	ND

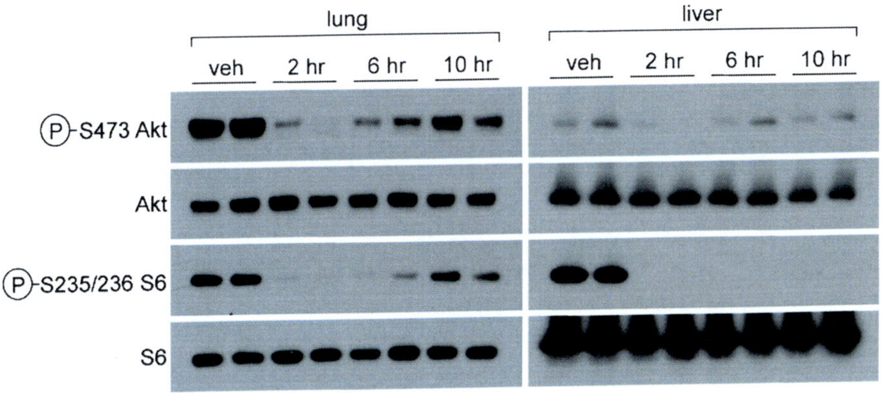

Fig. 5. Pharmacodynamic study of Torin1 (reproduced from ref. 17).

3. Tissues were collected and frozen on dry ice. The frozen tissue was thawed on ice and lysed by sonication in tissue lysis buffer (50 mM HEPES, pH 7.4, 40 mM NaCl, 2 mM EDTA, 1.5 mM sodium orthovanadate, 50 mM sodium fluoride, 10 mM sodium pyrophosphate, 10 mM sodium β-glycerophosphate, 0.1% SDS, 1.0% sodium deoxycholate, and 1.0% Triton, supplemented with protease inhibitor cocktail tablets [Roche]).
4. The concentration of clear lysate was measured using the Bradford assay and samples were subsequently normalized by protein content and analyzed by SDS-polyacrylamide gel electrophoresis (SDS-PAGE) and immunoblotting. The result of PD study of Torin1 was exemplified in Fig. 5.

3.9. Animal Xenograft Study (U87 Model)

1. 2×10^6 U87MG glioblastoma cells were resuspended in 100 μl of media that had been premixed with matrigel, and was injected subcutaneously in the upper flank region of mice (6 weeks old immunodeficient) that had been anesthetized with isoflurane. Tumors were allowed to grow to 1 cm^3 in size (see Note 6).
2. Animals were divided into two treatment groups randomly for vehicle and Torin1. Torin1 was formulated as 5 mg/ml in a solution of *N*-methyl-2-pyrrolidone:PEG400:water

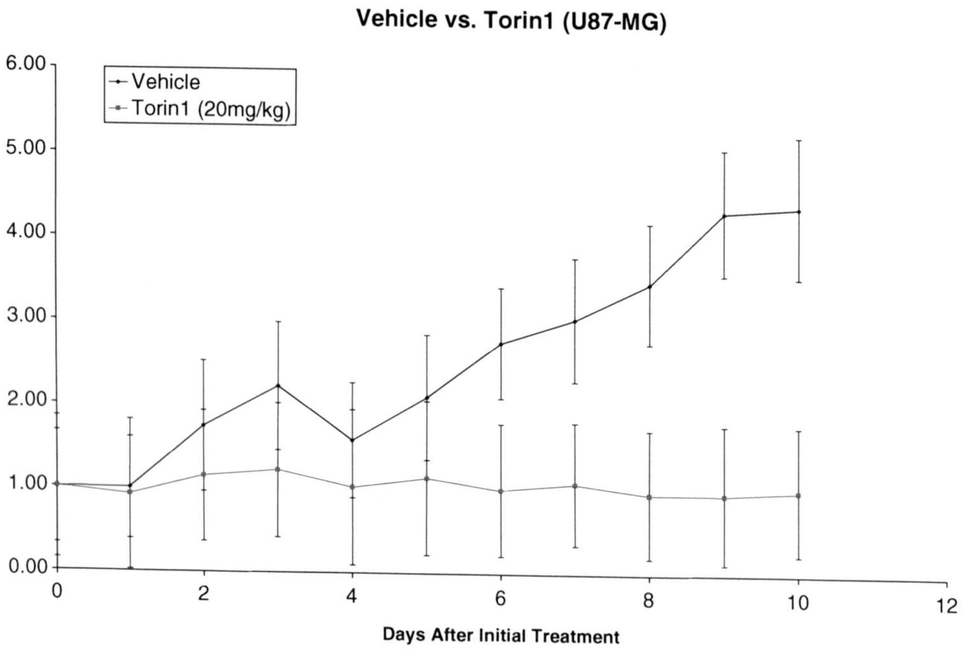

Fig. 6. U87MG tumor xenograft model study of Torin1 (reproduced from ref. 17).

(20%: 40%:40%). Both vehicle and Torin1 were delivered by IP injection at a dosage of 20 mg/kg (q.d.)

3. Tumors were measured with calipers in two dimensions every other day. Tumor volumes were estimated with the formula: volume = $(2a \times b)/2$, where a = short and b = long tumor axes, respectively, in millimeters. The result of Torin1's antitumor activity was exhibited in Fig. 6.

4. Notes

1. Some series of compound **5** would require $CHCl_3$/iPrOH (4:1) extraction system due to a poor solubility in EtOAC.
2. LAH should be added at 0°C to avoid facile over reduction of the benzyl alcohol to the toluene-like product. The reaction is typically monitored carefully by LC-MS to avoid over reduction after warming up to room temperature. A 0°C ice-water bath is preferred during the quenching procedure and the methanol should be added drop-wise due to the exothermic reaction. Extra EtOAc would be used to rinse the filtrate to avoid loss of the product in the celite.
3. Complete removal of the EtOH under vacuum is needed to avoid complications with the purification process.

4. Some compound **8** analogs are very polar and require reverse phase column chromatography with MeOH/water or CH_3CN/water solvent system.
5. All blood samples were collected in microcentrifuge tubes containing K2EDTA as an anticoagulant. At each time point, plasma samples from a set of three mice were separated by centrifugation and stored below −70°C until bioanalysis.
6. Immunodeficient mice (NCR nude, nu/nu; Taconic Laboratories) were maintained in a pathogen-free facility and were given autoclaved food and water ad libitum. All animal studies were performed according to the official guidelines from the MIT Committee on Animal Care and the American Association of Laboratory Animal Care.

Acknowledgements

We thank Life Technologies Corporation for SelectScreen® Kinase Profiling Service and Ambit Bioscience for performing KinomeScan™ profiling. We also thank SAI Advantium Pharma Limited Inc. (India) for the pharmacokinetic study.

References

1. Sarbassov, D. D., Ali, S. M., Sabatini, D. M. (2005) Growing roles for the mTOR pathway. *Curr. Opin. Cell Biol.* **17**, 596–603.
2. Sehgal, S. N.; Baker, H.; Vezina, C. (1975) Rapamycin (AY-22989), a new antifungal antibiotic. II. Fermentation, isolation and characterization. *J. antibiotics (Tokyo)*, **28**, 727–732.
3. Rao, R. D.; Buckner, J. C.; Sarkaria, J. N. (2004) mammalian Target of Rapamycin (mTOR) Inhibitors as Anti-Cancer Agents. *Curr. Cancer Drug. Targets.* **4**, 621–635.
4. Guertin, D. A., Sabatini, D.M. (2007) Defining the role of mTOR in cancer. *Cancer cell*, **12**, 9–22.
5. Molinolo, A. A., Hewitt, S. M., Amornphimoltham, P., Keelawat, S., Rangdaeng, S., Meneses , A. et al (2007) Dissecting the Akt/ mammalian target of rapamycin signaling network: emerging results from the head and neck cancer tissue array initiative. *Clin Cancer Res,* **13**, 4964–4973.
6. Karbowniczek, M., Spittle, C.S., Morrison, T., Wu, H., Henske, E.P. (2008) mTOR is activated in the majority of malignant melanomas. *J Invest Dermatol* **128**, 980–987.
7. Meric-Bernstam, F., Gonzalez-Angulo, A. M. (2009) Targeting the mTOR signaling network for cancer therapy. *J Clin Oncol*, **27**, 2278–2287.
8. Thoreen, C. C., Kang, S. A., Chang, J. W., Liu, Q., Zhang, J., Gao, Y., et al. (2009) An ATP-competitive mammalian target of rapamycin inhibitor reveals rapamycin-insensitive functions of mTORC1. *J Biol Chem.* **284**, 8023–8032.
9. Wan, X., Harkavy, B., Shen, N., Grohar, P., Helman, L. J. (2007) Rapamycin induces feedback activation of Akt signaling through an IGF-1R-dependent mechanism. *Oncogene*, 26, 1932–1940.
10. Walker, E. H., Pacold, M. E., Perisic, O., Stephens, L., Hawkins, P. T., Wymann, M. P., et al. (2000) Structural determinants of phosphoinositide 3-kinase inhibition by wortmannin, LY294002, quercetin, myricetin, and staurosporine. *Mol. Cell*, **6**, 909–919.

11. Hayakawa, M., Kaizawa, H., Moritomo, H., Koizumi, T., Ohishi, T., Yamano, M., et al. (2007) Synthesis and biological evaluation of pyrido[3',2':4,5]furo[3,2-d]pyrimidine derivatives as novel PI3 kinase p110alpha inhibitors. *Bioorg. Med. Chem. Lett.*, 17, 2438–2442.

12. Park, S., Chapuis, N., Bardet, V., Tamburini, J., Gallay, N., Willems, L., et al. (2008) PI-103, a dual inhibitor of class IA phosphatidylinositide 3-kinase and mTOR, has antileukemic activity in AML. *Leukemia*, **22**, 1698–1706.

13. García-Martínez, J. M., Moran, J., Clarke, R. G., Gray, A., Cosulich, S. C., Chresta, C. M., et al. (2009) Ku-0063794 is a specific inhibitor of the mammalian target of rapamycin (mTOR). *Biochem J*, **421**, 29–42.

14. Yu, K., Toral-Barza, L., Shi, C., Zhang, W. G., Lucas, J., Shor, B., et al. (2009) Biochemical, cellular, and in vivo activity of novel ATP-competitive and selective inhibitors of the mammalian target of rapamycin. *Cancer Res.* **69**, 6232–6240.

15. Liu, Q., Thoreen, C. C., Wang, J., Sabatini, D. M., Gray, N. S. (2009) mTOR medicated anti-cancer drug discovery. *Drug. Discov. Today: Therapeutic Strategies.* **6**, 47–55.

16. Ding, S., Gray, N. S., Wu, X., Ding, Q., Schultz, P. G. (2002) A combinatorial scaf-fold approach toward kinase-directed heterocycle libraries. *J. Am. Chem. Soc.*, **124**, 1594–1596.

17. Liu, Q., Chang, J., Wang, J., Kang, S. A., Thoreen, C.C., Markhard, A., Hur, W., Zhang, J., Sim, T., Sabatini, D. M., Gray, N. S. (2010) Discovery of 1-(4-(4-propionylpiperazin-1-yl)-3-(trifluoromethyl)phenyl)-9-(quinolin-3-yl) benzo[h][1,6]naphthyridin-2(1 H)-one as a highly potent, selective mTOR inhibitor for the treatment of cancer. *J. Med. Chem.* DOI: 10.1021/jm101144f

18. Guertin, D. A., Stevens, D. M., Thoreen, C. C., Burds, A. A., Kalaany, N. Y., Moffat, J., Brown, M., Fitzgerald, K. J., Sabatini, D. M. (2006) Ablation in mice of the mTORC components *raptor*, *rictor*, or *mLST8* reveals that mTORC2 is required for signaling to Akt-FOXO and PKCalpha, but not S6K1. *Dev. Cell*, **11**, 859–871

19. Trott, O., Olson, A. J. (2010) Autodock vina: Improving the speed and accuracy of docking with a new scoring function, efficient optimization, and multithreading. *J. Comput. Chem.*, **31**, 455–461.

Index

Thomas Weichhart (ed.), *mTOR: Methods and Protocols*, Methods in Molecular Biology, vol. 821, DOI 10.1007/978-1-61779-430-8, © Springer Science+Business Media, LLC 2012

E

F

G

H

I

K

L

M

N

O

P

R